THE ECOLOGY
OF
AQUATIC INSECTS

THE ECOLOGY
OF
AQUATIC INSECTS

Edited by
Vincent H. Resh
and
David M. Rosenberg

PRAEGER SPECIAL STUDIES • PRAEGER SCIENTIFIC

New York • Philadelphia • Eastbourne, UK
Toronto • Hong Kong • Tokyo • Sydney

Library of Congress Cataloging in Publication Data
Main entry under title:

The Ecology of aquatic insects.

Bibliography: p.
Includes index.
1. Insects, Aquatic—Ecology.   2. Insects—Ecology.
I. Resh, Vincent H.   II. Rosenberg, David M.
QL463.E19   1984   595.7'05263   83-21199
ISBN 0-03-059684-X

Published in 1984 by Praeger Publishers
CBS Educational and Professional Publishing
a Division of CBS Inc.
521 Fifth Avenue, New York, NY 10175 USA

© 1984 Praeger Publishers

456789 052 98765432

Printed in the United States of America
on acid-free paper

CONTENTS

PREFACE

This book has two main goals. The first is to present a contemporary overview of aquatic insect ecology. The second is to highlight research needs and avenues of investigation in aquatic insect ecology that currently seem most promising. To meet these goals, all of the contributors have attempted to present their subjects in a manner that will be informative to both beginning students in aquatic insect ecology and researchers already active in this field.

In planning the content of this book, we wanted a broad coverage of aquatic insect ecology, but one that avoided duplication with topics treated in existing, but still relevant, review articles (see Chapter 1). After choosing topics for individual chapters, we listed potential authors whom we felt would be especially suited to cover these topics. We approached our first-choice authors for each chapter and, in what may be a record for a volume of contributed articles, all accepted our invitation!

Aquatic entomology has its roots in natural history studies of freshwater insects by amateur entomologists. However, as can be seen from the topics covered in this book, ecological studies of these organisms now include approaches drawn from a variety of disciplines such as entomology, limnology, fisheries biology, and hydrology. We thank our contributors for their considerable efforts in producing the excellent treatments of their topics, and we thank the following individuals who served as chapter reviewers: Norman Anderson, Steven Balling, Ernest Benfield, Clifford Berg, Andrew Bodaly, Lazar Botosaneaneau, John Brittain, Joshua Collins, Colbert Cushing, Hugh Danks, Don Erman, James Gore, Roger Green, Ronald Hall, Charles Hawkins, Edwin Herricks, Laurent LeSage, Joseph McAuliffe, Eric McElravy, Jack Mollard, Harold Mundie, Stuart Neff, Dennis Newbold, Colin Paterson, Donald Price, Gordon Pritchard, Christopher Pugsley, Seth Reice, David Schindler, Andrew Sih, John Spence, Bernhard Statzner, Harry Tolkamp, David White, and John Wood. We greatly appreciate the assistance of Kay Sorg, Elizabeth Rodgers, Albert Hendricks, and especially Allen Wiens. The Rockefeller Foundation provided a one-month residency at the Villa Serbelloni in Bellagio, Italy, to one of us (V.H.R.) to complete this book; the support of the Freshwater Institute and the University of California at Berkeley is also gratefully acknowledged.

We are pleased to dedicate our efforts on this book to Ann, Lewis, Becky, and Ralph.

V. H. R. and D. M. R.

Bellagio, Italy
Berkeley, California, U.S.A.
Winnipeg, Manitoba, Canada

CONTRIBUTORS

J. David Allan
Department of Zoology
University of Maryland
College Park, Maryland 20742

David R. Barton
Department of Biology
University of Waterloo
Waterloo, Ontario N2L 3G1
Canada

Arthur C. Benke
School of Biology
Georgia Institute of Technology
Atlanta, Georgia 30332

Thomas M. Burton
Departments of Zoology, and Fisheries
and Wildlife
Michigan State University
East Lansing, Michigan 48824

Malcolm G. Butler
Department of Zoology
North Dakota State University
Fargo, North Dakota 58105

Kenneth W. Cummins
Department of Fisheries and Wildlife
Oregon State University
Corvallis, Oregon 97331

Michael Healey
Department of Fisheries and Oceans
Pacific Biological Station
P.O. Drawer 100
Nanaimo, British Columbia V9R 5K6
Canada

H. B. N. Hynes
Department of Biology
University of Waterloo
Waterloo, Ontario N2L 3G1
Canada

Steven L. Kohler
School of Natural Resources

University of Michigan
Ann Arbor, Michigan 48109

Gary A. Lamberti
Department of Entomological Sciences
University of California
Berkeley, California 94720

Richard W. Merritt
Department of Entomology
Michigan State University
East Lansing, Michigan 48824

G. Wayne Minshall
Department of Biology
Idaho State University
Pocatello, Idaho 83209

James W. Moore
Alberta Environmental Centre
Bag 4000
Vegreville, Alberta T0B 4L0
Canada

Robert W. Newbury
Freshwater Institute
501 University Crescent
Winnipeg, Manitoba R3T 2N6
Canada

Barbara L. Peckarsky
Department of Entomology
Cornell University
Ithaca, New York 14853

Vincent H. Resh
Department of Entomological Sciences
University of California
Berkeley, California 94720

David M. Rosenberg
Freshwater Institute
501 University Crescent
Winnipeg, Manitoba R3T 2N6
Canada

Andrew L. Sheldon
Department of Zoology
University of Montana
Missoula, Montana 59812

Stephen M. Smith
Department of Biology
University of Waterloo
Waterloo, Ontario N2L 3G1
Canada

Bernard W. Sweeney
Stroud Water Research Center
Philadelphia Academy of Natural Sciences
Avondale, Pennsylvania 19311

James V. Ward
Department of Zoology and Entomology

Colorado State University
Fort Collins, Colorado 80523

Torgny Wiederholm
The National Swedish Environment
Protection Board
Water Quality Laboratory Uppsala
Box 8043, S-750 08 Uppsala, Sweden

Michael J. Wiley
Aquatic Biology Section
Illinois Natural History Survey
Urbana, Illinois 61801

D. Dudley Williams
Division of Life Sciences
Scarborough College
University of Toronto
West Hill, Ontario M1C 1A4, Canada

THE ECOLOGY
OF
AQUATIC INSECTS

chapter 1
INTRODUCTION
Vincent H. Resh
David M. Rosenberg

THE GROWTH OF AQUATIC INSECT ECOLOGY

When the renowned freshwater ecologist T. T. Macan published an article in the *Annual Review of Entomology* that had the same title as this book his treatment of aquatic insect ecology was completed in 27 pages (Macan 1962). This book is several hundred pages longer than Macan's article, yet both the book and the article probably reflect the relative amount of information on the ecology of aquatic insects available at their respective times. Very likely, the number of papers on aquatic insect ecology published in the last decade exceeds the number of papers published on the topic in all previous decades.

The growth of interest in aquatic insect ecology can be seen by examining the increase in membership of the North American Benthological Society (NABS), from its inception with 13 members in 1953 as the Midwest Benthological Society to its 1983 membership of over 1,300 (Fig. 1.1). During the period of NABS's most rapid growth, the Aquatic Insect Subsection of the Entomological Society of America was also formed (Stewart 1977), and several international symposia on Odonata, Plecoptera, Ephemeroptera, Trichoptera, and Chironomidae were held, attracting participants from many different nations. Newsletters from these groups (*Selysia, Perla, Eatonia, The Trichoptera Newsletter, Chironomus*) are currently circulated worldwide.

Aquatic entomology courses are offered at over 70 North American universities (Morse 1979). In addition, workshops on the taxonomy and ecology of groups such as the Ephemeroptera, Trichoptera, and Chironomidae have been periodically taught to government regulatory personnel in the United States. On a different level, university extension courses in aquatic entomology for anglers have been an effective way of involving the interested public in conservation and research efforts.

The increased interest in aquatic entomology within the context of the entire field of entomology can be seen by examining the proportion of chapters in the *Annual Review of Entomology* that have been devoted to aquatic entomology subjects. For example, in the first 14 volumes of the *Review* (1956–69), less than 1 percent of all articles dealt with the ecology of aquatic insects. However, since 1970, more than 7 percent of the articles have dealt with this topic.

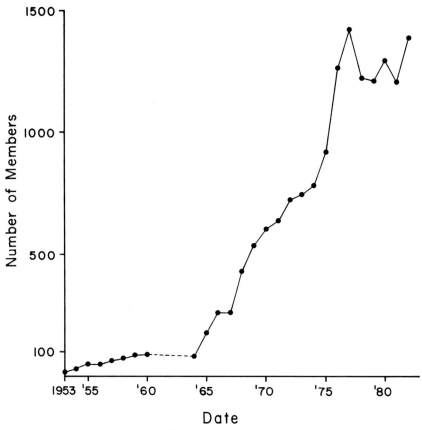

Figure 1.1 Growth of membership in the North American Benthological Society; *dashed line* (1961–63) indicates approximate numbers.

THE LITERATURE OF AQUATIC INSECT ECOLOGY

If, as has been suggested by historians of science, the first appearance of a journal can be used to determine the beginning of a new scientific discipline, then the journal *Aquatic Insects* would be a significant milestone in the growth of aquatic insect ecology. *Aquatic Insects* was first published in 1979 and is the only journal devoted entirely to the taxonomy and ecology of aquatic insects. The late Professor Joachim Illies gave the following reason for the creation of this new journal: "Aquatic entomology is a well established line of research with growing numbers of students, important for entomology as well as limnology, but without a publishing organ of its own" (Illies 1979, p. 1). Prior to *Aquatic Insects,* journals dealing with individual groups of aquatic insects had been published. For example, *Tombo* and *Odonatologica* were established in 1958 and 1972, respectively, for publication of original papers in the field of odonatology. Several journals (e.g. *Mosquito News*, established in 1941) have also been forums for ecological papers on pestiferous aquatic insects.

A book written ten or even five years ago on the ecology of aquatic insects would probably have had a very different table of contents than the one presented here. There are two reasons for this: First, a book prepared earlier would likely have had chapters on each of the orders of aquatic insects. However, several excellent reviews on the ecology of major groups have been published in the last few years. Thus, recent reviews of the biology of the Odonata (Corbet 1980), Plecoptera (Hynes 1976), Ephemeroptera (Brittain 1982), Trichoptera (Mackay and Wiggins 1979), Chironomidae (Oliver 1971), and Tipulidae (Pritchard 1983) make individual treatment of these orders unnecessary. Second, the availability of several excellent reviews has allowed expansion and synthesis, rather than initial review, of several topics that are presented in this book. For example, reviews of ecological processes involving aquatic insects—such as stream drift (Waters 1972; Müller 1974), filter feeding (Wallace and Merritt 1980), trophic relationships (Cummins 1973), detritus processing (Anderson and Sedell 1979), and animal-microbial interactions (Cummins and Klug 1979)—are still relevant, and have provided our contributors with a base upon which to expand and synthesize these topics.

Descriptive studies of aquatic insect ecology are probably more popular now than ever before. The immense literature concerning this topic generally falls into three areas: descriptions of life cycles and aspects of aquatic insect life histories, estimations of secondary production, and responses of benthic populations and communities to environmental alterations. Each of these topics is treated in individual chapters of this book, and expanded upon in others.

Bibliographic compilations of literature on a particular subject are useful for locating references to pertinent studies. The specialized newsletters for individual aquatic insect orders mentioned above (e.g. *Eatonia, Perla*) provide lists of references to studies dealing with those orders and, since 1959, NABS has published annual bibliographies for each of the aquatic insect orders, techniques for benthic studies, toxicology, and other relevant topics. In addition, major bibliographies on the Chironomidae, an important aquatic insect family, have recently been published (Fittkau et al. 1976; Hoffrichter and Reiss 1981).

Difficulties encountered in the identification of aquatic insects, particularly the immature stages, have long hindered ecological studies of aquatic insects. Fortunately, there have been recent compilations of generic keys to the North American fauna (e.g. Pennak 1978 and the revised edition of Merritt and Cummins 1978, which is in preparation), and to the faunas of specific regions (e.g. Brigham et al. 1982). These books have followed the pioneering works of Ward and Whipple (1918), Pennak (1953), and Usinger (1956). In addition, detailed treatments of the immature stages of several aquatic insect orders (e.g. Edmunds et al. 1976; Wiggins 1977; Menke 1980), orders containing aquatic representatives (e.g. McAlpine et al. 1981), aquatic insect families (e.g. Oliver and Roussel 1983), and genera of aquatic insects (e.g. Schuster and Etnier 1978) have facilitated ecological studies. A bibliography of such keys for North America is available in the individual chapters of Merritt and Cummins (1978).

The aquatic insect fauna of Europe is better known than that of North America (e.g. Bertrand 1954; see also discussion in Hynes, Chapter 2). Useful bibliographies of keys are available for several areas in Europe (e.g. Central Europe: Mauch 1966; France: Ginet 1980; Great Britain: Armitage et al. 1979). A major advance in the identification of insects from the European continent will be an updated edition of the classical work of Brauer (1909), which will appear in several volumes over the next decade. Recent keys also are available for the aquatic insect fauna of Africa (Durand and Lévêque 1981), New Zealand (Winterbourn and Gregson 1981), and Australia (Williams 1980). Identification keys to the Chironomidae of the Holarctic Region (Wiederholm, in press) will be a significant development in our ability to identify members of this group.

The systematics of aquatic insects is an especially important subject in ecological studies that attempt an evolutionary approach. Resh and Solem (1978) have provided a brief overview of the phylogenetic relationships of aquatic insects, and detailed phylogenetic treatments are available for the Odonata (Fraser 1957), Ephemeroptera (Edmunds 1972), Plecoptera (Illies 1965), Trichoptera (Ross 1967), and Chironomidae (Brundin 1967).

An earlier version of this book might have also included a series of chapters based on aquatic insect habitats. Recent treatments have reviewed insects in marine systems (Cheng 1976), regulated streams (Ward and Stanford 1979), temporary pools (Wiggins et al. 1980), and urban habitats (Resh and Grodhaus 1983). With a few exceptions, the emphasis in each of the chapters in this book has been on discussing the respective topics in terms of both lentic and lotic systems, and in habitats ranging from small seeps to large lakes.

We have not included a chapter on techniques and procedures for sampling aquatic insects, although the chapters by Newbury (Chapter 11), Barton and Smith (Chapter 15), and Allan (Chapter 16) will be of interest in this regard. A number of recent reviews on various aspects of sampling are available (e.g. sampling equipment: Merritt et al. 1978; artificial substrates: Rosenberg and Resh 1982; sampling variability: Resh 1979), and International Biological Program Handbook Number 17, which contains several chapters related to benthic sampling, has been revised (Downing and Rigler, in press).

Most books on aquatic ecology treat aquatic insects only superficially. However, important exceptions include Hynes (1960, 1970) and Macan (1974). Miall (1895) is an early book on aquatic insect ecology, and Wesenberg-Lund (1943) is a classical treatment of this subject. Although the European literature tends to be underused in English-speaking countries, Wesenberg-Lund (1943) and monographic treatments such as *Chironomus* (Thienemann 1954) are indispensable aids for researchers in aquatic insect ecology.

Several books such as Usinger (1956), Edmondson (1959), Pennak (1978), Merritt and Cummins (1978), Durand and Lévêque (1981), and Brigham et al. (1982) are primarily identification guides, but they provide information on aquatic insect ecology as well. Several of the chapters in Barnes and Minshall (1983) relate studies of stream benthos to ecological theory, and Resh and Rosenberg (1979, 1981) give examples of ecological approaches for use in teaching aquatic entomology.

Finally, far more books about aquatic insects have been written from the perspective of the angler and the hungry trout than from that of the aquatic entomologist. This large body of anecdotal information is usually ignored by aquatic insect ecologists, possibly because some of the information contained in these treatments, particularly the older ones, is inaccurate. The recent book by McCafferty (1981) is a major improvement over most of the previous literature on this subject.

We will not try to predict the direction of future developments and advances in aquatic insect ecology, since each contributor to this book has identified several possible areas within each of the topics covered. We do predict, however, that interest in the ecology of aquatic insects will continue to flourish. We hope that the information contained in this book will be a useful source of information for students and researchers who examine the ecology of these fascinating animals.

References

Anderson, N. H. and J. R. Sedell. 1979. Detritus processing by macroinvertebrates in stream ecosystems. Annual Review of Entomology 24:351–77.

Armitage, P. D., M. T. Furse, and J. F. Wright. 1979. A bibliography of works for the identification of freshwater invertebrates in the British Isles. Freshwater Biological Association Occasional Publication Number 5. 18 pp.

Barnes, J. R. and G. W. Minshall (eds.). 1983. Stream ecology: application and testing of general ecological theory. Plenum Publishing Co., New York, NY. (In press).

Bertrand, H. 1954. Les insectes aquatiques d'Europe. Encyclopédie Entomologique 30: 1–566, 31:1–547.

Brauer, A. 1909. Die Süsswasserfauna Deutschlands: Eine Exkursionsfauna. Gustav Fischer, Jena, Germany. Vols. 2–9.

Brigham, A. R., W. U. Brigham, and A. Gnilka (eds.). 1982. Aquatic insects and oligochaetes of North and South Carolina. Midwest Aquatic Enterprises, Mahomet, IL. 822 pp.

Brittain, J. E. 1982. Biology of mayflies. Annual Review of Entomology 27:119–47.

Brundin, L. 1967. Insects and the problem of austral disjunctive distribution. Annual Review of Entomology 12:149–68.

Cheng, L. (ed.). 1976. Marine insects. North Holland Publishing Co., Amsterdam, The Netherlands. 581 pp.

Corbet, P. S. 1980. Biology of Odonata. Annual Review of Entomology 25:189–217.

Cummins, K. W. 1973. Trophic relations of aquatic insects. Annual Review of Entomology 18:183–206.

Cummins, K. W. and M. J. Klug. 1979. Feeding ecology of stream invertebrates. Annual Review of Ecology and Systematics 10:147–72.

Downing, J. A. and F. H. Rigler (eds.). In press. A manual on methods for the assessment of secondary productivity in fresh waters. 2nd ed. International Biological Program Handbook Number 17. Blackwell Scientific Publications, Oxford, England.

Durand, J. -R. and C. Lévêque (eds.). 1981. Flore et faune aquatiques de l'Afrique sahelo-soudanienne. Vol. 2. Office de la Recherche Scientifique et Technique Outre-Mer, Paris, France. pp. 391–873.

Edmondson, W. T. (ed.) 1959. Fresh-water biology. 2nd ed. John Wiley and Sons Inc., New York, NY. 1,248 pp.

Edmunds, G. F., Jr. 1972. Biogeography and evolution of Ephemeroptera. Annual Review of Entomology 17:21–42.

Edmunds, G. F., Jr., S. L. Jensen, and L. Berner. 1976. The mayflies of North and Central America. University of Minnesota Press, Minneapolis. 330 pp.

Fittkau, E. J., F. Reiss, and O. Hoffrichter. 1976. A bibliography of the Chironomidae. Det Kongelige Norske Videnskabers Selskab, Museet, Gunneria 26:1–177.

Fraser, F. C. 1957. A reclassification of the order Odonata. Royal Zoological Society of New South Wales Handbook Number 12. 133 pp.

Ginet, R. 1980. Bases d'une bibliographie pour la détermination de la fauna des eaux douces. Association Francaise de Limnologie, Hors-série, 24 pp.

Hoffrichter, O. and F. Reiss. 1981. Supplement 1 to "A bibliography of the Chironomidae." Det Kongelige Norske Videnskabers Selskab, Museet, Gunneria 37:1–68.

Hynes, H. B. N. 1960. The biology of polluted waters. Liverpool University Press, Liverpool, England. 202 pp.

Hynes, H. B. N. 1970. The ecology of running waters. University of Toronto Press, Toronto. 555 pp.

Hynes, H. B. N. 1976. Biology of Plecoptera. Annual Review of Entomology 21:135–53.

Illies, J. 1965. Phylogeny and zoogeography of Plecoptera. Annual Review of Entomology 10:117–40.

Illies, J. 1979. Editorial. Aquatic Insects 1:1.

Macan, T. T. 1962. Ecology of aquatic insects. Annual Review of Entomology 7:261–88.

Macan, T. T. 1974. Freshwater ecology. 2nd ed. John Wiley and Sons Inc., New York, NY. 338 pp.

Mackay, R. J. and G. B. Wiggins. 1979. Ecological diversity in Trichoptera. Annual Review of Entomology 24:185–208.

Mauch, E. 1966. Bestimmungsliteratur für Wasserorganismen in mitteleuropäischen Gebiet. Schriftenreihe des Vereins für Wasser-, Boden-, und Lufthygiene 24:1–22.

McAlpine, J. F., B. V. Peterson, G. E. Shewell, H. J. Teskey, J. R. Vockeroth, and D. M. Wood (coordinators). 1981. Manual of Nearctic Diptera. Volume 1. Monograph Number 27, Research Branch, Agriculture Canada, Ottawa. 674 pp.

McCafferty, W. P. 1981. Aquatic entomology. The fishermen's and ecologists' illustrated guide to insects and their relatives. Science Books International, Boston. 496 pp.

Menke, A. S. (ed.). 1980. The semiaquatic and aquatic Hemiptera of California (Heteroptera: Hemiptera). Bulletin of the California Insect Survey 21:1–166.

Merritt, R. W. and K. W. Cummins (eds.). 1978. An introduction to the aquatic insects of North America. Kendall/Hunt Publishing Co., Dubuque, IA. 441 pp.

Merritt, R. W., K. W. Cummins, and V. H. Resh. 1978. Collecting, sampling, and rearing methods for aquatic insects, pp. 13–28. *In*: R. W. Merritt and K. W. Cummins (eds.). An introduction to the aquatic insects of North America. Kendall/Hunt Publishing Co., Dubuque, IA. 441 pp.

Miall, L. C. 1895. The natural history of aquatic insects. Macmillan and Co., London, England. 395 pp.

Morse, J. C. 1979. Techniques used in teaching aquatic insect taxonomy in North American colleges and universities, pp 3–13. *In*: V. H. Resh and D. M. Rosenberg (eds.). Innovative teaching in aquatic entomology. Canadian Special Publication of Fisheries and Aquatic Sciences 43. 118 pp.

Müller, K. 1974. Stream drift as a chronobiological phenomenon in running water ecosystems. Annual Review of Ecology and Systematics 5: 309–23.

Oliver, D. R. 1971. Life history of the Chironomidae. Annual Review of Entomology 16: 211–30.

Oliver, D. R. and M. E. Roussel. 1983. The genera of larval midges of Canada. Diptera: Chironomidae. The insects and arachnids of Canada. Part 11. Publication Number 1746, Research Branch, Agriculture Canada, Ottawa.

Pennak, R. W. 1953. Fresh-water invertebrates of the United States. The Ronald Press Co., New York, NY. 769 pp.

Pennak, R. W. 1978. Fresh-water invertebrates of the United States. 2nd ed. John Wiley and Sons Inc., New York, NY. 803 pp.

Pritchard, G. 1983. Biology of Tipulidae. Annual Review of Entomology 28:1–22.

Resh, V. H. 1979. Sampling variability and life history features: basic considerations in the design of aquatic insect studies. Journal of the Fisheries Research Board of Canada 36:290–311.

Resh, V. H. and G. Grodhaus. 1983. Aquatic insects in urban environments. *In*: G. W. Frankie and K. W. Koehler (eds.). Urban entomology: interdisciplinary perspectives. Praeger Publishers, New York, NY. (In press).

Resh, V. H. and D. M. Rosenberg (eds.). 1979. Innovative teaching in aquatic entomology. Canadian Special Publication of Fisheries and Aquatic Sciences 43. 118 pp.

Resh, V. H. and D. M. Rosenberg (eds.). 1981. Enseignement novateur de l'entomologie aquatique. Publication Spéciale Canadienne des Sciences Halieutiques et Aquatiques 43F. 122 pp.

Resh, V. H. and J. O. Solem. 1978. Phylogenetic relationships and evolutionary adaptations of aquatic insects, pp. 33–42. *In*: R. W. Merritt and K. W. Cummins (eds.). An introduction to the aquatic insects of North America. Kendall/Hunt Publishing Co., Dubuque, IA. 441 pp.

Rosenberg, D. M. and V. H. Resh. 1982. The use of artificial substrates in the study of freshwater benthic macroinvertebrates, pp. 175–235. *In*: J. Cairns, Jr. (ed.). Artificial substrates. Ann Arbor Science Publishers Inc., Ann Arbor, MI. 279 pp.

Ross, H. H. 1967. The evolution and past dispersal of the Trichoptera. Annual Review of Entomology 12:169–206.

Schuster, G. A. and D. A. Etnier. 1978. A manual for the identification of the larvae of the caddisfly genera *Hydropsyche* and *Symphitopsyche* Ulmer in Eastern and Central North America (Trichoptera: Hydropsychidae). EPA-600/4-78-060. 129 pp.

Stewart, K. W. 1977. Formation of the Aquatic Insects Subsection, ESA. Bulletin of the Entomological Society of America 23:245–46.

Thienemann, A. 1954. Chironomus. Leben, Verbreitung und wirtschaftliche Bedeutung der Chironomiden. Die Binnengewässer 20:1–834.

Usinger, R. L. (ed.). 1956. Aquatic insects of California with keys to North American genera and California species. University of California Press, Berkeley, CA. 508 pp.

Wallace, J. B. and R. W. Merritt. 1980. Filter-feeding ecology of aquatic insects. Annual Review of Entomology 25:103–32.

Ward, J. V. and J. A. Stanford (eds.). 1979. The ecology of regulated streams. Plenum Publishing Co., New York, NY. 398 pp.

Ward, H. B. and G. C. Whipple (eds.). 1918. Fresh-water biology. John Wiley and Sons Inc., New York, NY. 1,111 pp.

Waters, T. F. 1972. The drift of stream insects. Annual Review of Entomology 17:253–72.

Wesenberg-Lund, C. 1943. Biologie der Süsswasserinsekten. Springer, Berlin. 682 pp.

Wiederholm, T. (ed.). In press. Chironomidae of the Holarctic Region. Keys and diagnoses. Part 1. Larvae. Entomologica Scandinavica Supplement.

Wiggins, G. B. 1977. Larvae of the North American caddisfly genera (Trichoptera). University of Toronto Press, Toronto. 401 pp.

Wiggins, G. B., R. J. Mackay, and I. M. Smith. 1980. Evolutionary and ecological strategies of animals in annual temporary pools. Archiv für Hydrobiologie Supplement 58: 97–206.

Williams, W. D. 1980. Australian freshwater life. Macmillan Co. of Australia Pty. Ltd., Melbourne, Australia. 321 pp.

Winterbourn, M. J. and K. L. D. Gregson. 1981. Guide to the aquatic insects of New Zealand. Bulletin of the Entomological Society of New Zealand 5:1–80.

chapter 2

THE RELATIONSHIPS BETWEEN THE TAXONOMY AND ECOLOGY OF AQUATIC INSECTS

H. B. N. Hynes

INTRODUCTION

Insects that live in or on water are a remarkably diverse assemblage drawn from almost one-half of the orders of the arthropod class Insecta. It is clear therefore that despite their terrestrial origin (as shown, for example, by their tracheal systems, their almost universally aerial adults, and their waxy cuticles) the life-form of the Insecta is fairly easily adjustable to an aquatic existence. However, despite their notable success in inland waters, insects have made few inroads into the oceans, where they are represented mainly by a few species of Gerridae and some Chironomidae, all of which are undoubtedly derived from freshwater ancestors.

One may conclude that this failure to dominate the seas, as they have the inland waters, results from competitive exclusion by animals that were already resident in marine environments when the insects evolved about 350 million years ago. We can infer this because many insects inhabit inland waters that are saline and some species do occur in the sea; thus we know that salinity alone presents no insurmountable barrier to their physiology. Similarly, the physical conditions of the seas are no barrier; the wave-swept shores of the Laurentian Great Lakes in North America are rich in insect species (Barton and Hynes 1978), as are the sediments far from shore. Also, although the plankton is normally dominated by Crustacea, that of Victoria Nyanza in Africa is so densely inhabited by the larvae of *Chaoborus* (the phantom midge) that the adults often occur as great clouds like waterspouts over the lake. At times they blow ashore to cover the Ugandan town of Entebbe in an unpleasant grey coat of flies. (Thus it is appropriate than an inn there is named the Lake Fly Hotel!)

An alternative explanation for the failure of insects to dominate the seas (put forward by Hinton 1976) is that the violence of the intertidal zone has tended to exclude them from the oceans. This, however, seems implausible, as there are many types of insects in torrents, and many parts of the tidal zone are sheltered. It would seem more reasonable to accept the point, also shown by other habitats such as temporary pools, that where a group of organisms is firmly established in a habitat, and is occupying most of the ecological niches in a well-adapted way, it is more difficult for another group—even if it is an evolutionarily more advanced model of the same life-form—to invade the occupied territory.

PHYLOGENETIC RELATIONSHIPS

We see the above point illustrated by the aquatic insects in that they consist of a series of whole orders (Ephemeroptera, Odonata, Plecoptera, Megaloptera, Trichoptera) or suborders (Hydrocorisae, Hydradephaga, Nematocera) that are completely, or are very largely, aquatic. We may call these the primary invaders. There are also other groups that are relatively small taxonomical entities drawn from larger orders, most members of which are terrestrial. We may call these the secondary invaders. Such are the four families of water striders (Gerridae, Hydrometridae, Veliidae, and Mesoveliidae), many varied families of beetles and higher Diptera (Brachycera), a single family of Mecoptera (Nannochoristidae), some Lepidoptera, and even a few species of parasitic Hymenoptera. Within these groups, particularly the beetles, there is evidence that some of the aquatic families are closely related and thus may have diversified in the aquatic habitat after they invaded it (e.g. the Hydrophilidae and Hydraenidae, and the several families of the Dryopoidea). However, in most instances these groups consist of single families or even odd genera or species that clearly have entered the water quite independently.

In addition, there are some groups that have a different association with water in that they live in damp places along banks and shores, or even run on the water surface. These include springtails (Collembola) of several families, the pigmy mole crickets (Tridactylidae), the shore bugs (Saldidae), some staphylinid beetles, and representatives of many dipteran families. In tropical America there is even a cockroach that spends much of its time submerged in the pools of water collected in the leaf axils of the epiphytic Bromeliaceae. The occurrence of several other orthopteroid insects in aquatic situations in various parts of the tropics is discussed by Wesenberg-Lund (1943).

If we examine these various groups we see that each of the whole orders or suborders in our primary group is, in an evolutionary sense, a primitive one, whereas nearly all the secondary groups are representatives of evolutionarily more advanced taxa, which have somehow managed to infiltrate the aquatic habitat despite its presumably earlier occupancy by the less evolved groups. In this respect the secondary invaders resemble the rather few groups of insects that have managed to invade the oceans despite the competition there from older, established invertebrate groups. Indeed we know from the imperfect fossil record that Odonata, Ephemeroptera, and Plecoptera already had aquatic nymphs 250 or more million years ago, and it is easy to demonstrate that all of these primitive, primary groups have become very highly adapted to aquatic life.

The ingenuity and quality of these adaptations have long fascinated entomologists, and near the turn of the century Miall (1895) published a semipopular book describing and discussing many of them. Since that time they have become part of the lore of every entomological textbook, and much anatomical, behavioral, and physiological research has been done on them. It is appropriate to review them briefly here.

Locomotion

Among hydradephagan beetles and hydrocorisan bugs, both groups in which the adults remain aquatic, the adaptations to an aquatic existence include great streamlining of the body and modifications of the legs to form paddles. These adult insects are therefore efficient swimmers, but, like the mammalian seals, they have lost the ability to move satisfactorily on land. Similarly, among larval stages, one finds either hair-fringed legs that can be used as paddles (as in the bugs and some beetle larvae), or the ability to swim, which is performed by movements of the whole trunk. It is interesting to note that both the up and down movement of trunk and tail, as used by whales, and the side-to-side movement, as used by fishes or seals, have been adopted by aquatic insects. The former is well shown by mayfly larvae and some dytiscid beetles, and the latter by several groups, including stoneflies and damselflies. Usually the fins are formed by rows of hairs on terminal appendages, but in the damselflies they are formed by flattened cerci and filum terminalis, and, at first glance, the swimming larvae of some genera appear remarkably similar to larval fishes.

Larval stages of other groups have adopted modes of swimming unknown among the vertebrates. One mode is the rectal jet-propulsion of the anisopteran dragonflies, which is similar in principle to the swimming of a squid; another is the strange-looking wiggling motion that is so characteristic of mosquito larvae. In mosquitoes, and their relatives the Chaoboridae, this motion is aided by a fin formed by a median ventral line of hairs at the posterior end of the body, and as a result the larvae swim fairly well. However, in many other dipterans, such as the Chironomidae and many Ceratopogonidae (biting midges), and also in the Nannochoristidae (scorpionflies), no fin is present and the larvae are not very effective swimmers. It seems, however, that this motion is used in nature only to enable migration from adverse conditions, and it functions primarily as a means of suspension in the water column while currents cause dispersal. Perhaps, therefore, swimming is not really the correct term to use for this clumsy and apparently inept movement of larvae that lack fins. Probably, as in the nematodes that move in the same manner, this sinuous motion was primarily developed for moving through loose aquatic sediments, and thus represents a different adaptation to aquatic life.

Respiration

Respiratory mechanisms are also greatly modified in aquatic insects. Adult beetles and bugs continue to breathe air, and their spiracles open into bubbles that are held somewhere on the body—either under the wings, or by arrangements of unwettable hairs that keep the bubble against the body. Most such adult insects renew the air at intervals. Associated with this one finds a whole series of devices, mostly based on unwettable hairs, that penetrate the air-water interface to allow rapid gas exchange. Many of these devices were excellently described by early students of aquatic insects, and even recent studies with the scanning electron microscope have added little to our further understanding of their mechanisms.

An additional advantage to the insect is that a bubble held under air-saturated water has the valuable property of functioning as a so-called physiological gill. As oxygen is withdrawn from it and replaced by expired carbon dioxide, the very soluble CO_2 goes into solution in the water. Then, because the oxygen/nitrogen ratio in the bubble has become lower than it is in the surrounding solution, more oxygen enters the bubble. Thus the available supply of oxygen is greatly enhanced, and the bubble can deliver much more oxygen than it originally contained. Of course the nitrogen in the bubble will dissolve eventually even though the water is air-saturated, and then the insect must resurface to replenish its air supply. However, if some means of holding the bubble open can be evolved, as by surface tension between unwettable hairs or by enclosure in a rigid tracheal system, permanent underwater respiration is possible. We find the occurrence of unwettable hairs among some bugs; for example, *Aphelocheirus*, on which the classic work on so-called plastron respiration was done (Thorpe and Crisp 1947), and also on some beetles. Such hairs do not, however, occur on the primitive Hydradephaga, even though many species of this suborder can exploit their bubbles in this way for long periods, as under ice or during hibernation. Essentially, a plastron is a close pile of unwettable hairs on the body surface, among the bases of which the often elaborately branched spiracles open. However, a closed tracheal system, some parts of which lie close to the surface, functions in the same manner, and it is among larval stages that this mechanism has been particularly exploited.

A few larvae of the primary aquatic groups, notably mosquitoes and dytiscid beetles, retain an open tracheal system, although its openings are restricted to a pair of spiracles at the extreme posterior tip of the body, which can be brought to the water surface. Such devices, as in the adult beetles, have mechanisms for breaking the surface film. In the overwhelming majority of the primary aquatic groups, however, the tracheal systems of the larvae have become closed, but they are nevertheless filled with air and thus function as physiological gills. The tracheal system becomes filled with gas shortly after the larva hatches, by a mechanism that is not clearly understood, and thereafter it remains as a closed bubble, until the molt to a pupal or adult form with an open tracheal system puts it into contact with the atmosphere.

In all these so-called apneustic insects (i.e. those with a closed tracheal system), many fine branches of the tracheae lie near the surface, and often they extend out into thin-walled tubular, branched, or platelike tracheal gills. Such gills occur throughout the representatives of the primary aquatic groups and on various parts of the bodies of the insects. Clearly they have arisen independently on many occasions, and they are often important aids in identification. In several groups (Ephemeroptera, Megaloptera, and some Plecoptera, Zygoptera, and Coleoptera) the gills are segmentally arranged on the abdomen, and it has been suggested that they may represent abdominal appendages persisting from primitive insect ancestors. In most mayflies they are articulated and can be moved, and they are used to waft a current of water over the body or through a burrow. Similar irrigation currents that facilitate gas exchange are produced by undulations of the body in a tube or burrow, and are aided by fringes of hairs along the sides of the body of caddisfly larvae or by lateral shelflike welts along the sides of the abdominal segments in some chironomid larvae. Also, as already mentioned, anisopteran dragonfly larvae take water into the rectum, which is where the tracheal gills of this group occur. It is therefore apparent that in these and many other aquatic larvae, active respiratory movements are made that are quite different from the normal abdominal pumping respiratory movements used by terrestrial insects.

One should, however, be careful to note that not all the thin-walled projections on the bodies of aquatic insects are respiratory structures, even though they are often referred to as gills. For example, Wigglesworth (1933) showed long ago that the anal "gills" of mosquito larvae are really concerned with the uptake of salts. Also, the fact that the posterior gills of Nannochoristidae house the tips of the Malpighian tubules indicates that these structures may be concerned with water uptake (Pilgrim 1972).

A further development is the occurrence in some groups of the respiratory pigment hemoglobin. This is most unusual in insects, but we find it in the blood of many Chironomidae and also in special organs in one subfamily of back-swimmers, the Anisopinae. These bugs are unusual in that they lurk in the middle depths of the water, unlike their relatives who must either swim or hold onto something to prevent being floated up by their bubble. The hemoglobin of the Anisopinae stores much of the oxygen that they need, and it is also used to maintain the size of the bubble. The bubble can therefore be small, sufficient only to adjust the density of the insect to that of the water, and it can be maintained at constant volume for long periods (Wells et al. 1981). This enables these bugs to remain at a level that allows them to feed on planktonic microcrustacea, a food resource that is not available to most aquatic insects. It is interesting to note that the only other insect exploiters of this resource, the Chaoboridae, also have a modified respiratory system that gives them neutral buoyancy. In these phantom midges, two conchoidal bubbles remain fore and aft in each of the major tracheal trunks, and thus function like fish bladders to adjust the density of the insect. The rest of the tracheal system is filled with fluid until late in development, and so is presumably nonfunctional. Further aspects of aquatic insect respiration have been reviewed by Eriksen et al. (in press).

Terrestrial Life of Aquatic Insects

The adult stages of aquatic insects are usually terrestrial. Even the exceptions, the hemipteran suborder Hydrocorisae, the coleopteran suborder Hydradephaga and a few other families of beetles (e.g. Hydrophilidae, Hydraenidae, and Elmidae), are capable of leaving the water and flying to new habitats. It should, however, be noted that flightless specimens, which presumably do not leave the water, have been recorded as components of the populations of many species. Completely aquatic insects are a rarity, and among the beetles with aquatic adults, the pupae are terrestrial, or at least confined to an air pocket under water. Thus, aquatic insects have not severed their connection with the aerial environment except in a few very special cases. Examples of these are the wingless females of the moth *Acentropus*, whose males nevertheless do fly (Berg 1941), and the Lake Tahoe, California and Nevada, stonefly, whose adults have been found in deep water but not on land (Jewett 1963). Of the latter, however, we know so little that our interpretation of the known facts may be erroneous; the Lake Tahoe stonefly may indeed have at least some aerial adults.

Be that as it may, the great majority of the adult aquatic insects are certainly terrestrial; they disperse and mate on land and only the beetles and bugs already mentioned mate in the water. For many species the adult life is brief and is primarily reproductive; the adults of many groups do not even feed (e.g. mayflies, caddisflies, and many stoneflies and Nematocera). This means that much of the selective pressure exerted on the species has been on the immature stages, and it is probably because of this that the adults of, particularly, mayflies, stoneflies, and caddisflies are far less obviously morphologically diverse than are their larvae. In these groups one can usually identify larvae at least as to their family at arm's length with the naked eye; adults usually require much closer scrutiny.

In general, therefore, adults show less obvious variety within groups than do the larvae, and while some have obvious adaptations to such things as hunting (dragonflies) or feeding (mosquitoes, black flies), many are just breeding machines. However, perhaps because there has been no evolutionary pressure for change, we find that Ephemeroptera and Nematocera have elaborate nuptial dances, presumably retained from remote ancestors, whereas in other groups different mating rituals have allowed greater flexibility of form. Thus there are some flightless caddisflies, such as the winter form of *Dolophilodes* in North America and the Lake Baikal caddisflies of the subfamily Baicalini (Kozhov 1963). Among stoneflies, flightlessness, brachyptery, or even aptery, are not rare. Flightlessness, although it prevents dispersal, does have the advantage of keeping adults from being blown far from isolated or unique habitats, and it thus permits extreme species specialization. This is perhaps one reason why stoneflies have proved to be such interesting subjects for zoogeographical study (Illies 1965; Zwick 1980).

Feeding

Lastly, among our group of primary colonizers of inland waters, a major fact, and one that must represent evolutionary adaptation, is that many species feed on fine particles. The birthright of primitive insect groups is chewing mouthparts, and these together with the apparently early evolved piercing mouthparts of the bugs (first known from fossils about 250 million years old), are well adapted to carnivory. Indeed many of the more primitive (plesiomorphic, to use modern taxonomical jargon) families of our primary aquatic groups are, or at least include, carnivores. Such are the Siphlonuridae among the mayflies (Edmunds 1972), the Eustheniidae among the stoneflies (Zwick 1980), and the Rhyacophilidae among the trichopterans (Ross 1956, 1967). The Odonata, the Megaloptera, and most of the Hydrocorisae have all remained carnivorous. However, among the other families a great variety of feeding habits has developed. Many mayfly, stonefly, caddisfly, and nematoceran larvae scrape algae off stones, whereas others feed on small detrital particles that are on the substratum or that are sifted from the water with special appendages (e.g. mosquitoes).

In running water, elaborate mechanisms have been evolved for straining suspended solids from the current (Hynes 1970; Wallace and Merritt 1980; Merritt and Wallace 1981). Wooton (1972) noted that few aquatic insects feed on the tissues of higher plants, and he concluded that this may be because such material did not become widely available until the Tertiary period, long after these insects had become aquatic. Indeed, nearly all of those that do eat higher plant tissue do so after it has died, and very often it is terrestrial litter that has fallen into the water. These insects are the shredders of Cummins (1973) and they are mostly Trichoptera and Plecoptera. Even among the bugs, we find that the Corixidae, the most abundant family, have taken to sucking up fine particles in a manner unique among the Hemiptera (Hungerford 1948).

TAXONOMY AND ECOLOGICAL DIVERSITY

We can conclude from this brief survey that the primary aquatic groups have become highly adapted to life in water. Within these groups there has developed a wide diversity of niche, and most of the groups are represented in almost every type of aquatic habitat. Among the Ephemeroptera and Trichoptera in particular, diversity of form and way of life are very wide (Illies 1968; Edmunds 1972; Edmunds et al. 1976; Mackay and Wiggins 1979). Among other groups there has been some conservatism (e.g. carnivory in some families as mentioned previously and the necessity for cool well-oxygenated water among Plecoptera), but even in these groups there is great diversity in habitat and life-style (Corbet 1962; Zwick 1980).

One may note also that the "difficult" aquatic habitats have been colonized by representatives of these ancient aquatic groups. Thus Chironomidae dominate the deep, often deoxygenated, sediments of lakes; and torrential streams are populated almost entirely by Ephemeroptera, Plecoptera, Trichoptera, and

Nematocera, many of which are elaborately adapted to this peculiar environment (Hynes 1970).

In contrast to these primary orders of aquatic insects, the more taxonomically restricted, secondary groups belonging to orders that are predominantly terrestrial show far less diversity of structure and habitat, although many of them have clear adaptations to aquatic life. Thus, the four families of water striders have basically only an unwettable pile of hairs and a particularly waxy cuticle as an adaptation to life on the water surface, and they have, in fact, not moved from that restricted habitat. Indeed, the Hydrometridae, Mesoveliidae, and many small species of Veliidae live very much at the edge of the water and so are only slightly committed to it. However, one should note that the unwettability of the genus *Halobates,* which has a cuticle as waxy as the skin of a plum, enables it to survive ocean breakers, and that some genera of Veliidae, most notably *Rhagovelia,* have highly modified tarsi that give them a foothold even on the surface of turbulent streams.

Similarly, aquatic Lepidoptera, primarily members of the family Pyralidae, are most commonly associated with higher plants as they are on land. In this respect they are unusual among aquatic insects, and their caterpillars show few modifications from those of their terrestrial relatives. However, a few genera have developed tracheal gills and some have penetrated into rapidly running water where their ability to spin silk allows them to remain stationary. This invasion of running water habitats involved also a considerable change in their diet, from higher plants to algae scraped off the stones. But such unusual diets are by no means rare among caterpillars; one has only to think of wood borers and clothes moths. It is also perhaps significant that stream-dwelling caterpillars are the dominant animals only on some isolated Pacific islands where competition is minimal because the older aquatic groups have not become widely dispersed onto oceanic islands.

The Mecoptera, represented only by the Nannochoristidae of the Southern Hemisphere, seem to occupy only one niche, submerged silt in small streams and on lake shores. The carnivorous larvae are elongate and move like nematodes in the soft substrate, but apart from their closed spiracles, they show little adaptation to aquatic life. Certainly the group seems to have diversified little after its initial move into the water, which must date from before the breakup of Gondwanaland.

The same may be said of many of the families of Brachycera that have aquatic larvae. By and large, each of the families involved has remained as inhabitants of the water margins or of shallow water where aerial respiration is possible. It should be recalled that universal features of dipteran larvae are that they have reduced numbers of spiracles and that the posterior ones are the most important. They are therefore well adapted to foraging below a fluid surface while remaining in contact with the air. The addition of a mechanism around the spiracles for breaking the surface film, as is shown so well by the Stratiomyidae, or a snorkel-like tube as found in such Syrphidae as *Eristalis*, merely perfects this process, and one finds, in fact, that all of these families occupy fairly restricted habitat ranges.

It should be noted though, that this life-style, with the head directed downward for feeding and the posterior spiracles directed upward to the air, does enable many higher Diptera to thrive in fluids that would be uninhabitable if the larvae were totally submerged. Such is the case in the soup of rotting tissue in which flesh-fly maggots live, and, in the aquatic situation, the brines and seeps of mineral oil that have been invaded by some Ephydridae. Also, the Athericidae (sometimes included in the Rhagionidae) and some genera of other dipteran families have been able to break away from contact with the water surface and move out into the erosional beds of turbulent streams. Unlike other Diptera in such habitats, they retain their spiracles, but it seems that their respiration in this well-oxygenated habitat resembles that of insects with closed tracheal systems. They do not, however, become numerically dominant in such places, presumably because of competitive inferiority to the well-established, more ancient, stream insects.

Among the higher beetles we have somewhat more variety of habitat and adaptation, but even here most families are fairly uniform in life-style. Some, such as the weevils (Curculionidae), occur only at or very near the surface on aquatic plants, as do the Chrysomelidae (plant beetles)—except for the genus *Donacia*, whose larvae live on tubers of cattails. Associated with this is the modification of an abdominal spiracle that enables them to tap the aerenchyma in the plant for an air supply. Although the adults of Hydrophilidae swim fairly well and have an efficient mechanism operated by the antennae for breaking the surface film, both they and their larvae live in marginal and vegetation-rich habitats, and thus near the fringes of the aquatic world. The rather similar Hydraenidae live in somewhat the same way, but their larvae are confined more closely to the water's edge. In contrast to the Hydrophilidae, in most of which only the posterior spiracles are open, hydraenids have functional spiracles all along the abdomen. Also, although they do not swim, adults of *Hydraena* crawl out onto the beds of small stony streams, a habitat where one rarely finds a member of the Hydrophilidae. Presumably their bubble, which covers most of the ventral surface, functions as a plastron.

The flattened larvae of helodid beetles occur frequently in woodland ponds and have little obvious adaptation to the water apart from their posterior spiracles and the expanded internal airsacs that keep them at the surface. Some genera, however, occur in stony streams, and little is known about them. The adults are terrestrial and are rarely collected in the water. The reverse is true of the Dryopidae in that their larvae are terrestrial or marginally aquatic and the adults live under water, where they clamber on emergent plants without any clear adaptation to aquatic life. In North America, adults of the genus *Helichus*, like some Hydraenidae, crawl out onto stream beds.

The remaining three families of beetles (Elmidae, Ptilodactylidae, and Psephenidae) can be regarded as the most adapted to aquatic life among these secondary aquatic families, in that many of their larvae have tracheal gills and are not confined to the water's edge. Also, the adults of the Elmidae are aquatic, and although they do not swim, they have a well-developed plastron for respiration. Elmidae occur commonly and widely on lake shores and in running water, and their creeping larvae range in form from flattened to cylindrical. The variety of habitat types occupied by members of this family is comparable to that of some families of the Hydradephaga, such as the Haliplidae or the Gyrinidae. The adults of the Ptilodactylidae and Psephenidae are terrestrial, but their larvae live in streams and on stony lake shores. Larvae of the Psephenidae, especially *Psephenus*, the water penny, are highly modified for attachment to stones in swift water (Hynes 1970).

Lastly, we have the few parasitoid Hymenoptera that are very little modified to aquatic life and enter water only to find their hosts (Hagen 1978), and representatives of several groups already mentioned (Collembola, Saldidae, Staphylinidae, Dictyoptera, Tridactylidae, and some other Orthoptera) that live at the water's edge, but mostly on the shore. No particularly striking adaptations to aquatic environments are found here, although it is worth noting that the staphylinid *Dianous* secretes a substance from glands at its abdominal tip that enables it to return very quickly to the bank should it fall onto the water. In this respect it resembles the water strider *Velia* that can use a salivary secretion in the same way (Hynes 1970).

Reviewing all these points we can conclude that the more ancient, primary aquatic groups are closely and diversely adapted to their habitat, and have become ecologically and often morphologically diverse. The more recent, secondary invaders are representatives of higher taxa that evolved on land, and are generally much more restricted in their habitat. In a few secondary groups (e.g. Psephenidae and Elmidae), however, there has been much morphological adaptation and some diversification of habitat. This probably indicates that such families have been aquatic for long periods, especially when, as in these cases, the families have worldwide distribution. On the other hand changes such as the development of tracheal gills need not indicate a very old commitment to the aquatic habitat. We see this in the Lepidoptera, a group that appeared during the Tertiary and has probably had aquatic representatives for only about 50 million years.

Taxonomy and ecology tend therefore to be closely related in that the narrower, higher, and presumably more recently aquatic taxa are less ecologically diverse than the wider, lower, and more anciently aquatic taxa. This is, however, not an unexpected conclusion because the more recent groups each consist of a single, or, at most, a few families. It would seem that ecological diversity occurs more between families than it does within them.

To a great extent this is also true of the ancient groups. For instance, in the

Ephemeroptera the Baetidae are all streamlined swimmers, in the Plecoptera the Perlodidae and Eustheniidae are carnivorous inhabitants of stony substrata, and so on. However, there are many families of which this is not true at all. For example, within each of the following orders and families, and the list is by no means exhaustive, there is a wide range of life-style and habitat, mostly observable in the larval stages: Siphlonuridae and Leptophlebiidae of the Ephemeroptera; Aeschnidae and Agrionidae of the Odonata; Naucoridae and Corixidae of the Hydrocorisae; Corydalidae of the Megaloptera; Dytiscidae of the Coleoptera; Leptoceridae and Limnephilidae of the Trichoptera; and Culicidae and Chironomidae of the Nematocera. This reflects the fact that our classification of insects is based on adults, but, as we have seen, the force of selection has been primarily on the larvae and the adults tend to be conservative. Among the secondary invaders, only in the Elmidae do we find a family that has comparable ecological diversity to those listed above, although in each instance such a comparison would favor the ecological diversity of the family in the above listing.

AQUATIC INSECT TAXONOMY

Therefore, one can say that often, indeed usually, family boundaries define those of some ecological niches, but that there are very many exceptions—particularly among the ancient primary groups. This brings us to the question, How good is our taxonomic knowledge of aquatic insects? The answer varies greatly according to the group and to the geographical location.

In general, the biting flies (such as the Culicidae, Simuliidae, and Tabanidae) are well known almost all over the planet because of their medical and veterinary importance as carriers of disease and promoters of discomfort. In the first two families the larval stages have been well described and are identifiable to species at least in the later instars, but this is not true of the Tabanidae. In other groups much depends upon the location.

Study of aquatic insects began in Western Europe about a century ago, and one can say that much of the fauna of that region is fairly well known, even to the point of its being possible to map out the broad distributional patterns of most of the species (Illies 1978). This is particularly true of such well-studied areas as the British Isles, France, Scandinavia, and Germany. The larval stages of most groups have also been extensively studied in Europe. One can usually identify them at least to genus and often, in the areas mentioned above, to species. Much of this detailed work has been done during the past forty years, but even before that identification manuals for larval stages had begun to appear (Brauer 1909; Rousseau 1921). Even so, a continuous trickle of new species continues to be described, mostly from the peripheral parts of the region, and the taxonomy of certain difficult groups, such as the Baetidae, and, in particular, the ecologically important Chironomidae, remains far from clear. It should also be noted that the study of aquatic insects in Europe has been favored not only by its long-standing and concentrated population of interested entomologists, but also by the fact that it has a more restricted fauna than most comparable areas. This is because it was particularly affected by the Pleistocene glaciations, caused by the east to west orientation of the mountains in mid-latitudes. This resulted in a much more extensive area of severe climate than occurred, for example, in North America or Northern Asia.

Much work has been done in North America, but although a comprehensive work appeared early (Ward and Whipple 1918), the fauna is much more extensive. It is probably fair to say that the continent is 30 to 40 years behind Europe in the knowledge of its native fauna, and that there are few well-worked areas that can be equated with the European regions listed earlier. California is a possible exception because of the excellent work of Usinger (1956). In general, larval stages of aquatic insects can be identified only to genus (Merritt and Cummins 1978), and new species, and even genera, are constantly being found. Also, much of the southern area of North America has not been examined in any systematic way. This also applies to Central and South America, although some groups (e.g. Ephemeroptera, Plecoptera, Hemiptera, and Coleoptera) have received some particular attention.

Similarly, Southeast Asia has received only rather cursory attention, largely by visiting specialists in certain groups, who, as in South and Central America, have done rather little work on immature stages. Northern Asia has been somewhat more thoroughly worked, but there is no indication that definitive works, except for isolated groups or regions, are likely to appear for many years. Most publications merely describe new species and extend known distributions; in most instances, immature stages are poorly known.

This is to some extent also true of Africa and Australasia, although in both these biogeographical regions there are small areas that have been studied more thoroughly than the rest. In South Africa the early work of Barnard (1931, 1932, 1934a, 1934b) has led to several studies on particular groups. For New Zealand there are two handbooks for identification (Pendergrast and Cowley 1966; Winterbourn and Gregson 1981) and some more recent work on particular groups. These areas are therefore perhaps comparable to North America in their state of knowledge of the aquatic insects.

Rapid developments are, however, occurring elsewhere in these regions. For West Africa we now have a handbook (Durand and Lévêque 1981) comparable with the early European comprehensive works mentioned above. This will begin to open up our knowledge of the whole of tropical Africa. There is also a similar work for Australia (Williams 1980). If one can judge from experience in Europe and North America, the appearance of such preliminary aids to identification leads to a rapid increase in knowledge and an ever increasing effort to complete it; so we may say that the prospects are promising.

However, it should be emphasized that even in Europe the immature stages of some groups are poorly known, and that without the ability to identify larvae, ecological work is severely limited. This is particularly true of the large and very important family Chironomidae, but it also applies to many other groups, such as several families of Coleoptera and Diptera, especially the semiaquatic ones, and the Hemiptera. There is still much to be done and it is research that has to be performed in close association with the habitat. It is clearly not an activity that can be carried out effectively in distant museums.

We should not allow ourselves to become complacent about taxonomy. Quite often taxonomists and ecologists disagree as to what constitutes a species, the former maintaining that a small morphological difference between two entities is definitive, while the latter needs to see some ecological or geographical difference to be convinced. This is an argument that will persist, and through it we hope to arrive at the truth. However, even when taxonomists and ecologists do agree, they may be wrong, as has been shown recently by the discovery of so-called cytospecies in *Simulium* that are separable only by the banding of the chromosomes in the giant salivary glands of their larvae. In *S. damnosum*, the African vector of the nematode that causes river blindness (onchocerciasis), the various cytospecies differ subtly in ecology and in their ability to transmit the worm (Walsh et al. 1979). This is of great practical significance and the different species are now recognized as such even though one cannot yet distinguish the adults.

This raises the awesome question as to how widely this sort of thing occurs in aquatic insects. We know that it also occurs in mosquitoes (e.g. *Anopheles maculipennis* populations, which lay different eggs and vary in their ability to transmit malaria) but what about stoneflies and beetles? Also, should we not be using a trinomial nomenclature to cope with this problem so that the morphological identities of cryptospecies are not lost in the terminology? Similarly, should we not use trinomials for very closely allied forms separated by a narrow strait or a valley between two mountains? Such a practice would, for instance, not only reduce the apparent endemism of Corsica, but it would greatly clarify the biogeographical picture for the *general biologist*, a term that, in this context, includes all ecologists. Clearly, although we have come a long way in our understanding of the taxonomy and ecology of aquatic insects, an enormous amount remains to be done, and some current ideas and practices need some rethinking.

References

Barnard, K. H. 1931. The Cape alder-flies (Neuroptera, Megaloptera). Transactions of the Royal Society of South Africa 19:169–84.

Barnard, K. H. 1932. South African may-flies (Ephemeroptera). Transactions of the Royal Society of South Africa 20:201–59.

Barnard, K. H. 1934a. South African stone-flies (Perlaria) with descriptions of new species. Annals of the South African Museum 30:511–48.

Barnard, K. H. 1934b. South African caddis-flies (Trichoptera). Transactions of the Royal Society of South Africa 21:291–394.

Barton, D. R. and H. B. N. Hynes. 1978. Wave-zone macrobenthos of the exposed Canadian shores of the St. Lawrence Great Lakes. Journal of Great Lakes Research 4:27–45.

Berg, K. 1941. Contributions to the biology of the aquatic moth *Acentropus niveus* (Oliv.). Videnskabelige Meddelelser f. Dansk naturhistorisk Forening i Kjøbenhavn 105: 57–139.

Brauer, A. 1909. Die Süsswasserfauna Deutschlands: Eine Exkursionsfauna. Gustav Fischer Verlag, Jena, Germany. Vols. 2–9.

Corbet, P. S. 1962. A biology of dragonflies. Witherby, London, 247 pp.

Cummins, K. W. 1973. Trophic relations of aquatic insects. Annual Review of Entomology 18: 183–206.

Durand, J. -R. and C. Lévêque. 1981. Flore et faune aquatiques de l'Afrique sahelo-soudanienne. Vol. 2, Office de la Recherche Scientifique et Technique Outre-Mer, Paris, France. pp. 391–873.

Edmunds, G. F., Jr. 1972. Biogeography and evolution of Ephemeroptera. Annual Review of Entomology 17:21–42.

Edmunds, G. F., Jr., S. L. Jensen, and L. Berner. 1976. The mayflies of North and Central America. University of Minnesota Press, Minneapolis. 330 pp.

Eriksen, C. H., V. H. Resh, S. S. Balling, and G. A. Lamberti. In press. Aquatic insect respiration. *In*: R. W. Merritt and K. W. Cummins (eds.). An introduction to the aquatic insects of North America. 2nd ed. Kendall/Hunt Publishing Company, Dubuque, IA.

Hagen, K. S. 1978. Aquatic Hymenoptera, pp 233–39. *In*: R. W. Merritt and K. W. Cummins (eds.). An introduction to the aquatic insects of North America. Kendall/Hunt Publishing Company, Dubuque, IA. 441 pp.

Hinton, H. E. 1976. Enabling mechanisms, pp. 71–83. *In*: Proceedings of the Fifteenth International Congress of Entomology, Washington, DC. 824 pp.

Hungerford, H. B. 1948. The Corixidae of the Western Hemisphere (Hemiptera). University of Kansas Science Bulletin 32:5–827.

Hynes, H. B. N. 1970. The ecology of running waters. University of Liverpool Press, Liverpool, England. 555 pp.

Illies, J. 1965. Phylogeny and zoogeography of the Plecoptera. Annual Review of Entomology 10:117–40.

Illies, J. 1968. Ephemeroptera (Eintagsfliegen). Handbuch der Zoologie. Walter de Gruyter, Berlin, Germany. 4 (2) 2/5: Lief.7. 63 pp.

Illies, J. (ed.). 1978. Limnofauna Europaea. Gustav Fischer Verlag, Stuttgart, Germany. 532 pp.

Jewett, S. G. 1963. A stonefly aquatic in the adult stage. Science 139:484–85.

Kozhov, M. 1963. Lake Baikal and its life. Monographiae Biologicae. Vol. 11. Dr. W. Junk Publishers, The Hague, The Netherlands. 344 pp.

Mackay, R. J. and G. B. Wiggins. 1979. Ecological diversity in Trichoptera. Annual Review of Entomology. 24:185–208.

Merritt, R. W. and K. W. Cummins. (eds.). 1978. An introduction to the aquatic insects of North America. Kendall/Hunt Publishing Company, Dubuque, IA. 441 pp.

Merritt, R. W. and J. B. Wallace. 1981. Filter-feeding insects. Scientific American 244: 132–44.

Miall, L. C. 1895. The natural history of aquatic insects. Macmillan and Co., London, England. 395 pp.

Pendergrast, J. G. and D. R. Cowley. 1966. An introduction to New Zealand freshwater insects. Collins Bros. and Co. Ltd., Auckland, New Zealand. 100 pp.

Pilgrim, R. L. C. 1972. The aquatic larva and the pupa of *Choristella philpotti* Tillyard, 1917 (Mecoptera: Nannochoristidae). Pacific Insects 14:151–68.

Ross, H. H. 1956. Evolution and classification of the mountain caddisflies. University of Illinois Press, Urbana, IL. 213 pp.

Ross, H. H. 1967. The evolution and past dispersal of the Trichoptera. Annual Review of Entomology 12:169–206.

Rousseau, E. 1921. Les larves et nymphes aquatiques des insectes d'Europe. (Morphologie, biologie, systématique). Lebègue and Co., Brussels, Belgium. 967 pp.

Thorpe, W. H. and D. J. Crisp. 1947. Studies on plastron respiration. I. The biology of *Aphelocheirus* (Hemiptera, Aphelocheiridae (Naucoridae)) and the mechanism of plastron retention. Journal of Experimental Biology 24:227–69.

Usinger, R. L. (ed.). 1956. Aquatic insects of California. University of California Press, Berkeley, CA. 508 pp.

Wallace, J. B. and R. W. Merritt. 1980. Filter-feeding ecology of aquatic insects. Annual Review of Entomology 25:103–32.

Walsh, J. F., J. B. Davies, and R. le Berre. 1979. Entomological aspects of the first five years of the Onchocerciasis Control Programme in the Volta River Basin, West Africa. Tropenmedizin und Parasitologie 30:328–44.

Ward, H. B. and G. C. Whipple (eds.). 1918. Fresh-water biology. John Wiley and Sons, Inc. New York, NY. 1,111 pp.

Wells, R. M. G., M. J. Hudson, and T. Brittain. 1981. Function of the hemoglobin and the gas bubble in the backswimmer *Anisops assimilis* (Hemiptera: Notonectidae). Journal of Comparative Physiology 142:515–22.

Wesenberg-Lund, C. 1943. Biologie der Süsswasserinsekten. Verlag von Julius Springer, Berlin, Germany. 682 pp.

Wigglesworth, V. B. 1933. The function of the anal gills of the mosquito larva. Journal of Experimental Biology 10:16–26.

Williams, W. D. 1980. Australian freshwater life. Macmillan Co. of Australia Pty. Ltd. Melbourne, Australia. 321 pp.

Winterbourn, M. J. and K. L. D. Gregson. 1981. Guide to the aquatic insects of New Zealand. Bulletin of the Entomological Society of New Zealand 5:1–80.

Wooton, R. J. 1972. The evolution of insects in fresh water ecosystems, pp 69–82. *In*: R. B. Clark and R. J. Wooton. (eds.). Essays in hydrobiology. University of Exeter, Exeter, England. 136 pp.

Zwick, P. 1980. Plecoptera (Steinfliegen). Handbuch der Zoologie. Walter de Gruyter, Berlin, Germany. 4 (2) 2/7: Lief. 26. 111 pp.

chapter 3

LIFE HISTORIES OF AQUATIC INSECTS

Malcolm G. Butler

INTRODUCTION

Life-history information is of fundamental importance for virtually all ecological studies of freshwater invertebrates. Descriptive natural history has traditionally dominated aquatic insect ecology, and has provided a literature on life histories too vast to review in a single volume, much less in one chapter. Yet we still lack critical life-history knowledge that is essential to contemporary research on the structure and function of aquatic communities and ecosystems (Rosenberg 1979). Although the generation of new life-history information through population-level studies has great value, the accretion of such knowledge is often quite slow and is unlikely to increase in pace in the near future. A way to increase the efficiency of our research in this area may be to look for categories of life-history variables that are common to all taxa of aquatic insects, and then to ask why the particular life-history traits that we observe for a given variable have come to exist. The answers to such questions may be evolutionary (i.e. due to adaptation or phylogeny), ecological (involving physiological or population responses), or very often both. By testing hypotheses about life-history patterns with both experimental and comparative methods, we may hope to generalize both the reasons behind the observed patterns and the consequences of variations in life history for other questions in aquatic ecology.

The goal of this chapter is to develop a conceptual approach to the study of life histories that is appropriate to aquatic entomology and is consistent with historical and contemporary concepts in other areas of biology. The approach is to identify major categories of life-history variables, to explore the range of patterns known for aquatic insects, and to identify some particular questions that are in need of additional study.

WHAT IS A "LIFE HISTORY"?

What biological features are covered by the term "life history"? Oliver (1979, p. 319) defined a life history as "events that govern the reproduction (and survival) of a species or a population," including fecundity, development, longevity, and behavior. Waters (1979, p. 343) listed method of birth, pattern and rate of growth, feeding, locomotory and social behavior, length of life, selected habitat, response to environmental factors, mode of reproduction, and mode of death as factors involved in the "history of [a species'] life." Very little of an organism's biology is omitted from such inclusive definitions, and the task of organizing life-history information quickly becomes intractable. In addition, the term "life cycle" is often used almost interchangeably with "life history" in much of the literature on aquatic insects.

Calow (1978, p. 1) defines a life cycle as "the cycle of events which enables the products of reproduction themselves to reproduce." I will use "life cycle" to mean the sequence of morphological stages and physiological processes that link one generation to the next. The components of this cycle will be the same for all members of a species, and can generally be described qualitatively. In contrast, the qualitative and quantitative details of the variable events that are associated with the life cycle make up a "life history," which can vary among individuals or populations of one species. Thus, the life cycle of a mayfly includes egg, larval, subimago, and adult stages, incomplete metamorphosis, and feeding that is restricted to the larval stage. The life history of a particular population may involve asynchronous hatching of eggs, negative exponential mortality through the larval stage, and parthenogenetic reproduction. A given life cycle may be completed twice per year by a population with a bivoltine life history (*not* life cycle). A population with a semivoltine life history might complete the same qualitative sequence of events (life cycle) only once in two years, due to genetic or environmental differences or both.

LIFE HISTORIES AND ADAPTATION

Why do aquatic insects show such a diversity of life history patterns? We can look for determinants on two levels. First, there are the proximate effects (i.e. the physiological mechanisms that shape a life history in response to the environment), and these are treated by Sweeney in Chapter 4. Second, there are the ultimate causes, the morphological, behavioral, and physiological potentials that set general limits to a life history. Why do species and populations differ in these characteristics? For example, why (not how) does synchronous emergence occur in some aquatic insect populations and not others? Why is the potential for semivoltinism (or longer life cycles) apparently rare among mayflies (Brittain 1982), but common in other hemimetabolous orders such as odonates (Corbet 1980) and stoneflies (Hynes 1976)? This chapter deals primarily with such ultimate questions regarding life-history patterns.

The problem of distinguishing ultimate cause from proximate effect is a developing concern in evolutionary biology, and certainly applies to life-history studies of aquatic insects. The literature abounds with references to life-history features as "adaptations" that suit species of aquatic insects to the particular environmental conditions in which they are observed. Over-reliance on adaptationism to explain observations in nature has been recently criticized (Gould and Lewontin 1979; Gould 1980, 1982), and Gould and Vrba (1982) propose that we distinguish between characteristics that have arisen through natural selection (adaptations) and other beneficial features that may exist for historical reasons (exaptations). Thus, it may be no more useful to consider all apparently beneficial life-history characteristics as adaptations than to see the presence of six legs on a caddisfly larva as an adaptation to its particular mode of life. The caddisfly has six legs because it is an insect, not because there has been selection for that number. Although the six legs are certainly beneficial to the larva, and specific ways in which they are modified or used may be the adaptive result of selection for a particular habit, the number of legs should not be seen as an adaptation. Similarly, life-history features such as spring emergence, synchronous meta-morphosis, or reduced adult dispersal can have secondary, beneficial results that may not have been involved in the genesis of the feature.

It will often be difficult to determine whether a life-history feature is a true adaptation, or an exaptation that has resulted from cooptation of some previously existing characteristic. However, awareness of these two possibilities may lead us away from the simplistic approach of attributing every example of ecological success to adaptive natural selection.

LIFE-CYCLE PATTERNS

Among orders of aquatic insects, life cycles vary in the way processes such as feeding, growth, development, dormancy, dispersal, and reproduction are distributed among the various life-cycle stages (Table 3.1). Some major life-cycle features follow phylogenetic patterns. In orders with incomplete metamorphosis (Ephemeroptera, Odonata, Plecoptera, Hemiptera), growth and development occur in an immature larval stage, frequently termed a nymph or naiad. The remaining six aquatic orders have complete metamorphosis. Here the larval stage includes both growth and development, but in the nonfeeding pupal stage only development occurs.

All larvae must feed in order to grow, but growth of somatic tissues is rare among adult insects (Borror et al. 1981). Adult feeding has been retained in many groups, but has been lost in mayflies and megalopterans, and is reduced or lost in many stoneflies, caddisflies, aquatic lepidopterans, and some dipteran families like the Chironomidae, Tipulidae, and Chaoboridae. Lack of strict phylogenetic trends in adult feeding suggests that we look for ecological factors that may have influenced loss or retention of this behavior. Feeding in the adult stage may allow increased longevity and/or fecundity, thus potentially enhancing reproductive success. Loss of adult feeding might be expected in situations where females can obtain material for egg production more efficiently as aquatic larvae than as

TABLE 3.1: Distribution of Life-cycle Processes among Life-cycle Stages* for Ten Orders of Aquatic Insects

METAMORPHOSIS	ORDER	GROWTH	DEVELOPMENT	FEEDING	DORMANCY	REGIONAL DISPERSAL	REPRODUCTION
Incomplete	Ephemeroptera	L	L	L	E L	SI A	(SI) A
	Odonata	L	L	L A	E L	A	A
	Plecoptera	L	L	L (A)	(E) L	A	A
	Hemiptera	L	L	L A	E L	A	A
Complete	Megaloptera	L	L P	L	L	A	A
	Neuroptera	L	L P	L A	PP	A	A
	Coleoptera	L	L P	L A	(E) L (P) A	A	A
	Trichoptera	L	L P	L (A)	E PP	A	A
	Lepidoptera	L	L P	L A	L	A	A
	Diptera	L	L P	L (A)	E L,PP A	A	(P) A

* E = egg
L = larva
PP = prepupa
P = pupa
SI = subimago
A = adult
Parentheses indicate occurrence in a minority of species

foraging terrestrial adults. In mosquitoes, protein from the adult blood meal permits a greater increase in fecundity relative to egg production than would result from larval reserves alone. When environmental conditions limit the probability of obtaining a blood meal, as in the Arctic (Danks 1979), many biting flies show autogenous reproduction (i.e. oviposition without a blood meal). Autogeny can be either facultative, as in the mosquito *Aedes* (Corbet 1967), or obligate, as in tundra species of the ceratopogonid *Culicoides* and most arctic black flies (Downes 1965).

Occurrence of a dormant stage distinguishes heterodynamic life cycles, which are typical of most temperate zone aquatic insects, from homodynamic life cycles, where development is continuous and no regular period of dormancy occurs (Borror et al. 1981). As a life-history variable, dormancy is often critical to proper temporal scheduling of the life cycle in relation to environmental seasonality. Developmental arrests may correspond to cold (hibernation), heat or drying (aestivation), food shortage, or other forms of environmental adversity. Dormancy can be simply a period of quiescence directly caused by environmental factors (e.g. too hot or cold, too little food or oxygen to permit normal activity) or it can be a genetically programmed "true diapause" that is initiated in a particular life-cycle stage by an environmental cue (Danks 1979; Tauber and Tauber 1976, 1981).

The presence of dormancy and the stage at which it occurs vary greatly within and among populations. True diapause has been reported in the egg stage of several aquatic insect orders. Delayed hatching of some mayfly eggs was proposed by Macan (1958) as an explanation for the prolonged seasonal occurrence of small larvae even when adults emerge synchronously, and egg diapause has since been demonstrated for several mayflies (Bohle 1972). However, many examples of delayed appearance or prolonged seasonal occurrence of small mayfly larvae probably result from inadequate sampling (Brittain 1982) or slow growth (Elliott and Humpesch 1980) of the early instars.

Larvae usually make up the greater part of the life cycle and, consequently, animals in this stage are most likely to encounter seasonally adverse conditions. Larval dormancy occurs in all orders, usually as a means of passing cold and sometimes food-poor winter periods, or warm summer periods when lack of oxygen or even water may interrupt the life cycle. Larvae of some chironomids are able to withstand dehydration (Hinton 1951; Jones 1975), and larvae of some chironomid and trichopteran species are known to tolerate freezing (Scholander et al. 1953; Olsson 1981). Dormancy as a prepupa (i.e. a larva late in its final instar) is known in Trichoptera (Wiggins 1977), Chironomidae (Oliver 1968; Butler 1982c), and apparently Neuroptera (Pennak 1978). Delayed development during pupation appears to be uncommon, although some aquatic Coleoptera overwinter in this stage. Many water beetles and bugs, as well as some mosquitoes and chironomids, overwinter as adults, and some Trichoptera, especially those that live in temporary pools and streams, aestivate as adults during the dry summer (Wiggins 1977; Wiggins et al. 1980).

Dispersal may occur in both immature and adult life-cycle stages. Within aquatic habitats, substantial dispersal results from larval swimming, crawling, phoresy, and drift (see Wiley and Kohler, Chapter 5; and Sheldon, Chapter 13). Since both lentic and lotic freshwater habitats are frequently isolated from one another, the winged adult stage will be most effective in regional dispersal. Habitat quality can vary in both time and space, and diapause and dispersal are respective means of tracking favorable, or escaping unfavorable, conditions (Southwood 1977). Temporal changes in habitat quality will not always be correlated among the different intrahabitat patches, hence dispersal may be useful in both dimensions (Horn 1978). Insect dispersal is considered from an adaptive, life-history perspective by Dingle (1978) and Denno and Dingle (1981), and Sheldon (Chapter 13) gives many examples of dispersal among aquatic insects and discusses the related life-history phenomena of brachyptery and wing polymorphism; hence this topic is not treated further in this chapter.

The adult is the reproductive stage of the life cycle in most taxa, but there are some exceptions. In at least one mayfly the female oviposits as a subimago and no true adult stage exists (Edmunds et al. 1976). Paedogenesis (i.e. parthenogenetic reproduction by the immature stages) is also known in the Chironomidae, and midges that complete their life cycle entirely within water mains can present formidable control problems (Krüger 1941; Resh and Grodhaus 1983). Details of reproduction vary greatly between species; some major life-history variables relating to reproduction are considered herewith.

LIFE-HISTORY PARAMETERS

How is life-history information obtained? Both in the field and laboratory, all techniques can be classed as one of four empirical processes: enumeration, measurement, categorization, and observation. With these processes we determine densities, sizes, life-cycle stages, and behavior of organisms at any point in time. When carried out over a period of time (continuously or, more commonly, discontinuously) the values obtained for these variables provide information on various life-history parameters. When these parameters are considered in a temporal context, two additional life-history features emerge: voltinism and phenology (Fig. 3.1). The former relates to the frequency with which a life cycle is completed; the latter involves both the *seasonal timing* of life-cycle processes and the population *synchrony* of these processes. The following discussion treats most of the life-history features in Figure 3.1 from the standpoint of their ecological significance, the ultimate causes of life-history diversity seen among aquatic insects, and methods for approaching important questions that relate to each topic.

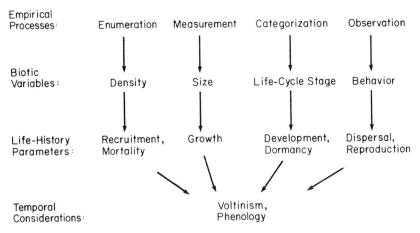

Figure 3.1 Processes and variables involved in the study of life histories.

Reproductive Recruitment

The recruitment of a new cohort through oviposition and subsequent hatching marks the beginning of a new generation. Like mortality, this life-history feature is a population phenomenon rather than an individual characteristic. Recruitment is a reflection of reproductive success at the population level, and many aquatic insect populations may be regulated at this point in the life cycle. Other aquatic invertebrates such as crustaceans and rotifers usually spend their entire life cycle in the aquatic habitat. When favorable conditions prevail for several generations, these animals may approach exponential population growth. In contrast, aquatic insect populations are frequently reduced between generations by mortality in the comparatively harsh terrestrial environment through which most species must pass.

Poor recruitment due to adult reproductive failure may be a major feature in the dynamics of many aquatic insect populations. Welch (1976) showed that emergence failure caused by late ice melt could dramatically influence subsequent recruitment of chironomid larvae in a polar lake. Kajak (1964) and Jónasson (1972) attributed recruitment success of profundal chironomids to weather conditions during adult emergence and oviposition. Fluctuations in water levels were found to delay the departure of newly hatched *Heteroplectron americanum* (Trichoptera) larvae from their egg masses, as well as to cause embryonic mortality of the eggs that were oviposited along the stream banks; this influenced both the timing and magnitude of reproductive recruitment (Patterson and Vannote 1979). In the tropics, reproductive failure caused by intense predation on adults may have been a major selective factor leading to compensatory life-history features such as short adult longevity (Edmunds and Edmunds 1979).

Despite the potential importance of this bottleneck in population continuity, recruitment is one of the more difficult life-history variables to study. Adult emergence can be monitored by a number of means (Corbet 1964), including qualitative trapping with lights or quantitative sampling with emergence traps (Davies, in press). However, these methods do not indicate how many females return to a particular habitat and successfully lay eggs. Sticky traps or pan traps for collecting females may be useful for species with certain oviposition behaviors, but most estimates of recruitment success are based on benthic sampling for early instar larvae. These new recruits are both small and frequently difficult to sample; hence, estimation of abundance is usually not feasible until a later stage, at which point some mortality will have taken place. Direct sampling of eggs or egg masses will be complicated by oviposition behaviors and identification problems, but may be useful in certain cases (Welton and Bass 1980; Wood et al. 1982). Comparison of recruitment levels between different years, generations, or populations may be made by means of standardized, representative samples of spent females, which are collected during oviposition periods. This can be accomplished with drift nets in lotic habitats, surface-towed plankton nets in lentic habitats, or with samples taken from eddies or windrows.

Mortality

Mortality can be viewed as both a life-history parameter and a determinant of other life-history features. A mortality schedule is a population parameter, although an individual within a population can be considered to have a corresponding probability schedule for survivorship through any point in its life cycle. This schedule obviously has a tremendous influence on the probability of reproductive success, and hence evolutionary fitness. We might therefore expect mortality patterns to play a major role in the evolutionary shaping of life-history patterns; conversely, certain life-history alternatives may influence mortality. For example, many odonates oviposit endophytically (Corbet 1980), and these eggs probably suffer lower mortality than eggs that are unprotected by plant stems. But if egg concealment is obtained at some cost such as lowered fecundity or delayed maturation (Denno and Dingle 1981), there could be selection for exophytic oviposition with consequent higher mortality (see also Wiley and Kohler, Chapter 5).

Mortality to aquatic insects resulting from vertebrate or invertebrate predation may constitute an important selective force in the evolution of life-history characteristics such as body size, life-cycle scheduling, and allocation of reproductive effort. Extensive research on freshwater zooplankton has confirmed the importance of vertebrate and invertebrate predation in influencing both zooplankton community composition and life-history features of planktonic crustaceans and rotifers (O'Brien 1979; Kerfoot 1980; Zaret 1980). Little research with aquatic insects has directly addressed the relationships between predation mortality and adaptive life-history evolution. Predatory relationships among aquatic insects are reviewed by Peckarsky (1982; and in Chapter 8 of this volume), in which she describes a host of morphological, behavioral, and chemical features of aquatic insects, many of which appear to serve defensive functions. Peckarsky (1982) concluded, however, that we do not know what contribution insect predators make to the total mortality of prey species. Extrapolating the results of zooplankton studies, we might expect invertebrate predation to elicit adaptive changes in such life-history features as growth pattern, body size, and fecundity, in addition to its effects on morphological and behavioral features.

Reice (1983) and Allan (Chapter 16) have independently conducted fish-removal experiments to look for influence of fish predation on stream insect populations. Neither could document any statistically significant effects on insect abundance. However, when trout were experimentally introduced into fish-free stream pools (S. Cooper, unpublished data) some taxa were rapidly eliminated whereas other insect populations appeared to be unaffected. Many morphological, behavioral, and life-history features of aquatic insects that normally coexist with predatory fish may reflect what O'Brien (1979, p. 572) has termed the "evolutionary thrust and parry of the predator-prey relationship." Among coadapted predator and prey populations, predation mortality may seldom be detectable but nonetheless may maintain a consistent, directional, selective pressure that shapes aquatic insect life histories (see also Healey, Chapter 9).

By knowing the schedule of mortality through the life cycle we may better understand how life histories have been shaped by factors such as predation. For example, Edmunds and Edmunds (1979) suggest that heavy adult mortality caused by predatory odonates and birds has led to the evolution of nocturnal emergence and very short adult life spans among tropical mayflies. Among aquatic stages, we may also expect that differences in mortality risk may also influence life-history evolution.

These mortality schedules are described by life tables, which are age-specific summaries of mortality rates within a population (Krebs 1978; Southwood 1978). There are two kinds of life tables, which differ in how the data (i.e. numbers surviving to age x) are obtained. The first is an age-specific (or cohort) life table, in which a group of individuals of roughly equal age is followed through the life cycle from recruitment to either death or emergence of the last individual. This is usually done by sampling. In addition to the many technical problems associated with benthic sampling (Resh 1979), sample data can never reflect the true survivorship schedule unless a cohort develops with perfect synchrony (Benke, Chapter 10). For an age-specific life table, one must be able to distinguish more or less discrete cohorts, which is a serious problem with populations that do not show reasonable synchrony throughout most of the life cycle.

A second type, the time-specific (or static) life table, is based on examination of a cross section of the population at one point in time and requires that the population be at equilibrium with regard to recruitment and mortality. It is also necessary that the generation time exceed the frequency of recruitment so that more than one age class is present when the population is sampled. This will be true for populations that show total asynchrony of recruitment, development, and emergence, as well as for synchronously reproducing populations that are very long-lived (Butler 1982b). Although time-specific life tables rarely have been used directly to describe mortality in aquatic insect populations, the size-frequency method for calculating secondary production incorporates a modification of the same concept (Benke and Waide 1977; Benke, Chapter 10).

Development

Invertebrate life-history studies often fail to distinguish between growth and development. Growth is simply an increase in organism size, whereas development is the physiological and morphological progression toward reproductive maturity; usually there will be a correlation between these processes, but frequently they can be separated.

Development cannot be quantified directly through linear measurement, but instead categorical variables must be used to document progress through the life cycle. The various life-cycle stages (i.e. the egg, a series of immature instars, a pupa in holometabolous orders, and the adult) are the basic developmental categories for aquatic insects. With the exception of diapausing eggs and long-lived adults in some Hemiptera and Coleoptera, the larval stages usually account for the largest portion of the life cycle in aquatic insects. A common approach to studying development is to analyze the progression of larvae through the sequence of instars.

The value of instars as developmental categories varies among orders, since as few as three instars to as many as fifty are known for aquatic insects (Table 3.2). More important is whether the number of instars is fixed within a species. Instar number is known to be indeterminate in mayflies (Brittain 1982), stoneflies (Hynes 1976), and odonates (Pritchard and Pelchat 1977). For most other groups it is apparently constant at least at the species level. When a species has a fixed number of instars, these can be used as a categorical variable to indicate developmental age, or progress toward metamorphosis. This may not be the same as temporal age, for development will rarely be constant through time. For taxa with an indeterminant number of instars, knowledge that an individual is in instar x really tells us rather little about its developmental state. Most of the methods that have been developed for determining instars in mayflies and stoneflies (Fink 1980, 1982) are based on linear measurement of body parts and tell us more about growth than about development (e.g. McClure and Stewart 1976).

TABLE 3.2: Number of Immature Instars Reported for Ten Orders of Aquatic Insects *

ORDER	NUMBER OF INSTARS	VARIATION WITHIN SPECIES
Ephemeroptera	(10) 15–25 (50)	Yes
Odonata	(8) 10–12 (15)	Yes
Plecoptera	12–22+	Yes
Hemiptera	(4) 5	No
Megaloptera	10–11	?
Neuroptera	3	No
Coleoptera	3 (6)	No
Trichoptera	5 (6, 7)	Some
Lepidoptera	5–7+	Some
Diptera	4–7	No

*Normal number or range is shown, with exceptions or extremes in parentheses.

In taxa with a fixed number of molts, the degree of developmental resolution possible is directly proportional to the number of instars. Chironomids, for example, have only four instars (Ford 1959) and the final stadium may often constitute more than half of the entire life cycle (Wülker and Götz 1968); thus, individuals in instar IV may really be at very different developmental stages. Pritchard (1978, p. 2639) encountered a similar problem in the final instar of *Tipula sacra* and indicated that, "This lack of a reliable method for determining the absolute age of individual larvae remains, therefore, as the major barrier to progress in [elucidation of life history]."

Relative age rather than absolute age will usually be sufficient for cohort analyses. In holometabolous orders, recognition of prepupae on the basis of thoracic swelling is a frequent technique for separating older and younger individuals in the final instar. Wülker and Götz (1968) found that in *Chironomus* larvae primordial development of pupal and adult structures (wings, legs, genitalia, etc.) begins shortly after the molt to the final instar and continues throughout the stadium. Under constant conditions, they were able to divide this fourth stadium into nine phases representing equal developmental increments. Butler (1982a) used these developmental phases as a categorical variable to help separate cohorts that were coexisting in an arctic *Chironomus* population. Thoracic and genital primordia are visible within final instar larvae of many other dipterans and could be used in a similar way.

Sweeney and Vannote (1981) used the first appearance of wing buds and male genital primordia as a developmental indicator for *Ephemerella* mayflies, and Clifford et al. (1979) defined a sequence of four developmental stages for *Leptophlebia cupida* based on wing-pad development. Similar techniques for determining developmental age should be sought for other taxa, as they will permit us to study development independently from growth.

Growth, Development, and Size

The interactions between growth rate, development, and maximum size present an interesting set of questions regarding life-history patterns. For those aquatic insects that do not feed as adults, female size is often a good indicator of fecundity (Colbo and Porter 1979; Ratte 1979; Seward 1980; Sweeney and Vannote 1981). Since evolutionary fitness is closely linked to fecundity, we might expect these insects to grow as large as possible; the environment, however, may limit adult size by affecting both the rate and duration of growth prior to metamorphosis. For a given species, therefore, is growth terminated by a size threshold, by a developmental marker, or by some external, environmental cue? Although there is probably a minimum size for metamorphosis in most species, development rather than growth will determine the timing of metamorphosis. If seasonal timing of the life cycle is critical (see following), scheduling mechanisms may exist that operate through environmental controls or cues that affect development (temperature, photoperiod, etc.). Size variation will then result from differences in the amount of growth achieved before development is completed.

These relationships have been studied in detail for stream mayflies (Vannote and Sweeney 1980; Sweeney and Vannote 1978; Sweeney, Chapter 4), and the Thermal Equilibrium Hypothesis was developed to synthesize both the proposed mechanisms and the consequences of variation in adult body size and fecundity. Critical to these arguments is the distinctness of the processes of growth and development. Many of the mechanisms behind this hypothesis remain to be tested experimentally, even for mayflies, and they provide a number of intriguing research problems in the physiological ecology of aquatic insects.

One of the most dramatic sources of size variation in aquatic insects is sexual dimorphism, but this topic has received very little critical attention. Sexual dimorphism in size occurs in most insect groups, and, in species that do not feed as adults, female weight can often exceed male weight two-fold (e.g. Pritchard 1976; Svensson 1977; Sweeney 1978; Clifford et al. 1979; Patterson and Vannote 1979; Ratte 1979; Butler 1982c). Although eggs in the female abdomen account for most of this difference, how this extra growth is achieved presents an interesting question, for males and females of most aquatic insects live in the same habitat, under identical food and temperature conditions, and for approximately the same amount of time. To put the question another way, why do males not grow larger?

Atchley (1971) considered this problem for *Chironomus* and proposed that female larvae may feed at higher rates or with greater efficiency. There is as yet no evidence for ruling out sexual differences in metabolic efficiency, as Atchley did, although it is hard to imagine why males should be less efficient at feeding or metabolism. Weight dimorphism appears quite early in the final instar of *Chironomus tardus* larvae (Butler 1982c), and there may be a sexual difference in the scheduling of larval growth relative to development (e.g. males may simply decrease or terminate feeding earlier than females, yet continue with development). This implies that males may optimize, whereas females maximize, their exploitation of food resources (Calow 1981). If this hypothesis is true, mature males should show lower variance in weight than do females. This may often be the case (e.g. Clifford et al. 1979; Patterson and Vannote 1979), but few studies report sex-specific mean weights, much less associated variances. More detailed studies of sex differences in feeding and growth may cause us to re-evaluate some ideas about the nature and consequences of resource limitation.

Reproduction

Reproductive behaviors such as sexual aggregation, territoriality, mate selection, copulation, and parental care are treated by Wiley and Kohler in Chapter 5. Here, I consider three additional reproductive life-history topics: sex ratio, parthenogenesis, and allocation of reproductive effort.

Sex Ratio. Sex ratios differing from 1:1 are sometimes reported for aquatic insects. Fisher (1930) explained why, at equilibrium, organisms should be expected to spend half of their reproductive effort in producing each sex. As long as males and females "cost" the same to produce, the initial sex ratio should be 1:1, a situation that should hold for nearly all aquatic insects.

Since it is often difficult to sex immature aquatic insects, most sex ratios in the literature are based on counts after emergence; unfortunately, different adult behaviors may bias the sample. Quantitative emergence-trap data are preferable, since sex ratios based on light-trap collections may even be influenced by climatic conditions (Resh and Sorg 1978). Deviations from a 1:1 adult sex ratio can arise from small sample size or from differential mortality prior to emergence, including that caused by parasitization (LeSage and Harrison 1980). Emergence samples that contain large numbers of individuals and are collected in a way that avoids the influence of sexual differences in emergence timing (see following), sometimes show significantly more males or more females. Flannagan and Lawler (1972) found a preponderance of females for most caddisfly species in a Manitoba lake, whereas males of the mayfly *Stenonema interpunctatum canadense* exceeded females by about 3:1. A 2:1 ratio favoring males of mature larval and adult *Tipula sacra* appeared to result from higher female mortality (Pritchard 1976). Collections of mature *Ephemera danica* nymphs usually consisted of 60 to 65 percent males, although in one cohort females showed better ability to survive a sudden freeze and males dropped to 45 percent of total emergence (Svensson 1977). Danks and Corbet (1973) trapped an excess of females of two *Aedes* species in arctic ponds. Differential mortality was implicated since temporary ponds consistently produced about 30 percent males, but equal or greater numbers of females emerged from permanent ponds. Some chironomid taxa emerging from other ponds at the same site had sex ratios near unity, but populations of Tanytarsini and some Orthocladiinae often had 60 to 65 percent females (Oliver and Danks 1972). In Alaskan arctic ponds, one *Tanytarsus* species regularly showed 65 to 70 percent females at emergence, but the sex ratio of a sympatric congener did not differ significantly from 1:1 (M. Butler, unpublished data). Similar unbalanced sex ratios have been reported for Finnish chironomid populations by Lindeberg (1971), who postulated parthenogenesis as a cause for the overabundance of females (see following). Genetically determined sex-ratio distortions are known in mosquitoes (Clements 1963; Hamilton 1967), but there is no evidence that deviations from 1:1 are adaptive for aquatic insects or for any other outbreeding, dioecious organisms (Williams 1979).

Parthenogenesis.   Parthenogenesis is widespread among insects (Suomalainen 1962) and is known or suspected in many species of mayflies (Brittain 1982), chironomids (Lindeberg 1971, 1974), and some caddisflies (Nielsen 1951; Corbet 1966) and stoneflies (Hynes 1976). Thelytokous parthenogenesis (in which females are produced from unfertilized eggs) can be of two kinds: automixis, in which meiosis is normal and haploid eggs must be restored to a diploid state by some cytological mechanism; and apomixis, in which meiosis is suppressed and diploid eggs are formed mitotically. Automixis is found in some insects but has not been confirmed for any aquatic species, hence apomixis is the probable mechanism when thelytoky is observed in aquatic insects.

Sporadic or occasional parthenogenesis (tychoparthenogenesis), in which a small proportion of unfertilized eggs hatch, is known in some mayflies (Brittain 1982), stoneflies (Hynes 1976), and chironomids (Grodhaus 1971). This is not likely to be an important mode of reproduction, however, because hatchlings produced in this way will be haploid, and it is unlikely that they could mature and reproduce. It would seem a great advantage for a female that is unable to find a mate to be able to lay her eggs parthenogenetically, but although such facultative parthenogenesis is known in a few terrestrial insects (White 1973), no aquatic insects have been shown to possess an automictic mechanism for restoring diploidy.

Occurrence of facultative parthenogenesis is occasionally suggested when populations of known bisexual species are found with sex ratios favoring females. McCafferty and Huff (1974) and Bergman and Hilsenhoff (1978) cite examples of mayflies where males are known, but females predominate and eggs may hatch without fertilization. In only one such case have larvae so produced been reared to maturity (Degrange 1960, cited in Gibbs 1977), and cytological mechanisms have not been investigated. Lindeberg (1974) hypothesized that mixed populations that consist of parthenogenetic and bisexual races may explain unbalanced sex ratios that have been observed in northern chironomid populations. When such parthenogenetic forms are known for other bisexual arthropods (Calow 1981), they frequently occur at higher latitudes or altitudes. Latitudinal variation in sex ratio is known in some Chironomidae (F. Reiss, personal communication), but pending confirmation of Lindeberg's (1971) hypothesis or documentation of true facultative parthenogenesis, such patterns remain enigmatic.

Allocation of Reproductive Effort. How might an aquatic insect best utilize its available resources for reproduction? Reproductive costs include not only the production of progeny themselves, but also the energy and materials invested in oviposition, maintenance, and defense of the eggs and young. Although we can expect natural selection to work toward the development of resource-apportionment strategies that maximize the number of offspring surviving to reproductive age (Price 1974), many alternatives will have been decided by the species' evolutionary history, which may essentially lock the species into certain phylogenetic patterns. Since these reproductive patterns are obviously still effective, they may be seen as long-standing adaptations. However, they may also represent constraints. For example, because of a short adult life, a mayfly cannot spread her oviposition over time, although the staggered hatching of eggs within a single clutch might be an alternative means of achieving the same result (see following). In addition to such phylogenetic legacies, other reproductive characteristics may represent adaptive responses to contemporary ecological conditions. For example, differences between two treehole mosquito (*Toxorhynchites*) species in the size of early instars may reflect an adaptive response to the size of prey available to newly hatched larvae (Lamb and Smith 1980).

There are several dimensions in which alternate patterns of reproductive investment are possible. For example, an ovipositing female can lay all her eggs in a single mass, or she can distribute them among a variety of habitats or habitat patches. An optimal choice will depend on the variance and predictability of habitat quality. If the best habitat (e.g. pond, riffle, treehole) can be determined, all eggs should be placed there. Otherwise, a female might "hedge her bets" by distributing the total investment among a variety of potential choices. Are the oviposition behaviors of aquatic insects simply the result of phylogenetic inertia, or do they represent adaptive responses to such a dilemma? Since we see a great deal of variety (or at least many exceptions to the basic pattern) within each order, many of these reproductive behaviors are probably adaptations. Indeed, selection of an oviposition site may be one of the single most important factors in reproductive success and will consequently be the object of intense selective pressure. Yet our understanding of the diversity of oviposition behaviors seen within aquatic insect taxa is scarcely beyond the anecdotal stage. Mayflies may scatter their eggs in small batches or lay a single mass on the water, but some *Baetis* females enter the water to place their eggs on stones (Brittain 1982). Most stoneflies lay single egg masses on the water surface, but one family, the Notonemouridae, includes species with ovipositors that are used to place eggs individually into wet crevices at streamside (Hynes 1976). Precise habitat selection and protective oviposition is widespread in Odonata (Corbet 1962), but there is great variability both among and within species (Corbet 1980). Oviposition by throwing individual eggs is seen by Machado and Martinez (1982) as an adaptation that permits the damselfly *Mecistogaster jocaste* to colonize treeholes in the Bolivian forest, whereas McCrae and Corbet (1982) suggest that attachment of egg masses to surfaces above the prevailing water level permits the equatorial African dragonfly *Tetrathemis polleni* to oviposit successfully in shallow, turbid pools.

Offspring can be distributed in time as well as in space. The great majority of aquatic insects show semelparous reproduction (i.e. they breed once and die). Species that feed as adults can be iteroparous, producing additional broods through an extended adult life. Again we see some general taxonomic patterns, in that all mayflies, megalopterans, chaoborids, tipulids, and chironomids appear to be obligately semelparous. However, within most other groups there is some potential for repeated reproduction, and, in fact, this is fairly common in odonates, hemipterans, and beetles.

It is generally held that ecological factors have been more important than phylogenetic constraints in the evolution of alternative reproductive strategies (Calow 1981). The ecological variables that select for each type of reproduction have been the subject of much theoretical work since Cole (1954) first considered the relative values of "big-bang" versus repeated breeding. In general, these analyses have suggested that although semelparity provides the fastest rate of population growth, adults will be selected to hedge their bets in a temporal dimension by extending their reproduction when juvenile mortality is high or unpredictable (Giesel 1976; Horn 1978; Calow 1978). The value of iteroparity will vary with adult survival ability (Charnov and Schaffer 1973), which of course depends partially on the level of reproductive effort at any time.

Price (1974) suggested that ephemeral insects offer simple cases for the study of reproductive strategies because they cannot shift distribution of their reproductive effort in time. This may not always be true if insect eggs exhibit the "relict seed phenomenon" suggested by Giesel (1976), whereby the parent effectively prolongs its reproductive life span by producing eggs with delayed (and perhaps asynchronous) development. There is some evidence for polymorphic egg diapause among mayflies (Brittain 1982), stoneflies (Hynes 1976), and odonates (Corbet 1980). Unfortunately, this topic is still poorly understood, in spite of the fact that asynchronous hatching could have great adaptive value in many environments. In addition, the degree of population synchrony at hatching can have an important impact on population structure and life-cycle analysis (see following).

Voltinism

Voltinism refers to the frequency with which life cycles are completed. The length of a life cycle will depend, in part, on factors that influence growth and development in all life-cycle stages. There may also be genetic constraints that limit rates of these processes or mandate dormancy at certain points in order to synchronize critical life-cycle events with environmental seasonality. A population may be multivoltine, trivoltine, bivoltine, or univoltine depending on the number of life cycles an individual and its progeny can complete in one year. In a semivoltine population one generation requires two years to complete the life cycle, and Pritchard (1983, p. 13) has recently used the term "merovoltine" to describe longer life cycles in which a cohort completes only a portion of a generation each year.

In a few cases, species that occur over a wide geographic range have been shown to have a consistent life-cycle duration, despite climatic variation. For example, species of *Leptophlebia* always appear to be univoltine (Brittain 1982). However, because of the potential for environmentally induced variation that probably exists for the majority of aquatic insects (Wallace and Merritt 1980), voltinism should be viewed as a life-history variable at the population level, and not as a characteristic of species.

Geographic variation in voltinism attributed to thermal or nutritional effects is commonly reported (Ward and Stanford 1982; Sweeney, Chapter 4). Yet even within populations there can be variability in life-cycle duration. Such "splitting" of cohorts can occur when a portion of a cohort reaches the adult stage within one season but the remainder is forced to add an additional year to the life cycle. Pritchard (1976, 1978) hypothesized such a situation for the cranefly *Tipula sacra* in a Canadian beaver pond, where univoltine and semivoltine portions of the population were separated by failure of most instar IV larvae to emerge after one year. *Chironomus anthracinus* in eutrophic lakes in Denmark (Jónasson 1972) and Northern Ireland (Carter 1980) showed similar cohort splitting at certain depths and in some years. Larvae that failed to achieve sufficient size by the spring emergence period continued to grow, as food and oxygen conditions permitted, until the next spring.

There are several examples of split cohorts from other aquatic insect orders as well. Two-thirds of a population of the caddisfly *Parachiona picicornis* had a one-year life cycle in a Danish spring; the remainder was semivoltine (Iversen 1976). In Southern England, the dragonfly *Anax imperator* was largely semivoltine, but a small portion of each cohort emerged after one year (Corbet 1957). A population of the stream-living mayfly *Ephemera danica* in Southern Sweden was divided between a two-year and a three-year life cycle (Svensson

1977). Another burrowing mayfly, *Hexagenia limbata*, appeared to have a life cycle longer than one year, but shorter than two years, in a Kansas reservoir (Horst and Marzolf 1975). Since emergence could only occur in summer, a mixture of univoltine and semivoltine individuals was suspected. In Lake Winnipeg, Manitoba, Flannagan (1979) proposed alternating life cycles of 14 and 22 months for this species as well as for *Hexagenia rigida*.

In multivoltine populations, cohort splitting takes a different form. Rosenberg et al. (1977) found that part of a *Cricotopus bicinctus* midge population produced three generations in one year and two in the next, whereas the portion that was bivoltine the first year could be trivoltine the following year due to a different overwintering stage.

Because life-cycle duration can be so variable, voltinism is only useful as a general descriptor; more precise knowledge of specific life-history variables will usually be preferable. For example, data on voltinism are commonly used to correct production estimates made by "short-cut" approaches such as the size-frequency and $P/\bar{B}$ methods (Waters 1977; Benke, Chapter 10). Actually, for this purpose one should know the length of the period when growth (and therefore production) may actually be occurring; this is the "cohort production interval" or CPI (Benke 1979, and in Chapter 10). A univoltine population may have a CPI considerably less than one year if egg and adult stages are of significant duration.

Emergence periodicity may reflect voltinism in some cases, but may just as often lead to an incorrect conclusion if no other life-history information is obtained. For example, life cycles greater than one year will generally show a single emergence period (e.g. Jónasson 1972; Butler 1982a), and bimodal emergence can occur in univoltine populations if emergence is interrupted by high or low temperatures (e.g. Kerst and Anderson 1974).

The length of a life cycle is usually determined by analyzing serial histograms of developmental stages. Unfortunately, size intervals rather than true developmental categories are often used, which, as we have seen above, means that sexual dimorphism and other sources of variance in growth can complicate the analysis. In this approach, modes in the distribution of developmental stages are followed through a temporal series of samples to determine the progression of a cohort through the life cycle. In such analyses, it is important that samples contain sufficient individuals to produce recognizable peaks on a histogram. The sampling interval must also be appropriate to the rate of development. Detection of multivoltinism may require weekly or even more frequent samples, whereas relatively infrequent sampling may suffice when other evidence points to a merovoltine life cycle (Butler 1982a). It is obviously preferable to deal with single species when determining a life cycle, but commonly two or more closely related, coexisting species cannot be distinguished in all aquatic stages. When taxonomic difficulties make it necessary to analyze such combined data, additional information must be sought to confirm the validity of presuming identical life cycles for all the species being considered.

Phenology

Seasonal Timing.   Even in "nonseasonal" tropical areas, natural environments exhibit some seasonality (McElravy et al. 1982), and life cycles must be temporally adjusted so that critical stages are aligned with appropriate environmental conditions. Factors that can vary seasonally and be important to aquatic insects include temperature (Vannote and Sweeney 1980; Ward and Stanford 1982), oxygen (Jónasson 1972; Nagell 1981), water level (Wiggins et al. 1980; Patterson and Vannote 1979), photoperiod (Sweeney, Chapter 4), and food (Jónasson 1972; Anderson and Sedell 1977; Anderson and Cummins 1979; Wallace and Merritt 1980). Biotic interactions will also change seasonally as populations of potential predators and competitors wax and wane.

Two taxa may show identical life-cycle stages and voltinism, yet their life histories can differ dramatically in seasonal scheduling. Landa (1968) and Clifford (1982) both used such seasonal differences in further subdividing univoltine and bivoltine categories in their classifications of mayfly life histories. Different seasonal patterns may even occur within a population, as reported by Resh (1976) for the caddisfly *Ceraclea transversa* in a Kentucky creek. In this population, a larval cohort hatched in the spring and fed on freshwater sponges. These larvae completed growth before sponge gemmulation in the autumn, and subsequently overwintered as prepupae. A second cohort hatched from eggs laid during a summer flight period. These larvae reached instar III or IV before gemmulation, overwintered as active larvae using other food sources, and completed their life cycle the following summer. A similar situation was reported by Cushman et al. (1977) for a univoltine population of *Diplectrona modesta*, a detritus-feeding caddisfly in a Tennessee woodland stream. A spring cohort hatched in June-July and pupated the following March-April, overlapping with a summer cohort that hatched in September-October and pupated the following July-August. Adults were not collected in this latter study. Since such cohorts may effectively be genetically isolated by their temporal segregation, one should always consider the possibility that cryptic, sibling species may be involved.

Are these phenological features of life histories adaptive for aquatic insects? The timing of life-cycle events may be constrained directly by environmental conditions. For example, larval growth of *Chironomus anthracinus* may cease in response to low oxygen levels (Jónasson 1972), and food shortage can influence the timing of chironomid emergence (Danks 1978). In contrast to such short-term effects, life-cycle events like diapause and emergence are often regulated by the environment through physiological responses to stimuli such as temperature and photoperiod (Sweeney, Chapter 4). Such intrinsic patterns may be adaptive responses to long-term environmental effects on life histories (Danks 1979).

The plethora of "adaptive" interpretations that can be applied to a single phenomenon is illustrated by the example of emergence seasonality. In aquatic insects, seasonal timing is often studied most conveniently at emergence, since individuals can be monitored as they pass through the aquatic-terrestrial interface (Corbet 1964; Danks 1979). Precise seasonal timing of emergence may be beneficial if the presence of newly emerged adults coincides with optimal conditions for dispersal, feeding, or reproduction. Alternatively, phenological control at emergence may also serve to schedule other portions of the life cycle where regulation is not so easily achieved. For example, emergence of *Chironomus anthracinus* from Lake Esrom is strictly limited to a short period in May every year, and this results in the small instar I and II larvae being present at the very time when small phytoplankton species are dominant (Jónasson 1972). Whether spring emergence is an adaptation to the feeding requirements of early instars is uncertain, for an additional selective pressure is presented by the hypolimnetic anoxia that is characteristic of eutrophic lakes during summer stratification. The ability of certain *Chironomus* species to thrive in such profundal environments appears to be related to their large size, which permits the larvae to ventilate their tubes during microstratification at the sediment-water interface (Brundin 1951). The necessity of reaching a minimum size prior to such microstratification in late summer could also have selected for a life cycle that begins with emergence and oviposition in early spring.

Emergence timing often differs slightly between males and females. Protandrous emergence, where males peak slightly prior to females, is best detected in highly synchronous populations and has been reported in many aquatic insect orders, including mayflies (Brittain 1982), stoneflies (Hynes 1976), and chironomids (Danks 1971; Danks and Oliver 1972; LeSage and Harrison 1980; Butler 1980, 1982c). Danks and Oliver (1972) suggest that protandry may assist in outbreeding, although the temporal difference is so slight that the sexes overlap broadly in most cases. In species with highly synchronous, short-lived adults, males that emerge slightly prior to peak female emergence should have the highest probability of mating successfully, whereas females that emerge before the expected peak abundance of males would be at a selective disadvantage, since they risk mortality while waiting for mates.

Synchrony.   In a totally synchronous population, all individuals would show indentical seasonal timing. In nature, life histories generally involve greater synchrony at some points in the life cycle than at others. Population synchrony may result both from environmentally imposed mortality on deviant individuals, and from regulation by physiological mechanisms that control the timing of life-history events such as hatching, development, diapause, and emergence. Life histories of synchronous populations are much easier to study because such processes as growth, development, and mortality can easily be determined from static samples. Consequently, the best studies of secondary production have been made on relatively synchronous populations (Benke, Chapter 10).

Synchrony (or asynchrony) can occur in any life-cycle stage, but, as with seasonal timing, this life-history feature has been studied most intensively at emergence. Adult emergence seems to be the most common point in the life cycle where synchrony is re-established within a population. North-temperate populations tend to show greater synchrony than tropical populations (e.g. McElravy et al. 1982), and emergence is often highly synchronous in the Arctic (Oliver 1968; Danks and Oliver 1972; Welch 1973; Butler 1980). Despite this apparent latitudinal trend, variation in emergence synchrony can be found among populations within a single location. Following Corbet's (1958) contrast between synchronous and protracted emergence patterns in two British dragon-flies, similar dichotomies have been proposed within communities of stoneflies (Harper and Pilon 1970), mayflies (Harper and Magnin 1971), and chironomids (Cloutier and Harper 1978).

Extended emergence (or egg hatching or any other life-cycle event) may result from a variety of life-history scenarios. But the high degree of population synchrony seen in many emergence patterns causes one to wonder whether this is an adaptive life-history feature. Synchronous emergence may in fact be an artifact of synchrony at some other point in the life cycle. If for some reason natural selection acted upon synchrony of egg hatching, and if larvae diverged little during growth and development, relatively synchronous emergence might follow as a consequence. Selection might also act not upon synchrony per se, but on the seasonal timing of an event. If individuals are selected to achieve fairly precise seasonal timing at some stage, population synchrony in that stage will also follow as a consequence.

There may also be selection for synchrony itself. Mass emergences may in some circumstances reduce adult mortality by swamping predators (Corbet 1957). The co-occurrence of large numbers of adults can also be critical to mating success, especially in short-lived and swarming species (Downes 1969; Savolainen 1978). Observations that parthenogenetic populations of a chironomid (Welch 1973) and a mayfly (Gibbs 1977) showed lower synchrony than sympatric bisexual populations support the idea that sexual aggregation is at least one of the driving forces behind emergence synchrony.

A number of mechanisms for synchronizing emergence have been proposed. Corbet (1964, p. 275) suggested a mechanism by which synchrony results from a "spring species" life history in which overwintering occurs in the final instar. Mature individuals accumulate in this stage in the autumn, and emerge synchronously in response to warmer temperatures in spring. Another hypothesis deals with synchronization of "summer species" that overwinter in a variety of instars. This mechanism involves a series of rising temperature thresholds for development in the final instars, and would operate only when temperatures are rising. A number of workers have tested and extended this hypothesis (e.g. Paulson and Jenner 1971; Lutz 1968, 1974). This work has shown the importance of photoperiod as well as temperature in regulating emergence, at least for Odonata. Vannote (1978) proposed a model, based on metabolic data from a spring-emerging mayfly, that provides a mechanism for synchronous emergence of individuals that have diverged in size during larval growth.

Species with the capacity to maintain population synchrony may not do so in all environments. For example, the caddisfly *Lepidostoma unicolor* had a synchronous life cycle and late summer emergence in one Oregon stream, but a population in a spring-fed Oregon river apparently lacked a temperature stimulus to control life-cycle phenology, and mature larvae were found through most of the year with emergence spanning five months (Grafius and Anderson 1980).

In contrast to the many synchronous life histories reported for aquatic insects in North America and Europe, New Zealand stream insects generally show flexible, poorly synchronized life histories with nonseasonal or weakly seasonal development, and extended periods of hatching and emergence (Winterbourn et al. 1981). These life-history patterns reflect the relatively nonseasonal and climatically unpredictable nature of New Zealand stream ecosystems, and have prompted these authors to question implications of the river continuum hypothesis (Vannote et al. 1980) that relate to aquatic insect life histories and benthic community structure.

Phenology and Community Structure.   Is there a relationship between life-cycle phenology and community structure? Throughout the literature one finds statements similar to the following: "In a given community, biologically similar species are often temporally separated, presumably the result of competitive interactions" (Danks 1979, p. 556). Reviews by Hynes (1976), Mackay and Wiggins (1979), Wallace and Merritt (1980), Brittain (1982), and Ward and Stanford (1982) list many studies in which sympatric populations of congeneric or functionally similar species show temporal differences in certain life-cycle events such as periods of maximum growth or adult emergence, and many authors imply that avoidance of competition has been the selective cause of divergent life-cycle phenologies.

Are competitive interactions between species really responsible for this temporal aspect of community structure? There are four possibilities worth considering when such a "temporal sequence of synchronized species replacement" (Vannote et al. 1980, p. 134) is observed.

1. *Random patterns.* It is possible that many apparent patterns are trivial and, in fact, nonexistent. However, biologists are inherently biased toward drawing such conclusions by the fact that they are looking for patterns in nature. Unintentional selectivity among data sets and creative presentation of data may lead us to see regularity where randomness may really exist. Appeals have been made for rigorous statistical testing of data against random models (Poole and Rathcke 1979; Connor and Simberloff 1979; Strong 1980), but such tests can be overly simplistic when the biological complexities of each case are considered (Stiles 1979; Grant and Abbott 1980).

2. *Physiological accommodation.* Life histories may be sufficiently labile that, should there indeed be a potential for competition, physiologically controlled shifts in the timing of life-cycle events could minimize temporal overlap. Brittain (1982, p. 131) mentions that European *Leptophlebia* species having similar life cycles are "out of step" when they occur together. Phenological information on populations in allopatry is required to determine whether an actual shift has resulted from sympatric occurrence. Cloutier and Harper (1978) observed that two *Arctopelopia* midge species showed different emergence patterns when collected from the same river section than when either species occurred alone. They assumed that this reflected alterations in other portions of the life cycle as well. This mechanism would require the ability to sense the presence of one or more potential competitors, and then to effect an appropriate physiological change in life history. This does not seem likely and has rarely been specifically hypothesized for aquatic insects.

3. *Adaptive coevolution.* Phenological differences in life history could conceivably also result from an evolutionary shift, in that the biotic environment (i.e. the presence of certain other populations) could constitute a selective pressure if individual fitness were affected by interspecific competition. This hypothesis requires that populations persist in sympatry over a sufficiently long time period for evolutionary change to occur. However, community composition can be expected to change frequently since insect populations can be quite ephemeral. Thus, an evolutionary response would be most reasonably expected within a pool of regionally co-occurring species, rather than among populations presently inhabiting a given waterbody or stretch of river.

A similar evolutionary argument has been made for the phenological patterns observed in communities of flowering plants that may be competing for pollinators (Heinrich 1976; Stiles 1977). Although rarely stated explicitly, an evolutionary shift in life-cycle phenology appears to be the implied mechanism behind some statements to the effect that "closely allied species avoid competition by spatial or temporal segregation" (Hynes 1976, p. 148).

Butler (1980) suggested that the emergence phenologies of some chironomids inhabiting arctic Alaskan tundra ponds may reflect such evolutionary accommodation. Swarming and mating success appears to be a major bottleneck for local population success in this cold and windy environment, and six midge species show morphological and behavioral modifications that allow them to swarm, mate, and oviposit without leaving the pond surface. The regular spacing of highly synchronous emergence periods within this group of species may have resulted from phenological shifts that were selected as a result of interspecific interference during swarming on the pond surface.

4. *Ecological assembly.* If phenological differences in life histories are indeed beneficial (*aptations* sensu Gould and Vrba 1982) with regard to interspecific interactions, they may in most cases be *exaptations* (i.e. resulting from historical reasons, rather than natural selection). In this scenario, temporal niche segregation has not resulted from evolutionary change within a persistent community, but from a process of community development in which coexisting groups of species are assembled from taxa with preexisting life-history characteristics. These life histories may have evolved in response to other biotic or abiotic pres-

sures, or may simply be consequences of historical and developmental constraints.

This is the mechanism invoked by Vannote and Sweeney (1980) to explain the temporal distribution of development within different functional groups of aquatic insects in a fourth-order stream in Pennsylvania. Phenologies are temperature-controlled according to species-specific physiological responses that were determined by the thermal environment of the glacial and interglacial periods when speciation is believed to have occurred. Although the region contains a pool of about 25 *Ephemerella* species that could potentially colonize such a stream, only about five species occur within any given reach. Competitive interactions are seen as restricting invasion by additional species, but not as the selective agent of evolution.

Much recent controversy surrounds the testing of similar competition-based hypotheses in bird community ecology (e.g. Connor and Simberloff 1979; Grant and Abbott 1980; Alatalo 1982). Accordingly, aquatic entomologists should be careful to state specifically the mechanisms believed to be responsible for an observed phenomenon, to entertain alternative hypotheses, and to test between them whenever possible (see also Allan, Chapter 16).

CONCLUSIONS

Our knowledge of aquatic insect life histories continues to increase in a rather piecemeal fashion, even as we become increasingly aware of how essential life-history information is to our attempts to understand the structure and productivity of aquatic communities. We need consistent definitions for concepts such as life cycle, life history, and life-history adaptation, and we need to identify particular categories of life-history variables about which we can ask general questions, and to which we can relate the results of research on any kind of aquatic insect (or even other freshwater invertebrates).

In this chapter I have presented one view of how these terminological and organizational problems may be approached, and have discussed a selected set of life-history topics. Throughout, I have tried to illustrate the interplay between ecological, evolutionary, and descriptive life-history questions. Aquatic entomologists can make a much more significant contribution to broader questions in modern biology than has been historically the case, but to do this we must keep informed in subject areas beyond our own subdiscipline.

A critical need in life-history research is the development of new techniques that will permit us to obtain greater precision in measurement of life-history variables as well as to discriminate between parameters that are often not distinguished, such as growth and development. A field study conducted in a new habitat or on a different species, but using traditional methods, may provide some new information; however, interpretation of results will suffer from traditional limitations. It is easy to appreciate the wisdom of recommendations that we sample all life-cycle stages over more than one seasonal cycle, use standardized techniques, and obtain large samples with many replicates. Likewise, in experimental studies we must provide adequate replication and control (Allan, Chapter 16). But we also need creative methodological improvements that can make such goals feasible in terms of time and funding. Finally, the value of our work derives ultimately from the quality of the questions we address; hence there is a need for the development of a more unified, conceptual approach to the study of aquatic insect life histories.

References

Alatalo, R. V. 1982. Bird species distributions in the Galapágos and other archipelagoes: competition or chance? Ecology 63:881–87.

Anderson, N. H. and K. W. Cummins. 1979. Influences of diet on the life histories of aquatic insects. Journal of the Fisheries Research Board of Canada 36:335–42.

Anderson, N. H. and J. R. Sedell. 1979. Detritus processing by macroinvertebrates in stream ecosystems. Annual Review of Entomology 24:351–77.

Atchley, W. R. 1971. Components of sexual dimorphism in *Chironomus* larvae (Diptera: Chironomidae). The American Naturalist 105:455–66.

Benke, A. C. 1979. A modification of the Hynes method for estimating secondary production with particular significance for multivoltine populations. Limnology and Oceanography 24:168–71.

Benke, A. C. and J. B. Waide. 1977. In defense of average cohorts. Freshwater Biology 7: 61–63.

Bergman, E. A. and W. L. Hilsenhoff. 1978. Parthenogenesis in the mayfly genus *Baetis* (Ephemeroptera: Baetidae). Annals of the Entomological Society of America 71: 167–68.

Bohle, H. W. 1972. Die Temperaturabhängigkeit der Embryogenese und der embryonalen Diapause von *Ephemerella ignita* (Poda) (Insecta, Ephemeroptera). Oecologia (Berlin) 10:253–68.

Borror, D. J., D. M. DeLong, and C. A. Triplehorn. 1981. An introduction to the study of insects. 5th ed. Saunders College Publishing, Philadelphia, PA. 827 pp.

Brittain, J. E. 1982. Biology of mayflies. Annual Review of Entomology 27:119–47.

Brundin, L. 1951. The relation of O_2-microstratification at the mud surface to the ecology of the profundal bottom fauna. Drottningholm Institute of Freshwater Research, Report 32:32–42.

Butler, M. G. 1980. Emergence phenologies of some arctic Alaskan Chironomidae, pp. 307–14. *In*: D. A. Murray (ed.). Chironomidae. Ecology, systematics, cytology and physiology. Pergamon Press, Oxford. 354 pp.

Butler, M. G. 1982a. A 7-year life cycle for two *Chironomus* species in arctic Alaskan tundra ponds (Diptera: Chironomidae). Canadian Journal of Zoology 60:58–70.

Butler, M. G. 1982b. Production dynamics of some arctic *Chironomus* larvae. Limnology and Oceanography 27:728–36.

Butler, M. G. 1982c. Morphological and phenological delimitation of *Chironomus prior* sp. n. and *C. tardus* sp. n. (Diptera: Chironomidae), sibling species from arctic Alaska. Aquatic Insects 4:219–35.

Calow, P. 1978. Life cycles. An evolutionary approach to the physiology of reproduction, development, and ageing. Chapman and Hall, London. 164 pp.

Calow, P. 1981. Invertebrate biology. A functional approach. Halsted Press, New York, NY. 183 pp.

Carter, C. E. 1980. The life cycle of *Chironomus anthracinus* in Lough Neagh. Holarctic Ecology 3:214–17.

Charnov, E. L. and W. M. Schaffer. 1973. Life history consequences of natural selection: Cole's result revisited. The American Naturalist 107:791–93.

Clements, A. N. 1963. The physiology of mosquitoes. Pergamon Press, Oxford. 393 pp.

Clifford, H. F. 1982. Life cycles of mayflies (Ephemeroptera), with special reference to voltinism. Quaestiones Entomologicae 18:15–90.

Clifford, H. F., H. Hamilton, and B. A. Killins. 1979. Biology of the mayfly *Leptophlebia cupida* (Say) (Ephemeroptera: Leptophlebiidae). Canadian Journal of Zoology 57: 1026–45.

Cloutier, L. and P. P. Harper. 1978. Phénologie de Tanypodinae de ruisseaux des Laurentides (Diptera: Chironomidae). Canadian Journal of Zoology 56:1129–39.

Colbo, M. H. and G. N. Porter. 1979. Effects of the food supply on the life history of Simuliidae (Diptera). Canadian Journal of Zoology 57:301–6.

Cole, L. C. 1954. The population consequences of life history phenomena. The Quarterly Review of Biology 29: 103–37.

Connor, E. F. and D. Simberloff. 1979. The assembly of species communities: chance or competition? Ecology 60:1132–40.

Corbet, P. S. 1957. The life-history of the emperor dragonfly *Anax imperator* Leach (Odonata: Aeshnidae). Journal of Animal Ecology 26:1–69.

Corbet, P. S. 1958. Temperature in relation to seasonal development of British dragonflies (Odonata). Proceedings of the Xth International Congress of Entomology, Montreal 2:755–57.

Corbet, P. S. 1962. A biology of dragonflies. H. F. and G. Witherby Ltd., London. 247 pp.

Corbet, P. S. 1964. Temporal patterns of emergence in aquatic insects. The Canadian Entomologist 96:264–79.

Corbet, P. S. 1966. Parthenogenesis in caddisflies (Trichoptera). Canadian Journal of Zoology 44:981–82.

Corbet, P. S. 1967. Facultative autogeny in arctic mosquitoes. Nature (London) 215: 662–63.

Corbet, P. S. 1980. Biology of Odonata. Annual Review of Entomology 25:189–217.

Cushman, R. M., J. W. Elwood, and S. G. Hildebrand. 1977. Life history and production dynamics of *Alloperla mediana* and *Diplectrona modesta* in Walker Branch, Tennessee. American Midland Naturalist 98:354–64.

Danks, H. V. 1971. Life history and biology of *Einfeldia synchrona* (Diptera: Chironomidae). The Canadian Entomologist 103:1597–1606.

Danks, H. V. 1978. Some effects of photoperiod, temperature, and food on emergence in three species of Chironomidae (Diptera). The Canadian Entomologist 110:289–300.

Danks, H. V. 1979. Characteristic modes of adaptation in the Canadian insect fauna, pp. 548–66. *In*: H. V. Danks (ed.). Canada and its insect fauna. Memoirs of the Entomological Society of Canada 108:1–573.

Danks, H. V. and P. S. Corbet. 1973. Sex ratios at emergence of two species of high-arctic *Aedes* (Diptera: Culicidae). The Canadian Entomologist 105:647–51.

Danks, H. V. and D. R. Oliver. 1972. Seasonal emergence of some high arctic Chironomidae (Diptera). The Canadian Entomologist 104:661–86.

Davies, I. J. 1983. Sampling aquatic insect emergence. *In*: J. A. Downing and F. H. Rigler (eds.). A manual on methods for the assessment of secondary productivity in fresh waters. IBP Handbook Number 17. 2nd ed. Blackwell Scientific Publications, Oxford, England (In press).

Degrange, C. 1960. Recherches sur la reproduction des Ephéméroptères. Travaux Laboratoire d'Hydrobiologie et de Pisciculture Université de Grenoble 51:7–193.

Denno, R. F. and H. Dingle (eds.). 1981. Insect life history patterns: habitat and geographic variation. Springer-Verlag, New York, NY. 226 pp.

Dingle, H. (ed.). 1978. Evolution of insect migration and diapause. Springer-Verlag, New York, NY. 284 pp.

Downes, J. A. 1965. Adaptations of insects in the arctic. Annual Review of Entomology 10:257–74.

Downes, J. A. 1969. The swarming and mating flight of Diptera. Annual Review of Entomology 14:271–98.

Edmunds, G. F., Jr. and C. H. Edmunds. 1979. Predation, climate, and emergence and mating of mayflies, pp. 277–85. *In*: J. F. Flannagan and K. E. Marshall (eds.). Advances in Ephemeroptera biology. Plenum Press, New York, NY. 552 pp.

Edmunds, G. F., Jr., S. L. Jensen, and L. Berner. 1976. The mayflies of North and Central America. University of Minnesota Press, Minneapolis. 330 pp.

Elliott, J. M. and U. H. Humpesch. 1980. Eggs of Ephemeroptera. Freshwater Biological Association Annual Report 48:41–52.

Fink, T. J. 1980. A comparison of mayfly (Ephemeroptera) instar determination methods, pp. 367–80. *In*: J. F. Flannagan and K. E. Marshall (eds.). Advances in Ephemeroptera biology. Plenum Press, New York, NY. 552 pp.

Fink, T. J. 1982. Why the simple frequency and Janetschek methods are unreliable for determining instars of mayflies: an analysis of the number of nymphal instars of *Stenonema modestum* (Banks) (Ephemeroptera: Heptageniidae) in Virginia, USA. Aquatic Insects 4:67–71.

Fisher, R. A. 1930. The genetical theory of natural selection. Clarendon Press, Oxford. 272 pp.

Flannagan, J. F. 1979. The burrowing mayflies of Lake Winnipeg, Manitoba, Canada, pp. 103–113. *In*: K. Pasternak and R. Sowa (eds.). Proceedings of the Second International Conference on Ephemeroptera. Panstwowe Wydawnictwo Naukowe, Warsaw, Poland. 312 pp.

Flannagan, J. F. and G. H. Lawler. 1972. Emergence of caddisflies (Trichoptera) and mayflies (Ephemeroptera) from Heming Lake, Manitoba. The Canadian Entomologist 104:173–83.

Ford, J. B. 1959. A study of larval growth, the number of instars, and sexual differentiation in the Chironomidae (Diptera). Proceedings of the Royal Entomological Society of London Series A 34:151–60.

Gibbs, K. E. 1977. Evidence for obligatory parthenogenesis and its possible effect on the emergence period of *Cloeon triangulifer* (Ephemeroptera: Baetidae). The Canadian Entomologist 109:337–40.

Giesel, J. T. 1976. Reproductive strategies as adaptations to life in temporally heterogeneous environments. Annual Review of Ecology and Systematics 7:57–79.

Gould, S. J. 1980. Is a new and general theory of evolution emerging? Paleobiology 6: 119–30.

Gould, S. J. 1982. Darwinism and the expansion of evolutionary theory. Science 216: 380–87.

Gould, S. J. and R. C. Lewontin. 1979. The spandrels of San Marco and the Panglossian paradigm: a critique of the adaptationist program. Proceedings of the Royal Society of London Series B 205:581–98.

Gould, S. J. and E. S. Vrba. 1982. Exaptation—a missing term in the science of form. Paleobiology 8:4–15.

Grafius, E. and N. H. Anderson. 1980. Population dynamics and role of two species of *Lepidostoma* (Trichoptera: Lepidostomatidae) in an Oregon coniferous forest stream. Ecology 61:808–16.

Grant, P. R. and I. Abbott. 1980. Interspecific competition, island biogeography and null hypotheses. Evolution 34:332–41.

Grodhaus, G. 1971. Sporadic parthenogenesis in three species of *Chironomus* (Diptera). The Canadian Entomologist 103:338–40.

Hamilton, W. D. 1967. Extraordinary sex ratios. Science 156:477–88.

Harper, F. and E. Magnin. 1971. Émergence saisonnière de quelques éphéméroptères d'un ruisseau des Laurentides. Canadian Journal of Zoology 49:1209–21.

Harper, P. P. and J. -G. Pilon. 1970. Annual patterns of emergence of some Quebec stoneflies (Insecta: Plecoptera). Canadian Journal of Zoology 48:681–94.

Heinrich, B. 1976. Flowering phenologies: bog, woodland, and disturbed habitats. Ecology 57:890–99.

Hinton, H. E. 1951. A new chironomid from Africa, the larva of which can be dehydrated without injury. Proceedings of the Zoological Society of London 121:371–80.

Horn, H. S. 1978. Optimal tactics of reproduction and life-history, pp. 411–29. *In:* J. R. Krebs and N. B. Davies (eds.). Behavioral ecology: an evolutionary approach. Sinauer Associates Ltd., Sunderland, MA. 494 pp.

Horst, T. J. and G. R. Marzolf. 1975. Production ecology of burrowing mayflies in a Kansas reservoir. Internationale Vereinigung für Theoretische und Angewandte Limnologie Verhandlungen 19:3029–38.

Hynes, H. B. N. 1976. Biology of Plecoptera. Annual Review of Entomology 21:135–53.

Iversen, T. M. 1976. Life cycle and growth of Trichoptera in a Danish spring. Archiv für Hydrobiologie 78:482–93.

Jónasson, P. M. 1972. Ecology and production of the profundal benthos in relation to phytoplankton in Lake Esrom. Oikos Supplementum 14:1–148.

Jones, R. E. 1975. Dehydration in an Australian rockpool chironomid larva (*Paraborniella tonnoiri*). Journal of Entomology Series A 49:111–19.

Kajak, Z. 1964. Remarks on conditions influencing the appearance of new generations of Tendipedidae larvae. Ekologia Polska Seria A 12:173–83.

Kerfoot, W. C. (ed.). 1980. Evolution and ecology of zooplankton communities. University Press of New England, Hanover, NH. 793 pp.

Kerst, C. D. and N. H. Anderson. 1974. Emergence patterns of Plecoptera in a stream in

Oregon, USA. Freshwater Biology 4:205–12.

Krebs, C. J. 1978. Ecology: the experimental analysis of distribution and abundance. 2nd ed. Harper and Row, New York, NY. 678 pp.

Krüger, F. 1941. Parthenogenetische *Stylotanytarsus*-larven als Bewohner einer Trinkwasserleitung. (*Tanytarsus*-Studien III: Die Gattung *Stylotanytarsus*). Archiv für Hydrobiologie 38:214–53.

Lamb, R. J. and S. M. Smith. 1980. Comparison of egg size and related life-history characteristics for two predaceous tree-hole mosquitoes (*Toxorhynchites*). Canadian Journal of Zoology 58:2065–70.

Landa, V. 1968. Developmental cycles of central European Ephemeroptera and their interrelations. Acta Entomologica Bohemoslovaca 65:276–84.

LeSage, L. and A. D. Harrison. 1980. The biology of *Cricotopus* (Chironomidae: Orthocladiinae) in an algal-enriched stream: Part I. Normal biology. Archiv für Hydrobiologie Supplement 57:375–418.

Lindeberg, B. 1971. Parthenogenetic strains and unbalanced sex ratios in Tanytarsini (Diptera: Chironomidae). Annales Zoologici Fennici 8:310–17.

Lindeberg, B. 1974. Parthenogenetic and normal populations of *Abiskomyia virgo* Edw. (Diptera, Chironomidae). Entomologisk tidskrift 95 (Supplementum): 157–61.

Lutz, P. E. 1968. Effects of temperature and photoperiod on larval development in *Lestes eurinus* (Odonata: Lestidae). Ecology 49:637–44.

Lutz, P. E. 1974. Effects of temperature and photoperiod on larval development in *Tetragoneuria cynosura* (Odonata: Libellulidae). Ecology 55:370–77.

Macan, T. T. 1958. Causes and effects of short emergence periods in insects. Internationale Vereinigung für Theoretische und Angewandte Limnologie Verhandlungen 13: 845–49.

McCafferty, W. P. and B. L. Huff. 1974. Parthenogenesis in the mayfly *Stenonema femoratum* (Say) (Ephemeroptera: Heptageniidae). Entomological News 85:76–80.

McClure, R. G. and K. W. Stewart. 1976. Life cycle and production of the mayfly *Choroterpes (Neochoroterpes) mexicanus* Allen (Ephemeroptera: Leptophlebiidae). Annals of the Entomological Society of America 69:134–44.

McCrae, A. W. R. and P. S. Corbet. 1982. Oviposition behaviour of *Tetrathemis polleni* (Selys): a possible adaptation to life in turbid pools (Anisoptera: Libellulidae). Odonatologica 11:23–31.

McElravy, E. P., H. Wolda, and V. H. Resh. 1982. Seasonality and annual variability of caddisfly adults (Trichoptera) in a "non-seasonal" tropical environment. Archiv für Hydrobiologie 94:302–17.

Machado, A. B. M. and A. Martinez. 1982. Oviposition by egg-throwing in a zygopteran, *Mecistogaster jocaste* Hagen, 1869 (Pseudostigmatidae). Odonatologica 11:15–22.

Mackay, R. J. and G. B. Wiggins. 1979. Ecological diversity in Trichoptera. Annual Review of Entomology 24:185–208.

Nagell, B. 1981. Overwintering strategy of two closely related forms of *Cloeon (dipterum?)* (Ephemeroptera) from Sweden and England. Freshwater Biology 11:237–44.

Nielsen, A. 1951. Spring fauna and speciation. Internationale Vereinigung für Theoretische und Angewandte Limnologie Verhandlungen 11:261–63.

O'Brien, W. J. 1979. The predator-prey interaction of planktivorous fish and zooplankton. American Scientist 67:572–81.

Oliver, D. R. 1968. Adaptations of arctic Chironomidae. Annales Zoologici Fennici 5: 111–18.

Oliver, D. R. 1979. Contribution of life history information to taxonomy of aquatic insects. Journal of the Fisheries Research Board of Canada 36:318–21.

Oliver, D. R. and H. V. Danks. 1972. Sex ratios of some high arctic Chironomidae (Diptera). The Canadian Entomologist 104:1413–17.

Olsson, T. I. 1981. Overwintering of benthic macroinvertebrates in ice and frozen sediment in a North Swedish river. Holarctic Ecology 4:161–66.

Patterson, J. W. and R. L. Vannote. 1979. Life history and population dynamics of *Heteroplectron americanum*. Environmental Entomology 8:665–69.

Paulson, D. R. and C. E. Jenner. 1971. Population structure in overwintering larval Odonata in North Carolina in relation to adult flight season. Ecology 52:96–107.

Peckarsky, B. L. 1982. Aquatic insect predator-prey relations. BioScience 32:261–66.

Pennak, R. W. 1978. Fresh-water invertebrates of the United States. 2nd ed. John Wiley and Sons, New York, NY. 803 pp.

Poole, R. W. and B. J. Rathcke. 1979. Regularity, randomness, and aggregation in flowering phenologies. Science 203:470–71.

Price, P. W. 1974. Energy allocation in ephemeral adult insects. Ohio Journal of Science 74:380–87.

Pritchard, G. 1976. Growth and development of larvae and adults of *Tipula sacra* Alexander (Insecta: Diptera) in a series of abandoned beaver ponds. Canadian Journal of Zoology 54:266–84.

Pritchard, G. 1978. Study of dynamics of populations of aquatic insects: the problem of variability in life history exemplified by *Tipula sacra* Alexander (Diptera; Tipulidae). Internationale Vereinigung für Theoretische und Angewandte Limnologie Verhandlungen 20:2634–40.

Pritchard, G. 1983. Biology of Tipulidae. Annual Review of Entomology 28:1–22.

Pritchard, G. and B. Pelchat. 1977. Larval growth and development of *Argia vivida* (Odonata: Coenagrionidae) in warm sulfur pools at Banff, Alberta. The Canadian Entomologist 109:1563–70.

Ratte, H. T. 1979. Tagesperiodische Vertikalwanderung in thermisch geschichteten Gewässern: Einfluss von Temperatur- und Photoperiode-Zyklen auf *Chaoborus crystallinus* de Geer (Diptera: Chaoboridae). Archiv für Hydrobiologie Supplement 57:1–37.

Reice, S. R. 1983. Predation and substratum: factors in lotic community structure, pp. 325–45. *In*: T. D. Fontaine, III and S. M. Bartell (eds.). Dynamics of lotic ecosystems. Ann Arbor Science Publishers, Ann Arbor, MI. 494 pp.

Resh, V. H. 1976. Life histories of coexisting species of *Ceraclea* caddisflies (Trichoptera: Leptoceridae): the operation of independent functional units in a stream ecosystem. The Canadian Entomologist 108:1303–18.

Resh, V. H. 1979. Sampling variability and life history features: basic considerations in the design of aquatic insect studies. Journal of the Fisheries Research Board of Canada 36:290–311.

Resh, V. H. and G. Grodhaus. 1983. Aquatic insects in urban environments. *In*: G. W. Frankie and K. W. Koehler (eds.). Urban entomology: interdisciplinary perspectives. Praeger Publishers, New York, NY. (In press).

Resh, V. H. and K. L. Sorg. 1978. Midsummer flight activity of caddisfly adults from a northern California stream. Environmental Entomology 7:396–98.

Rosenberg, D. M. (ed.). 1979. Freshwater benthic invertebrate life histories: current research and future needs. Journal of the Fisheries Research Board of Canada 36: 289–345.

Rosenberg, D. M., A. P. Wiens, and O. A. Saether. 1977. Life histories of *Cricotopus (Cricotopus) bicinctus* and *C. (C.) mackenziensis* (Diptera: Chironomidae) in the Fort Simpson area, Northwest Territories. Journal of the Fisheries Research Board of Canada 34:247–53.

Savolainen, E. 1978. Swarming in Ephemeroptera: the mechanism of swarming and the effects of illumination and weather. Annales Zoologici Fennici 15:17–52.

Scholander, P. F., W. Flagg, R. J. Hock, and L. Irving. 1953. Studies on the physiology of frozen plants and animals in the Arctic. Journal of Cellular and Comparative Physiology 42:1–56.

Seward, R. M. 1980. A study of fecundity in the Chironomidae, with emphasis on intra- and inter-specific variability in primary follicle number. Unpublished Ph.D. thesis, University of Pittsburgh, Pittsburgh, PA. 198 pp.

Southwood, T. R. E. 1977. Habitat, the templet for ecological strategies? Journal of Animal Ecology 46:337–65.

Southwood, T. R. E. 1978. Ecological methods with particular reference to the study of insect populations. 2nd ed. Chapman and Hall, London. 524 pp.

Stiles, F. G. 1977. Coadapted competitors: the flowering seasons of hummingbird-pollinated plants in a tropical forest. Science 198:1177–78.

Stiles, F. G. 1979. Response to R. W. Poole and B. J. Rathcke. Science 203:471.

Strong, D. R., Jr. 1980. Null hypotheses in ecology. Synthese 43:271–85.

Suomalainen, E. 1962. Significance of parthenogenesis in the evolution of insects. Annual Review of Entomology 7:349–66.

Svensson, B. 1977. Life cycle, energy fluctuations, and sexual differentiation in *Ephemera danica* (Ephemeroptera), a stream-living mayfly. Oikos 29:78–86.

Sweeney, B. W. 1978. Bioenergetic and developmental response of a mayfly to thermal variation. Limnology and Oceanography 23:461–77.

Sweeney, B. W. and R. L. Vannote. 1978. Size variation and the distribution of hemimetabolous aquatic insects: two thermal equilibrium hypotheses. Science 200:444–46.

Sweeney, B. W. and R. L. Vannote. 1981. *Ephemerella* mayflies of White Clay Creek: bioenergetic and ecological relationships among six coexisting species. Ecology 62: 1353–69.

Tauber, C. A. and M. J. Tauber. 1981. Insect seasonal cycles: genetics and evolution. Annual Review of Ecology and Systematics 12:281–308.

Tauber, M. J. and C. A. Tauber. 1976. Insect seasonality: diapause maintenance, termination, and postdiapause development. Annual Review of Entomology 21:81–107.

Vannote, R. L. 1978. A geometric model describing a quasi-equilibrium of energy flow in populations of stream insects. Proceedings of the National Academy of Sciences of the United States of America 75:381–84.

Vannote, R. L., G. W. Minshall, K. W. Cummins, J. R. Sedell, and C. E. Cushing. 1980. The river continuum concept. Canadian Journal of Fisheries and Aquatic Sciences 37:130–37.

Vannote, R. L. and B. W. Sweeney. 1980. Geographic analysis of thermal equilibria: a conceptual model for evaluating the effect of natural and modified thermal regimes on aquatic insect communities. The American Naturalist 115:667–95.

Wallace, J. B. and R. W. Merritt. 1980. Filter-feeding ecology of aquatic insects. Annual Review of Entomology 25:103–32.

Ward, J. V. and J. A. Stanford. 1982. Thermal responses in the evolutionary ecology of aquatic insects. Annual Review of Entomology 27:97–117.

Waters, T. F. 1977. Secondary production in inland waters. Advances in Ecological Research 10:91–164.

Waters, T. F. 1979. Benthic life histories: summary and future needs. Journal of the Fisheries Research Board of Canada 36:342–45.

Welch, H. E. 1973. Emergence of Chironomidae (Diptera) from Char Lake, Resolute, Northwest Territories. Canadian Journal of Zoology 51:1113–23.

Welch, H. E. 1976. Ecology of Chironomidae (Diptera) in a polar lake. Journal of the Fisheries Research Board of Canada 33:227–47.

Welton, J. S. and J. A. B. Bass. 1980. Quantitative studies on the eggs of *Simulium (Simulium) ornatum* Meigen and *Simulium (Wilhelmia) equinum* L. in a chalk stream in southern England. Ecological Entomology 5:87–96.

Wood, J. R., V. H. Resh, and E. M. McEwan. 1982. Egg masses of Nearctic sericostomatid caddisfly genera (Trichoptera). Annals of the Entomological Society of America 75:430–34.

White, M. J. D. 1973. Animal cytology and evolution. 3rd ed. Cambridge University Press, London. 961 pp.

Wiggins, G. B. 1977. Larvae of the North American caddisfly genera (Trichoptera). University of Toronto Press, Toronto. 401 pp.

Wiggins, G. B., R. J. Mackay, and I. M. Smith. 1980. Evolutionary and ecological strategies of animals in annual temporary pools. Archiv für Hydrobiologie Supplement 58: 97–206.

Williams, G. C. 1979. The question of adaptive sex ratio in outcrossed vertebrates. Proceedings of the Royal Society of London Series B 205:567–80.

Winterbourn, M. J., J. S. Rounick, and B. Cowie. 1981. Are New Zealand stream ecosystems really different? New Zealand Journal of Marine and Freshwater Research 15:321–28.

Wülker, W. and P. Götz. 1968. Die Verwendung der Imaginalscheiben zur Bestimmung des Entwicklungszustandes von *Chironomus*-Larven (Diptera). Zeitschrift für Morphologie und Ökologie der Tiere 62:363–88.

Zaret, T. M. 1980. Predation and freshwater communities. Yale University Press, New Haven, CT. 187 pp.

chapter 4

FACTORS INFLUENCING LIFE-HISTORY PATTERNS OF AQUATIC INSECTS

Bernard W. Sweeney

INTRODUCTION

Insects exhibit a wide diversity of life-history patterns in aquatic habitats. For a given species, the life-history pattern is a population level concept that is usually described in terms of the magnitude and timing of individual life-history parameters in a specific habitat. This review examines how various environmental factors have affected specific life-history parameters of representative aquatic insect species. Parameters selected for review include those having a high probability of being ecologically significant (i.e. with respect to maximizing individual fitness; see also Butler, Chapter 3) and for which a good data base exists for a large number of aquatic species. These parameters include rate and success of embryonic development; rate, efficiency, and magnitude of larval growth; timing of pupation and adult emergence; and adult size and fecundity.

Temperature, nutrition, and photoperiod are the most important environmental factors that are discussed. These factors were chosen because they are important to all species, and general principles can presently be offered from a relatively large data base. Other important abiotic (e.g. dissolved oxygen, pH, current, substrate) and biotic (e.g. predation, parasitism, competition) factors are not included here since many of them are considered in other chapters of this book.

The information presented here is organized by environmental factors. Interactions between factors are pointed out where possible because life-history patterns in nature are compromises between a complex of environmental demands. Although narrowly focused, the topic of this review is still a very extensive one. As a result, not all known references on a particular subject are listed, and preference is generally given to the most typical study, or to relevant reviews. A bias towards the North American literature is acknowledged.

TEMPERATURE

Seasonal temperature fluctuations characterize most natural lotic and lentic environments (Ward and Stanford 1982). Streams differ from lakes and ponds, however, in that: (1) the seasonal rate of temperature change in streams is often greater than in lakes, even though the annual temperature range is often less; (2) diurnal temperature changes are often greater in streams than in lakes, especially when streams are compared to nonlittoral areas of lakes; and (3) thermal stratification is rare in natural streams and persists for only short periods relative to lakes (Hynes 1970; Brittain 1976a).

Although streams and lakes exhibit thermal fluctuations, the annual temperature regime (i.e. both rate and magnitude) of a given habitat seems highly structured and predictable from year to year. For example, a statistical analysis of a continuous ten-year temperature record for White Clay Creek in Pennsylvania indicated significant predictability both within seasons and over the entire year (Vannote and Sweeney 1980). In addition, although both the seasonal pattern and magnitude of temperature for streams varies geographically, the amount of heat accumulated (i.e. degree days) on an annual basis appears to be predictable when tributaries of similar size are compared over a broad range of latitudes (Vannote and Sweeney 1980). Aquatic insects, therefore, inhabit environments with temperature patterns that vary, but they do so in a predictable fashion.

Although temperature has usually been measured in field studies of aquatic insects, the type and frequency of measurements (e.g. once daily, weekly) has often lacked enough detail to properly describe existing thermal conditions. The availability of submersible, long-term thermographs has greatly increased the potential for studying thermal effects, although the placement of these devices within a habitat is critical.

Most laboratory studies involving aquatic insects have emphasized constant temperatures in evaluating thermal effects. The thermal variation of natural systems is rarely included in experimental designs because of technical difficulties in producing suitable test environments, interpretation problems with variable temperature data, and the association of constant temperatures with a controlled experiment (Sweeney 1978; Beck 1983). Although constant temperature seems to present a more controlled experimental situation, manipulation of maximum temperature, minimum temperature, and the rate of change of temperatures, on a diel or seasonal basis, permits a more natural exposure of test animals to the experimental variable.

Pradhan (1945) presented an instructive theoretical analysis on how and why activity rates (e.g. rates of growth, development, feeding) that are associated with insect life histories might differ when these rates are studied at constant as compared to variable temperatures. He argued that when data were obtained over a broad range of constant temperatures, most insects exhibit a sigmoid relationship between activity and temperature (e.g. see Fig. 4.1). In this hypothetical case, if temperature remains constant then activity will remain constant. However, temperature fluctuations above or below a given constant temperature will generally accelerate or retard activity, respectively. The activity rate for a fluctuating regime will be the net sum of these effects. Whether this sum is positive or negative depends on the range of temperatures encompassed by the

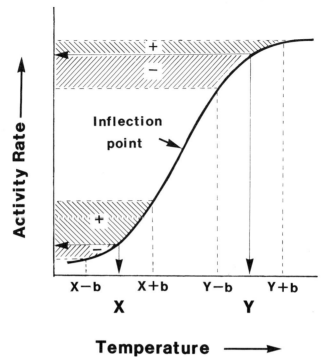

Figure 4.1 Hypothetical example showing the sigmoid relationship between activity rate and temperature that is often observed for insects. The relative increase or decrease in activity rate associated with two fluctuating thermal regimes having different mean temperatures (i.e. X and Y), but the same magnitude of fluctuation (i.e. $\pm b$), is demonstrated. Note that when temperature fluctuations are approximately symmetrical around the mean, then fluctuating regime $X \pm b$ has a net stimulatory effect on activity relative to a constant regime of X, whereas fluctuating regime $Y \pm b$ has a net retardative effect relative to a constant regime Y (modified from Pradhan 1945).

fluctuating regime, and the shape of the curve that depicts activity rate as a function of temperature. In the hypothetical example, if a variable regime fluctuates symmetrically about a given mean temperature (i.e. thermal accumulation above and below the mean is identical), then the net effect would be accelerative at temperatures below the inflection point (Fig. 4.1, $X \pm b$) and retardative above the inflection point (Fig. 4.1, $Y \pm b$) relative to the equivalent constant temperature. The magnitude of the net effect should be most pronounced near the lower and upper ends of the curve.

Two related questions need to be addressed experimentally in order to improve our understanding of insect activity rates in constant versus variable temperature regimes. First, does a given temperature have the same effect on activity when it is experienced by the insect constantly or when it is part of an array of cooler and warmer temperatures? This is an important, but perhaps untestable, assumption underlying all historical discussions on this subject. Second, when the main intent is to compare biological response among experiments, is mean temperature the best available statistic for characterizing a fluctuating temperature regime? This is important because the response of aquatic insects to thermal variation remains a significant area of research.

Most life-history parameters (especially larval growth, adult size, and fecundity) discussed herewith are affected significantly by both temperature and nutrition (see Anderson and Cummins 1979; Vannote and Sweeney 1980 for reviews). It is often difficult to differentiate experimentally between the relative importance of temperature and nutrition, because temperature affects larval bioenergetics as well as the quantity and/or quality of food materials. For example, temperature can affect larval growth directly by its influence on rates of feeding, assimilation, and respiration; food conversion efficiencies; enzymatic kinetics; and endocrine processes (Vannote and Sweeney 1980; Sweeney and Vannote 1981), or indirectly by altering the quantity (e.g. density and/or productivity of periphyton) and quality (e.g. microbial populations associated with detritus) of available food material (Cummins and Klug 1979). Likewise, temperature and photoperiod are often confounded in most experimental designs. Beck (1983) has pointed out that although both thermoperiod and photoperiod can greatly influence the ecological adaptations of insects, there has been little research on thermal periodicity alone. Daily thermoperiods in water are less variable than in air and form a seasonally predictable pattern by which life-history events could be timed. The dynamic nature of environmental temperatures must be incorporated into experimental studies of growth, development, diapause induction, and emergence timing. Thus, the following discussion will focus on studies where temperature was the only experimental variable measured, or where its effects can be differentiated clearly from other factors.

Egg Development

The duration of egg development in nondiapausing aquatic insects is often inversely related to temperature (e.g. Hanec and Brust 1967; Elliott 1972; Brittain 1977; Humpesch 1980; Humpesch and Elliott 1980). This inverse relationship is adequately described by a power function for most species (e.g. see Humpesch and Elliott 1980 for data on 12 species of Ephemeroptera). In contrast, hatching success often remains uniformly high at intermediate temperatures and then decreases at the low and/or high temperature extremes (e.g. Brittain 1977; Elliott and Humpesch 1980). Also, individual developmental time is more variable at lower temperatures (Elliott 1972).

Extension of constant-temperature laboratory data to field development of eggs has sometimes been successful (Elliott 1972), but it also has underestimated or overestimated developmental time (Humpesch 1978). The rate of egg development in fluctuating thermal regimes depends largely on the magnitude of the diel temperature pulse for some species, and mean daily temperatures may underestimate the developmental stimulus provided by the warm part of the cycle (Sweeney and Schnack 1977; Sweeney 1978). For both aquatic and terrestrial species, the optimum constant temperature for egg development (i.e. the shortest developmental period and highest percent hatch) usually occurs near the upper lethal limit. Howe (1967) suggests that these optima may have arisen from selection for rapid development in variable temperature environments.

Many aquatic insects exhibit an embryonic diapause that is considered an adaptation for surviving periods of drought (Needham 1903; Corbet 1963), cold weather (Corbet 1958, 1963), or other unfavorable conditions. The stage in development where diapause begins is consistent within a species but varies among different species (Hynes 1970). The effect of temperature on diapause depends on the stage of development with respect to diapause (e.g. prediapause, diapause, or postdiapause) and the species. For example, temperature clearly affects: (1) the duration of the prediapause period (Bohle 1972; Sawchyn and Church 1973; Sawchyn and Gillott 1974); (2) the length of the diapause (Corbet 1956a, 1956b; Britt 1962; Bohle 1972; Sawchyn and Church 1973; Sawchyn and Gillott 1974; Snellen and Stewart 1979); and (3) the duration of postdiapause development (Britt 1962; Bohle 1972). Low or seasonally decreasing temperatures often terminate diapause for some aquatic species (Snellen and Stewart 1979), especially eggs that diapause in a more advanced embryonic stage (Britt 1962; Bohle 1972).

Temperatures may also induce "quiescence" in species where embryonic diapause is not obligatory (Howe 1967). This type of response is found in representatives of most insect orders. For these species, development slows to reduced rates at low temperatures, regardless of the embryonic stage. Available data indicate that: (1) quiescent eggs remain viable for a long time, but not indefinitely (Hanec and Brust 1967); (2) the incubation period at optimal temperatures equals the sum of prequiescence and postquiescence intervals at the same temperatures (e.g. Hunt 1951); and (3) mortality during quiescence increases with decreased temperatures below the thermal threshold, and also with exposure time at a specific temperature (Hanec and Brust 1967). For terrestrial insects, death during quiescence has been attributed to the depletion of stored energy reserves (Richards 1957), but this possibility has not been tested for aquatic species.

Intraspecific variation in developmental time at a given temperature has been studied for geographically separated populations, but results do not show any consistent pattern among species. For example, Humpesch (1980) found that eggs of two mayfly species obtained from populations in cold environments took longer to hatch at a given temperature than eggs from warmer habitats. Significant differences in hatching time at identical temperatures have also resulted when lake and stream populations were compared (Khoo 1968b; Humpesch 1980). Although other aquatic species do not exhibit significant geographic variation in the response of egg development time to temperature

(Pajunen 1975; Humpesch 1980), it has been shown that the temperature resulting in the maximum developmental rate of eggs decreases as latitude or elevation increases for certain other aquatic invertebrates and terrestrial insects (Taylor 1981).

Larval Development and Growth

The magnitude of larval growth, in terms of size and/or biomass, depends largely on the total length of time available for growth (i.e. the developmental time) and how fast a larva can accumulate biomass during that period (i.e. the growth rate). A given temperature, or change in temperature, does not always affect a species' developmental time and growth rate to the same extent. For this reason, developmental time and growth rate are discussed separately.

The duration of the larval period is an important life-history parameter because it is a critical component of most life-table and production estimates for aquatic insects, and temperature corrections are often applied to these types of data. The developmental period for aquatic insects, when carefully studied at several constant temperatures, usually exhibits an inverse relationship to temperature (e.g. Huffaker 1944; Konstantinov 1958; Brust 1967; Hanec and Brust 1967; Lutz 1968a, 1968b; Harvey 1971; Becker 1973; Pajunen and Sundbäck 1973; Pajunen 1975). For some species, however, the larval period decreases with increased temperature to a certain point and then either decreases (Brust 1967) or remains unchanged (Nebeker 1973) with subsequent increments of temperature.

The faster development at high temperatures largely results from a decrease in the duration of each of the larval stages (Headlee 1940; Huffaker 1944; Bar-Zeev 1958; Brust 1967; Hanec and Brust 1967; Lutz 1974b), or from a decrease in both the duration and number of instars (Ross and Merritt 1978). Some species, however, actually have more instars at higher temperatures even though the duration of the larval period decreases (Pritchard and Pelchat 1977). For those species where the number of growth stages remains unchanged, temperature affects the duration of each instar but not the percentage of total larval development time spent in a given instar (Huffaker 1944; Brust 1967). The temperature resulting in the shortest developmental time remains constant for all instars in some (Huffaker 1944; Brust 1967), but not all species (Bar-Zeev 1958).

Many recent, but less-detailed studies (i.e. fewer temperatures or narrower range of temperatures) on winter- and spring-growing species have shown that exposure of larvae to unseasonably warm temperatures results in shorter developmental periods and premature metamorphosis (Brittain 1976a, 1976b; Sweeney and Vannote 1978, 1981; Vannote and Sweeney 1980). For multivoltine species, developmental time is significantly shorter for summer cohorts (Thibault 1971; Benech 1972a, 1972b; Fahy 1973; Clifford and Boerger 1974; Sweeney 1978; Ross and Merritt 1978; Illies 1979; Vannote and Sweeney 1980).

The larval developmental period of a population can often be predicted using summation of thermal units (i.e. degree days). This method requires an accurate estimate of the daily mean temperatures of the habitat and the developmental threshold temperature for the species. The threshold temperature refers theoretically to a specific cold temperature at which the rate of development is at or near zero. Thus, the degree-day accumulation for any given day represents the difference (in degrees) between the mean daily temperatures and the developmental threshold. The "sum of degree days" rule suggests that the number of degree days needed to complete development for a given species is a constant. This generalization can and has been applied to development of any life-history stage, but here we will focus only on larval development.

Although many studies on aquatic insects report the number of degree days needed to complete larval development in a specific habitat or laboratory experiment, only a few studies have evaluated degree-day accumulation for a given species in more than one habitat or in an experiment having different thermal regimes. For aquatic insects, the number of degree days needed to complete larval development seems to be constant regardless of rearing temperature in some species (Headlee 1941; Konstantinov 1958), but not in others (Huffaker 1944; Bar-Zeev 1958; Becker 1973; Sweeney and Schnack 1977). Additional studies are needed to evaluate the usefulness of this concept for aquatic insects. For example, no one has measured degree-day requirements of populations that are separated geographically along a natural thermal gradient, or that occupy habitats having the same total number of degree days per year but differing in the seasonal distribution and/or annual range of temperatures.

Most studies concerning temperature and aquatic insect development have been made at constant temperatures. A serious problem centers on whether results obtained under constant temperatures can be reliably applied to the fluctuating thermal regimes that are prevalent in the field. For some aquatic species, developmental times in fluctuating regimes have been predicted accurately from constant temperature data (e.g. Headlee 1941; Pajunen and Sundbäck 1973), whereas for other species, development was either accelerated (Huffaker 1944) or retarded (Headlee 1940) in fluctuating regimes that were related to a constant temperature equivalent (which is usually taken as the mean). Although the mean of a fluctuating regime is generally assumed to be the suitable constant temperature for such a comparison, biological data sometimes fail to support this assumption (Taylor 1982).

Besides influencing the length of the larval growth period, temperature also affects the ingestion and assimilation of food. Ingestion rates either increase uniformly with increased temperature (Iversen 1979) or increase and then decrease when a very broad range of temperatures are studied (Monakov 1972; Johannsson 1980). Some aquatic insects exhibit more or less constant ingestion rates at low to intermediate temperatures and then, depending on the species, either increase (e.g. Lacey and Mulla 1979) or decrease (e.g. Gallepp 1977) at higher temperatures.

The effects of temperature on larval digestion has only been reported for one aquatic insect, the cranefly *Tipula abdominalis*. For this species, Martin et al. (1980) show that protein digestion (i.e. proteolytic activity) is highly temperature dependent, increasing by a factor of 7 to 10 for each 10°C rise in temperature. This area of research should be pursued further.

Available data indicate an inverse relationship between ingestion rate and assimilation efficiency for several aquatic insects (see McCullough 1975 for review). Since ingestion rate is usually greater at higher temperatures, the assimilation rate should be lower. Few studies, however, have estimated assimilation efficiency over a broad range of temperatures. Heiman and Knight (1975) demonstrated that increased temperature resulted in lower assimilation efficiency but higher ingestion rates for the predatory stonefly *Acroneuria californica*. Although this is consistent with data on at least one other aquatic invertebrate (Elwood and Goldstein 1975), more studies are needed on the effect of temperature on assimilation efficiency values for insects that represent both primary and secondary consumer levels, and that have different growth patterns during various seasons.

Temperature can influence larval growth by simultaneously altering the relationship among several bioenergetic factors. It is not surprising, therefore, that numerous workers have observed high positive correlations between larval growth rates of natural populations and water temperature (e.g. Brittain 1976b; Sweeney 1978; Humpesch 1979; Mackay 1979; Johannsson 1980), and that laboratory studies have confirmed this as well (e.g. Brittain 1976b; Mokry 1976; Gallepp 1977; Mackey 1977; Sweeney and Schnack 1977; Sweeney 1978; Ward and Cummins 1979; Vannote and Sweeney 1980). For some species, however, abnormally high temperatures result in decreased growth rates (Konstantinov 1958; Brust 1967; Nebeker 1973; Heiman and Knight 1975) and/or reduced net growth efficiencies (Heiman and Knight 1975). Thermal effects on larval growth efficiency are also implied when temperatures resulting in fastest development fail to maximize absolute growth (Konstantinov 1958; Brust 1967).

Corbet (1957) proposed a model to describe how odonates could use seasonal changes in temperature to synchronize their life history. His model pertains to species that overwinter in several larval instars (i.e. "summer species"), and have a positive correlation between the thermal threshold of growth and individual larval size. Thus, the smallest larvae resume growth earlier than larger larvae as water temperatures rise in the spring. Size variation among larvae is gradually reduced, and the population becomes synchronized for adult emergence. Life-history data on several odonates support this model (Lutz 1963, 1968a, 1968b; Procter 1973). For other hemimetabolous aquatic insects, especially mayflies and stoneflies, the model fails to explain the high variance in adult male and female sizes that is often associated with species having synchronous emergence.

Sweeney and Vannote (1981) proposed an alternate model for the mayfly

Ephemerella subvaria that also exhibits a synchronous adult emergence despite having a large size range of larvae present throughout the growth period. They proposed that, during the spring, water temperatures rise and eventually exceed the threshold temperature for initiating the development of adult tissues (i.e. nonlarval structures such as wing buds, genitalia, gonads) by altering the metabolism and/or the production of insect hormones. Adult tissue development, and hence the propensity to proceed toward metamorphosis, continues uninterrupted as long as temperatures remain above the threshold. Thus, synchronous adult emergence and high variance in adult size are attributed to: (1) thermal stimulation of adult tissue development in both small and large larvae on or about the same date; and (2) no appreciable difference among small or large larvae in the effect of temperature on rate of adult tissue development once the water warms above the threshold. This model predicts a synchronous emergence, but not similar adult size, and is consistent with experimental data on both field and experimental populations of several other mayfly species. However, it remains to be verified for other hemimetabolous insects.

For most aquatic insects, especially those without an egg diapause, the duration of the larval stage is generally equal to or greater than the duration of all other stages (egg, pupa, adult) combined. Since temperature often determines the rate of larval development, it has been implicated as the key environmental factor determining the duration of aquatic insect life histories. For example, the number of generations per year and/or the number of years per generation for a given species have been shown to differ when populations were compared at different latitudes (Zahar 1951; Corbet 1980), altitudes (Zahar 1951; Hildrew 1978; Brittain 1978), in cold versus warm streams in the same geographic province (Lehmkuhl 1973; Newell and Minshall 1978b; Mackay 1979), in cold versus warm parts of the same lake (Jansson and Scudder 1974), and in cold upstream areas relative to warmer downstream reaches of the same river (Harper 1973; Hildrew 1978; Mackay 1979). Ingram and Jenner (1976) have shown for the damselfly *Ischnura aspersuma* (Odonata) that larvae hatching from eggs in early spring are bivoltine, while those hatching from eggs in late spring are univoltine. For the Chironomidae, which contain many multivoltine species at temperate latitudes, all life histories are greater than one year in the arctic (Oliver 1971), and some are even considerably longer (e.g. Butler 1982). While many stoneflies (Plecoptera) have multi-year life histories in cold water, warm water tends to eliminate species rather than to permit bivoltine development.

Pupation and/or Adult Emergence

Laboratory studies have shown that artificially raising or lowering the normal pattern of temperature consistently causes premature or delayed emergence, respectively, for an array of aquatic insect species (e.g. see Sweeney and Vannote 1981 for references). Similarly, adult emergence of natural populations occurs earlier or later during years that are abnormally warm or cool, relative to average conditions (Gledhill 1960; Illies 1971; Langford 1975; Hynes 1976). Natural populations also emerge earlier in streams that are warmer due to lower latitude (Thibault 1971; Lehmkuhl 1974), lower altitude (Nebeker 1971c; Wise 1980), or the warm-water effluent of power plants (Coutant 1967; Langford 1975; Mattice and Dye 1978). Delayed adult emergence has been observed for populations inhabiting colder tributaries of a given river system (Ide 1935, 1940; Sprules 1947; Clifford 1969; Wise 1980), cooler areas of a given lake (Moore 1980), and rivers that are seasonally cooled by hypolimnetic water from large reservoirs (Pearson et al. 1968).

Adult Size and Fecundity

The size of an individual insect at maturity (i.e. its pupal or adult length or weight) depends largely on two temperature dependent phenomena—the duration of the larval development and the rate of growth during that developmental period. Theoretically, if growth rates and duration of larval development were affected to the same relative degree by a given change in water temperature, then intraspecific variation in pupal or adult size would be very low. However, growth and duration of development do not seem to respond similarly because pupal and adult sizes vary significantly for a wide variety of species, depending upon rearing temperatures (e.g. Konstantinov 1958; Heuvel 1963; Brust 1967; Hagstrum and Workman 1971; Mackey 1977; Sweeney and Schnack 1977; Vannote and Sweeney 1980).

Variation in size of individuals in natural populations has also been correlated with either temporal or spatial temperature gradients. For example, studies on multivoltine mayfly species consistently show that adults of winter-spring cohorts are significantly larger than those of summer cohorts (e.g. see Merritt et al. 1982; Rhame and Stewart 1976; Ross and Merritt 1978; Illies 1979; Kondratieff and Voshell 1980; Vannote and Sweeney 1980; and Wise 1980 for recent examples). Also, populations in rivers that are warmed by power-plant effluent have shown either slight (Langford 1975) or extensive (Obrdlík et al. 1979) modification of adult size, depending on the species and the degree of warming. However, it remains to be demonstrated to what extent temperature affects adult size in natural populations either directly through physiological and developmental processes or indirectly by altering the quantity or quality of food and habitat.

The ecological significance of variation in adult size, especially with respect to aquatic insect life histories, lies in the positive correlation between size and fecundity (e.g. see Clifford and Boerger 1974; Colbo and Porter 1979; Kondratieff and Voshell 1980; Sweeney and Vannote 1981). In 1978, Sweeney and Vannote reported that adult size and fecundity for ten species of hemimetabolous aquatic insects depended largely on thermal conditions during the larval growth period. They hypothesized that the geographic distribution of some aquatic insects may be limited at least partially by lowered fecundity in habitats of increasingly cold or warm temperature cycles. Vannote and Sweeney (1980) expanded and restated this hypothesis to suggest that population stability within the geographic range of a species was thought to reflect a dynamic equilibrium between temperature and factors such as growth, development, metabolism, reproductive potential, and generation time.

The thermal equilibrium hypothesis, as stated above, appears superficially to be too simplistic because numerous factors (biotic and abiotic) other than temperature are undoubtedly involved in population stability. The hypothesis, however, provides a conceptual framework for analyzing and organizing a rather large and heretofore disjointed mass of field and laboratory data concerning temperature and aquatic insect populations. In addition, the hypothesis emphasizes the need for studies aimed at quantifying the amount of intraspecific geographic variation that occurs in the life-history characteristics of aquatic insects, and for determining if observed patterns represent a phenotypic response to environmental conditions or genetic differentiation among populations. Thus, although life-history studies on a variety of species are important contributions, data comparing geographically separated populations of a given species are also needed. Comparative studies should ideally include a broad array of life-history variables, a detailed analysis of potential environmental constraints and influences, and an assessment of genetic identity among populations (e.g. through laboratory breeding and rearing experiments, field transplant studies, isozyme analysis).

Concluding Remarks on Temperature

Research has indicated that temperature may play a major role in influencing life-history patterns of aquatic insects. Table 4.1 summarizes, according to taxa, studies that examined the response of certain life-history characteristics to temperature (see also Ward and Stanford 1982 for references on temperature factors not considered here). A significant aspect of Table 4.1 lies in the conspicuous absence of data on particular life-history characteristics for several groups. For example, although numerous studies on mayflies (Ephemeroptera) have focused on egg development at various temperatures, data on other important aquatic groups are either very limited (e.g. Plecoptera, Diptera [Culicidae], Odonata) or lacking altogether (e.g. Trichoptera, Diptera [Chironomidae, Simuliidae], Megaloptera, Neuroptera). Similarly, a disproportionate number of studies involving Diptera (especially Chironomidae) have focused on the effects of temperature on larval development.

Most of the studies referenced in Table 4.1 are based on laboratory analyses. Many field studies have also correlated seasonal patterns in life-history characteristics with temperature, but few such studies have compared populations in two or more habitats that differ in temperature. Thus, it is presently unknown whether aquatic insect populations along natural thermal gradients exhibit as much intraspecific variation in life-history characteristics as might be predicted *a priori* based on laboratory or field studies of one local population.

We also lack fundamental data in several other areas. For example, growth and development of aquatic insects is regulated largely by hormones (e.g. ecdysone, juvenile hormone), but little is known about the general mechanism by which temperature can alter endocrine processes. Thus, does premature adult emergence at abnormally high temperatures result because juvenile hormone production ceases, or because the hormone is catabolized by enzymes more rapidly or efficiently at high temperatures? Is the effect of temperature on endocrine processes the principal factor involved in limiting the number of aquatic species that exhibit adult emergence during periods of decreasing, rather than increasing, temperatures (e.g. autumn)?

Few studies have examined the effects of temperature on the quantitative and qualitative changes in enzymes, especially those critical to digestive and assimilative processes, basal metabolism, etc. For example, Schott and Brusven (1980) demonstrated that certain gene loci of the damselfly *Argia vivida* were stimulated, activated, or deactivated by shifts in temperature, and this resulted in changes in the electromorph characteristics of certain enzymes. The genetic component, however, is usually not considered when the relationship between temperature and intraspecific variation in life-history characteristics is interpreted. This seems inexcusable, especially since many natural populations of invertebrates, including insects, probably contain a substantial amount of genetic variation.

NUTRITION

In most natural aquatic ecosystems, both living and detrital organic matter are available to consumer insects throughout the year. The relative availability, however, of detritus and primary production food sources can be very seasonal. Autumn-winter food chains in temperate areas often depend largely on detritus, whereas autotrophic sources form the major food base in spring and summer months (Wetzel 1975; Minshall 1978; Vannote et al. 1980).

Various aspects of nutrition, diet, and other food-related processes have been examined in terms of their influence on aquatic insect life histories (see Anderson and Cummins 1979 for recent review). Studies of natural populations often implicate food quantity and/or quality as an important factor affecting life-history dynamics, but interpretation is often complicated by other abiotic factors. The array of food resources for aquatic insects is large and includes (in approximate descending order of nutritional importance): animal tissues, living algae (filamentous and diatoms), decomposing vascular hydrophytes, fine

TABLE 4.1: **Summary of Laboratory and Field Experiments on Aquatic Insects where Certain Life-history Characteristics Have Been Studied at Two or More Temperatures, or in Two or More Habitats that Differ Significantly in Temperature***

ORDER FAMILY GENUS SPECIES	Egg r, h, d	Larva r, t, g	Adult s, f	REFERENCES
Ephemeroptera				
Baetidae				
Baetis rhodani	X X		X	Benech 1972a, 1972b; Elliott 1972; Fahy 1973
Centroptilum rufostrigatum		X X	X X	Sweeney & Vannote 1978; Vannote & Sweeney 1980
Caenidae				
Caenis simulans		X X	X X	Sweeney & Vannote 1978; Vannote & Sweeney 1980
Ephemerellidae				
Ephemerella dorothea	X			Sweeney & Vannote 1981
Ephemerella ignita	X X			Bohle 1972; Elliott 1978
Ephemerella subvaria	X	X X	X X	Vannote & Sweeney 1980; Sweeney & Vannote 1981
Eurylophella funeralis	X	X X	X X	Vannote & Sweeney 1980; Sweeney & Vannote 1981
Serratella deficiens	X			Sweeney & Vannote 1981
Ephemeridae				
Ephemera simulans		X		Britt 1962
Hexagenia limbata	X			Hunt 1951
Hexagenia rigida	X X			Friesen et al. 1979
Heptageniidae				
Ecdyonurus dispar	X X			Humpesch 1980
Ecdyonurus insignis	X X			Humpesch 1980

Taxon			Reference
Ecdyonurus picteti	X X		Humpesch 1978, 1980
Ecdyonurus torrentis	X X		Humpesch 1980
Ecdyonurus venosus	X X		Humpesch 1980
Rhithrogena cf. *hybrida*	X X		Humpesch & Elliott 1980
Rhithrogena loyolaea	X X		Humpesch & Elliott 1980
Rhithrogena semicolorata	X X		Humpesch & Elliott 1980
Leptophlebiidae			
Leptophlebia cupida	X X	X X	Sweeney & Vannote 1978; Vannote & Sweeney 1980
Leptophlebia vespertina	X X	X X	Brittain 1976b
Polymitarcyidae			
Ephoron album	X		Britt 1962
Tortopus incertus	X		Tsui & Peters 1974
Siphlonuridae			
Ameletus ludens	X X	X X	Sweeney & Vannote 1978; Vannote & Sweeney 1980
Isonychia bicolor	X	X X	X X	Sweeney 1978; Vannote & Sweeney 1980
Tricorythidae			
Tricorythodes atratus	X	X X	Sweeney & Vannote 1978; Vannote & Sweeney 1980
Tricorythodes minutus	X	X X	Newell & Minshall 1978a
Odonata			
Aeshnidae			
Anax imperator	X		Corbet 1956a
Anax junius	X X		Beesley 1972
Coenagrionidae			
Argia vivida	X		Pritchard & Pelchat 1977
Corduliidae			
Tetragoneuria cynosura	X		Lutz 1974a, 1974b
Lestidae			
Lestes congener	X X	X	Sawchyn & Church 1973; Sawchyn & Gillott 1974
Lestes disjunctus	X X	X	Sawchyn & Church 1973
Lestes dryas	X X	X	Sawchyn & Church 1973

TABLE 4.1: Continued

ORDER FAMILY GENUS SPECIES	LIFE STAGE			REFERENCES
	Egg r, h, d	Larva r, t, g	Adult s, f	
Lestes eurinus		X		Luz 1968a, 1968b
Lestes sponsa	X X X			Fisher 1958; Corbet 1956a
Lestes unguiculatus	X X			Sawchyn & Church 1973
Plecoptera				
Nemouridae				
Nemoura avicularis	X	X		Brittain 1973, 1976a
Perlidae				
Acroneuria californica		X		Heiman & Knight 1975
Perlesta placida	X			Snellen & Stewart 1979
Pteronarcidae				
Pteronarcys dorsata		X	X	Nebeker 1971a, 1971b
Taeniopterygidae				
Taeniopteryx nebulosa	X X			Brittain 1977
Hemiptera				
Corixidae				
Arctocorisa carinata	X	X		Pajunen & Sundbäck 1973; Pajunen 1975
Callicorixa producta	X	X		Pajunen & Sundbäck 1973
Sigara alternata	X	X X	X	Sweeney & Schnack 1977; Vannote & Sweeney 1980

Diptera

Chironomidae

Species					Reference
Ablabesmyia monilis	X	X			Monakov 1972; Mackey 1977
Chironomus annularius	X	X	X		Konstantinov 1958
Chironomus decorus	X				Mackey 1977
Chironomus dorsalis	X	X	X	X	Konstantinov 1958
Chironomus heterodentatus	X	X			Konstantinov 1958
Chironomus plumosus	X	X			Konstantinov 1958; Harvey 1971
Chironomus staegeri	X				Danks 1978
Cladotanytarsus atridorsum	X	X			Mackey 1977
Cricotopus gr. *algarum*	X	X			Konstantinov 1978
Cricotopus bicinctus	X	X	X	X	Mackey 1977
Cricotopus sylvestris	X	X	X		Konstantinov 1958
Cricotopus sylvestris	X	X			Mackey 1977
Cryptochironomus pararostratus	X	X			Konstantinov 1958
Endochironomus nigricans	X				Danks 1978
Endochironomus tendens	X	X			Konstantinov 1958
Glyptotendipes pallens	X	X			Konstantinov 1958
Glyptotendipes paripes	X				Harvey 1971
Limnochironomus gr. *nervosus*	X	X			Konstantinov 1958
Limnochironomus pulsus	X	X			Mackey 1977
Metriocnemus hirticollis	X	X			Mackey 1977
Microcricotopus bicolor	X	X			Mackey 1977
Parachironomus biannulatus	X	X			Mackey 1977
Paratendipes albimanus		X			Ward & Cummins 1979
Phaenopsectra flavipes	X	X			Mackey 1977
Polypedilum convictum	X	X			Mackey 1977
Polypedilum nubeculosum	X	X			Konstantinov 1958

TABLE 4.1: Concluded

ORDER FAMILY GENUS SPECIES	Egg r, h, d,	Larva r, t, g	Adult s, f	REFERENCES
Rheotanytarsus photophilus		X X		Mackey 1977
Synorthocladius semivirens		X X		Mackey 1977
Tanytarsus dissimilis		X X		Nebeker 1973
Culicidae				
Aedes aegypti		X X	X	Hadlee 1940, 1941; Bar-Zeev 1958; Heuvel 1963
Aedes nigromaculis		X X	X X	Brust 1967
Aedes vexans		X X	X X	Brust 1967; Slater & Pritchard 1979
Anopheles quadrimaculatus	X	X X		Huffaker 1944
Culex tarsalis		X X	X	Hagstrum & Workman 1971
Culiseta inornata	X	X X	X X	Brust 1967; Hanec & Brust 1967
Simuliidae				
Cnephia dacotensis		X X		Ross & Merritt 1978
Prosimulium fuscum		X X X		Ross & Merritt 1978
Prosimulium mixtum		X X X		Ross & Merritt 1978
Simulium venustum		X		Mokry 1976
Simulium vittatum		X X		Becker 1973; Lacey & Mulla 1979
Stegopterna mutata		X X		Ross & Merritt 1978
Trichoptera				
Brachycentridae				
Brachycentrus americanus		X		Gallepp 1977
Brachycentrus occidentalis		X		Gallepp 1977

LIFE STAGE

Hydropsychidae
  Diplectrona felix   X   Edington & Hildrew 1973
  Hydropsyche fulvipes   X   Edington & Hildrew 1973
Lepidostomatidae
  Lepidostoma quercina   X   Grafius & Anderson 1979
Limnephilidae
  Potamophylax cingulatus   X   Otto 1974

* r = rate of development
  h = hatch success
  d = diapause development
  t = thermal summation (degree days) for development
  g = growth rate
  s = pupal or adult size
  f = fecundity

particulate organic matter, terrestrial leaf litter, and wood (Anderson and Cummins 1979). Aside from the nutritional gradient among these categories (see Lamberti and Moore, Chapter 7), there can exist a gradient of food quality within a given category (e.g. among species of deciduous leaves). The nutritional quality of a given detritus food often seems to depend on the quantity and type of associated microflora (bacteria and fungi). However, the relative importance of microbes to biomass production and the developmental dynamics of aquatic insect larvae has only recently been appreciated.

Egg Development

The amount of nutrients (e.g. lipids, carbohydrates, protein) available for egg production in some aquatic groups depends completely (e.g. all Ephemeroptera), or almost completely (e.g. Plecoptera), on materials assimilated and stored during the larval stage. Other groups (e.g. Odonata, Culicidae, Simuliidae, and some other Diptera) obtain most of their reproductive energy through adult feeding. Although the rate and success of egg development has been shown to vary with egg size in terrestrial insects (Richards and Kolderie 1957), few if any data are available for aquatic insects. It seems, however, that variation in the quantity and quality of nutrients available for use in egg production could potentially have an indirect effect on egg development of aquatic insects.

Intraspecific variation (either within or among individuals) in egg quality (e.g. size, weight, proportion of lipids, fats, carbohydrates) has apparently not been examined in detail for any species of aquatic insect. However, limited studies on a few mayfly species have shown that although egg size is relatively constant for multivoltine species, adults vary considerably in size and fecundity (McClure and Stewart 1976; Sweeney 1978), and a linear relationship exists between fecundity and adult dry weight (e.g. Harvey et al. 1979; Vannote and Sweeney 1980; Sweeney and Vannote 1981). These data therefore suggest, but do not prove, that environmental factors largely affect egg quantity rather than egg quality. Clearly, more studies that are based on a wide variety of insects are needed.

Larval Growth

Although the total amount of potential food rarely seems limiting in natural ecosystems, the actual amount of high quality or preferred food may be limiting to certain species at specific times of the year or in different habitats. Just over a decade ago, Hynes (1970) concluded that the relative importance of autochthonous compared to allochthonous organic matter to aquatic insect growth was generally unknown. Although more data are still needed, studies on the nutritive importance of these food resources have received increased emphasis in recent years (see Cummins and Klug 1979 and Anderson and Cummins 1979 for reviews).

Most laboratory studies on the response of larval growth to food quality have focused on insects that feed on whole leaf substrates (i.e. shredder species having life histories that are keyed to the seasonal availability of deciduous leaves). In general, the magnitude of larval growth for shredders can vary by a factor of two or more, depending on the species of leaf provided as food (Cummins et al. 1973; Iverson 1974; Otto 1974; Anderson and Cummins 1979). Many species exhibit selective or preferential feeding when offered a variety of leaf types (Wallace et al. 1970; Iversen 1974; Otto 1974; Grafius and Anderson 1979; Anderson and Sedell 1979). For the aquatic cranefly *Tipula abdominalis*, larval growth rate was highest on leaf species that were preferentially eaten (when given a choice of nine species), and decreased among less preferred deciduous leaves (Vannote, unpublished data). In this last study, no significant difference in food conversion efficiencies was observed, and differences in growth rates were attributed to higher feeding rates on preferred leaf types. Anderson and Cummins (1979) suggest that some shredders increase feeding rates on lower quality leaf species in order to keep growth rates uniform. More information is needed to test this hypothesis.

The high feeding preference and growth rates exhibited by a species for certain species of leaves may be related to the amount of associated bacteria and fungi. Thus, leaves characterized by a dense microbial flora, high nitrogen (or ATP) content, or high respiration rate are usually fed on preferentially (e.g. Anderson and Cummins 1979; Ward and Cummins 1979). Few growth studies have attempted to assess the relative nutritional value of microbes compared to the leaf itself. Klug and Cummins (unpublished data cited in Cummins and Klug 1979) reported for *Tipula abdominalis* that only 8.3 percent of larval growth could be attributed to microbial biomass, even when assimilation efficiency was assumed to be 100 percent. They also reported that *T. abdominalis* had nearly the same growth rate on sterile leaves when compared to leaves that were heavily colonized with microbes. Findlay and Tenore (1982) suggest that the role of microbes in the nutrition of marine detritivores may vary and may depend largely on the source of detritus. Their experimental approach of feeding various combinations of ^{15}N-labeled detritus and microbes to detritivores seems highly suited for freshwater studies and should be examined further. Thus, the role of microbes in the nutrition of insect detritivores (especially leaf shredders) is a rapidly evolving area of research, and conclusions at this time are very tentative.

Larval growth studies on insect species that are unable to eat whole leaf detritus (e.g. grazers, collectors) also demonstrate the importance of food quality. For example, Ward and Cummins (1979) clearly showed that substrates with high microbial activity (i.e. respiration) and biomass (i.e. ATP) resulted in greater larval growth rates for *Paratendipes albimanus*, a collector midge. They also demonstrated that the relationship between growth rate and food quality remained similar over a 10°C range of temperatures. Although Ward and Cummins (1979) included only detrital diets, other studies have shown that larval growth rates of detritivores were increased significantly when detritus was supplemented with either algae (Fuller and Mackay 1981) or animal material (Mackey 1977; Anderson 1976; Anderson 1978; Anderson and Cummins 1979; Fuller and Mackay 1981).

Laboratory studies have focused largely on the response of insect growth to gradients of food quality. Food quantity, however, can also exert a significant effect. Available data indicate that growth decreases proportionally with decreased food levels for detritivores (Gallepp 1977; Colbo and Porter 1979), herbivores (Collins 1980), and predators (Hosseinie 1976; Macan 1977; Fox and Murdoch 1978; Lawton et al. 1980).

The above discussion emphasizes a growing awareness that natural insect populations can be, and probably often are, significantly affected by the type, quantity, and quality of food resources. For example, populations of species that are normally univoltine have been reported to be semivoltine in habitats that are characterized as having limited amounts of food (Lillehammer 1975, Macan 1977) or lacking high-quality food (Brittain 1974, 1978; Lavandier and Pujol 1975), although temperatures may be complicating the interpretation of these trends. Conversely, the normal life cycle of certain species has been accelerated considerably when food levels were artificially increased by industrial effluents (Azam and Anderson 1969), sewage effluents (Learner and Potter 1974), or the discharge of lakes (Carlsson et al. 1977).

Pupation and/or Adult Emergence

Reduced food levels cause slower larval growth and often result in delayed pupation (holometabolous forms) or adult emergence (hemimetabolous forms) of aquatic insects (e.g. Hosseinie 1976; Colbo and Porter 1979; Collins 1980). For some species, moderate reduction of food levels does not greatly alter the initial emergence pattern, although severe food reduction ultimately results in marked retardation of the pattern (Danks 1978; Colbo and Porter 1979). Delayed emergence has also been observed when species were reared on suboptimal diets (Cianciara 1979; Anderson and Cummins 1979). Attempts, however, to accelerate normal emergence time by providing unlimited quantities of normal food material, while keeping other environmental parameters similar to field conditions, have been unsuccessful (Brittain 1976a).

Size at Maturation

Few studies have included the rearing of test species through to the adult stage in examining the effect of varied food levels or the quality of available food. Hosseinie (1976) and Colbo and Porter (1979) showed that reduced food levels prolonged the duration of the larval period of predaceous hydrophilid beetles and ultimately resulted in smaller adults of filter-feeding black flies. Although the mayfly *Cloeon dipterum* required almost 50 percent longer to complete development feeding on algae compared to feeding on detritus, maximum larval size did not differ significantly among diets (Cianciara 1979). In contrast, Anderson and Cummins (1979) showed that mature pupae of the caddisfly *Clistoronia magnifica* grew to only half the size on an alder leaf diet as compared to animals fed an alder leaf diet that was supplemented with enchytraeid worms.

Food quality has been suspected to influence adult size in natural populations. For example, larvae of the caddisfly *Ceraclea transversa* that developed late in the summer fed largely on detritus and produced smaller adults relative to early summer larvae that fed mainly on freshwater sponge (Resh 1976). For multivoltine species, summer generation adults are often smaller than winter generation adults (Benech 1972b; LeSage and Harrison 1980; Sweeney and Vannote 1981). However, for those species where instantaneous larval growth rates of summer generations are equal to or greater than those of winter generation larvae, the possibility that food quality or quantity are limiting seems unlikely (Vannote and Sweeney 1980).

Concluding Remarks on Nutrition

The preceding discussion indicates that aquatic insects, like other animals, are significantly affected by the quantity and quality of food resources. The potential interactions between organic detritus, its associated microbial flora, and the nutrition of detritivores has received increased attention in recent years, but more work is needed. Similarly, the results from supplemental feeding experiments (e.g. where algae are added to a detritivore diet, or animal tissue to that of a herbivore) have yielded valuable insights into the role of diet in influencing life-history patterns.

Experimental studies on an array of species that represent all trophic levels and various feeding types are needed to differentiate the effects of food type and abundance from other factors (e.g. temperature, photoperiod) with respect to insect life histories, especially the rate and magnitude of larval growth. The most productive experimental approach will be to measure growth or developmental response to two or more variables simultaneously. The experimental range for each variable within the context of the experiment should be made equitable, if possible. Thus, if food quality ranges from an extremely poor food source to an extremely nutritious one (e.g. low vs. high levels of protein, ATP), then the array of food types should be presented over a broad range of temperatures (e.g. 5, 10, 15, 20, 25°C). Only then will the relative importance of each variable be evident from the results. For example, the study of larval growth over a wide range of food types, but a narrow range of temperatures, would bias the interpretation of results towards the relative importance of food.

Table 4.2 summarizes, according to taxa, studies (mostly laboratory-based) that examined the effects of nutrition on certain insect life-history characteristics. Although most of the major aquatic insect orders are represented, additional

TABLE 4.2: Summary of Laboratory and Field Experiments Concerning the Effects of Food Quantity and/or Quality on Various Life-history Characteristics of Aquatic Insects*

ORDER FAMILY GENUS SPECIES	LIFE STAGE Larva r, g, i, fp, ae, l	Pupa or Adult p, a, f	REFERENCES
Ephemeroptera			
Baetidae			
Cloeon dipterum	X X		Gianciara 1979
Tricorythidae			
Tricorythodes minutus	X		McCullough et al. 1979
Odonata			
Coenagrionidae			
Ischnura elegans	X X		Lawton et al. 1980
Pyrrhosoma nymphula	X		Macan 1977
Plecoptera			
Peltoperlidae			
Peltoperla maria			Wallace et al. 1970
Hemiptera			
Notonectidae			
Notonecta hoffmanni	X X		Fox & Murdoch 1978
Coleoptera			
Hydrophylidae			
Tropisternus nimbatus	X X	X	Hosseinie 1976

Taxon			Reference
Megaloptera			
Sialidae			
Sialis californica	X		Azam 1969 (cited by Anderson & Cummins 1979)
Sialis rotunda	X		Azam 1969 (cited by Anderson & Cummins 1979)
Diptera			
Chironomidae			
Ablabesmyia monilis	X		Mackey 1977
Chironomus decorus	X		Danks 1978
Chironomus staegeri	X		Danks 1978
Endochironomus nigricans	X		Danks 1978
Parabornniella tonmoiri	X X		Jones 1974
Paratendipes albimanus	X	X	Ward & Cummins 1979
Ephydridae			
Ephydra cinerea	X	X X	Collins 1980
Simuliidae			
Simulium austeni	X		Ladle & Hansford 1981
Simulium venustum	X		Fredeen 1964
Simulium verecundum	X	X X	Colbo & Porter 1979
Simulium vittatum	X	X X	Fredeen 1964; Colbo & Porter 1979
Tipulidae			
Tipula abdominalis	X X		Anderson & Cummins 1979; Cummins & Klug 1979
Trichoptera			
Brachycentridae			
Brachycentrus americanus	X X		Gallepp 1977
Brachycentrus occidentalis	X X		Gallepp 1977
Hydropsychidae			
Hydropsyche betteni	X		Fuller & Mackay 1981
Hydropsyche slossonae	X		Fuller & Mackay 1981
Hydropsyche sparna	X		Fuller & Mackay 1981

TABLE 4.2: Concluded

ORDER FAMILY GENUS SPECIES	LIFE STAGE		REFERENCES
	Larva r, g, i, fp, ae, l	*Pupa or Adult* p, a, f	
Lepidostomatidae			
Lepidostoma quercina	X X X		Grafius & Anderson 1979
Limnephilidae			
Clistoronia magnifica	X	X	Anderson 1976; Anderson & Cummins 1979
Potamophylax cingulatus	X X X		Otto 1974
Sericostomatidae			
Sericostoma personatum	X X X		Iversen 1974

* r = rate of development
   g = growth rate
   i = ingestion rate
  fp = feeding preference
  ae = assimilation efficiency
   l = maximum larval weight
   p = pupal weight
   a = adult weight
   f = fecundity

species need to be studied. In addition to more laboratory studies, field experiments are needed to assess the importance of food quantity or quality to the life-history characteristics of natural populations. These studies need not involve laborious, expensive manipulations of streams or lakes since human activities have altered the landscape significantly and created a mosaic of trophic conditions in aquatic environments of many areas. For example, it is not unusual for small streams to flow alternately through woodlots and meadows, with the quantity and quality of organic inputs changing in parallel. Humans have also altered the species composition of woodlots so that leaf litter inputs to aquatic habitats may range from a mixed deciduous composition to mostly leaves from a single species of deciduous or coniferous tree. Species composition may vary from one side of a lake to another depending on exposure, human activities, and a variety of other factors. Thus, many field experiments on nutrition are presently underway, but unfortunately, only a few of these have been recognized, and even fewer are being monitored.

PHOTOPERIOD

Photoperiodic control of insect life histories has been reported for numerous terrestrial species, but fewer accounts are available for aquatic forms (Saunders 1976; Beck 1980). Photoperiod should be equally important to insects in both aquatic and terrestrial habitats due to its predictability and its reliable signaling of forthcoming seasonal changes, especially in temperate regions. The importance of photoperiod, relative to other environmental signals, for synchronizing life histories in a given habitat (terrestrial or aquatic) depends in part on (1) the insects' ability to perceive and measure absolute and/or relative changes in daylength; and (2) the availability of other environmental signals, or cues.

Since insects respond to very low light levels (Saunders 1976; Beck 1980), light detection is probably not a problem for them in most aquatic habitats. It should be recognized, however, that (1) light is attenuated by water in deep lakes and rivers (especially during floods); (2) many lentic and lotic habitats are naturally dark due to dissolved humic substances, suspended sediment, ice and snow cover, and other factors; and (3) certain life-history stages occur in microhabitats where light is significantly reduced or altogether lacking (e.g. eggs are often deposited beneath rocks, deposited inside vascular plants and submerged wood, or carried by the current to depositional areas where silt covers them). These considerations obviously do not apply to all species in all habitats, but they should be considered in the experimental design and subsequent data interpretation in photoperiodic studies of aquatic insects.

Although it is impossible to predict the *actual* significance of photoperiod to the life history of a given species, the *potential* significance might be inferred *a priori* by evaluating certain known habitat characteristics. For example, many headwater streams or springbrooks have a very constant thermal regime, even in temperate regions. The constant temperature regime is predictable, but it does not reliably signal the seasonal changes that are critical to certain life-history stages (e.g. air temperatures suitable for adult flight, feeding, reproduction). One might expect that photoperiod would provide a more reliable and informative signal than temperature in this constant-temperature habitat and, therefore, might not only affect more species, but also influence them to a greater extent.

Experimental studies on aquatic insects are needed to broaden the data base concerning the effects of photoperiod on the induction or termination of diapause in the egg, larval, or pupal stage; the rate and efficiency of larval growth; and the timing of adult emergence. These experiments should be designed with an awareness that little information presently exists to determine whether aquatic insects (1) perceive daylength as being either long or short (e.g. Do daylengths shorter than a critical photoperiod elicit one response whereas daylengths longer than a critical photoperiod elicit an alternative response?); (2) perceive the actual duration of photophase (or scotophase), and show a quantitative response to seasonally changing daylengths; or (3) perceive and respond to the direction of daylength changes, regardless of the actual daylength (Tauber and Tauber 1975).

Egg Development

The effects of photoperiod on the rate of embryonic development have only been studied for a few aquatic species, mostly Ephemeroptera. These limited results consistently show that most intraspecific variation in hatching time can be attributed to temperature, even when eggs were subjected to abnormal photoperiods (e.g. continuous light: Humpesch 1978; Humpesch and Elliott 1980; or continuous darkness: Brittain 1977) relative to natural or quasi-natural conditions. For other species, especially mosquitoes (Diptera: Culicidae), photoperiod seems to play a critical role in either inducing or terminating an embryonic diapause (see Beck 1980 for review). Because of the paucity of data, generalizations concerning photoperiodic effects on embryonic development are inappropriate at this time.

Larval Growth and Development

To my knowledge, no study has been published concerning photoperiodic influences on larval growth, in terms of biomass, for any aquatic insect. However, there have been several studies on Odonata in which the rate of larval development and the timing of adult emergence have been measured as a response to both temperature and photoperiod. These results are somewhat inconclusive because most species were studied only during the final instars. For some species, however, it appears that photoperiod either induced a larval diapause (Corbet 1956b) or affected the amount of time necessary to complete larval development and emerge as an adult (Lutz 1968b, 1974a, 1974b; Ingram

1976; Ingram and Jenner 1976). For other species, experiments indicate little or no photoperiodic effect (Procter 1973). However, interpretation of these experiments is difficult, because the experiments often lacked controls entirely (i.e. larvae were not kept at either natural temperatures or photoperiods for comparison—Corbet 1956b; Ingram 1976; Ingram and Jenner 1976) or only had partial controls (i.e. larvae were kept at natural temperatures but not at natural photoperiods—Lutz 1968b, 1974a, 1974b). The lack of adequate controls could be a critical shortcoming because experiments that showed a significant photoperiodic effect at high temperatures usually failed to show any photoperiodic effect on larvae kept at natural temperatures (Lutz 1968b, 1974a, 1974b).

Although the temperature-photoperiod interaction for odonates requires further study, we should remember that larval development of many of these species was positively correlated with temperature and, within a temperature, development was fastest at long-day photoperiods. One recent hypothesis concerning these results is that increased daylength at a given temperature increases the overall feeding period of odonates, especially since most are visual predators and feeding activity may be reduced or less efficient in darkness (Sweeney and Vannote 1981). This hypothesis is untested but it is interesting to note that, for a given temperature, significant photoperiodic effects were only observed for odonates that were provided with a continuous food supply and were permitted to feed according to their own schedule (Corbet 1956b; Lutz 1968b, 1974a, 1974b; Ingram 1976; Ingram and Jenner 1976). However, Procter (1973), using this approach with the damselfly *Enallagma boreale*, failed to show significant photoperiodic effects.

Given that many species of aquatic insects have been shown to feed either largely during the day (Mecom 1970; Monakov 1972; Gallepp 1977) or exclusively at night (Chapman and Demory 1963; Elliott 1969; Thorup and Iversen 1974; Meier and Bartholomae 1980; Ploskey and Brown 1980), it seems intuitive that variation in photoperiod could potentially have an important modifying effect on patterns of larval growth and development. Thus, an increase in daylength from 11 to 14 hours not only increases the feeding period for a day-active species by 27 percent, but it also reduces the amount of time each day that the insect relies on reserve energy for maintenance metabolism. These considerations could take on added significance when geographic dispersal is considered. For example, in day-active species, prolonged photoperiods at higher latitudes during certain seasons might compensate partially for the reduced feeding rates associated with the low temperatures of more northern habitats.

Pupation and/or Adult Emergence

Photoperiodic control of adult metamorphosis has been reported for several species of Diptera, especially in the Chironomidae and Culicidae (see Danks 1978; Trimble and Smith 1979). Daylength, however, might also be important in the phenology of other groups of aquatic insects. For example, Malicky (1981) recently reported that continuous illumination of a mountain stream disrupted the normal emergence pattern of several caddisfly species. Since data for most species indicated that adult emergence occurred over a longer part of the year than normal, the response to increased photoperiod is difficult to explain.

In contrast, Nebeker's (1971a) study of ten species of stream insects (including Ephemeroptera, Trichoptera, and Plecoptera) indicated that photoperiod had little or no measurable effect with respect to the timing of adult emergence. Brittain (1976a) also demonstrated that larvae of a mayfly and a stonefly species that were kept in total darkness would emerge at the normal time if kept at ambient field temperatures, but emerged several weeks prematurely if water temperatures were artificially elevated. These data indicate that hemimetabolous aquatic insects may be less responsive to photoperiodic change than holometabolous species (e.g. the caddisflies in Malicky's study).

Concluding Remarks on Photoperiod

Table 4.3 summarizes, according to taxa, studies that examined the effects of photoperiod on various life-history characteristics of aquatic insects. This summary indicates that (1) a disproportionate number of studies have involved aquatic Diptera (>40%); (2) most studies have been concerned with the effects of photoperiod on either embryonic, larval, or adult diapause; and (3) very few studies have focused on photoperiodic effects on the rate and magnitude of growth. The lack of studies concerning photoperiod and pupal diapause is also noteworthy, especially since a significant response has been demonstrated or reasonably inferred for about 60 species of terrestrial insects (see Table 14 in Beck 1980). A comparison of Table 4.3 with data summarized in Beck (1980, Tables 5, 12, 13, 14, 15) indicates that only about 24 percent ($\sim 16\%$ if Diptera are not included) of the taxa used in studies that examined the influence of photoperiod on insect life-history characteristics are aquatic. It appears, therefore, that photoperiodic studies of aquatic insects have not kept pace with those of terrestrial species.

Although photoperiod has been known and used to stimulate reproduction and body growth of vertebrates for more than 60 years (Tucker and Ringer 1982), the importance of photoperiod to individual growth (i.e. biomass production) and reproduction has not been studied for any species of aquatic insect. The role of photoperiod in the induction and termination of larval diapause has been examined for several taxa of Odonata and Diptera, but few studies have focused on the photoperiodic response of developmental time in nondiapausing species. Thus, photoperiodic effects on developmental time, growth, and reproductive output, which are primary fitness characters (Istock 1981), remain unexamined for many groups of aquatic insects.

FINAL CONSIDERATIONS

I have attempted here to examine the effects of three important environmental factors on various life-history parameters in aquatic insects. Although considerable knowledge has accumulated in recent years, it is clear that the task at hand still remains formidable, and many areas remain largely unexplored. Multivariate experimental studies will certainly play a critical role in evaluating the relative importance of environmental factors on life-history characteristics of natural populations. For example, critical daylengths for diapause determination, and degree day temperature requirements for seasonal growth, are both involved in determining the number of generations per year of many insects (Beck 1980). The relationships between these two adaptations, however, have not been worked out for any aquatic or terrestrial insect species.

At this point we can only begin to assess the degree to which environmentally induced changes in life-history parameters are adaptive, although researchers have already speculated in this area. For example, Collins (1980) has recently hypothesized that the degree to which a life-history parameter changes in response to realistically applied, naturally occurring environmental variables should be an inverse measure of the importance of that parameter to individual fitness. Ricklefs (1969), however, cautioned that when one observes relative uniformity in the face of ecological diversity, internal restraints might be involved. Thus, limits to the amount of change possible for a given life-history parameter are frequently internal, and involve relationships between parts of the organism rather than between the organism and the environment.

In this review, components of an insect's life history have been viewed as a collection of coadapted traits that are shaped by natural selection. Although life-history differences among conspecific populations are generally thought to be genetic, and are thought to reflect differences in selective pressures among environments, we presently lack sufficient data to completely justify these assumptions. Berven (1982) cautions that ignoring the variation induced by the environment could lead to false conclusions of adaptiveness. Thus, environmental influences could mask the underlying genetic differences between populations, or have a nonadditive effect on different genotypes. Furthermore, the acquisition of life-history traits that allow individuals to complete their life histories under an array of biotic and abiotic conditions may, in itself, be an important adaptation (Berven 1982). It is now clear that meaningful interpretation of life-history patterns necessitates that a distinction be made between genotypic variation and environmentally induced phenotypic variation. As Denno and Dingle (1981) suggest, future studies must include more detailed analyses of environmental constraints and environmental influences on a broader range of life-history variables, as well as a more sophisticated assessment of the genetic structure of aquatic insect life histories.

TABLE 4.3: Summary of Laboratory and Field Experiments on Aquatic Insects where Certain Life-history Characteristics Have Been Studied at Two or More Photoperiodic Regimes, or in Two or More Geographic Localities that Differ Significantly in Photoperiod*

ORDER FAMILY GENUS SPECIES	LIFE STAGE Egg r, e	Larva r, l	Adult a, m	REFERENCES
Ephemeroptera				
Heptageniidae				
Ecdyonurus dispar	X			Humpesch 1980
Ecdyonurus insignis	X			Humpesch 1980
Ecdyonurus picteti	X			Humpesch 1978
Ecdyonurus torrentis	X			Humpesch 1980
Ecdyonurus venosus	X			Humpesch 1980
Rhithrogena cf. *hybrida*	X			Humpesch & Elliott 1980
Rhithrogena loyolaea	X			Humpesch & Elliott 1980
Rhithrogena semicolorata	X			Humpesch & Elliott 1980
Leptophlebiidae				
Leptophlebia vespertina		X	X	Brittain 1976a
Odonata				
Aeshnidae				
Aeshna cyanea		X		Schaller 1965 (cited by Beck 1980)
Anax imperator		X		Corbet 1956b
Coenagrionidae				
Argia vivida		X X		Pritchard & Pelchat 1977
Enallagma aspersum		X X		Ingram 1976; Ingram & Jenner 1976

Taxon						Reference
Enallagma boreale		X				Procter 1973
Enallagma hageni		X	X			Ingram 1975; Ingram & Jenner 1976
Corduliidae						
Tetragoneuria cynosura		X	X			Lutz & Jenner 1960, 1964, Lutz 1974a, 1974b
Lestidae						
Lestes disjunctus				X		Sawchyn & Church 1973
Lestes dryas				X		Sawchyn & Church 1973
Lestes eurinus			X			Lutz 1968a, 1968b
Lestes inguiculatus				X		Sawchyn & Church 1973
Libellulidae						
Leucorrhinia glacialis		X				Procter 1973
Libellula quadrimaculatus		X				Procter 1973
Neotetrum pulchellum			X			Montgomery & Macklin 1962 (cited by Beck 1980)
Plecoptera						
Capniidae						
Capnia bifrons		X				Khoo 1968a
Nemouridae						
Nemoura avicularis	X		X			Brittain 1976a
Perlodidae						
Diura bicaudata				X		Khoo 1968b
Taeniopterygidae						
Brachyptera risi				X		Khoo 1968c
Taeniopteryx nebulosus					X	Brittain 1977
Hemiptera						
Gerridae						
Gerris lacustris	X					Vepsäläinen 1974a
Gerris lateralis	X					Vepsäläinen 1974a
Gerris odontogaster	X					Vepsäläinen 1974b
Notonectidae						
Notonecta undulata	X					Vanderlin & Streams 1977

TABLE 4.3: Concluded

ORDER FAMILY GENUS SPECIES	LIFE STAGE Egg r, e	Larva r, l	Adult a, m	REFERENCES
Diptera				
Chironomidae				
Chironomus decorus		X	X	Danks 1978
Chironomus nuditarsis		X		Fischer 1974
Chironomus plumosus		X		Fischer 1974
Chironomus salinarius		X		Koskinen 1968
Chironomus staegeri		X	X	Danks 1978
Chironomus tentans		X		Englemann & Shappirio 1965
Einfeldia synchrona		X		Danks 1978
Endochironomus nigricans		X	X	Danks 1978
Metriocnemus knabi		X		Faris & Jenner 1959
Culicidae				
Aedes atropalpus	X			Anderson 1968; Beach & Craig 1977
Aedes campestris	X			Tauthong & Brust 1977 (cited by Beck 1980)
Aedes canadensis	X			Finger & Eldridge 1977
Aedes sierrensis		X		Jordan 1980
Aedes togoi	X		X	Vinogradova 1960 (cited by Beck 1980)
Aedes triseriatus	X	X X		Kappus & Venard 1967; Beck 1980
Anopheles barberi		X		Baker 1935
Anopheles bifurcatus		X		Danilevsky 1965
Anopheles freeborni			X	Depner & Harwood 1966
Anopheles maculipennis			X	Vinogradova 1960 (cited by Beck 1980)

	r	e	l	a	m	Reference
Anopheles plumbeus			X			Vinogradova 1960 (cited by Beck 1980)
Anopheles pulcherrimus		X				Danilevsky 1965
Anopheles superpictus			X			Vinogradova 1960 (cited by Beck 1980)
Chaoborus americanus			X			Bradshaw 1969
Culex pipiens pipiens				X		Spielman & Wong 1973 (cited by Beck 1980)
Culiseta inornata				X		Hudson 1977 (cited by Beck 1980)
Orthopodomyia signifera			X			Bradshaw 1973
Psorophora ferox	X					Pinger & Eldridge 1977
Toxorhynchites rutilus			X	X		McCrary & Jenner 1965; Bradshaw & Holzapfel 1977
Toxorhynchites septentrionalis	X		X	X		Trimble & Smith 1979
Wyeomyia smithii			X			Smith & Brust 1971; Bradshaw & Lounibos 1972
Trichoptera						
Limnephilidae						
Acrophylax zerberus					X	Malicky 1981
Allogamus uncatus					X	Malicky 1981
Limnephilus spp.			X			Novák & Sehnal 1963
Lithax niger					X	Malicky 1981
Psychomyiidae						
Tinodes dives					X	Malicky 1981
Rhyacophilidae						
Rhyacophila aurata					X	Malicky 1981
Rhyacophila stigmatica					X	Malicky 1981
Rhyacophila tristis					X	Malicky 1981
Rhyacophila vulgaris					X	Malicky 1981

* r = rate of development
e = embryonic diapause
l = larval diapause
a = adult diapause
m = metamorphosis or adult emergence

Acknowledgments

Preparation of this manuscript was supported in part by the U.S. Department of Energy (contract No. DE-AC-02-79EV10259), the National Science Foundation (contract No. DAR 78-18589), the Stroud Foundation, and the Frances Boyer Research Endowment. Valuable comments on the manuscript were provided by three anonymous reviewers.

References

Anderson, J. F. 1968. Influence of photoperiod and temperature on the induction of dia pause in *Aedes atropalpus* (Diptera: Culicidae). Entomologia experimentalis et applicata 11:321–30.

Anderson, N. H. 1976. Carnivory by an aquatic detritivore, *Clistoronia magnifica* (Trichoptera: Limnephilidae). Ecology 57:1081–85.

Anderson, N. H. 1978. Continuous rearing of the limnephilid caddisfly *Clistoronia magnifica* (Banks), pp. 317–29. *In*: M. I. Crichton (ed.). Proceedings of the Second International Symposium on Trichoptera. Dr. W. Junk b.v. Publishers, The Hague, The Netherlands. 359 pp.

Anderson, N. H. and K. W. Cummins. 1979. Influences of diet on the life histories of aquatic insects. Journal of the Fisheries Research Board of Canada 36:335–42.

Anderson, N. H. and J. R. Sedell. 1979. Detritus processing by macroinvertebrates in stream ecosystems. Annual Review of Entomology 24:351–77.

Azam, K. M. and N. H. Anderson. 1969. Life history and habits of *Sialis rotunda* and *S. californica* in western Oregon. Annals of the Entomological Society of America 62:549–58.

Baker, F. C. 1935. The effect of photoperiodism on resting, treehole mosquito larvae. The Canadian Entomologist 67:149–53.

Bar-Zeev, M. 1958. The effect of temperature on the growth rate and survival of the immature stages of *Aedes aegypti* (L). Bulletin of Entomological Research 49:157–63.

Beach, R. F. and G. B. Craig, Jr. 1977. Night length measurements by the circadian clock controlling diapause induction in the mosquito *Aedes atropalpus*. Journal of Insect Physiology 23:865–70.

Beck, S. D. 1980. Insect photoperiodism. 2nd ed. Academic Press, New York, NY. 387 pp.

Beck, S. D. 1983. Insect thermoperiodism. Annual Review of Entomology 28:91–108.

Becker, C. D. 1973. Development of *Simulium (Psilozia) vittatum* Zett. (Diptera: Simuliidae) from larvae to adults at thermal increments from 17.0 to 27.0°C. The American Midland Naturalist 89:246–51.

Beesley, C. 1972. Investigations of the life history and predatory capacity of *Anax junius* Drury (Odonata: Aeschnidae). Unpublished Ph.D. thesis, University of California, Riverside, CA. 114 pp.

Benech, V. 1972a. La fécondité de *Baetis rhodani* Pictet. Freshwater Biology 2:337–54.

Benech, V. 1972b. Le polyvoltinisme chez *Baetis rhodani* Pictet (Insecta, Ephemeroptera) dans un ruisseau a truites des Pyrénées-Atlantiques, le Lissuraga. Annales d'Hydrobiologie 3:141–71.

Berven, K. A. 1982. The genetic basis of altitudinal variation in the wood frog *Rana sylvatica*. I. An experimental analysis of life history traits. Evolution 36:962–83.

Bohle, H. W. 1972. Die Temperaturabhängigkeit der Embryogenese und der embryonalen Diapause von *Ephemerella ignita* (Poda) (Insecta, Ephemeroptera). Oecologia 10: 253–68.

Bradshaw, W. E. 1969. Major environmental factors inducing the termination of larval diapause in *Chaoborus americanus* Johannsen (Diptera: Culicidae). The Biological Bulletin 136:2–8.

Bradshaw, W. E. 1973. Homeostasis and polymorphism in vernal development of *Chaoborus americanus*. Ecology 54:1247–59.

Bradshaw, W. E. and C. M. Holzapfel. 1977. Interaction between photoperiod, temperature, and chilling in dormant larvae of the tree-hole mosquito, *Toxorhynchites rutilus* Coq. The Biological Bulletin 152:147–58.

Bradshaw, W. E. and L. P. Lounibos. 1972. Photoperiodic control of development in the pitcher-plant mosquito, *Wyeomyia smithii*. Canadian Journal of Zoology 50:713–19.

Britt, N. W. 1962. Biology of two species of Lake Erie mayflies, *Ephoron album* (Say) and *Ephemera simulans* Walker. Bulletin of the Ohio Biological Survey 1:1–70.

Brittain, J. E. 1973. The biology and life cycle of *Nemoura avicularis* Morton (Plecoptera). Freshwater Biology 3:199–210.

Brittain, J. E. 1974. Studies on the lentic Ephemeroptera and Plecoptera of southern Norway. Norsk Entomologisk Tidsskrift 21:135–54.

Brittain, J. E. 1976a. The temperature of two Welsh lakes and its effect on the distribution of two freshwater insects. Hydrobiologia 48:37–49.

Brittain, J. E. 1976b. Experimental studies on nymphal growth in *Leptophlebia vespertina* (L.) (Ephemeroptera). Freshwater Biology 6:445–49.

Brittain, J. E. 1977. The effect of temperature on the egg incubation period of *Taeniopteryx nebulosa* (Plecoptera). Oikos 29:302–5.

Brittain, J. E. 1978. Semivoltinism in mountain populations of *Nemurella pictetii* (Plecoptera). Oikos 30:1–6.

Brust, R. A. 1967. Weight and development time of different stadia of mosquitoes reared at various constant temperatures. The Canadian Entomologist 99:986–93.

Butler, M. G. 1982. A 7-year life cycle for two *Chironomus* species in arctic Alaskan tundra ponds (Diptera: Chironomidae). Canadian Journal of Zoology 60:58–70.

Carlsson M., L. M. Nilsson, Bj. Svensson, S. Ulfstrand, and R. S. Wotton. 1977. Lacustrine seston and other factors influencing the blackflies (Diptera: Simuliidae) inhabiting lake outlets in Swedish Lapland. Oikos 29:229–38.

Chapman, D. W. and R. L. Demory. 1963. Seasonal changes in the food ingested by aquatic insect larvae and nymphs in two Oregon streams. Ecology 44:140–46.

Cianciara, S. 1979. Some study on the biology and bioenergetics of *Cloeon dipterum* (L.) Ephemeroptera (preliminary data), pp. 175–92. *In*: K. Pasternak and R. Sowa (eds.). Proceedings of the Second International Conference on Ephemeroptera. Panstwowe Wydawnictwo Naukowe, Warsaw, Poland. 312 pp.

Clifford, H. F. 1969. Limnological features of a northern brown-water stream, with special reference to the life histories of the aquatic insects. The American Midland Naturalist 82:578–97.

Clifford, H. F. and H. Boerger. 1974. Fecundity of mayflies (Ephemeroptera), with special reference to mayflies of a brown-water stream of Alberta, Canada. The Canadian Entomologist 106:1111–19.

Colbo, M. H. and G. N. Porter. 1979. Effects of the food supply on the life history of Simuliidae (Diptera). Canadian Journal of Zoology 57:301–6.

Collins, N. C. 1980. Developmental responses to food limitation as indicators of environmental conditions for *Ephydra cinerea* Jones (Diptera). Ecology 61:650–61.

Corbet, P. S. 1956a. The influence of temperature on diapause development in the dragonfly *Lestes sponsa* (Hansemann) (Odonata: Lestidae). Proceedings of the Royal Entomological Society of London (A) 31:45–48.

Corbet, P. S. 1956b. Environmental factors influencing the induction and termination of diapause in the emperor dragonfly, *Anax imperator* Leach (Odonata: Aeshnidae). Journal of Experimental Biology 33:1–14.

Corbet, P. S. 1957. The life-histories of two spring species of dragonfly (Odonata: Zygoptera). Entomological Gazette 8:79–89.

Corbet, P. S. 1958. Temperature in relation to seasonal development of British dragonflies (Odonata). Proceedings of the Xth International Congress of Entomology, Montreal 2:755–57.

Corbet, P. S. 1963. A biology of dragonflies. Quadrangle Books, Chicago, 247 pp.

Corbet, P. S. 1980. Biology of Odonata. Annual Review of Entomology 25:189–217.

Coutant, C. C. 1967. Effect of temperature on the developmental rate of bottom organisms, pp. 11–12. *In*: Biological effects of thermal discharges. Annual report Pacific North West Laboratory of the United States Atomic Energy Commission for Development, Biology, and Medicine. Report number BNWL-714.

Cummins, K. W. and M. J. Klug. 1979. Feeding ecology of stream invertebrates. Annual Review of Ecology and Systematics 10:147–72.

Cummins, K. W., R. C. Petersen, F. O. Howard, J. C. Wuycheck, and V. I. Holt. 1973. The utilization of leaf litter by stream detritivores. Ecology 54:336–45.

Danilevsky, A. S. 1965. Photoperiodism and seasonal development of insects. Oliver and Boyd, London. 283 pp.

Danks, H. V. 1978. Some effects of photoperiod, temperature, and food on emergence in three species of Chironomidae (Diptera). The Canadian Entomologist 110:289–300.

Denno, R. F. and H. Dingle. 1981. Considerations for the development of a more general life history theory, pp. 1–6. *In*: R. F. Denno and H. Dingle (eds.). Insect life history patterns: habitat and geographic variation. Springer-Verlag, New York, NY. 225 pp.

Depner, K. R. and R. F. Harwood. 1966. Photoperiodic responses of two latitudinally diverse groups of *Anopheles freeborni* (Diptera: Culicidae). Annals of the Entomological Society of America 59:7–11.

Edington, J. M. and A. H. Hildrew. 1973. Experimental observations relating to the distribution of net-spinning Trichoptera in streams. Internationale Vereinigung für Theoretische und Angewandte Limnologie Verhandlungen 18:1549–58.

Elliott, J. M. 1969. Life history and biology of *Sericostoma personatum* Spence (Trichoptera). Oikos 20:110–18.

Elliott, J. M. 1972. Effect of temperature on the time of hatching in *Baëtis rhodani* (Ephemeroptera: Baëtidae). Oecologia 9:47–51.

Elliott, J. M. 1978. Effect of temperature on the hatching time of eggs of *Ephemerella ignita* (Poda) (Ephemeroptera: Ephemerellidae). Freshwater Biology 8:51–58.

Elliott, J. M. and U. H. Humpesch. 1980. Eggs of Ephemeroptera. Freshwater Biological Association Annual Report 48:41–52.

Elwood, J. W. and R. A. Goldstein. 1975. Effects of temperature on food ingestion rate and absorption, retention, and equilibrium burden of phosphorous in an aquatic snail, *Goniobasis clavaeformis* Lea. Freshwater Biology 5:397–406.

Englemann, W. and D. G. Shappirio. 1965. Photoperiodic control of the maintenance and termination of larval diapause in *Chironomus tentans*. Nature (London) 207:548–49.

Fahy, E. 1973. Observations on the growth of Ephemeroptera in fluctuating and constant temperature conditions. Proceedings of the Royal Irish Academy 73B:133–49.

Findlay, S. and K. Tenore. 1982. Nitrogen source for a detritivore: detritus substrate versus associated microbes. Science 218:371–73.

Fischer, J. 1974. Experimentelle Beiträge zur Ökologie von *Chironomus* (Diptera). I. Dormanz bei *Chironomus nuditarsis* und *Ch. plumosus.* Oecologia 16:73–95.

Fischer, Z. 1958. Influence exerted by temperature on the development of the eggs of *Lestes sponsa* Leach. Ekologia Polska Seria B 4:1–5.

Fox, L. R. and W. W. Murdoch. 1978. Effects of feeding history on short-term and long-term functional responses in *Notonecta hoffmanni.* Journal of Animal Ecology 47: 945–59.

Fredeen, F. J. H. 1964. Bacteria as food for blackfly larvae (Diptera: Simuliidae) in laboratory cultures in natural streams. Canadian Journal of Zoology 42:527–48.

Friesen, M. K., J. F. Flannagan, and S. G. Lawrence. 1979. Effects of temperature and cold storage on development time and viability of eggs of the burrowing mayfly *Hexagenia rigida* (Ephemeroptera: Ephemeridae). The Canadian Entomologist 111: 665–73.

Fuller, R. L. and R. J. Mackay. 1981. Effects of food quality on the growth of three *Hydropsyche* species (Trichoptera: Hydropsychidae). Canadian Journal of Zoology 59:1133–40.

Gallepp, G. W. 1977. Responses of caddisfly larvae *(Brachycentrus* spp.) to temperature, food availability and current velocity. American Midland Naturalist 98:59–84.

Gledhill, T. 1960. The Ephemeroptera, Plecoptera and Trichoptera caught by emergence traps in two streams during 1958. Hydrobiologia 15:179–88.

Grafius, E. and N. H. Anderson. 1979. Population dynamics, bioenergetics, and role of *Lepidostoma quercina* (Trichoptera: Lepidostomatidae) in an Oregon woodland stream. Ecology 60:433–41.

Hagstrum, D. W. and E. B. Workman. 1971. Interaction of temperature and feeding rate in determining the rate of development of larval *Culex tarsalis* (Diptera, Culicidae). Annals of the Entomological Society of America 64:668–75.

Hanec, W. and R. A. Brust. 1967. The effect of temperature on the immature stages of *Culiseta inornata* (Diptera: Culicidae) in the laboratory. The Canadian Entomologist 99:59–64.

Harper, P. P. 1973. Life histories of Nemouridae and Leuctridae in Southern Ontario (Plecoptera). Hydrobiologia 41:309–56.

Harvey, R. S. 1971. Temperature effects on the maturation of midges (Tendipedidae) and their sorption of radionuclides. Health Physics 20:613–16.

Harvey, R. S., R. L. Vannote, and B. W. Sweeney. 1979. Life history, developmental processes, and energetics of the burrowing mayfly *Dolania americana,* pp. 211–30. *In:* J. F. Flannagan and K. E. Marshall (eds.). Advances in Ephemeroptera biology. Proceedings of the Third International Conference on Ephemeroptera. Plenum Press, New York, NY. 552 pp.

Headlee, T. J. 1940. The relative effects on insect metabolism of temperatures derived from constant and variable sources. Journal of Economic Entomology 33:361–64.

Headlee, T. J. 1941. Further studies of the relative effects on insect metabolism of temperatures derived from constant and variable sources. Journal of Economic Entomology 34:171–74.

Heiman, D. R. and A. W. Knight. 1975. The influence of temperature on the bioenergetics of the carnivorous stonefly nymph, *Acroneuria californica* Banks (Plecoptera: Perlidae). Ecology 56:105–16.

Heuvel, M. J. 1963. The effect of rearing temperature on the wing length, thorax length, leg length and ovariole number of the adult mosquito, *Aedes aegypti* (L.). Transactions of the Royal Entomological Society of London 115.197–216.

Hildrew, A. G. 1978. Ecological aspects of life history in some net-spinning Trichoptera, pp. 269–81. *In:* M. I. Crichton (ed.). Proceedings of the Second International Symposium on Trichoptera. Dr. W. Junk b.v. Publishers, The Hague, The Netherlands. 359 pp.

Hosseinie, S. O. 1976. Effects of the amount of food on duration of stages, mortality rates, and size of individuals in *Tropisternus lateralis nimbatus* (Say) (Coleoptera: Hydrophylidae). Internationale Revue der gesamten Hydrobiologie 61:383–88.

Howe, R. W. 1967. Temperature effects on embryonic development in insects. Annual Review of Entomology 12:15–42.

Huffaker, C. B. 1944. The temperature relations of the immature stages of the malarial mosquito, *Anopheles quadrimaculatus* Say, with a comparison of the developmental power of constant and variable temperatures in insect metabolism. Annals of the Entomological Society of America 37:1–27.

Humpesch, U. H. 1978. Preliminary notes on the effect of temperature and light-condition on the time of hatching in some Heptageniidae (Ephemeroptera). Internationale Vereinigung für Theoretische und Angewandte Limnologie Verhandlungen. 20: 2605–11.

Humpesch, U. H. 1979. Life cycles and growth rates of *Baetis* spp. (Ephemeroptera: Baetidae) in the laboratory and in two stony streams in Austria. Freshwater Biology 9:467–79.

Humpesch, U. H. 1980. Effect of temperature on the hatching time of eggs of five *Ecdyonurus* spp. (Ephemeroptera) from Austrian streams and English streams, rivers and lakes. Journal of Animal Ecology 49:317–33.

Humpesch, U. H. and J. M. Elliott. 1980. Effect of temperature on the hatching time of eggs of three *Rhithrogena* spp. (Ephemeroptera) from Austrian streams and an English stream and river. Journal of Animal Ecology 49:643–61.

Hunt, B. P. 1951. Reproduction of the burrowing mayfly, *Hexagenia limbata* (Serville), in Michigan. Florida Entomologist 34:59–70.

Hynes, H. B. N. 1970. The ecology of running waters. University of Toronto Press, Toronto, Canada. 555 pp.

Hynes, H. B. N. 1976. Biology of Plecoptera. Annual Review of Entomology 21:135–53.

Ide, F. P. 1935. The effect of temperature on the distribution of the mayfly fauna of a stream. University of Toronto Studies, Biological Series 39:1–76.

Ide, F. P. 1940. Quantitative determination of the insect fauna of rapid water. University of Toronto Studies, Biological Series 47:1–20.

Illies, J. 1971. Emergenz 1969 im Breitenbach. Schlitzer productionsbiologische Studien (1). Archiv für Hydrobiologie 69:14–59.

Illies, J. 1979. Annual and seasonal variation of individual weights of adult water insects. Aquatic Insects 1:153–63.

Ingram, B. R. 1976. Effects of photoperiod and temperature on abnormal wing-pad development in two species of Odonata. Canadian Journal of Zoology 54:1103–10.

Ingram, B. R. and C. E. Jenner. 1976. Influence of photoperiod and temperature on developmental time and number of molts in nymphs of two species of Odonata. Canadian Journal of Zoology 54:2033–45.

Istock, C. A. 1981. Natural selection and life history variation: theory plus lessons from a mosquito, pp. 113–28. *In*: R. F. Denno and H. Dingle (eds.). Insect life history patterns: habitat and geographic variation. Springer-Verlag, New York, NY. 225 pp.

Iversen, T. M. 1974. Ingestion and growth in *Sericostoma personatum* (Trichoptera) in relation to the nitrogen content of ingested leaves. Oikos 25:278–82.

Iversen, T. M. 1979. Laboratory energetics of larvae of *Sericostoma personatum* (Trichoptera). Holarctic Ecology 2:1–5.

Jansson, A. and G. G. E. Scudder. 1974. The life cycle and sexual development of *Cenocorixa* species (Hemiptera, Corixidae) in the Pacific Northwest of North America. Freshwater Biology 4:73–92.

Johannsson, O. E. 1980. Energy dynamics of the eutrophic chironomid *Chironomus plumosus* f. *semireductus* from the Bay of Quinte, Lake Ontario. Canadian Journal of Fisheries and Aquatic Sciences 37:1254–65.

Jones, R. E. 1974. The effects of size-selective predation and environmental variation on the distribution and abundance of a chironomid, *Paraborniella tonnoiri* Freeman. Australian Journal of Zoology 22:71–89.

Jordan, R. G. 1980. Geographic differentiation in the development of *Aedes sierrensis* (Diptera: Culicidae) in nature. The Canadian Entomologist 112:205–10.

Kappus, K. D. and C. E. Venard. 1967. The effects of photoperiod and temperature on the induction of diapause in *Aedes triseriatus* (Say). Journal of Insect Physiology 13:1007–19.

Khoo, S. G. 1968a. Experimental studies on diapause in stoneflies. I. Nymphs of *Capnia bifrons* (Newman). Proceedings of the Royal Entomological Society of London (A) 43:40–48.

Khoo, S. G. 1968b. Experimental studies on diapause in stoneflies II. Eggs of *Diura bicaudata* (L.). Proceedings of the Royal Entomological Society of London (A) 43:49–56.

Khoo, S. G. 1968c. Experimental studies on diapause in stoneflies III. Eggs of *Brachyptera risi* (Morton). Proceedings of the Royal Entomological Society of London (A) 43:141–46.

Kondratieff, B. C. and J. R. Voshell, Jr. 1980. Life history and ecology of *Stenonema modestum* (Banks) (Ephemeroptera: Heptageniidae) in Virginia, USA. Aquatic Insects 2:177–89.

Konstantinov, A. S. 1958. The effect of temperature on growth rate and development of chironomid larvae. Doklady Akademii Nauk SSSR. Seriya Biologiya 20:506–9.

Koskinen, R. 1968. Seasonal and diel emergence of *Chironomus salinarius* Kieff. (Dipt., Chironomidae) near Bergen, Western Norway. Annales Zoologici Fennici 5:65–70.

Lacey, L. A. and M. S. Mulla. 1979. Factors affecting feeding rates of black fly larvae. Mosquito News 39:315–19.

Ladle, M. and R. G. Hansford. 1981. The feeding of the larva of *Simulium austeni* Edwards and *Simulium (Wilhelmia)* spp. Hydrobiologia 78:17–24.

Langford, T. E. 1975. The emergence of insects from a British river warmed by power station cooling-water. Part II. The emergence patterns of some species of Ephemeroptera, Trichoptera, and Megaloptera in relation to water temperature and river flow, upstream and downstream of the cooling-water outfalls. Hydrobiologia 47: 91–133.

Lavandier, P. and J. -Y. Pujol. 1975. Cycle biologique de *Drusus rectus* (Trichoptera) dans les Pyrénées centrales: influence de la température et de l'enneigement. Annales de Limnologie 11:255–62.

Lawton, J. H., B. A. Thompson, and D. J. Thompson. 1980. The effects of prey density on survival and growth of damselfly larvae. Ecological Entomology 5:39–51.

Learner, M. A. and D. W. B. Potter. 1974. The seasonal periodicity of emergence of insects from two ponds in Hertfordshire, England, with special reference to the Chironomidae (Diptera: Nematocera). Hydrobiologia 44:495–510.

Lehmkuhl, D. M. 1973. A new species of *Baetis* (Ephemeroptera) from ponds in the Canadian arctic, with biological notes. The Canadian Entomologist 105:343–46.

Lehmkuhl, D. M. 1974. Thermal regime alteration and vital environmental physiological signals in aquatic organisms, pp 116–222. *In*: J. W. Gibbons and R. R. Sharitz (eds.). Thermal ecology. National Technical Information Service Conference 730505. United States Atomic Energy Commission. 670 pp.

LeSage, L. and A. D. Harrison. 1980. The biology of *Cricotopus* (Chironomidae: Orthocladiinae) in an algal-enriched stream: Part I. Normal biology. Archiv für Hydrobiologie Supplement 57:375–418.

Lillehammer, A. 1975. Norwegian stoneflies. IV. Laboratory studies on ecological factors influencing distribution. Norwegian Journal of Entomology 22:99–108.

Lutz, P. E. 1963. Seasonal regulation in nymphs of *Tetragoneuria cynosura* (Say). Proceedings of the North Central Branch, Entomological Society of America 18:135–38.

Lutz, P. E. 1968a. Life-history studies on *Lestes eurinus* Say (Odonata). Ecology 49: 576–79.

Lutz, P. E. 1968b. Effects of temperature and photoperiod on larval development in *Lestes eurinus* (Odonata: Lestidae). Ecology 49:637–44.

Lutz, P. E. 1974a. Effects of temperature and photoperiod on larval development in *Tetragoneuria cynosura* (Odonata: Libellulidae). Ecology 55:370–77.

Lutz, P. E. 1974b. Environmental factors controlling duration of larval instars in *Tetragoneuria cynosura* (Odonata). Ecology 55:630–37.

Lutz, P. E. and C. E. Jenner. 1960. Relationship between oxygen consumption and photoperiodic induction of the termination of diapause in nymphs of the dragonfly *Tetragoneuria cynosura*. Journal of the Elisha Mitchell Science Society 76:192–93.

Lutz, P. E. and C. E. Jenner. 1964. Life history and photoperiodic responses of nymphs of *Tetragoneuria cynosura* (Say). The Biological Bulletin 127:304–16.

Macan, T. T. 1977. The influence of predation on the composition of fresh-water animal communities. Biological Reviews of the Cambridge Philosophical Society 52: 45–70.

Mackay, R. J. 1979. Life history patterns of some species of *Hydropsyche* (Trichoptera: Hydropsychidae) in southern Ontario. Canadian Journal of Zoology 57:963–75.

Mackey, A. P. 1977. Growth and development of larval Chironomidae. Oikos 28:270–75.

Malicky, H. 1981. Artificial illumination of a mountain stream in lower Austria: effect of constant daylength on the phenology of the caddisflies (Trichoptera). Aquatic Insects 3:25–32.

Martin, M. M., J. S. Martin, J. J. Kukor, and R. W. Merritt. 1980. The digestion of protein and carbohydrate by the stream detritivore, *Tipula abdominalis* (Diptera, Tipulidae). Oecologia (Berlin) 46:360–64.

Mattice, J.S. and L. L. Dye. 1978. Effect of a stream electric generating station on the emergence timing of the mayfly, *Hexagenia bilineata* (Say). Internationale Vereinigung für Theoretische und Angewandte Limnologie Verhandlungen 20:1752–58.

McClure, R. G. and K. W. Stewart. 1976. Life cycle and production of the mayfly *Choroterpes (Neochoroterpes) mexicanus* Allen (Ephemeroptera: Leptophlebiidae). Annals of the Entomological Society of America 69:134–44.

McCrary, A. B. and C. E. Jenner. 1965. Influence of day length on sex ratio in the giant mosquito, *Toxorhynchites rutilus*, in nature. American Zoologist 5:206. (Abstract).

McCullough, D. A. 1975. The bioenergetics of three aquatic insects determined by radioisotopic analyses. Battelle Pacific Northwest Laboratories, B.N.W.L. 1928, special distribution VC-48. 225 pp.

McCullough, D. A., G. W. Minshall, and C. E. Cushing. 1979. Bioenergetics of lotic filter-feeding insects *Simulium* spp. (Diptera) and *Hydropsyche occidentalis* (Trichoptera) and their function in controlling organic transport in streams. Ecology 60:585–96.

Mecom, J. O. 1970. Evidence of diurnal feeding activity in Trichopteran larvae. Journal of the Graduate Research Center, Southern Methodist University, Dallas 38:44–57.

Meier, P. G. and P. G. Bartholomae. 1980. Diel periodicity in the feeding activity of *Potamanthus myops* (Ephemeroptera). Archiv fur Hydrobiologie 88:1–8.

Merritt, R. W., D. H. Ross, and G. J. Larson. 1982. Influence of stream temperature and seston on the growth and production of overwintering larval black flies (Diptera: Simuliidae). Ecology 63:1322–31.

Minshall, G. W. 1978. Autotrophy in stream ecosystems. BioScience 28:767–71.

Mokry, J. E. 1976. Laboratory studies on the larval biology of *Simulium venustum* Say (Diptera: Simuliidae). Canadian Journal of Zoology 54:1657–63.

Monakov, A. V. 1972. Review of studies on feeding of aquatic invertebrates conducted at the Institute of Biology of Inland Waters, Academy of Sciences, USSR. Journal of the Fisheries Research Board of Canada 29:363–83.

Moore, J. W. 1980. Factors influencing the composition, structure and density of a population of benthic invertebrates. Archiv für Hydrobiologie 88:202–18.

Nebeker, A. V. 1971a. Effect of high winter water temperatures on adult emergence of aquatic insects. Water Research 5:777–83.

Nebeker, A. V. 1971b. Effect of water temperature on nymphal feeding rate, emergence, and adult longevity of the stonefly *Pteronarcys dorsata*. Journal of the Kansas Entomological Society 44:21–26.

Nebeker, A. V. 1971c. Effect of temperature at different altitudes on the emergence of aquatic insects from a single stream. Journal of the Kansas Entomological Society 44:26–35.

Nebeker, A. V. 1973. Temperature requirements and life cycle of the midge *Tanytarsus dissimilis* (Diptera: Chironomidae). Journal of the Kansas Entomological Society 46:160–65.

Needham, J. G. 1903. Aquatic insects in New York State. Part 3. Life histories of Odonata, suborder Zygoptera. New York State Museum Bulletin 68:218–79.

Newell, R. L. and G. W. Minshall. 1978a. Effect of temperature on the hatching time of *Tricorythodes minutus* (Ephemeroptera: Tricorythidae). Journal of the Kansas Entomological Society 51:504–6.

Newell, R. L. and G. W. Minshall. 1978b. Life history of a multivoltine mayfly, *Tricorythodes minutus*: an example of the effect of temperature on the life cycle. Annals of the Entomological Society of America 71:876–81.

Novak, K. and F. Sehnal. 1963. The development cycle of some species of the genus *Limnephilus* (Trichoptera). Časopis Československé společnosti entomologické 60: 68–80.

Obrdlík, P., Z. Adámek, and J. Zahrádka. 1979. Mayfly fauna (Ephemeroptera) and the biology of the species *Potamanthus luteus* (L.) in a warmed stretch of the Oslava River. Hydrobiologia 67:129–40.

Oliver, D. R. 1971. Life history of the Chironomidae. Annual Review of Entomology 16: 211–30.

Otto, C. 1974. Growth and energetics in a larval population of *Potamophylax cingulatus* (Steph.) (Trichoptera) in a South Swedish stream. Journal of Animal Ecology 43: 339–61.

Pajunen, V. I. 1975. Effect of temperature on development in some populations of *Arctocorisa carinata* (Sahlb.) (Hemiptera, Corixidae). Annales Zoologici Fennici 12: 211–14.

Pajunen, V. I. and E. Sundbäck. 1973. Effect of temperature on the development of *Arctocorisa carinata* (Sahlb.) and *Callicorixa producta* (Reut.) (Hemiptera, Corixidae). Annales Zoologici Fennici 10:372–77.

Pandian, T. J., S. Mathavan, and C. P. Jeyagopal. 1979. Influence of temperature and body weight on mosquito predation by the dragonfly nymph *Mesogomphus lineatus*. Hydrobiologia 62:99–104.

Paris, O. H. and C. E. Jenner. 1959. Photoperiodic control of diapause in the pitcher-plant midge, *Metriocnemus knabi*, pp 601–24. *In*: Photoperiodism and related phenomena in plants and animals. American Association for the Advancement of Science, Washington, DC. 903 pp.

Pearson, W. D., R. H. Kramer, and D. R. Franklin. 1968. Macroinvertebrates in the Green River below Flaming Gorge Dam, 1964–65 and 1967. Proceedings of the Utah Academy of Sciences, Arts and Letters 45:148–67.

Pinger, R. R. and B. F. Eldridge. 1977. The effect of photoperiod on diapause induction in *Aedes canadensis* and *Psorophora ferox* (Diptera: Culicidae). Annals of the Entomological Society of America 70:437–41.

Ploskey, G. R. and A. V. Brown. 1980. Downstream drift of the mayfly *Baetis flavistriga* as a passive phenomenon. The American Midland Naturalist 104:405–9.

Pradhan, S. 1945. Insect population studies: II. Rate of insect development under variable temperature in the field. Proceedings of the National Institute of Sciences of India 11:74–80.

Pritchard, G. and B. Pelchat. 1977. Larval growth and development of *Argia vivida* (Odonata: Coenagrionidae) in warm sulphur pools at Banff, Alberta. The Canadian Entomologist 109:1563–70.

Procter, D. L. C. 1973. The effect of temperature and photoperiod on larval development in Odonata. Canadian Journal of Zoology 51:1165–70.

Resh, V. H. 1976. Life histories of coexisting species of *Ceraclea* caddisflies (Trichoptera: Leptoceridae): the operation of independent functional units in a stream ecosystem. The Canadian Entomologist 108:1303-18.

Rhame, R. E. and K. W. Stewart. 1976. Life cycles and food habits of three Hydropsychidae (Trichoptera) species in the Brazos River, Texas. Transactions of the American Entomological Society 102:65-99.

Richards, A. G. 1957. Cumulative effects of optimum and suboptimum temperatures on insect development, pp. 145-62. *In*: F. H. Johnson (ed.). Influence of temperature on biological systems. American Physiological Association, Washington, DC. 275 pp.

Richards, A. G. and M. Q. Kolderie. 1957. Variation in weight, developmental rate, and hatchability of *Oncopeltus* eggs as a function of the mother's age. Entomological News 68:57-64.

Ricklefs, R. E. 1969. Preliminary models for growth rates in altricial birds. Ecology 50: 1031-39.

Ross, D. H. and R. W. Merritt. 1978. The larval instars and population dynamics of five species of black flies (Diptera: Simuliidae) and their responses to selected environmental factors. Canadian Journal of Zoology 56:1633-42.

Saunders, D. S. 1976. Insect clocks. International Series in Pure and Applied Biology, Zoology Division, Vol. 54. Permagon Press, New York, NY. 279 pp.

Sawchyn, W. W. and N. S. Church. 1973. The effects of temperature and photoperiod on diapause development in the eggs of four species of *Lestes* (Odonata: Zygoptera). Canadian Journal of Zoology 51:1257-65.

Sawchyn, W. W. and C. Gillott. 1974. The life history of *Lestes congener* (Odonata: Zygoptera) on the Canadian prairies. The Canadian Entomologist 106:367-76.

Schott, R. J. and M. A. Brusven. 1980. The ecology and electrophoretic analysis of the damselfly, *Argia vivida* Hagen, living in a geothermal gradient. Hydrobiologia 69: 261-65.

Slater, J. D. and G. Pritchard. 1979. A stepwise computer program for estimating development time and survival of *Aedes vexans* (Diptera: Culicidae) larvae and pupae in field populations in southern Alberta. The Canadian Entomologist 111:1241-53.

Smith, S. M. and R. A. Brust. 1971. Photoperiodic control of the maintenance and termination of larval diapause in *Wyeomyia smithii* (Coq.) (Diptera: Culicidae) with notes on oogenesis in the adult female. Canadian Journal of Zoology 49:1065-73.

Snellen, R. K. and K. W. Stewart. 1979. The life cycle of *Perlesta placida* (Plecoptera: Perlidae) in an intermittent stream in northern Texas, USA. Annals of the Entomological Society of America 72:659-66.

Sprules, W. M. 1947. An ecological investigation of stream insects in Algonquin Park, Ontario. University of Toronto Studies, Biological Series 56:1-81.

Sweeney, B. W. 1978. Bioenergetic and developmental response of a mayfly to thermal variation. Limnology and Oceanography 23:461-77.

Sweeney, B. W. and J. A. Schnack. 1977. Egg development, growth, and metabolism of *Sigara alternata* (Say) (Hemiptera: Corixidae) in fluctuating thermal environments. Ecology 58:265-77.

Sweeney, B. W. and R. L. Vannote. 1978. Size variation and the distribution of hemimetabolous aquatic insects: two thermal equilibrium hypotheses. Science 200: 444-46.

Sweeney, B. W. and R. L. Vannote. 1981. *Ephemerella* mayflies of White Clay Creek:

bioenergetic and ecological relationships among six coexisting species. Ecology 62: 1353–69.

Tauber, M. J. and C. A. Tauber. 1975. Natural daylengths regulate insect seasonality by two mechanisms. Nature (London) 258:711–12.

Taylor, F. 1981. Ecology and evolution of physiological time in insects. The American Naturalist 117:1–23.

Taylor, F. 1982. Sensitivity of physiological time in arthropods to variation of its parameters. Environmental Entomology 11:573–77.

Thibault, M. 1971. Écologie d'un ruisseau a truites des Pyrénées-Atlantiques, le Lissuraga. II. Les fluctuations thermiques de l'eau; répercussion sur les périodes de sortie et la taille de quelques Éphéméroptères, Plécoptères et Trichoptères. Annales d'Hydrobiologie 2:241–74.

Thorup, J. and T. M. Iversen. 1974. Ingestion by *Sericostoma personatum* Spence (Trichoptera: Sericostomatidae). Archiv für Hydrobiologie 74:39–47.

Trimble, R. M. and S. M. Smith. 1979. Geographic variation in the effects of temperature and photoperiod on dormancy induction, development time, and predation in the tree-hole mosquito, *Toxorhynchites rutilus septentrionalis* (Diptera: Culicidae). Canadian Journal of Zoology 57:1612–18.

Tsui, P. T. P. and W. L. Peters. 1974. Embryonic development, early instar morphology, and behavior of *Tortopus incertus* (Ephemeroptera: Polymitarcidae). Florida Entomologist 57:349–56.

Tucker, H. A. and R. K. Ringer. 1982. Controlled photoperiodic environments for food animals. Science 216:1381–86.

Vanderlin, R. L. and F. A. Streams. 1977. Photoperiodic control of reproductive diapause in *Notonecta undulata*. Environmental Entomology 6:258–62.

Vannote, R. L., G. W. Minshall, K. W. Cummins, J. R. Sedell, and C. E. Cushing. 1980. The river continuum concept. Canadian Journal of Fisheries and Aquatic Sciences 37:130–37.

Vannote, R. L. and B. W. Sweeney. 1980. Geographic analysis of thermal equilibria: a conceptual model for evaluating the effect of natural and modified thermal regimes on aquatic insect communities. The American Naturalist 115:667–95.

Vepsäläinen, K. 1974a. Determination of wing length and diapause in water-striders (*Gerris* Fabr., Heteroptera). Hereditas 77:163–76.

Vepsäläinen, K. 1974b. Lengthening of illumination period is a factor in averting diapause. Nature (London) 247:385–86.

Wallace, J. B., W. R. Woodall, and F. F. Sherberger. 1970. Breakdown of leaves by feeding of *Peltoperla maria* nymphs (Plecoptera: Peltoperlidae). Annals of the Entomological Society of America 63:562–67.

Ward, G. M. and K. W. Cummins. 1979. Effects of food quality on growth of a stream detritivore, *Paratendipes albimanus* (Meigen) (Diptera: Chironomidae). Ecology 60: 57–64.

Ward, J. V. and J. A. Stanford. 1982. Thermal responses in the evolutionary ecology of aquatic insects. Annual Review of Entomology 27:97–117.

Wetzel, R. G. 1975. Limnology. W.B. Saunders Co., Philadelphia, PA. 743 pp.

Wise, E. J. 1980. Seasonal distribution and life histories of Ephemeroptera in a Northumbrian River. Freshwater Biology 10:101–11.

Zahar, A. R. 1951. The ecology and distribution of black-flies (Simuliidae) in south-east Scotland. Journal of Animal Ecology 20:33–62.

chapter 5

BEHAVIORAL ADAPTATIONS OF AQUATIC INSECTS

Michael Wiley
Steven L. Kohler

INTRODUCTION

One of the most fascinating characteristics of the aquatic insects as a group is their diverse and often intriguing behavior. The behavioral repertoire of each insect consists of an entire collection of interacting "ways of behaving" that have been accumulated genetically through generations of selection. Every individual behavior pattern has been retained because—alone or in concert—it has improved for its bearer the chances of surviving in a dangerous environment; it is in this sense that we refer in this chapter to behavior as an adaptation. The diversity of aquatic insect behaviors is a reflection of the diversity of challenges posed by the aquatic environment to insect life; as for the apparent ingenuity of insect behavior, this too is the product of natural selection operating in a milieu of harsh constraints and high reproductive potential.

Much of what has been written in the past about aquatic insect behavior is anecdotal in nature. This is unfortunate because it tends to de-emphasize the fundamental importance of behavioral adaptations and helps to promulgate the myth that insect behavior is somehow mechanical and uninteresting. We believe, on the contrary, that the study of behavior as a coherent system of adaptations is essential to a complete understanding of the ecology of aquatic insects.

Our intent in this chapter is not to provide a systematic catalogue of aquatic insect behavior. Any detailed listing would be an enormous undertaking far beyond the present scope of this book. We will instead present a selective sampling of behaviors and hope that these representative examples will suffice both to illustrate the ecological sophistication of these adaptations and to share our enthusiasm for their study. Our examination of behavioral adaptations focuses on several broad categories of behavior that embody responses to relatively distinct classes of problems faced by aquatic insects. These include (1) *regulatory behavior*, behavioral adaptations that increase the control that an individual exerts over its own metabolic status; (2) *foraging behavior*, behavior that involves the gathering and processing of energy and other key resources from the environment; and (3) *reproductive behavior*, behavior that is responsible for the successful continuation of life into the next generation. We also discuss in some detail the topic of *behavioral drift*, an important subject that has served as a focal point for behavioral investigations during the past two decades.

Throughout our treatment of these topics we endeavor to emphasize the role of natural selection in the development of insect behavior. We have tried in each of these topical areas to discuss both representative evolutionary problems and their behavioral solutions. Where possible we direct the reader to more comprehensive review articles. When this is not possible we hope that the material presented is sufficiently interesting to encourage the reader to investigate the available literature further.

REGULATORY BEHAVIOR

A fundamental principle in bioenergetics is that in order to grow and reproduce, an animal's energy intake must exceed its energy expenditure. Energy lost in the form of work or heat is replaced by energy produced from the oxidation of organic material (food) gathered from the environment. Behavior contributes to the maintenance of a favorable energy balance by (1) enabling the organism to efficiently acquire food resources from the environment, and (2) by regulating the rate of their metabolism via the behavioral regulation of temperature (thermoregulation) and/or of respiratory gas exchanges (O_2 and CO_2). Temperature regulates metabolic processes directly by affecting the rate at which biochemical reactions take place (Chapman 1971). Rates of gas exchange control metabolic processes indirectly by affecting the rate at which metabolic reactants (O_2) and/or products (CO_2) can be processed.

In terms of both gas exchange and thermal regulation, the aquatic environment poses a very different set of problems for insect life than does the terrestrial environment. For terrestrial insects, thermal regimes are highly variable, whereas the environmental partial pressure of both O_2 and CO_2 are relatively invariant. Under these conditions thermoregulation is the principle goal of insect regulatory behavior (Chapman 1971; Heinrich and Casey 1978). For aquatic insects, however, the situation is the reverse. The high specific heat of water promotes thermal stability; thus, diel and seasonal fluctuations in temperature are considerably moderated. On the other hand, in water there are large variations in the partial pressure of both O_2 and CO_2. As a result, behavioral adaptations to facilitate gas exchange are generally more important than thermoregulation in equipping aquatic insects for life in the water.

The Regulation of Gas Exchange

Aquatic insects have evolved a wide variety of techniques to facilitate respiration. These range from simple gas exchange through the cuticle (as occurs in many dipteran larvae), to the muscle-driven tracheal gill systems of mayfly nymphs (Ephemeroptera). Several excellent reviews of aquatic insect respiration are available elsewhere (Thorpe 1950; Crisp 1964; Chapman 1971; Eriksen et al., in press). We will restrict our discussion to behavioral strategies related to the physics of diffusion.

Fundamentally, all aquatic insects that do not respire at the water's surface exchange O_2 and CO_2 directly with the surrounding water by diffusion. In this category are included the Ephemeroptera, Plecoptera, Megaloptera, Odonata, Trichoptera, and many Coleoptera and Diptera. The rate at which a gas will diffuse across any exchange surface can be described in terms of Fick's Law. In the case of oxygen diffusing into the tissues of an aquatic insect, we can write

$$QO_2 = \Delta pO_2 \times k \times A/L, \tag{1}$$

where QO_2 is the rate at which O_2 can diffuse into an insect's tracheal system; ΔpO_2 is the difference between the internal partial pressure of O_2, and the partial pressure of O_2 in the water immediately adjacent to the exchange surface; A is the total area of the exchange surface (e.g. gills, plastron, integument) across which O_2 can diffuse; L is the distance the molecules must diffuse (thickness of integument, gills, etc.); and k is a diffusion constant characteristic of the material comprising the exchange surface. For any individual insect, the parameters A, L, and k are constant—they are set by the physical morphology of its respiratory system. The partial pressure of oxygen in the tracheal system is almost always constant and very low; consequently, the rate of oxygen uptake is primarily controlled by the dissolved oxygen concentration (DO) of the water. If the ambient DO drops, metabolic processes are forced to slow down because the rate of oxygen diffusion decreases. However, even when the pO_2 in the surrounding water is high, rates of diffusion will tend to decrease over time because the external O_2 concentration in the immediate vicinity of respiratory surfaces declines as O_2 molecules diffuse across them. As the water surrounding the exchange surface becomes more and more depleted of O_2 molecules, ΔpO_2, and consequently the rate of diffusion, continues to decline.

A common solution to this problem is behavioral ventilation (i.e. the replacement of O_2-depleted water in the vicinity of the respiratory exchange surface with nondepleted water; that is, water at ambient DO) by physically moving the exchange surface, the overlying water, or both. For ventilation, ephemeropteran nymphs beat their abdominal gills (Eastham 1937; Eriksen 1963), plecopteran nymphs wag their abdomens or perform "push-ups" by bobbing up and down in an attempt to force new water over their gills and other exchange surfaces (Eriksen 1966; Knight and Gaufin 1963), and trichopteran and dipteran larvae undulate their abdomens (Walshe 1950; Philipson and Moor-house 1974; Philipson 1978). For aquatic insects using plastron or gas-gill respiration (Hemiptera and Coleoptera), or those living in areas of high current velocity (e.g. the mayfly *Baetis tricaudatus*), ventilation is automatic because either the insects or the surrounding water are continually in motion. In such cases, the decline in diffusion rate due to depletion is offset by natural renewal rates of O_2 due to water turbulence.

Any lowering of ΔpO_2 by decreasing ambient DO or decreasing current velocity (natural ventilation) will lead to an increase in behavioral ventilation in the organism's attempt to raise the concentration gradient (Philipson 1954; Ambühl 1959; Eriksen 1963; Philipson and Moorhouse 1974, 1976). Likewise, any factor that causes an increase in metabolic demand (e.g. temperature or activity) also results in increased ventilation rates, because increasing the average ΔpO_2 is the only way the rate of oxygen uptake can be increased. Since ventilation of respiratory exchange surfaces involves significant muscular activity, it is energetically costly. For some taxa there is a double cost involved because ventilation cannot be carried out simultaneously with other activities; therefore it precludes feeding, retreat repair, or other energy profitable behaviors (e.g. *Chironomus plumosus*, Walshe 1950; hydropsychid caddisflies, Philipson and Moorhouse 1974). In this context, ambient current velocity is an especially important factor because if it is great enough, it can achieve the same effect as behavioral ventilation with little or no energetic cost to the organism.

Eriksen (1966) suggested that lotic mayflies (Ephemeroptera) may attempt to regulate O_2 consumption rates behaviorally with a minimum of physiological work. This could be achieved by preferentially remaining in those microhabitats with sufficient current velocity to ventilate exchange surfaces, or by temporarily moving to areas of higher velocity or DO when under respiratory stress. Overwintering populations of the mayfly *Leptophlebia vespertina* respond in just this way to falling DO concentrations in ice-covered ponds. Brittain and Nagell (1981) report that *L. vespertina* survive months of anoxic conditions in Norwegian humic ponds by migrating to the underside of the ice along the shore, the only location where oxygen concentrations are predictably high enough to permit survival. Extensive labortory experiments demonstrated that under high DO conditions nymphs were negatively phototactic. However, at low DO nymphs became very active, positively phototactic, and negatively thermotactic. Shifts in taxis orientation allowed the nymphs to move toward the pond's edge where temperature was lower, but oxygen concentrations were higher. A similar reversal in phototactic and thermotactic orientation in association with behavioral metabolic regulation has been reported for the mayfly *Cloeon dipterum* (Nagell 1977). Field studies with the caddisfly *Glossosoma nigrior* suggest that seasonal shifts in microhabitat use that are a result of behavioral regulation may also occur in other orders of insects (Kovalak 1976, 1979).

Short-term changes in DO probably elicit less extreme behavioral responses. Laboratory studies with several different species of lotic mayfly nymphs have demonstrated that, as oxygen levels decline, nymphs compensate by moving from sheltered locations to nearby positions with maximal exposure to current (Wiley and Kohler 1980). As would be expected, the propensity of nymphs to engage in this short-term regulatory behavior generally decreases with increasing ambient current velocity, and is inversely related to physiological exchange capacity (A/L, Equation 1).

Light rather than DO concentration appears to be the most commonly used directional cue for the positioning changes and taxis orientation that are related to respiratory regulation. Light intensity presumably is an accurate guide because it is associated with the water-atmosphere interface where current velocity and pO_2 are both most likely to be at a maximum, and pCO_2 is at a minimum. Scherer (1965), in an elegant laboratory experiment, demonstrated the relationship between phototactic response of the mayfly *Baetis rhodani,* and gas exchange requirements. By varying the CO_2 concentration in water, and thereby the rate at which nymphs could rid themselves of this metabolic by-product, negatively phototactic nymphs could be made to alternately move into and out of illuminated and darkened patches in an artificial stream. High CO_2 and low O_2 concentrations bring about a positive phototaxis (movement towards light) for the same reason; nymphs have a higher probability of encountering favorable partial-pressure gradients near the water's surface.

Alteration of microhabitat positioning is an effective way to reduce costs of ventilation, but it is not the only behavioral solution that aquatic insects have evolved to this problem. Among those groups using abdominal undulation to facilitate gas exchange through unsclerotized abdominal cuticle (Trichoptera, Megaloptera, Lepidoptera, some Coleoptera, and some Diptera), the mechanical efficiency of renewing the water around the body's surface can be greatly enhanced by surrounding the abdomen with a solid tube against which water can be pushed (e.g. the cases of most Trichoptera, and the cocoons of some Diptera and Lepidoptera). The result is a type of peristaltic pump, which can move large volumes of water over the abdomen with much less energy than would be required if ventilation were due to simple motion-induced turbulence alone. This decrease in cost per unit effort of ventilation has been proposed as a primary mechanism that allowed adaptive radiation of the Trichoptera from their ancestral habitat of cold, highly oxygenated mountain streams into warmer waters with relatively low oxygen concentrations (Ross 1956; Wiggins 1977). Cases and tube structures clearly fulfill other functions (defense, ballast, etc.) wherever they have evolved (see Wiggins 1977; Mackay and Wiggins 1979), but certainly their role in respiratory regulation is well established (Wiggins 1977; Walshe 1950). No doubt, this flexible utility in part accounts for the phylogenetically widespread development of such structures among aquatic insects.

Thermoregulation

Although thermal regimes in aquatic habitats are less variable than those in terrestrial habitats, considerable work with freshwater fishes suggests that behavioral thermoregulation remains an important adaptation for life in water (Brett 1956). Thermoregulation by submerged insects, however, has received little attention. Research by Nagell (1977) and Brittain and Nagell (1981) demonstrated that thermotaxis exists in ephemeropteran nymphs, and that it can be used to locate microhabitats with favorable DO concentrations. However, it is not clear whether thermotaxis is used directly to regulate body temperature. It is possible that the thermotactic response of *Leptophlebia vespertina* during the winter is in part an adaptation to lower body temperature and, consequently, metabolic demand for O_2 (Brittain and Nagell 1981). Larvae of the cranefly *Tipula plutonis* have been shown to move to preferred temperatures, and these preferences vary with a true circadian rhythm (Kavaliers 1981). Significantly higher temperatures are selected during dark hours than during light hours, and, once established, the cyclic pattern persists even in constant illumination. Since these larvae presumably forage only at night, preference for environments with higher temperatures during the foraging period may be an adaptation to increase activity rates and, as a result, increase caloric intake. In contrast, preferences for lower temperature environments during nonforaging periods could reduce metabolic costs and help to conserve energy for the next feeding cycle (e.g. see Kavaliers 1981).

Since many aquatic insects face serious energetic constraints, the energy savings from a behavioral strategy like that of the *Tipula* larvae described above should prove valuable. The fact that behavioral thermoregulation has also been reported in both freshwater gastropods and decapods (Crawshaw 1974; Casterlin and Reynolds 1977; Kavaliers 1980) suggests that it may be a more widespread adaptation among aquatic insects than we currently realize. Among adult stages of aquatic insects, thermoregulation by submersion in water warmer than air, a practice sometimes referred to as "underwater basking" (Spence et al. 1980), is known in at least two orders. Some gerrid water striders commonly spend the winter underwater where higher temperatures result in faster ovarian development and earlier reproduction in the spring (Spence et al. 1980). The stonefly *Zapada cinctipes* has been reported to submerge itself in mountain streams to escape freezing air temperatures on frosty spring mornings (Tozer 1979). Behavioral thermoregulation is also important in the energetics of adult flight (see Corbet 1962; Heinrich and Casey 1978, May 1978).

FORAGING BEHAVIOR

Aquatic systems are remarkably heterogeneous in space and time. The food resources for almost all aquatic insects are patchily distributed (e.g. periphyton: Jones 1978, Tett et al. 1978; detritus: Egglishaw 1964, Rabeni and Minshall 1977; prey: Ulfstrand 1967, Townsend and Hildrew 1979), and as a result the efficient location and utilization of food is an important problem facing insects in aquatic environments. Optimal foraging theory (reviewed by Pyke et al. 1977; Krebs 1978; Cowie and Krebs 1979; see also Peckarsky, this volume, Chapter 8) provides a useful conceptual framework for investigating behavioral solutions to this problem because it attempts to predict the kinds of behavior necessary to maximize foraging efficiency. A substantial body of theory has developed that deals with strategies for exploiting patchily distributed food, much of which is applicable to the foraging behavior of aquatic insects. Although there are a number of distinct hypotheses, basically all involve sets of behavioral rules that allow an individual to maximize the time it spends in high quality patches and to minimize the time it spends in low quality patches (Krebs 1978). Two specific topics in optimal foraging theory seem particularly useful in discussing the foraging behavior of aquatic insects. They are "area-restricted searching" (Hassell and May 1974; Krebs 1979), and the use of rules for "giving up" when foraging in discrete patches (Charnov 1976; Krebs 1978).

Area-Restricted Searching

Hassell and May (1974) have argued that animals foraging in a patchy environment allocate their time in various patches according to patch quality. The result is that, as quantitative models of optimal time allocation predict (Royama 1970), efficient foragers tend to aggregate in high quality resource patches and spend very little time in low quality patches. Area-restricted searching, a simple behavioral strategy which will allow animals to achieve this distribution (Krebs 1978), refers to an immediate adjustment of a search path in response to resource density. This "adjustment" generally involves increasing rates of turning and/or decreasing travel velocity. Singularly or together, these responses to resource density result in the insect "loitering" in areas with high resource density. Area-restricted searching is a widely used behavior among both vertebrates and invertebrates (Hassell and May 1973; Smith 1974; Krebs 1979), and recent studies suggest that it may be commonly used by aquatic insects as a simple adaptation to foraging in a patchy environment.

Field observations and experiments indicated that the limnephilid caddisfly *Dicosmoecus gilvipes* efficiently exploited patchily distributed periphyton food resources (Hart and Resh 1980; Hart 1981). In experiments in which periphyton patchiness was manipulated, larvae allocated most of their search effort (>80%) to areas containing periphyton, and effectively avoided the previously grazed patches, by (1) tending to move in a straight line between patches (a strategy that minimizes patch revisitation); and (2) moving more slowly when in high quality patches. As a result, larvae spent significantly less time in low quality compared with high quality patches (Hart 1981).

Time-lapse films (Wiley and Kohler 1981) of *Baetis tricaudatus (vagans)* foraging on natural substrates suggest that in this species too, allocation of search effort is largely restricted to distinct resource (in this case, periphyton) patches (Fig. 5.1*A*). Subsequent observations on substrates in which periphyton patches and periphyton-free areas were systematically created (Fig. 5.1*B*), confirmed the marked response of this species to small-scale resource heterogeneity. Within patches, movement velocities were greatly reduced and larvae tended to exhibit

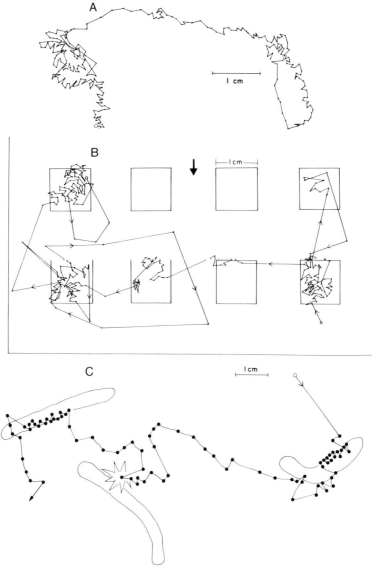

Figure 5.1 Some examples of area-restricted searching by baetid mayfly nymphs based upon time-lapse films. (*A*) Foraging movements of a *Baetis tricaudatus* nymph on natural substrate in Hunt Creek, Michigan, suggesting natural patchiness of periphytic food resources. Positions at 30-second intervals denoted by *dots*. (*B*) The pattern of *Baetis tricaudatus* foraging in an artificially produced patchy environment in Hunt Creek. Each square represents 1 cm² of periphyton at a density of approximately 300,000 cells/cm². The area surrounding the squares has been cleaned and is totally devoid of periphyton. The large *arrow* denotes direction of current, with lateral and bottom lines representing boundaries of the cement substrate; each position being recorded at 30-second intervals. (*C*) The mayfly *Paracloeodes* sp. foraging on patchily distributed midge tubes (Chironomidae) in the Vermilion River, Illinois. *Dots* represent position at 10-second intervals, with the star burst at the middle midge tube denoting an aggressive encounter between the foraging nymph and the resident midge larvae.

frequent sharp turns between moves; these are both characteristics of area-restricted search. Another mayfly, *Paracloeodes* sp., feeds on the periphyton growing on chironomid midge tubes in warm-water streams, and shows a similar area-restricted search strategy (Fig. 5.1C); that is, forward velocity decreases dramatically when nymphs encounter midge tubes, and turning angles become much more acute.

Area-restricted searching is a necessary condition for efficient foraging in patchy environments, but it is not sufficient to demonstrate that optimal foraging is occurring. When a group of individuals use area-restricted searching they will tend to aggregate in higher quality patches (the so-called aggregative response; Hassell and May 1974). However, just as area-restricted searching by an individual does not necessarily result in an optimal allocation of its time between patches of varying quality, the observation of aggregative responses at the population level does not provide conclusive evidence for optimal foraging by that population.

Rules for "Giving Up"

Once foraging in a patch, an individual insect may use some behavioral rule to decide how long to stay in that patch, and when to give up and move on. Two very different strategies are possible with respect to giving-up-time (GUT, i.e. the time between last prey capture and abandonment of a patch). Hassell and May (1974) have shown that a GUT that is constant (i.e. not dependent upon patch quality) will result in an aggregative response in much the same fashion as does area-restricted searching. The rule in this case is simple: abandon the patch X time units after the last food item is taken (or after the intake rate is below some constant threshold). As in area-restricted searching, there is no requirement for behavioral flexibility or any kind of long-term memory, but again—like area-restricted searching—it does not result in a truly optimal allocation of time within patches. The optimal solution to the problem of time allocation can only be achieved by allowing the GUT to vary with mean patch quality (Charnov 1976). In what has become known as the *Marginal Value Theorem*, Charnov (1976) demonstrated that an optimally foraging organism should forage in a patch just long enough to reduce the resource level of the patch to the average value for the habitat as a whole. In order to use this strategy, the organism must maintain information about its previous experiences in that habitat (i.e. memory) against which to evaluate its current rate of foraging success.

The GUT strategy of the polycentropodid caddisfly *Plectrocnemia conspersa* has been examined by Hildrew and Townsend (1980). This larva spins a net in slowly flowing waters to entangle drifting and wandering prey (Hildrew and Townsend 1977; Tachet 1977). Field studies indicated that larvae aggregate in areas of high prey density (Hildrew and Townsend 1977) as a result of (1) an increased probability of building a net at a site where prey are encountered frequently; and (2) an increased probability of remaining at a particular site where rates of prey capture are high. In subsequent laboratory experiments, Hildrew and Townsend (1980) found that the GUT at a particular site (i.e. habitat) was independent of prey-capture rate (i.e. habitat quality). Townsend and Hildrew (1980) then estimated the marginal capture rate for *P. conspersa* from the reciprocal of its GUT, and the average capture rate from average daily consumption rates in the field. Because the values were approximately equal, they concluded that *P. conspersa* foraged by using the simple constant GUT rule to achieve an efficient allocation of time to feeding sites of varying quality. This is inconsistent with Charnov's (1976) prediction that GUT for an optimally foraging predator should decrease with increasing mean patch quality. Thus, in this sense, *P. conspersa* did not forage in an optimal manner. Implicit in this interpretation is the suggestion that this particular insect does not utilize a memory of past experiences in making foraging decisions.

A somewhat different result was obtained by Cook and Cockrell (1978) working with the backswimmer *Notonecta glauca*. Because the ingestion rate of *N. glauca* feeding on individual mosquito larvae decreased over time, individual prey could be treated as distinct resource patches. Subsequent experiments demonstrated that increases in the average time between prey captures resulted in increases in the mean time spent feeding on individual mosquito larvae (implying increased GUT), which is in general agreement with the predictions of the Marginal Value Theorem (Charnov 1976; Krebs 1978). Similar results, also for notonectids feeding on mosquito larvae, were obtained by Sih (1980). In these cases, past experiences did modify current patch utilization, with the result that, in contrast to *Plectrocnemia* larvae, *Notonecta* foraging appears consistent with optimal foraging models.

TERRITORIALITY AND COMPETITION

Once a resource patch is located, the opportunity may exist for an individual to monopolize its use by actively defending the area against intrusion by potential competitors. Resource defense, leading to more or less exclusive use by a single individual, is termed territoriality, and is a common strategy used among birds and mammals (Wilson 1975; Wolf and Hainsworth 1977). Generally speaking, territoriality among insects is much less common (Wilson 1975; Baker 1983). Nevertheless, conditions under which many aquatic insects must forage favor territoriality from a theoretical point of view (Gill and Wolf 1978; Wolf and Hainsworth 1977; Meyers et al. 1981). These include (1) patchily distributed resources; (2) patch size small enough relative to body size to be controlled or defended without excessive traveling time; and (3) high resource depletion and renewal rates. Not surprisingly then, as the foraging behavior of aquatic insects receives more study, observations of resource defense and territoriality are becoming frequent.

Resource Theft

A rudimentary but common form of territoriality among filter-feeding predators is the defense of fixed feeding sites. We have already described the foraging nets of *Plectrocnemia* caddisfly larvae. As these predators aggregate in areas of high prey concentration, competition for a suitable net building site increases and larvae searching for a filtering site will often attempt to displace larvae that are already established (Hildrew and Townsend 1980). In such encounters, an intruding larva faces a resident larva at the mouth of the resident's net, and both larvae rear up and strike at each other with open mandibles. The contest ends when one larva retreats; either the intruder backs away or enters the drift, or the defender exits through the tube's rear entrance and leaves both the site and the net to the intruder.

A similar defense of filtering sites occurs among trichopteran larvae of the family Hydropsychidae. As with the polycentropodid *Plectrocnemia*, displaced larvae attempt to commandeer neighboring retreats in preference to expending their energy locating a suitable unoccupied site and constructing their own retreat. Larvae defending their retreat will physically attack intruders (Glass and Bovbjerg 1969; Schuhmacher 1970); the ensuing combat is occasionally fierce enough to result in mortal injury (Jansson and Vuoristo 1979). Furthermore, among the hydropsychids a new dimension to the contest occurs; that is, defending larvae produce auditory alarm signals by stridulating with their forelegs on ventrolateral striations of the head capsule (Johnstone 1964). Jansson and Vuoristo (1979) have demonstrated that stridulation increases the probability of successful defense against intruders. Defenders apparently can assess the relative size of the intruder from vibrations on the outside of the retreat and can respond accordingly. If the intruder is small relative to the defender, stridulation is less likely to occur, and the retreat is easily defended by physical attack alone. However, if the intruder appears larger, the defender is likely to stridulate in addition to physically attacking the intruder. In encounters where stridulation does not occur, large intruders have a significantly higher probability of displacing defenders than in similar contests in which defenders use stridulation. Stridulation, therefore, appears to be a signal that the defender is highly motivated, despite its size disadvantage. It is apparently advantageous for the intruder to heed the stridulation signal because the risk of injury in combat is substantial, and the intruder's probability of success, although greater than if the defender held a size advantage, is still relatively low (less than 13% according to Jansson and Vuoristo's calculations).

Invasion and defense of retreats is by no means limited to filter-feeding predators. Larvae of the chironomid midge *Cricotopus bicinctus* forage by grazing periphyton (LeSage and Harrison 1980), and they construct a central tube retreat from which they forage in a more or less elliptical territory surrounding the tube. These feeding territories are held until they are depleted, at which point the larva abandons the tube and seeks out a new site by commandeering a neighbor's tube, by renovating an abandoned tube, or—failing either of these options—by constructing a new tube in an unoccupied site. The feeding territories are used more or less exclusively by a single larva, in part because intruders are driven off if encountered, and in part because other larvae seem to avoid trespassing in neighboring territories. The mechanism involved in this avoidance is not known.

It is interesting to note that in all of these examples, and despite major differences in foraging mechanisms and trophic level, *resource theft* appears to be an important behavioral strategy. The commandeering of a neighbor's retreat is clearly preferable to spending time and energy (in terms of both mechanical work and silk production) in building a new retreat. Furthermore, if the commandeered retreat is occupied, the probability is high that the resource patch associated with the retreat is of high quality. Successful intruders thereby avoid the risk of building a new retreat or renovating an old one, only to find that the resource base at that location is depleted or insufficient.

The resolution of contests over retreat ownership also follows a typical pattern: defending larvae are more likely to be successful than intruding larvae, and larger larvae are more likely to succeed than smaller larvae. As a result, small intruders almost never displace large defenders, whereas large intruders have the greatest chance of securing a retreat from a small defender. We can imagine that there is some minimum probability of success that is required to make an attempt at retreat theft an adaptive behavioral strategy, and further, that this probability is likely to be inversely related to the potential energetic gain accrued by successfully stealing an occupied retreat. For example, a large hydropsychid larva has a rather slight chance of displacing a smaller defender. However, the behavior is apparently adaptive because, if successful, that larva can avoid constructing a rather elaborate and presumably energetically costly feeding retreat. On the other hand, *Cricotopus* larvae construct very simple, amorphous detrital tubes, and although the net energetic gain in commandeering such a tube is probably much less for these larvae, the probability of success is also much higher (60% for a large intruder attacking a small defender; Wiley, unpublished data).

Not all territory defense in aquatic insects is associated with a central tube or retreat. Macan (1973), using field data, has suggested that odonate nymphs may maintain and defend feeding territories that are important in preserving the balance between these predators and their prey populations. Recent behavioral experiments have supported his speculations (Crowley 1979; Baker 1980; Rowe 1980). Rowe (1980) found that nymphs of the damselfly *Xanthocnemis zealandica* in aquaria established territories on artificial "reed stems," which they defended from intruders by a combination of threat displays (consisting of flaring of the caudal lamellae and abdominal swinging), and direct physical assaults with the labium. Once again, defending larvae were much more likely to win in an aggressive interaction than were intruding larvae.

All resource theft does not involve territoriality. The true bug *Hydrometra championiana* has been reported to creep up behind feeding water striders (*Gerris* and *Velia*) and loot their prey. As inobtrusively as possible, *H. championiana* plunges its long beak into their captured prey, stealing sips from a dinner which its competitors have captured for themselves (Schaeffer 1978).

Interactions and Displacement

Territoriality is a behavioral adaptation to reduce exploitative competition from neighbors by maintaining exclusive use of a resource patch. When patches are not used exclusively, opportunities abound for physical confrontations and interactions among the various individuals and taxa. These interactions may be either *disruptive* (i.e. competitive in the broad sense) or *facilitative* in nature. In either case there is growing evidence that these behavioral interactions occur frequently enough to have substantial effects on the population dynamics of the species involved.

Filter-feeding black fly larvae commonly occur in feeding aggregations on the surfaces of rocks or other solid objects in areas with adequate current velocities. Biting or bumping, and local displacement of neighbors, are common occurrences in these aggregations (Gersabeck and Merritt 1979; Wiley and Kohler 1981) as larvae jockey for better filtering positions. Larger larvae routinely displace smaller larvae, which are then often forced to the downstream periphery of the aggregation, where flow is turbulent and filtering is less efficient. Individuals of other species, either grazing on periphyton or searching for filtering sites, also interact with black-fly larvae, often causing cascading chain reactions as each individual is locally displaced, and in turn displaces one or more weaker neighbors.

In situ studies using time-lapse cinematography indicate that the rate of these disruptive interactions can be surprisingly high. Interactions between simuliid larvae in a Michigan trout stream occurred at an average rate of 0.76 ± 0.18 (mean $\pm$ S.E.) interactions per larva-day of observation. If interactions between simuliids and other taxa are included, the observed interaction rate increased to 1.63 ± 0.71 (Wiley and Kohler 1981).

Disruptive interactions between individuals can involve a wide variety of taxa, and are not restricted to animals feeding at the same trophic level (Table 5.1). These interactions most certainly have energetic costs for individuals that are displaced, which occur both immediately (since feeding is interrupted) and over a longer period (since the insects must undertake the risky task of locating a new and adequate foraging site). These interactions are "competitive" in that even individuals that feed at different trophic levels are attempting to utilize the same foraging space. Individuals or populations that are consistently involved in interactions where they "lose" (i.e. are displaced), will likely suffer reduced reproductive capacity or growth, and eventually may become locally rare or even extinct.

Documented instances of competitive displacement among whole popula-

TABLE 5.1: **Types of Behavioral Interactions Observed In Situ in an Epilithic Stream Community[a] Using Time-lapse Cinematography[b]**

TAXA	TYPE OF INTERACTION	INTERACTIONS OBSERVED	
		Number	Percentage
Simuliidae vs. Simuliidae	Disruptive; competition for food and space	72	50
Glossosoma vs. Simuliidae	Disruptive; competition for food and space	19	13
Rhyacophila vs. Simuliidae	Predator-prey	19	13
Eukiefferiella vs. Simuliidae	Disruptive; competition for foraging space	14	10
Parapsyche vs. Simuliidae	Predator-prey	11	8
Brachycentrus vs. Simuliidae	Disruptive; competition for foraging space	4	3
Baetis vs. Simuliidae	Disruptive; competition for foraging space	3	2
Glossosoma vs. Rhyacophila	Predator-prey	1	<1
Baetis vs. Eukiefferiella	Disruptive; competition for food and space	1	<1
Glossosoma vs. Brachycentrus	Disruptive; competition for food and space	1	<1
TOTAL		145	100
Total interactions between competitors		114	79
Total interactions between predators and prey		31	21

[a]Hunt Creek, Montmorency County, MI
[b]Based on the analysis of 78 hours of film from September and October 1979 (Wiley and Kohler 1981). Notice that competitive interactions were more than three times as frequent as predator-prey interactions.

tions of aquatic insects are relatively rare, although this is in part due to the acknowledged difficulty of examining competitive processes in nature. Cantrell and McLachlan (1977) found evidence suggesting that *Chironomus* midge larvae displaced small tanytarsine midges during the colonization of an English reservoir. In the laboratory, *Chironomus* larvae also displaced tanytarsines, and caused them to emigrate from preferred to less preferred sediments.

REPRODUCTIVE BEHAVIOR

Although behavioral adaptations related to metabolic regulation and foraging are used on a daily basis by most aquatic insects, specialized reproductive behaviors come into play for only brief periods, usually near the end of the individual's life. Nevertheless, they are among the most important behavioral adaptations in an individual insect's repertoire since reproduction is, in a very real sense, the climax towards which regulatory and foraging behaviors have been striving.

Among aquatic insects, the distinction between foraging habitat and reproductive habitat is particularly strongly enforced. For example, although foraging and growth take place in or on the water, reproduction in many aquatic insects must occur in either aerial or terrestrial situations (exceptions being mainly in the aquatic families of the Coleoptera and Hemiptera). In all cases, sexually mature individuals face two principal problems for which behavioral solutions must be found: (1) they must physically locate potential mates; and (2) once potential mates are located they must individually compete with other suitors for copulation and subsequent fertilization.

Procuring a Mate

Aerial swarming is perhaps the most typical mating behavior of aquatic insects; it predominates in the archaic order Ephemeroptera, as well as among the Diptera and Trichoptera, and occurs sporadically among the odonates. Its adaptive function appears to be that of providing a staging ground for sexual rendezvous. Its essential feature is the "flight-station, an assembly point controlled by a landmark (referred to as a *swarm marker*) at which the short-range recognition and capture of the female can take place" (Downes 1969, p. 272). The mating swarm, regardless of its size, is generally a male-dominated gathering; females enter the swarm and are almost immediately recognized and mated, and then leave the swarm to oviposit (Downes 1960; Dahl 1965; Syrjämäki 1965, 1966). In a few species, separate female swarms have been reported, with females periodically leaving their own swarm to enter a nearby male swarm to copulate (Downes 1955, 1958). Bisexual swarms have been observed in the mosquito *Aedes* (McClelland 1959) and in some dance flies (Empididae) (Downes 1969).

As females enter the swarms they are approached and mated by waiting males. The males, having to discriminate between the females and other male swarm members, rely upon visual and/or acoustic cues. The eyes of the male in many Diptera and Ephemeroptera have specialized upper (or lower) ommatidia that improve resolution and tracking abilities, and assist in the visual recognition of erratically flying females that enter the swarm (Davies and Peterson 1956; Downes 1958, 1969). Among the culiciformes (e.g. Culicidae, Chironomidae), flight tone also appears to be used as a sexual cue, with the large plumose antennae of the males designed to pick out the slightly lower flight tone of the female (Downes 1969; Säwedal and Hall 1979).

Although swarming can be viewed as a behavioral mechanism to insure contact between potential mates, it also may function to isolate populations prior to copulation. Interspecific copulation attempts waste time, energy, and gametes. Since most female insects are reluctant to mate for some period after insemination (Parker 1970), mating with a member of the wrong species results, at best, in lower reproductive potential and, at worst, in reproductive failure. Swarm markers are usually species-specific, and apparently are selected on the basis of genetically determined preferences (Cameron 1922; Downes 1955, 1969; Peterson 1959; Lindeberg 1964; Solem 1978). The swarming sites of some species are so predictable that some empidid and ceratopogonid predators regularly seek out the typical swarm markers of their prey in an attempt to locate an easy meal (Frohne 1959; Downes 1960).

Mating Territories. Establishment of mating territories is common among the Odonata and some Diptera. Territoriality and swarming appear to be closely related behaviors. Downes (1969) suggests that a territory is a specialized derivative of the assembly station (i.e. a swarm marker defended by a solitary male). Pajunen (1980) argues conversely—that the swarming behavior of some insects may be the result of territorial breakdown at high densities. Decreases in intermale aggression and the formation of bands of nonterritorial males have been reported to occur at high densities in the damselfly *Calopteryx virgo* (Pajunen 1966a), and the dragonflies *Leuchorrhinia* (Pajunen 1966b) and *Sympetrum* (Uéda 1979). However, this type of swarming among odonates appears primarily related to intrasexual competition between males who aggregate at oviposition sites, whereas swarming in Ephemeroptera, Diptera, and Trichoptera is an adaptive behavior of both sexes to insure sexual rendezvous and copulation. Furthermore, in the latter orders, the swarming site is a neutral feature of the landscape and is not related to oviposition, whereas among odonates the swarm's location is not determined by a swarm marker with distinct characteristics, but by oviposition sites chosen by females.

Typically, mating territories among the Odonata are defended at prime oviposition sites. Males arrive early in the day and defend an area that will be attractive to the later-arriving females (Bick and Bick 1965; Pajunen 1966a, 1966b; Arai 1971; Heymer 1972; Pezalla 1979). Since oviposition sites are usually limited, only a certain portion of the male population is able to procure territories, presumably the more fit. The rest are relegated to a nonterritorial status, and must gain access to females either by intrusion or at the periphery of the breeding site. Males usually hold their territory for a single day or less, retreating at night to roosts further from the water. In some species, males come to the breeding site on just one day (e.g. *Libellula pulchella*; Pezalla 1979); in others, males may return two or more days in a row, occasionally, but not usually, occupying the same territory (Campanella 1972; Heymer 1972; Waage 1973; Campanella and Wolf 1974; Hassan 1978).

Territory defense involves both ritualized and actual aggression. Contests between *Libellula pulchella* males are often resolved by what Pezalla (1979) has called *circle flights,* an intense, aggressive interaction in which males make rapid circular flights at first around the territory in dispute, and—eventually—around the whole pond. During the flight the dominant male flies in a series of smaller circles around the subdominant male, so that it alternately matches and exceeds the speed of its rival. Dominance in this case is apparently decided by the relative agility and speed of the contestants. Physical attacks also occur but are much rarer, presumably because they risk permanent injury. Pezalla reported that physical assaults occurred in only 6 percent of the encounters between territorial males of *L. pulchella.* In encounters with nonterritorial males the frequency of assaults increased to 28 percent, presumably reflecting the persistence and high level of motivation of nonterritory holders.

The females' role in odonate territoriality is limited. In most species the female actively cooperates with territorial males by ovipositing in their territories and fleeing from competitors (which include neighboring territory holders or intruding nonterritorial males). The benefit derived from this cooperation appears to be that females are guaranteed access to males of proven fitness (Alcock 1979). Beyond this, females have little to gain from a system enforced by intrasexual competition between males, which is a topic we will return to shortly. They do, however, control the arena for this competition since male territories are chosen on the basis of their probable attractiveness as oviposition sites to females.

Intrasexual Signaling. The use of acoustic or other signals is another way in which potential mates are located by some aquatic insects. Many adult Plecoptera use their abdomen to "drum" species-specific mating calls that are then answered by receptive females (Rupprecht 1977; Ziegler and Stewart 1979; Szczytko and Stewart 1979). In some cases, the communication involves three steps with the male answering the female's return signal, although more often the interchange consists only of a male signal and a female answer. Females remain stationary during the interchange, while the male moves forward using her signal as a directional cue. After repeated rounds of drumming, the male locates the female and initiates copulation.

Corixid males (water boatmen) stridulate underwater both in confrontations with other males, and also in the presence of corixid females. If a female is ready to copulate, she answers with a distinct signal of her own, and the male will approach and grasp her. If the female is not receptive, she gives no answer but retreats quickly from the male (Jansson 1973, 1979). Among the Corixidae, acoustic signals are not used to locate the female, but visual contact and possibly chemical cues initiate stridulation by the male (Jansson 1979). The communication's primary functions appear to be to signal receptivity for copulation, and to serve as a mechanism for reproductive isolation. Each species (and each sex) has a distinctive stridulatory signal consisting of a set pattern in the number and timing of pulse trains (i.e. burst of pulses), but males and females only respond sexually to corixids that give the correct stridulatory "password."

Some water striders also use a signaling system to query receptivity and species identity in their reproductive encounters. Signals in this case, however, are transmitted by wave propagation on the surface of the water, and are inaudible to the human ear (Wilcox 1972, 1979).

Intrasexual Competition

Even though a male may have succeeded in copulating with a female, his paternity of her offspring is not ensured. This is because the eggs are often only fertilized at the time of oviposition, with the sperm being held in storage in the spermatheca and bursa copulatrix of the female. If the female mates again before oviposition, the sperm of the last male she mates with will have the greatest chance of fertilizing most of the eggs (Parker 1970; Alcock 1979). Recent work by Waage (1979), and Miller and Miller (1981) clarified just how great this "last male" advantage can be. In the zygopterans *Calopteryx maculata* and *Enallagma cyathigerum* (and probably many others as well), the greater part of copulation is involved in removing sperm from the bursa and spermatheca of the female. Only at the end of copulation is sperm transferred from the male to the female. In this way, competition from rival sperm is completely eliminated; and if the male can keep the female from mating again before she oviposits, his paternity is guaranteed.

The need to keep an inseminated female from mating with other males before oviposition has led to two distinct guarding strategies among territorial odonates. In some species, the male and female remain in tandem following copulation until the female has completed oviposition (Bick and Bick 1961, 1963, 1965; Sawchyn and Gillott 1974). The male, by engaging in "contact guarding," makes sure that no competing male can initiate copulation and displace his sperm. In other species, the male releases the female following copulation but then follows her closely, defending her from other males until she oviposits in what has become known as "noncontact guarding" (Alcock 1979). Which strategy a male uses apparently depends upon balancing the risk of sperm displacement by rival males against lost opportunities for his own multiple matings. Contact (tandem) guarding helps ensure the female will not mate or wander off before oviposition, but also keeps the male from further copulations during that time. When females are plentiful, noncontact guarding allows the male to defend his reproductive investment by watching over inseminated females, and at the same time gives him freedom to respond sexually to other females in his territory. Field studies of *Calopteryx maculata* by Alcock (1979) show that noncontact guarding males have achieved up to nine copulations in an afternoon. One male was observed to simultaneously guard five females ovipositing in his territory, all of which he had inseminated during a single 27-minute period.

Female giant water bugs of the subfamily Belostomatinae lay their eggs on the back of the males, who carry and brood the eggs (Voelker 1968; Smith 1976). Males not only spend substantial energy carrying and aerating the eggs, but expose themselves to increased risk of attack by predators (Smith 1979). This large reproductive investment by the male makes paternity assurance a matter of critical importance. Since polyandrous females, which includes most female insects, can store sperm from previous copulations, males may be tricked into investing energy in the care of eggs fertilized by their competitors. However, males of this subfamily have evolved a behavioral solution to this dilemma that may seem excessive, but certainly is effective. A male will not allow a female to oviposit on his back except immediately following copulation. Furthermore, the female is regularly interrupted during oviposition (often causing her to drop eggs from the ovipositor), and required to copulate again. The male enforces alternating bouts of copulation and oviposition, allowing an average of 1.4 eggs to be oviposited per copulation; thus, the entire mating process lasts for 24 to 48 hours (Smith 1979). Since females can carry upwards of 150 eggs at a time, males may demand and receive 100 or more copulations from a single female during this period.

DRIFT BEHAVIOR

From a historical perspective, perhaps no ecological topic has attracted as much attention to aquatic insect behavior as has the phenomenon of stream drift. The term *drift* refers broadly to the downstream transport by current of benthic animals normally living on or in bottom substrates. Early drift studies emphasized both biotic and abiotic influences on drift rates and their apparent diel periodicity (e.g. Müller 1954; Waters 1961). As the search for an "explanation" of the drift phenomenon continued, considerable controversy arose over the contribution of behavior to observed periodicities in drift density (see extensive reviews by Waters 1972 and Müller 1974). The terms "active drift" and "passive drift" came to epitomize two competing hypotheses about the causes and significance of drift, and much of the resulting discussion of these concepts has contained a strong element of semantic confusion. Waters (1965, 1972) classified drift into three categories: (1) catastrophic, (2) behavioral, and (3) constant. Despite mechanistic overtones, these terms only distinguish between the temporal properties that different types of drift exhibit (catastrophic = pulsed; behavioral = periodic; constant = continuous). Thus, the term "behavioral drift" refers to drift that exhibits a diel periodicity and may, or may not, be the result of *drift behavior*, which is a term we use here (following Müller 1974) to mean a behavioral adaptation involving voluntary entry into the water column.

At the heart of the *active* versus *passive* drift controversy lies our inability to generalize about the relative importance of *drift behavior* in accounting for observed periodicity. No one seriously contends that drift is either exclusively active or passive, but clearly it is a composite phenomenon that is the result of a myriad of independent, simultaneous, and to some extent unknown causes.

Modes of Drift Entry

Active entry into the water column by voluntary abandonment of the substrate appears to be a common behavioral response of aquatic insects to stress stimuli. Particularly in lotic environments, entry into the drift provides an immediate escape and an energetically inexpensive mode of local dispersal. For example, acute respiratory stress has repeatedly been shown to cause the active entry of mayfly nymphs (Hughes 1966; Minshall and Winger 1968; Wiley and Kohler 1980). The need to escape aggressive competitors or predators can also serve as a stimulus for drifting behavior. As noted earlier, *Plectrocnemia* caddisfly larvae, when losing ground in aggressive encounters with conspecifics, release their hold on the substrate and "push off" into the water column, rather than risk physical injury by continuing a conflict they are likely to lose (Hildrew and Townsend 1980). In time-lapse films of the predaceous caddisfly *Rhyacophila* foraging on black fly larvae, black flies released their hold on the substrate and drifted away at the first touch of *Rhyacophila*. "Drifting on contact" appears to be a very successful behavioral defense strategy against these slow-moving, tactile predators. In 2.6 hours of observed foraging, 9.5 black fly larvae escaped via drift for each larva that was successfully captured by *Rhyacophila* (Wiley and Kohler 1981).

Given sufficient current, any insect that loses its hold on the substrate will be immediately swept into the water column; we refer to such an event as "erosion." Therefore, passive entry is presumably the result of accidental or unintentional release, and subsequent erosion. Passive entry is difficult to quantify in the laboratory or field, primarily because it is difficult to assess the motivation behind an entry into the water column. In a particular instance, the absence of an observable "releaser" stimulus may suggest that an entry is accidental, but this can never be demonstrated. However, aquatic insects, like humans, must occasionally make mistakes, and given adequate current, these mistakes are likely to result in drift. Kovalak (1979) proposed a simple conceptual model of drift in which the probability of passive drift entry is the product of the probability of exposure to the current and the probability of erosion given this exposure. Thus, any behavior that increases either the probability of exposure to eroding currents, or the probability of erosion itself, can dramatically increase the chances (and for a population, the rate) of passive drift entry.

Ecological Antecedents to Drift

The specific mechanisms that give rise to drift cannot be divorced from the ecological contexts in which they occur. For example, the role of competitive displacement in producing observed drift densities has been a subject of considerable debate (e.g. Waters 1966; Dimond 1967; Bishop and Hynes 1969). Attempts at resolving such questions by unraveling the behavioral mechanisms that underlie drift entry require both controlled experiments and detailed behavioral observations.

Several recent papers have examined the influence of population density on drift rates. Walton et al. (1977) found that field densities of the stonefly *Acroneuria abnormis* varied with substrate particle size; densities were highest on stones, intermediate on cobble, and lowest on gravel. Laboratory drift rates varied inversely with apparent substrate preference (e.g. drift was the lowest from stones), and drift was density dependent (i.e. the proportion drifting increased with increasing density) on stones and cobble, but density independent on gravel. Based on these results, interference competition for interstitial spaces was judged to be the mechanism accounting for the observed density-dependent relationships.

Densities above which behavioral interactions result in a significant increase in the rate of drift entry can be estimated from this type of laboratory manipulation. Determination of the contribution of competitive displacements to the drift in a natural setting requires a comparison of laboratory-estimated threshold densities with observed densities in the field. For example, Wiley (1981) examined density-dependent emigration of two lotic chironomids from silt and sand substrates in a series of stream microcosm experiments. A comparison of threshold densities derived from these experiments, and extensive field density data, suggested that competitive interactions leading to displacement and drift occurred regularly in the sand communities of a Michigan trout stream.

Although mayflies often dominate stream drift, experiments have not shown density-dependent drift for several members of this order (Hildebrand 1974; Bohle 1978; Corkum 1978). Corkum (1978) found no indication of interactions between *Baetis tricaudatus (vagans)* individuals in still water, even at high densities; studies with in situ time-lapse films (Kohler, personal observation) support these observations. In general, the presence of competitors in laboratory experiments has not affected mayfly drift rates (Keller 1975; Corkum and Clifford 1980), even though total densities may have been greatly increased (Corkum and Clifford 1980).

The role of predation in drift has recently begun to receive attention. In several studies, drift rates of herbivore-detritivores increased in the presence of a stonefly predator (Corkum and Pointing 1979; Corkum and Clifford 1980; Walton 1980), and the timing of predator-influenced drift appeared largely to be a function of predator feeding rhythms (Corkum and Pointing 1979; Corkum and Clifford 1980). The intensity of response to predators varies between taxa, and between size groups within taxa. Drift of small *Baetis tricaudatus (vagans)* was not affected by the presence of the predaceous stonefly *Isogenoides elongatus*, although drift of mature nymphs was increased significantly by the presence of this predator (Corkum and Clifford 1980). Heptageniid mayflies appear less susceptible to predator-influenced drift than other taxa (Keller 1975; Walton 1978, 1980).

The effect of predators on the drift rates of prey must depend to a large extent on the ability of the prey to detect and evade approaching predators. Peckarsky (1980) found that mayfly taxa differed in their responses to contact encounters with stonefly predators. Activity of *Baetis* species and heptageniid taxa increased in response to predator contact, whereas *Ephemerella* responded with reduced locomotion. Heptageniids generally crawled away to evade predators, whereas *Baetis* usually entered the water column. Peckarsky (1980) suggested that such differential responses may contribute to differences in relative importance of these taxa in drift.

Wiley and Kohler (1981) filmed predator-prey interactions in an epilithic stream community. Results of the outcomes of these interactions were categorized as either no displacement, local displacement (i.e. displacement of several cm at most), or drift (i.e. emigration). Of the thirty interactions observed involving physical contact between black fly larvae and their predators, 37 percent resulted in black fly drift, but 63 percent resulted in only local displacement. The percentage of contacts with predators that resulted in black fly drift was substantially higher than that observed in contacts with either large nonpredators or other black flies (2.4% and 1.3%, respectively). These data suggest that black flies are able to distinguish physical contacts with predators from contacts with nonpredators, and behave accordingly.

Diel Periodicity

The reports of diel periodicities in drift by Tanaka (1960), Waters (1961), and Müller (1963) stimulated considerable interest. Many taxa drift, and most of these exhibit diel periodicities with nocturnal maxima (see Bournaud and Thibault 1973). Factors that serve as phase-setting agents in drift periodicities have been identified, and are thoroughly reviewed by Waters (1972) and Müller (1974). Mechanisms producing diel variation in propensity to drift, however, are poorly understood. At least two classes of mechanisms have been considered (i.e. those involving positioning changes, and those involving activity levels). Although these are not mutually exclusive, and in fact may act in concert, we will consider them separately.

Requirements for maintaining metabolic homeostasis can necessitate behaviors that may require changes in positioning relative to current. Elliott (1967) suggested that diel changes in positioning may be at least partly responsible for diel variation in propensity to enter the drift. Some mayflies and caddisflies move to more current-exposed surfaces of stones in response to decreased oxygen availability (Kovalak 1976; Wiley and Kohler 1980). Since oxygen levels typically reach late-night minima in streams, behavioral regulation of oxygen consumption via positioning changes may, in part, account for late-night drift peaks of some taxa.

The movement of insects to the top of stones to forage on periphyton at night has repeatedly been suggested as the mechanism responsible for diel changes in the positioning of negatively phototactic aquatic insects (e.g. see Elliott 1967; Bishop 1969; Waters 1972; Ploskey and Brown 1980). Although such changes undoubtedly occur, attempts to associate them with drift patterns have not been totally successful (Elliott 1967; Bailey 1981). Field studies indicate that changes in diel positioning are not nearly of the magnitude reported from laboratory studies (Kohler 1979). Also, taxa that dominate the drift (e.g. *Baetis, Pseudocloeon*) do not show marked diel positioning changes (Bohle 1978), and often occur largely on current-exposed surfaces throughout the day and night (Corkum et al. 1977). Clearly, in these instances, positioning behavior alone is insufficient to account for diel drift periodicities.

Positioning changes are often associated with increases in activity rates (Elliott 1967; Bailey 1981). Although increases in exposure may be sufficient to account for increases in drift probability (Elliott 1967; Kovalak 1979), activity alone may act to influence erosion probability by (1) affecting both intraspecific and interspecific encounter rates (Elliott 1967; Waters 1972); and (2) increasing the probability of an organism encountering erosion-inducing zones of increased current or turbulence (Kovalak 1979; Waters and Hokenstrom 1980). Diel activity patterns of aquatic insects are not well known, and they have often been erroneously inferred from drift patterns (e.g. Müller 1973; Fjellheim 1980). Although both activity and drift often exhibit nocturnal maxima (Elliott 1968, 1970; Müller 1974; Ploskey and Brown 1980), the relationship between the two is not always clear. For example, the mayfly *Atalophlebioides* exhibits both elevated drift and activity maxima at night, but drift rate peaks do not correspond with activity peaks (Bailey 1981). Bailey concluded that activity and positioning changes alone could not account for diel drift periodicities, and Elliott (1968) reached similar conclusions. In future studies, planned manipulations of positioning changes and activity, with simultaneous monitoring of drift, may be useful in separating the influences of these two factors on diel drift periodicities.

Allan (1978) has suggested that fish predation on drift may act as an effective selective mechanism influencing drift patterns. He observed that brook trout that fed on drifting *Baetis bicaudatus* during the day were strongly size-selective for larger nymphs. At night, this size selection was not as pronounced. He argued that since large nymphs should be subject to greater mortality risk by drifting during the day, natural selection should act to suppress day drift of large mayflies. In subsequent field studies, he found that large nymphs were more prone to drift at night than small nymphs, and concluded that the diel drift periodicity with a nocturnal maxima of large nymphs is an adaptation to avoid predation.

Progress in our attempts to understand the drift phenomenon depends largely on the further elaboration and development of testable hypotheses. The work of Allan (1978) and of Kovalak (1979) is particularly noteworthy in this respect since they both present concrete hypotheses about the mechanisms controlling drift entry. At the same time, it is important to remember that in nature drift is very likely to be a composite phenomenon, involving both active and passive components. These components undoubtedly vary in their relative importance, depending upon specific ecological conditions. No single hypothesis will suffice to explain the drift of all taxa at all times. In this regard, multiple hypotheses about the origins of drift are entirely appropriate.

FUTURE DIRECTIONS

Our knowledge of the behavior of aquatic insects is still in its infancy. For the vast majority of species we know only the broadest outlines of their behavioral strategies, and we are entirely ignorant of most of the relevant details. Even among those few aquatic insects that have received detailed study (most of which we have touched upon here), our information seldom extends beyond a single facet of their behavioral repertoire (e.g. foraging, reproduction). We conclude our discussion of behavioral adaptations by suggesting some important questions and research topics that we believe deserve increased attention in the future. In each of the topical areas that we have discussed, there remains much to be done.

1. *Regulatory behavior* clearly exists among aquatic insects and is easily demonstrated in the laboratory. However, we currently have little information of its relative importance in the field. Are positioning changes and thermotaxic migrations extraordinary events, or are they regular responses to diel fluctuations in temperature and dissolved oxygen? Do they disrupt foraging efficiencies and/or increase mortality rates? Are such behaviors ecologically relevant? There are energetic reasons to suspect that the importance of regulatory behavior may differ substantially between warm-water and cold-water environments. Similarly, because of differences in flow and thermal regimes, regulatory behavior may vary as a function of stream order, although to date there has been no attempt to examine regulatory behavior in a geographic context.

2. The *foraging behavior* of aquatic insects also has received little comparative treatment, either geographically or taxonomically. Comparative studies of foraging strategies would, in our view, prove extremely valuable in elucidating the key evolutionary pressures insects face while foraging in aquatic environments. Understanding the role of learning is an important future task. Work with terrestrial insects has demonstrated that learning from past experience can be a key factor in the formulation of individual foraging strategies (Heinrich

1979). Also critical to this whole topic is a better understanding of relative re-source levels, and the scale and magnitude of patchiness in aquatic habitats. For example, is food or foraging space often in short supply? In communities of aquatic insects, is resource competition the exception or the rule?

3. Although we know a great deal about *reproductive behavior*, our knowledge is derived largely from the intense study of a relatively small number of taxa. Furthermore, there are several areas about which we have almost no useful in-formation. Insects exhibit a wide variety of tactics during oviposition. Eggs may be laid singly or in masses, in riffles or pools, at the center or near the edge of a body of water. How are such strategies formulated, and what ecological factors affect their development? Sperm competition has been shown to be important among the odonates; how important is it among other groups? Do reproductive swarms have any real social structure, as recently suggested by Thornhill (1976), or is the classical view of a noninteracting aggregation correct?

4. Considerable progress in understanding the phenomenon of *behavioral drift* has been made since it was first described several decades ago. An understanding of the causes of drift and drift periodicity will require an integrated approach that recognizes that the catch taken in drift nets is the result of an array of simul-taneous and often independent events. Further work should emphasize the elucidation of the ecological mechanisms underlying these component events. A rigorous experimental approach, which has generally been lacking, will be essen-tial in this regard.

We should like to expand on this last point and stress that the need for a rigorous quantitative and experimental approach in the study of aquatic insect behavior transcends any particular topical area. Our understanding of the behavioral ecology of aquatic insects can progress only as rapidly as we can frame and unambiguously test appropriate hypotheses. While the study of any behavior necessarily requires a strong foundation of detailed observation, tests of behavioral and evolutionary hypotheses require carefully crafted experimental designs. Here, as elsewhere in science, experiment refines theory. The develop-ment of a coherent theory of behavior (evolutionary or otherwise) is essential to an integrated view of behavioral adaptations and, therefore, ultimately to a comprehensive understanding of the ecology of aquatic insects.

Acknowledgments

We are very grateful to Jana Waite for her patience and effort in the typing and editing of this chapter. We also would like to thank Beth Kohler, Sue Peratt, and Susan Wiley for cheerful technical assistance, and R. Gorden, W. Larimore, J. McAuliffe, B. Peckarsky, J. Spence, and J. Unzicker for helpful reviews and comments.

References

Alcock, J. 1979. Multiple mating in *Calopteryx maculata* (Odonata: Calopterygidae) and the advantage of noncontact guarding by males. Journal of Natural History 13: 439–46.

Allan, J. D. 1978. Trout predation and the size composition of stream drift. Limnology and Oceanography 23:1231–37.

Ambühl, H. 1959. Die Bedeutung der Strömung als ökologischer Faktor. Schweizerische Zeitschrift für Hydrologie 21:133–264.

Arai, Y. 1971. Reproductive behavior of *Orthetrum albistylum speciosum* and *O. triangulare melania* (Libellulidae). Tombo 17:13–17. (In Japanese with English summary).

Bailey, P. C. E. 1981. Diel activity patterns in nymphs of an Australian mayfly *Atalophlebioides* sp. (Ephemeroptera: Leptophlebiidae). Australian Journal of Marine and Freshwater Research 32.121–31.

Baker, R. L. 1980. Use of space in relation to feeding areas by zygopteran nymphs in captivity. Canadian Journal of Zoology 58:1060–65.

Baker, R. R. 1983. Insect territoriality. Annual Review of Entomology 28:65–89.

Bick, G. H. and J. C. Bick. 1961. An adult population of *Lestes disjunctus australis* Walker (Odonata: Lestidae). Southwestern Naturalist 6:111–37.

Bick, G. H. 1963. Behavior and population structure of the damselfly, *Enallagma civile* (Hagen) (Odonata: Coenagriidae). Southwestern Naturalist 8:57–84.

Bick, G. H. 1965. Demography and behavior of the damselfly, *Argia apicalis* (Say) (Odonata: Coenagriidae). Ecology 46:461–72.

Bishop, J. E. 1969. Light control of aquatic insect activity and drift. Ecology 50:371–80.

Bishop, J. E. and H. B. N. Hynes. 1969. Downstream drift of the invertebrate fauna in a stream ecosystem. Archiv für Hydrobiologie 66:56–90.

Bohle, H. W. 1978. Beziehungen zwischen dem Nahrungsangebot, der Drift und der raumlichen Verteilung bei Larven von *Baetis rhodani* (Pictet) (Ephemeroptera: Baetidae). Archiv für Hydrobiologie 84:500–25.

Bournaud, M. and M. Thibault. 1973. La dérive des organismes dans les eaux courantes. Annales d'Hydrobiologie 4:11–49.

Brett, J. R. 1956. Some principles in the thermal requirements of fishes. Quarterly Review of Biology 31:75–87.

Brittain, J. E. and B. Nagell. 1981. Overwintering at low oxygen concentrations in the mayfly *Leptophlebia vespertina*. Oikos 36:45–50.

Cameron, A. E. 1922. The morphology and biology of a Canadian cattle-infesting black fly, *Simulium simile* Mall. Canadian Department of Agriculture Entomology Bulletin 20. 26 pp.

Campanella, P. J. 1972. The evolution of mating systems in dragonflies (Odonata: Anisoptera). Unpublished Ph.D. thesis, Syracuse University, Syracuse, NY.

Campanella, P. J. and L. L. Wolf. 1974. Temporal leks as a mating system in a temperate zone dragonfly (Odonata: Anisoptera). I. *Plathemis lydia* (Drury). Behaviour 51: 49–87.

Cantrell, M. A. and A. J. McLachlan. 1977. Competition and chironomid distribution patterns in a newly flooded lake. Oikos 29:429–33.

Casterlin, M. E. and W. W. Reynolds. 1977. Behavioral fever in crayfish. Hydrobiologia 56:99–101.

Chapman, R. F. 1971. The insects: structure and function. Elsevier North Holland, Inc., New York, NY. 819 pp.

Charnov, E. L. 1976. Optimal foraging: the marginal value theorum. Theoretical Population Biology 9:129–36.

Cook, R. M. and B. J. Cockrell. 1978. Predator ingestion rate and its bearing on feeding time and the theory of optimal diets. Journal of Animal Ecology 47:529–47.

Corbet, P. S. 1962. A biology of dragonflies. H. F. and G. Witherby Ltd., London. 247 pp.

Corkum, L. D. 1978. The influence of density and behavioural type on the active entry of two mayfly species (Ephemeroptera) into the water column. Canadian Journal of Zoology 56:1201–6.

Corkum, L. D. and P. J. Pointing. 1979. Nymphal development of *Baetis vagans* McDunnough (Ephemeroptera: Baetidae) and drift habits of large nymphs. Canadian Journal of Zoology 57:2348–54.

Corkum, L. D., P. J. Pointing, and J. J. H. Ciborowski. 1977. The influence of current velocity and substrate on the distribution and drift of two species of mayflies (Ephemeroptera). Canadian Journal of Zoology 55:1970–77.

Corkum, L. D. and H. F. Clifford. 1980. The importance of species associations and substrate types to behavioural drift, pp. 331–42. *In*: J. F. Flannagan and K. E. Marshall (eds.). Advances in ephemeroptera biology. Plenum Press, New York, NY 434 pp.

Cowie, R. J. and J. R. Krebs. 1979. Optimal foraging in patchy environments, pp. 183–205. *In*: R. M. Anderson, B. D. Turner, and L. R. Taylor (eds.). Population dynamics. Blackwell Scientific Publications, Oxford. 434 pp.

Crawshaw, L. T. 1974. Temperature selection and activity in the crayfish *Orconectes immunis*. Journal of Comparative Physiology 95:315–22.

Crisp, D. J. 1964. Plastron respiration. Recent Progress in Surface Science 2:277–425.

Crowley, P. H. 1979. Behavior of zygopteran nymphs *Ischnura verticalis* in a simulated weed bed. Odonatologica (Utrecht) 8:91–102.

Dahl, C. 1965. Studies on swarming activity in Trichoceridae (Diptera) in Southern Sweden. Opuscula Entomologica Supplementum 27:1–68.

Davies, D. M. and B. V. Peterson. 1956. Observations on the mating, feeding, ovarian development, and oviposition of adult blackflies. Canadian Journal of Zoology 34: 615–55.

Dimond, J. B. 1967. Evidence that drift of stream benthos is density related. Ecology 48: 855–57.

Downes, J. A. 1955. Observations on the swarming flight and mating of *Culicoides* (Diptera: Ceratopogonidae). Transactions of the Royal Entomological Society of London 106:213–36.

Downes, J. A. 1958. Assembly and mating in the biting Nematocera. Proceedings of the Xth International Congress of Entomology 2:425–34.

Downes, J. A. 1960. Feeding and mating, and their interrelationship in the insectivorous Ceratopogoninae. Proceedings of the XIth International Congress of Entomology 1: 418.

Downes, J. A. 1969. The swarming and mating flight of Diptera. Annual Review of Entomology 14:271–98.

Eastham, L. E. S. 1937. The gill movements of nymphal *Ecdyonurus venosus* (Ephemeroptera) and the currents produced by them in water. Journal of Experimental Biology 14:219–28.

Egglishaw, H. J. 1964. The distributional relationship between the bottom fauna and plant detritus in streams. Journal of Animal Ecology 33:463–76.

Elliott, J. M. 1967. Invertebrate drift in a Dartmoor stream. Archiv für Hydrobiologie 63: 202–37.

Elliott, J. M. 1968. The daily activity patterns of mayfly nymphs (Ephemeroptera). Journal of Zoology (London) 155:201–21.

Elliott, J. M. 1970. The diel activity patterns of caddis larvae (Trichoptera). Journal of Zoology (London) 160:279–90.

Eriksen, C. H. 1963. Respiratory regulation in *Ephemera simulans* Walker and *Hexagenia limbata* (Serville) (Ephemeroptera). Journal of Experimental Biology 40:455–67.

Eriksen, C. H. 1966. Benthic invertebrates and some substrate-current-oxygen interrelationships. Pymatuning Laboratory of Ecology Special Publication 4:98–115.

Eriksen, C. H., V. H. Resh, S. S. Balling, and G. A. Lamberti. In press. Aquatic insect respiration. *In*: R. W. Merritt and K. W. Cummins (eds.). Introduction to the aquatic insects of North America. 2nd ed. Kendall/Hunt Publishing Co., Dubuque, IA.

Fjellheim, A. 1980. Differences in drifting of larval stages of *Rhyacophila nubila* (Trichoptera). Holarctic Ecology 3:99–103.

Frohne, W. C. 1959. Predation of dance flies (Diptera: Empididae) upon mosquitoes in Alaska, with especial reference to swarming. Mosquito News 19:7–11.

Gersabeck, E. F. Jr. and R. W. Merritt. 1979. The effect of physical factors on the colonization and relocation behavior of immature black flies (Diptera: Simuliidae). Environmental Entomology 8:34–39.

Gill, F. B. and L. L. Wolf. 1978. Nonrandom foraging by sunbirds in a patchy environment. Ecology 58:1284–96.

Glass, L. W. and R. V. Bovbjerg. 1969. Density and dispersion in laboratory populations of caddisfly larvae *(Cheumatopsyche,* Hydropsychidae). Ecology 50:1082–84.

Hart, D. D. 1981. Foraging and resource patchiness: field experiments with a grazing stream insect. Oikos 37:46–52.

Hart, D. D. and V. H. Resh. 1980. Movement patterns and foraging ecology of a stream caddisfly larva. Canadian Journal of Zoology 58:1174–85.

Hassan, A. T. 1978. Reproductive behavior of *Acisoma panorpoides inflatum* Selys (Anisoptera: Libellulidae). Odonatologica (Utrecht) 7:237–45.

Hassell, M. P. and R. M. May. 1973. Stability in insect host-parasite models. Journal of Animal Ecology 42:693–726.

Hassell, M. P. and R. M. May. 1974. Aggregation of predators and insect parasites and its effect on stability. Journal of Animal Ecology 43:567–94.

Heinrich, B. 1979. Bumblebee economics. Harvard University Press, Cambridge, MA. 245 pp.

Heinrich, B. and T. M. Casey. 1978. Heat transfer in dragonflies: 'fliers' and 'perchers.' Journal of Experimental Biology 74:17–36.

Heymer, A. 1972. Comportements socials et territorials des Calopterygidae (Odon. Zygoptera). Annales de la Societe Entomologique de France 8:3–53.

Hildebrand, S. G. 1974. The relation of drift to benthos density and food level in an artificial stream. Limnology and Oceanography 19:951-57.

Hildrew, A. G. and C. R. Townsend. 1977. The influence of substrate on the functional response of *Plectrocnemia conspersa* (Curtis) larvae (Trichoptera: Polycentropodidae). Oecologia (Berlin) 31:21-26.

Hildrew, A. G. and C. R. Townsend. 1980. Aggregation, interference and foraging by larvae of *Plectrocnemia conspersa* (Trichoptera: Polycentropodidae). Animal Behaviour 28:553-60.

Hughes, D. A. 1966. On the dorsal light response in a mayfly nymph. Animal Behaviour 14:13-16.

Jansson, A. 1973. Stridulation and its significance in the genus *Cenocorixa* (Hemiptera, Corixidae). Behaviour 46:1-36.

Jansson, A. 1979. Geographic variation in the stridulatory signals of *Arctocorisa carinata* (C. Sahlberg) (Heteroptera, Corixidae). Annales Zoologici Fennici 16:36-43.

Jansson, A. and T. Vuoristo. 1979. Significance of stridulation in larval Hydropsychidae (Trichoptera). Behaviour 71:167-86.

Johnstone, G. W. 1964. Stridulation by larval Hydropsychidae (Trichoptera). Proceedings of the Royal Entomological Society of London 39:146-150.

Jones, J. G. 1978. Spatial variation in epilithic algae in a stony stream (Wilfin Beck) with particular reference to *Cocconeis placentula*. Freshwater Biology 8:539-46.

Kavaliers, M. 1980. A circadian rhythm of behavioural thermoregulation in a freshwater gastropod *Helisoma trivolis*. Canadian Journal of Zoology 58:2152-55.

Kavaliers, M. 1981. Rhythmic thermoregulation in larval cranefly (Diptera: Tipulidae). Canadian Journal of Zoology 59:555-58.

Keller, A. 1975. Die Drift und ihre ökologische Bedeutung. Experimentelle Untersuchung an *Ecdyonurus venosus* (Fabr.) in einem Fliesswassermodell. Schweizerische Zeitschrift für Hydrologie 37:294-331.

Knight, A. W. and A. R. Gaufin. 1963. The effect of water flow, temperature, and oxygen concentration on the Plecoptera nymph, *Acroneuria pacifica* Banks. Proceedings of the Utah Academy of Science, Arts and Letters 40:175-84.

Kohler, S. L. 1979. Substrate positioning and drift of stream insects. Unpublished M.S. thesis, University of Michigan, Ann Arbor, MI.

Kovalak, W. P. 1976. Seasonal and diel changes in the positioning of *Glossosoma nigrior* Banks (Trichoptera: Glossosomatidae) on artificial substrates. Canadian Journal of Zoology 54:1585-94.

Kovalak, W. P. 1979. Day-night changes in stream benthos density in relation to current velocity. Archiv für Hydrobiologie 87:1-18.

Krebs, J. R. 1978. Optimal foraging: decision rules for predators, pp. 23-63. *In*: J. R. Krebs and N. B. Davies (eds.). Behavioral ecology. An evolutionary approach. Blackwell Scientific Publications, Oxford. 494 pp.

Krebs, J. R. 1979. Foraging strategies and their social significance, pp. 225-70. *In*: P. Marler and J. G. Vandenbergh (eds.). Handbook of behavioral neurobiology. Vol. 3, Social behavior and communication. Plenum Press, New York, NY. 411 pp.

LeSage, L. and A. D. Harrison. 1980. The biology of *Cricotopus* (Chironomidae: Orthocladiinae) in an algal enriched stream. 1. Normal biology. Archiv für Hydrobiologie Supplementband 57:375-418.

Lindeberg, B. 1964. The swarm of males as a unit for taxonomic recognition in the Chironomids (Diptera). Annales Zoologici Fennici 1:72–76.

McClelland, G. A. H. 1959. Observations on the mosquito, *Aedes (Stegomyia) aegypti* (L.) in East Africa. Bulletin of Entomological Research 50:227–35.

Macan, T. T. 1973. Ponds and lakes. George Allen and Unwin Ltd., London. 148 pp.

Mackay, R. J. and G. B. Wiggins. 1979. Ecological diversity in Trichoptera. Annual Review of Entomology 24:185–208.

May, M. L. 1978. Thermal adaptations of dragonflies. Odonatologica (Utrecht) 7:27–47.

Meyers, J. P., P. G. Conners, and F. A. Pitelka. 1981. Optimal territory size and the sonderling: compromise in a variable environment, pp. 135–58. *In*: A. C. Kamile and F. D. Sargent (eds.). Foraging behavior: ecology, ethological and psychological approaches. Garland STPM Press, New York, NY. 534 pp.

Miller, P. L. and C. A. Miller. 1981. Field observations on copulatory behavior in Zygoptera, with an examination of the structure and activity of the male genitalia. Odonatologica (Utrecht) 10.201–18.

Minshall, G. W. and P. V. Winger. 1968. The effect of reduction in stream flow on invertebrate drift. Ecology 49:580–82.

Müller, K. 1954. Investigations on the organic drift in North Sweden streams. Report Drottingholm Institute for Freshwater Research 35:133–48.

Müller, K. 1963. Tag-Nachtrhythmus von Baetidenlarven in der "Organischen Drift." Naturwissenschaften 50:161.

Müller, K. 1973. Circadian rhythms of locomotor activity in aquatic organisms in the subarctic summer. Aquilo Ser Zoologica 14:1–18.

Müller, K. 1974. Stream drift as a chronobiological phenomenon in running water ecosystems. Annual Review of Ecology and Systematics 5:309–23.

Nagell, B. 1977. Phototactic and thermotactic responses facilitating survival of *Cloeon dipterum* (Ephemeroptera) larvae under winter anoxia. Oikos 29:342–47.

Pajunen, V. I. 1966a. Aggressive behaviour and territoriality in a population of *Calopteryx virgo* L. (Odonata: Calopterygidae). Annales Zoologici Fennici 3:201–14.

Pajunen, V. I. 1966b. The influence of population density on the territorial behaviour of *Leucorrhinia rubicunda* L. (Odonata: Libellulidae). Annales Zoologici Fennici 3: 40–52.

Pajunen, V. I. 1980. A note on the connexion between swarming and territorial behaviour in insects. Annales Entomologici Fennici 46:53–55.

Parker, G. A. 1970. Sperm competition and its evolutionary consequences in the insects. Biological Reviews of the Cambridge Philosophical Society 45:525–67.

Peckarsky, B. L. 1980. Predator-prey interactions between stoneflies and mayflies: behavioral observations. Ecology 61:932–43.

Peterson, B. V. 1959. Observations on mating, feeding, and oviposition of some Utah species of black flies. The Canadian Entomologist 91:147–55.

Pezalla, V. M. 1979. Behavioral ecology of the dragonfly *Libellula pulchella* Drury (Odonata: Anisoptera). American Midland Naturalist 102:1–22.

Philipson, G. N. 1954. The effect of water flow and oxygen concentration on six species of caddis fly (Trichoptera) larvae. Proceedings of the Zoological Society, London 124: 547–64.

Philipson, G. N. 1978. The undulatory behaviour of larvae of *Hydropsyche pellucidula* Curtis and *Hydropsyche siltalai* Döhler, pp. 241–47. *In*: M. I. Crichton (ed.) Proceedings of the Second International Symposium on Trichoptera. Dr. W. Junk b.v. Publishers, The Hague, Netherlands. 359 pp.

Philipson, G. N. and B. H. S. Moorhouse. 1974. Observations on ventilatory and net-spinning activities of larvae of the genus *Hydropsyche* Pictet (Trichoptera, Hydropsychidae) under experimental conditions. Freshwater Biology 4:525–33.

Philipson, G. N. and B. H. S. Moorhouse. 1976. Respiratory behaviour of larvae of four species of the Family Polycentropodidae (Trichoptera). Freshwater Biology 6: 347–53.

Ploskey, G. R. and A. V. Brown. 1980. Downstream drift of the mayfly *Baetis flavistriga* as a passive phenomenon. American Midland Naturalist 104:405–9.

Pyke, G. H., H. R. Pulliam, and E. L. Charnov. 1977. Optimal foraging: a selective review of theory and tests. The Quarterly Review of Biology 52:137–54.

Rabeni, C. F. and G. W. Minshall. 1977. Factors affecting microdistribution of stream benthic insects. Oikos 29:33–43.

Ross, H. H. 1956. Evolution and classification of the mountain caddisflies. University of Illinois Press, Urbana, IL. 213 pp.

Royama, T. 1970. Factors governing the hunting behaviour and selection of food by the great tit (*Parus major* L.). Journal of Animal Ecology 39:619–68.

Rowe, R. J. 1980. Territorial behaviour of a larval dragonfly *Xanthocnemis zealandica* (McLachlan) (Zygoptera: Coenagrionidae). Odonatologica (Utrecht) 9:285–92.

Rupprecht, R. 1977. Nachweis von Trommelsignalen bei einem europäischen Vertreter der Steinfliegen—Familie Leuctridae (Plecoptera). Entomologica Germanica 3: 333–36.

Sawchyn, W. W. and C. Gillott. 1974. The life histories of three species of *Lestes* (Odonata: Zygoptera) in Saskatchewan. The Canadian Entomologist 106:1283–93.

Säwedal, L. and R. Hall. 1979. Flight tone as a taxonomic character in Chironomidae (Diptera). Entomologica Scandinavica Supplement 10:139–44.

Schaeffer, C. W. 1978. Hydrometrid-gerrid interactions: a further note (Hemiptera). Journal of the Kansas Entomological Society 51:286.

Scherer, E. 1965. Zur Methodik experimenteller Fliesswasser-Ökologie. Archiv für Hydrobiologie 61:242–48.

Schuhmacher, H. 1970. Untersuchungen zur Taxonomie, Biologie und Ökologie einiger Kocherfliegenarten der Gattung *Hydropsyche* Pict. (Insecta, Trichoptera). Internationale Revue der gesamten Hydrobiologie 55:511–57.

Sih, A. 1979. Stability and prey behavioural responses to predatory density. Journal of Animal Ecology 48:79–89.

Sih, A. 1980. Optimal foraging: partial consumption of prey. The American Naturalist 116:281–90.

Smith, J. N. M. 1974. The food searching behaviour of two European thrushes. I. Description and analysis of search paths. Behaviour 48:276–302.

Smith, R. L. 1976. Male brooding behavior of the water bug *Abedus herberti* (Heteroptera: Belostomatidae). Annals of the Entomological Society of America 69:740–47.

Smith, R. L. 1979. Paternity assurance and altered roles in the mating behaviour of a giant water bug, *Abedus herberti* (Heteroptera: Belostomatidae). Animal Behaviour 27:716–25.

Solem, J. O. 1976. Studies on the behavior of adults of *Phryganea bipunctata* and *Agrypnia obsoleta* (Trichoptera). Norwegian Journal of Entomology 23:23–28.

Solem, J. O. 1978. Swarming and habitat segregation in the Family Leptoceridae (Trichoptera). Norwegian Journal of Entomology 25:145–48.

Spence, J. R., D. H. Spence, and G. G. E. Scudder. 1980. Submergence behavior in *Gerris*: underwater basking. American Midland Naturalist 103:385–91.

Syrjämäki, J. 1965. Laboratory studies on the swarming behaviour of *Chironomus strenzkei* Fittkau in litt. Annales Zoologici Fennici 2:145–52.

Szczytko, S. W. and K. W. Stewart. 1979. Drumming behavior of four western nearctic *Isoperla* (Plecoptera) species. Annals of the Entomological Society of America 72: 781–86.

Tachet, H. 1977. Vibrations and predatory behaviour of *Plectrocnemia* larvae (Trichoptera). Zeitschrift für Tierpsychologie 45:61–74.

Tanaka, H. 1960. On the daily change of the drifting of benthic animals in stream, especially on the types of daily change observed in taxonomic groups of insects. Bulletin of the Freshwater Fisheries Research Laboratory, Tokyo 9:13–24. (In Japanese with English summary).

Tett, P., C. Gallegos, M. G. Kelly, G. M. Hornberger, and B. J. Cosby. 1978. Relationships among substrate, flow, and benthic microalgal pigment density in the Mechums River, Virginia. Limnology and Oceanography 23:785–97.

Thornhill, R. 1976. Sexual selection and parental investment in insects. The American Naturalist 110:153–63.

Thorpe, W. H. 1950. Plastron respiration in aquatic insects. Biological Reviews of the Cambridge Philosophical Society 25:344–90.

Townsend, C. R. and A. G. Hildrew. 1979. Resource partitioning by two freshwater invertebrate predators with contrasting foraging strategies. Journal of Animal Ecology 48:909–20.

Townsend, C. R. and A. G. Hildrew. 1980. Foraging in a patchy environment by a predatory net-spinning caddis larva: a test of optimal foraging theory. Oecologia (Berlin) 47:219–21.

Tozer, W. 1979. Underwater behavioural thermoregulation in the adult stonefly, *Zapada cinctipes*. Nature (London) 281:566–67.

Uéda, T. 1979. Plasticity of the reproductive behaviour in a dragonfly, *Sympetrum parvulum* Barteneff, with reference to the social relationship of males and the density of territories. Researches on Population Ecology (Kyoto) 21:135–52.

Ulfstrand, S. 1967. Microdistribution of benthic species (Ephemeroptera, Plecoptera, Trichoptera, Diptera: Simuliidae) in Lapland streams. Oikos 18:293–310.

Voelker, J. 1968. Untersuchungen zu Ernährung, Fortpflanzungsbiologies und Entwicklung von *Limnogeton fieberi* Mayr (Belostomatidae, Hemiptera) als Beitrag zur Kenntnis von naturlichen Feiden tropischer Susswasserschnecken. Entomologische Mitteilungen aus dem Zoologischen Staatsinstitut und Zoologischen Museum Hamburg 3:1–24.

Waage, J. K. 1973. Reproductive behavior and its relation to territoriality in *Calopteryx maculata* (Beauvois) (Odonata: Calopterygidae). Behaviour 47:240–56.

Waage, J. K. 1979. Dual function of the damselfly penis: sperm removal and transfer. Science 203:916–18.

Walshe, B. M. 1950. The function of haemoglobin in *Chironomus plumosus* under natural conditions. Journal of Experimental Biology 27:73–95.

Walton, O. E., Jr. 1978. Substrate attachment by drifting aquatic insect larvae. Ecology 59:1023–30.

Walton, O. E., Jr. 1980. Invertebrate drift from predator-prey associations. Ecology 61: 1486–97.

Walton, O. E., Jr., S. R. Reice, and R. W. Andrews. 1977. The effects of density, sediment particle size and velocity on drift of *Acroneuria abnormis* (Plecoptera). Oikos 28: 291–98.

Waters, T. F. 1961. Standing crop and drift of stream bottom organisms. Ecology 42: 532–37.

Waters, T. F. 1965. Interpretation of invertebrate drift in streams. Ecology 46:327–34.

Waters, T. F. 1966. Production rate, population density and drift of a stream invertebrate. Ecology 47:595–604.

Waters, T. F. 1972. The drift of stream insects. Annual Review of Entomology 17:253–72.

Waters, T. F. and J. C. Hokenstrom. 1980. Annual production and drift of the stream amphipod *Gammarus pseudolimnaeus* in Valley Creek, Minnesota, USA. Limnology and Oceanography 25:700–10.

Wiggins, G. B. 1977. Larvae of the North American caddisfly genera (Trichoptera). University of Toronto Press, Toronto, Canada. 401 pp.

Wilcox, R. S. 1972. Communication by surface waves. Mating behavior of a water strider (Gerridae). Journal of Comparative Physiology 80:255–66.

Wilcox, R. S. 1979. Sex discrimination in *Gerris remigis*: role of a surface wave signal. Science 206:1325–27.

Wiley, M. J. 1981. Interacting influences of density and preference on the emigration rates of some lotic chironomid larvae (Diptera: Chironomidae). Ecology 62:426–38.

Wiley, M. J. and S. L. Kohler. 1980. Positioning changes of mayfly nymphs due to behavioral regulation of oxygen consumption. Canadian Journal of Zoology 58: 618–22.

Wiley, M. J. and S. L. Kohler. 1981. An assessment of biological interactions in an epilithic stream community using time-lapse cinematography. Hydrobiologia 78: 183–88.

Wilson, E. O. 1975. Sociobiology: the new synthesis. Belknap Press of Harvard University Press, Cambridge, MA. 697 pp.

Wolf, L. L. and F. R. Hainsworth. 1977. Temporal patterning of feeding by hummingbirds. Animal Behaviour 25:976–89.

Ziegler, D. D. and K. W. Stewart. 1977. Drumming behavior of eleven Nearctic stonefly (Plecoptera) species. Annals of the Entomological Society of America 70:495–505.

chapter 6

THE ROLE OF AQUATIC INSECTS IN THE PROCESSING AND CYCLING OF NUTRIENTS

Richard W. Merritt

Kenneth W. Cummins

Thomas M. Burton

INTRODUCTION

Nutrient processing in aquatic systems involves complex physical and biological interactions. In this chapter, nutrient processing is defined as the factors and activities that influence the cycling of material within and between organic and inorganic phases. In the past, emphasis on nutrient processing in aquatic ecosystems has focused on carbon, phosphorus, and nitrogen, largely because carbonaceous material is the end product of photosynthesis, and phosphorus and nitrogen are the elements most likely to limit its production (e.g. see Hutchinson 1957; Odum 1971; Wetzel 1975).

Kitchell et al. (1979) reviewed the ways that aquatic and terrestrial consumers influence nutrient cycling through physical-chemical processes not directly reflected in energy flow. They suggested that consumers influence nutrient cycling in three ways: (1) by transforming nutrients from one part of an ecosystem to another; (2) by transforming nutrients through mineralization, uptake of inorganic nutrients with conversion to organic material, changes in surface-to-volume ratio, etc.; and (3) by storing nutrients in consumer standing crop. The role of aquatic insects in the processing of nutrients occurs mainly through (1) their feeding activities, which change surface-to-volume relationships (Kitchell et al. 1979; Zimmerman and Wissing 1980); (2) bioturbation, which consists of activities that move nutrients from sediments to overlying water (Gallepp et al. 1978; Nalepa et al. 1980); (3) the uptake, excretion, and organic mineralization of phosphorus (Nalepa et al. 1980); and (4) filtration of material from the water column (Wallace et al. 1977; Wallace and Merritt 1980).

Most studies of aquatic insects and their role in nutrient cycling have dealt with either changes in the distribution and/or abundance of aquatic insects in response to the amount of a particular food resource present, which results in correlations between food quantity (and quality) and animal abundance (for review, see Hynes 1970a, 1970b), or the rates of organic matter processing by various aquatic insects in a variety of habitats (for review, see Petersen and

134

Cummins 1974). Few studies, however, have actually assessed the role of aquatic insects in nutrient processing, storage, or cycling (for a general review of consumers' role, see Kitchell et al. 1979). Thus, the major objectives for the remainder of this chapter are (1) to compare and contrast the nutrient-processing roles of aquatic insects in streams and lakes; and (2) to evaluate the qualitative and quantitative roles these insects play in nutrient storage, processing, and cycling.

INFLUENCE OF STREAM ORDER AND LAKE TROPHIC STATUS ON NUTRIENT CYCLING

The "River Continuum Concept," a holistic view of a river system (Vannote et al. 1980), describes the structure and function of lotic communities from headwaters to mouth (Fig. 6.1). According to this concept, the distribution of stream invertebrates reflects the shift in location and types of food resources available in different-sized streams. This shift is summarized in Table 6.1, which also compares different stream types with other aquatic habitats. The morphological and behavioral feeding adaptations of stream invertebrates reflect this shift as well. Thus, large particle "shredders" (see Table 6.2 for definitions of these functional groups) are common in headwater streams where most energy is derived from coarse particulate organic matter (CPOM, i.e. >1 mm; e.g. leaves, twigs, wood) that enters the stream from the adjacent watershed, but becomes rare in larger (i.e. higher order) streams since organic matter is reduced to fine particle size, and periphyton becomes more important as a food source (Table 6.1). These biological processes, combined with physical abrasion, reduce the particle size of organic matter as it goes through the transport and storage process. Therefore, particle size decreases as stream order increases. The accompanying shifts in stream community composition result in progressively more efficient utilization of smaller particles, although tributary effects can modify this pattern (Minshall et al. 1983).

According to the continuum concept, downstream communities in higher order streams depend on the inefficiency, or "leakage," of food resources from the preceding orders (Cummins 1980a), with many of these resources derived from upstream processing of CPOM. However, New Zealand streams, and perhaps others, lack large particle feeders because large organic particles are not efficiently retained due to high gradients, unpredictable flooding, and other factors (Winterbourn et al. 1981). Therefore, the unpredictable recruitment of fine particulate organic material (i.e. particles $0.5~\mu m < FPOM < 1$ mm) from sources such as windblow and bank erosion may be the major source of energy for stream invertebrates. If true, Winterbourn et al. (1981) suggest that instream processing of coarse particulate organic matter may be less important than is suggested by the river continuum concept.

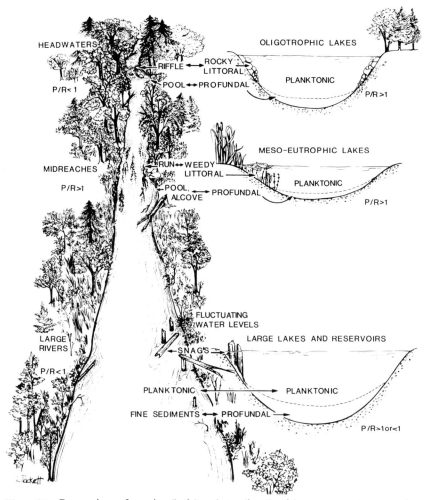

Figure 6.1 Comparison of running (lotic) and standing (lentic) water ecosystems, emphasizing the similarities in organic nutritional resource categories and habitats
used by aquatic insects in the two types of systems.

Regardless of the source of material, many open, flowing-water ecosystems
use nutrients through a storage-cycle-release pattern. This pattern is the basis for
the "Nutrient Spiralling Concept" (Webster 1975; Newbold et al. 1981, 1982;
Elwood et al. 1983). The cascading of nutrients in the particulate and dissolved
phase, and the subsequent, repeated recycling through the particulate phase as
nutrients "spiral" downstream, is a fundamental feature of lotic ecosystems that is
tied to the flow regime and the physical and chemical retention features of a given
stream reach.

TABLE 6.1: Resource Utilization by Insects in Several Aquatic Habitats[1]

		NUTRIENT RESOURCES			
HABITATS[2]	Algae	Vascular Plants	FPOM[3]	Leaf Litter (CPOM)[4]	Wood
Headwater streams (orders 1–3) (100–250 species)	S; some scraper species: Ephemeroptera Plecoptera Trichoptera Coleoptera Diptera	A (except mosses and liverworts); possibly a few species using mosses: Ephemeroptera Trichoptera	C; many collector species: Ephemeroptera Trichoptera Diptera	C; many shredder species: Trichoptera Coleoptera Diptera	C; few shredder species: Trichoptera Coleoptera Diptera
Oligotrophic lakes (25–100 species)	C (primarily littoral); moderate numbers of species, many are stream riffle analogs: Ephemeroptera Trichoptera Diptera	A (or S); few if any species	S (primarily in profundal zone); few species: Ephemeroptera Coleoptera Diptera	S (or A) (seasonally); few shredder species: Trichoptera Diptera	S (or A); no known species
Midreach rivers (orders 3–6) (200–500 species)	C; many species: Ephemeroptera Trichoptera Diptera	C (eutrophic lake littoral zone analogs); few species: Trichoptera Lepidoptera Coleoptera Diptera	C (especially in transport); many collector species: Ephemeroptera Trichoptera Coleoptera Diptera	S; a few shredder species in protected areas seasonally or localized at entrance of low order streams: Trichoptera Diptera	S (or C), distributions very clumped; few species: Diptera

TABLE 6.1: Concluded

HABITATS[2]	Algae	Vascular Plants	FPOM[3]	Leaf Litter (CPOM)[4]	Wood
			NUTRIENT RESOURCES		
Eutrophic lakes (10–50 species)	S (epiphytes on plants in littoral zone); few species: Ephemeroptera Diptera	C (littoral zone); some live plant tissue shredders: Trichoptera Lepidoptera Coleoptera Diptera	C; few gathering collector species some at high density: Ephemeroptera Diptera	S (littoral zone); few shredder species: Trichoptera Diptera	S (or A); no known species?
Large rivers (orders >6) (10–50 species)	S; very few species: Ephemeroptera Diptera	S (or A); few if any species: Diptera	C (especially in transport) (mostly Annelida, Mollusca); few species of insects at high densities: Ephemeroptera Trichoptera Coleoptera Diptera	A (or S in protected areas); shredders rare or absent as in midreach rivers: Trichoptera Diptera	S (or A), very clumped distribution; few if any species: Ephemeroptera (Povilla burrows in wood) Diptera
Temporary ponds (and streams) (50–75 species)	S (attached to plants, etc.); few species: Ephemeroptera Diptera	S (or C seasonally); few species (eutrophic lake analogs): Coleoptera	C; few species (may be limited by anaerobic conditions or moisture at times):	S (or C, depending on presence of riparian vegetation); some shredder species:	S (or C in some headwater intermittent streams); few species: Diptera

		Diptera	Ephemeroptera Diptera	Plecoptera Trichoptera Diptera	
Terrestrial (400–600 species woodland; 1600–1800 species old field)	S (in lichens, damp locations); few species: Coleoptera Diptera	C; many species have adapted to feed on specific plants or parts of plants: Orthoptera Hemiptera Lepidoptera Hymenoptera Coleoptera Diptera	C; few gathering collector species (mostly Annelida, Isopoda, Acarina): Collembola Diptera	C (seasonally); few shredder species (mostly Annelida, Isopoda, Mollusca): Diptera	C; many species, including those with social behavior: Thysanura Dictyoptera Psocoptera Isoptera Hymenoptera Coleoptera Diptera
Marine (beach and intertidal) (1–20 species)	C; similar to FPOM: Diptera	A (except certain marsh plants); very few species: Orthoptera Homoptera Diptera	C; very few collector species (mostly Annelida, Crustacea, and Mollusca): Diptera	A (except related to vascular aquatic plant marshes and river inputs); no known species	C (in estuaries, some beach and inshore areas); few species (mostly Annelida and Mollusca): Diptera

[1] The relative dominance of each resource type for a given habitat is shown at the beginning of each entry (C = Common, S = Sparse, A = Absent). Below each habitat category, a hypothetical range of insect species richness is given parenthetically.

[2] Ordering of habitats follows sequence in Figure 6.1.

[3] FPOM: Fine Particulate Organic Matter ($0.5\ \mu m < FPOM < 1\ mm$)

[4] CPOM: Coarse Particulate Organic Matter ($> 1\ mm$)

Data for this table were taken primarily from the following sources: Hynes 1961; Evans and Murdoch 1968; Janzen and Schoener 1968; Coffman et al. 1971; Hilsenhoff et al. 1972; Coffman 1973; Janzen 1973; Wetzel and Rich 1973; Cheng 1976; Teskey 1976; Williams and Hynes 1976; Merritt and Cummins 1978; Merritt and Lawson 1979; Wiggins et al. 1980; McAlpine et al. 1981; Williams 1983.

TABLE 6.2: General Classification System for Aquatic Insect Trophic Relations*

FUNCTIONAL GROUP (GENERAL CATEGORY BASED ON FEEDING MECHANISM)	SUBDIVISION OF FUNCTIONAL GROUP		GENERAL PARTICLE SIZE RANGE OF FOOD (MICRONS)
	Dominant Food	Feeding Mechanism	
Shredders	Living vascular hydrophyte plant tissue	Herbivores—chewers and miners	
	Decomposing vascular plant tissue and wood—coarse particulate organic matter (CPOM)	Detritivores—chewers, wood borers, and gougers	$>10^3$
Collectors	Decomposing fine particulate organic matter (FPOM)	Detritivores—filterers or suspension feeders	
		Detritivores—gatherers or deposit (sediment) feeders (includes surface film feeders)	$<10^3$

Scrapers	Periphyton—attached algae and associated material	Herbivores—grazing scrapers of mineral and organic surfaces	$<10^3$
	Living vascular hydrophyte cell and tissue fluids or filamentous (macroscopic) algal cell fluids	Herbivores—pierce tissues or cells and suck fluids	$>10^2$–10^3
Piercers	Living animal tissue	Carnivores—attack prey and pierce tissues and cells and suck fluids	$>10^3$
Engulfers (Predators)	Living animal tissue	Carnivores—whole animals (or parts)	$>10^3$

*Modified from Merritt and Cummins, 1978

Lakes are also categorized according to a "continuum," but this pattern represents a chronological sequence associated with aging or nutrient loading (Wetzel 1975). A lake continuum proceeds from oligotrophy to mesotrophy to eutrophy (Fig. 6.1). Climatic conditions, morphological characteristics, nutrient loading, and lake age influence the transition from one lake type to another. An increase or a reduction in nutrient inputs can either speed up or slow down the transition from one trophic state to the next or, in some cases, reverse the trophic classification of lentic ecosystems (Wetzel 1975).

The trophic status of a lake will drastically affect the diversity, distribution, and abundance of its insect fauna (Table 6.1; also see Wiederholm, Chapter 17). Thus, the influence of insects on nutrient cycling in lakes is closely associated with trophic status. However, insects in streams are influenced by allochthonous inputs of energy associated with stream width, discharge, gradient, type and density of riparian vegetation, and other factors. These factors change in a predictable sequence as stream order increases, and the aquatic insect fauna generally reflect changes in these factors (Table 6.1). The manner in which vegetation and morphological characteristics affect the trophic status of lakes and streams is summarized in Figure 6.1.

SOURCES OF NUTRIENTS IN LAKE AND STREAM ECOSYSTEMS

Organic Matter Inputs

Organic matter in lentic and lotic ecosystems may come from auto-chthonous (produced within the ecosystem) or allochthonous (transported into the system from elsewhere) sources. In *lakes*, the primary source of particulate organic matter (POM) is autochthonous organic matter production (e.g. see Wetzel 1975; Goldman and Kimmel 1978; Loucks and Odum 1978; Odum and Prentki 1978), with secondary inputs of allochthonous material from shoreline vegetation and fluvial inputs. Although littoral production by vascular hydro-phytes can be important in some lakes, phytoplankton production is often the primary source of autochthonous organic matter (Wetzel 1975; Goldman and Kimmel 1978). While some phytoplankton is cycled through zooplankton or other herbivore populations, and then through secondary and tertiary con-sumers, before entering the detritivore food chains, other phytoplankton-produced carbon directly enters the detritus pathway by natural cell mortality (e.g. Moore 1980).

Dissolved organic matter (DOM, i.e. particles < 0.5 μm) produced by "leakage" of living phytoplankton cells, from lysis of dead cells, or from microbial activities, may be taken up directly by microbial cells and provide a significant nutrient resource for aquatic insects (e.g. Wetzel and Rich 1973; Dahm 1981).

Nonbiological formation of FPOM may also occur by physical-chemical flocculation of dissolved organic compounds (Wetzel 1975). These particles then become a nutrient resource for filter-feeding zooplankton, aquatic insects, and other organisms.

In contrast to the dominant pattern of autochthonous planktonic production of organic matter in lentic systems, small-sized to medium-sized, forested, lotic ecosystems (stream orders 1–3) receive large inputs of allochthonous matter from the surrounding watershed (Hynes 1963, 1970a, 1975; Ross 1963; Minshall 1967; Fisher and Likens 1973; Cummins 1974, 1977). Up to 99 percent of the annual energy requirements for primary consumers in woodland streams can come from forest organic matter, with autumnal leaf fall comprising most of these inputs (Fisher and Likens 1973).

Events in the riparian ecosystem largely determine the quantity, quality, timing, and retention of allochthonous organic material received by streams (Nelson 1969; Webster and Patten 1979; Sedell et al. 1974, 1978; Hynes 1975; Cummins et al. 1980b). Following the introduction of coarse particulate organic matter into streams, the soluble DOM is leached from the CPOM, and microorganisms begin to colonize CPOM surfaces (Cummins 1974). The pool of FPOM is derived from (1) feeding activities of macroinvertebrates ("shredders" of leaf material, "gougers" of wood, etc.) on CPOM and subsequent release of feces; (2) abrasion of CPOM by sediments in transport; (3) scouring of algal cells from substrate surfaces; (4) uptake of DOM by microbes; and (5) flocculation of DOM by physical-chemical processes (Lush and Hynes 1978; Anderson and Sedell 1979; Cummins and Klug 1979; Wallace and Merritt 1980; Dahm 1981). FPOM can also enter the stream from adjacent terrestrial areas through windblow, surface runoff, bank erosion (Winterbourn et al. 1981), or from rain dripping through the forest canopy, which leaches particles into the stream (throughfall; Fisher and Likens 1973). If present, vascular hydrophytes tend to enter the processing cycle through the detrital food chain rather than through the feeding activities of stream herbivore-shredders (Welton 1980).

In addition to leaching from CPOM, DOM may enter the stream through other pathways, such as groundwater seepage, throughfall from the canopy, and surface runoff. DOM is a continual input of energy and nutrients to the stream throughout the year, but this is only available to most stream insects after microbial uptake or flocculation to FPOM.

The floodplain along flowing waters contains sediments in quasi-equilibrium. Erosion during flooding and runoff balances sediment deposition during flood recession, with additional inputs of organic matter to the floodplain resulting from litter fall and lateral movement (Moon 1939). Leaf litter on the floodplain is processed in a manner similar to that occurring in a stream. A period of physical-chemical and microbial conditioning is required before invertebrate feeding begins, and microorganisms and soil invertebrates combine to process the material (Merritt and Lawson 1979). While most allochthonous material enters deciduous woodland streams as leaf fall in autumn (e.g. Kaushik and Hynes 1968; Sedell et al. 1974; Mahan 1980), leaves falling onto the floodplain may not actually enter the stream until major flooding occurs in the spring. This often coincides with the highest rate of litter processing by terrestrial invertebrates (Merritt and Lawson 1979). Thus, spring flooding may serve as a mechanism to transport relatively high quality organic material (i.e. either conditioned CPOM or recently generated FPOM) into the stream ecosystem when the energy reserves from the stream-processed litter of autumn are significantly diminished. Thus, the floodplain may serve as a "storage" or preprocessing area for allochthonous material before its ultimate input into streams (Merritt and Lawson 1979; Cummins et al. 1983).

Inorganic Inputs

Sources of nitrogen, phosphorus, and other inorganic nutrients in lentic ecosystems include direct precipitation, streamflow and groundwater inputs, litter from shoreline vegetation and, for nitrogen, fixation by blue-green algae or microbial populations (Likens and Loucks 1978). Precipitation can be especially important for nitrogen. For example, 26 percent of the total input of nitrogen to Findley Lake, Washington, and 54 percent of total nitrogen input to Mirror Lake, New Hampshire, was derived from precipitation (Likens and Loucks 1978). Lake Jackson, Florida, received 66 percent of its nitrogen from direct rainfall (Burton et al. 1978). Precipitation is less important for phosphorus (i.e. 3–10% of total input: Burton et al. 1978; Likens and Loucks 1978). The importance of direct precipitation as an input source decreases as the size of the watershed area draining into the lake increases, and/or as anthropogenic sources increase.

Direct precipitation inputs into streams are rarely important since the open area of stream is generally small compared to the watershed area. Thus, major inputs in streams come from groundwater seepage, overland flow, and leaching and breakdown of the allochthonous organic matter that has been transported into the stream. Minor inputs of nutrients are also derived from throughfall.

Inputs of inorganic nutrients to lakes and streams are highly influenced by land use within the watershed (see reviews by Beaulac 1980; Omernik 1976, 1977; Stewart et al. 1975, 1976). The highest inputs coincide with intensive agricultural practices, such as row cropping and orchards, and with urban land uses; the lowest nutrient inputs are from undisturbed forest or grassland. Less intensive agricultural land uses that involve minimum tillage and fertilization give intermediate nutrient inputs. Streams that drain abandoned farm fields usually have low nutrient inputs that are only slightly higher than undisturbed forests, especially after several years of abandonment (Burton and Hook 1979).

Gibbs (1970) identified three primary processes that influence the chemistry of large rivers: (1) *atmospheric precipitation*, which is especially important for streams draining highly leached soils (primarily tropical); (2) *rock-weathering,* especially for rivers on geologically young or easily leached substrates (and important particularly in mountainous regions); and (3) *evaporation-crystallization,* especially for streams in arid regions. The type 3 streams above are usually rock-weathering influenced (i.e. type 2 above) in the headwaters, but become type 3 as the stream enters arid basins. In this classification, the dominant sources of inorganic ions in streams come from atmospheric inputs and/or from rock-weathering. On a local scale, factors such as the amount and type of vegetation (especially in the riparian zone), soil type, and land-use patterns in the watershed determine weathering and erosion rates, and influence nutrient inputs into streams.

QUALITATIVE AND QUANTITATIVE ROLES OF AQUATIC INSECTS IN NUTRIENT CYCLING

Aquatic insects play potentially major qualitative roles in the processing and turnover of nutrients in freshwater ecosystems. They may also have a minor but significant quantitative impact on nutrient storage and cycling. Although there is an extensive literature on the general association between the distribution of aquatic insects and various nutritional resources (e.g. Cushing 1963; Egglishaw 1964; Maciolek 1966; Minshall 1967; Malmqvist et al. 1978; Ward and Dufford 1979; Haefner and Wallace 1981), the actual roles played by insects in nutrient turnover have received much less attention. Several of these roles are reviewed herewith.

Role of Aquatic Insect Feeding Activities in Nutrient Transformations and Turnover Rates

Aquatic insects may influence nutrient turnover rates in streams and lakes by filtering particles from the water column, by changing surface-to-volume relationships and the chemical composition of their food substrate, and by activities such as bioturbation.

The role of filter-feeding insects in flowing waters has been reviewed by Wallace et al. (1977) and Wallace and Merritt (1980). Filter-feeding insects have evolved an array of mechanisms that enable different species to consume different particle sizes and types of materials, resulting in the efficient use of particulate resources in streams. This usage reduces export and enhances storage of these particulates, thereby increasing the efficiency of organic input utilization within a stream reach. By increasing the cycling rate of inorganic constituents, such as phosphorus and heavy metals that are readily sorbed to the surfaces of these particulates, filter-feeding insects can potentially affect movement of inorganic ions in streams. Such increased cycling of phosphorus in a phosphorus-limited system, as in Walker Branch, Tennessee (Elwood et al. 1981, 1983), would thereby increase the productivity of those streams. However, the efficiency of filter-feeding insects in removing and processing detritus varies among localities, and our comprehension of their role is limited (Wallace and Merritt 1980; Merritt and Wallace 1981).

Filter-feeding aquatic insects probably have little effect in removing FPOM from the water column in most streams and rivers (e.g. McCullough et al. 1979a, 1979b; Oswood 1979; Benke and Wallace 1980; Cudney and Wallace 1980; Wallace and Merritt 1980; Newbold et al. 1981; Haefner and Wallace 1981; Georgian and Wallace 1981). McCullough et al. (1979a) estimated that simuliids and hydropsychids removed only about 0.01 percent of the transported POM each day. The availability of suitable attachment sites for filter-feeders in large rivers is undoubtedly a severe limitation (e.g. Cudney and Wallace 1980). However, the low removal rate can change when attachment is possible; for example, Maciolek and Tunzi (1968) estimated a 60 percent removal of algal cells by black flies in the stream outfall of a lake.

Detritus is a major food source for macroinvertebrates, and their role in processing this material has been extensively studied (see reviews by Berrie 1976; Anderson et al. 1978; Anderson and Sedell 1979; Cummins and Klug 1979). Aquatic insects are especially important processors of microbially colonized CPOM. Their feeding on CPOM leads to quantitative and qualitative alteration of the material, and consequently, the production of FPOM, which is utilized by a variety of FPOM-feeding insects (collector-filterers and collector-gatherers) (McDiffett 1970; Cummins et al. 1973; Short and Maslin 1977; Grafius and Anderson 1979; Wallace and Merritt 1980).

The production of shredder fecal material is often correlated with collector ingestion. For example, Grafius and Anderson (1979) noted that the maximum fecal output of the shredder caddisfly *Lepidostoma* and the maximum production of the collector black fly *Simulium* corresponded, and they calculated that fecal ingestion by the collector could account for one-fourth to one-half of its observed production. Although the production of the shredder was only 10 percent that of the collector, its impact on the collector's nutritional resource could be as high as 50 percent. Black flies may also affect the transformation of particulates, since they probably increase FPOM particle size by ingesting very fine particulates, which are then compacted into larger feces (Ladle 1972). Simuliids in high density populations also may reingest their feces (Wotton 1980). Other such delays in nutrient transfer may accompany aquatic invertebrate activity. For example, the decreased mobilization of dissolved inorganic nitrogen was associated with leaf-litter feeding by a stream snail (McDiffett and Jordan 1978). These processes can be considered to constitute a recycling mechanism that increases the efficiency of organic input utilization in an aquatic ecosystem.

The transformation of CPOM to FPOM involves a series of abiotic and biotic processes involving aquatic insects and microorganisms. Even though a process-oriented model exists for this transformation (Boling et al. 1975), the process is still not completely understood (Anderson and Sedell 1979).

Fresh leaf litter is not readily consumed by stream insects, but rather requires days to weeks (depending on leaf species and temperature) of conditioning to render this substrate palatable to detritivores (Kaushik and Hynes 1971). In the first 24 hours, litter is leached of soluble organics such as carbohydrates, amino acids, and polyphenols, which amount to about 5 to 30 percent of the loss of the initial dry weight of leaves (Kaushik and Hynes 1971; Cummins et al. 1972; Petersen and Cummins 1974). Readily leached inorganic ions, such as potassium (K), are also removed, and up to 80 percent is lost in the first 24 hours (Killingbeck et al. 1982). In fact, insect detritivores in streams may have some mechanism for dissolved-K uptake from the water (Webster and Patten 1979), since they appear to ingest more K than is present in the organic matter in the stream. Calcium (Ca) is not so readily leached from newly fallen leaf material (Killingbeck et al. 1982), and high detritivore ingestion efficiencies for Ca appear to be linked to reingestion of egested materials (Webster and Patten 1979). Woodall and Wallace (1975) suggested that feeding activities by aquatic insects probably release inorganic ions such as Ca^{++} and magnesium (Mg^{++}) from detritus.

After initial leaching, microbial colonization and the degradative activities of fungi and bacteria further transform leaf litter (Suberkropp et al. 1976). This process physically softens and chemically modifies the leaf, which can then be consumed and more efficiently assimilated than newly fallen leaves (Triska 1970; Kaushik and Hynes 1971; Bärlocher and Kendrick 1973a, 1973b, 1981; Cummins et al. 1973; Petersen and Cummins 1974). Increases in nitrogen content with detrital conditioning involve not only microbially induced changes, but also the progressive formation of recalcitrant humic substances (Rice 1982). It is not yet known whether the feeding preferences demonstrated for microbially colonized leaves are based on the nutritive value of the degraded leaf, the associated microbial biomass, or some chemical by-product of one or both of these factors.

Clearly, the microorganisms on these leaves can be readily assimilated (42–97% assimilation rate: Hargrave 1970), and remain potential sources of carbon and nitrogen (Fenchel 1970; Berrie 1976; Hargrave 1976; Lopez et al. 1977; Cammen 1980). However, microbial biomass is a small percentage (0.03–10%) of the diet of detritivores (Iversen 1973; Baker and Bradnam 1976; Cummins and Klug 1979). For example, microbial biomass contributes only about 8.3 percent to the growth of the stream shredder *Tipula abdominalis* (Cummins and Klug 1979).

Despite the adaptive feeding and digestive mechanisms demonstrated by aquatic insects, the assimilation efficiency for detritus is low—6 to 35 percent (Berrie 1976). Since assimilation is low, aquatic insects must further modify the detrital particles with their own digestive mechanisms (potentially including enzymes from ingested microorganisms and resident gut flora) and/or process relatively large quantities of detritus to obtain sufficient nutrients for growth (Findlay and Tenore 1982). Although all of the above processes can affect the nutrient recycling rate, their specific role has yet to be quantified in any detail.

Many aquatic insects cannot digest the major structural carbohydrates of leaf litter without microbial enzymes (Bjarnov 1972; Nielsen 1962, 1963; Martin et al. 1980; 1981a, 1981b). Many herbivorous aquatic insects have cellulase in their guts (see Lamberti and Moore, Chapter 7), but it is not known whether this enzyme is endogenous or microbial in origin. Microorganisms associated with leaf litter have the enzymatic capacity to degrade many of the refractory carbohydrates such as cellulose, hemicellulose, and pectin, and this microbial degradation may produce more digestible by-products that can be used by detritivores. Digestion of cellulose in the stonefly *Pteronarcys* and the cranefly *Tipula* has been detected (Benfield, personal communication), but an intra-intestinal microbial source of the enzyme has not been ruled out.

Nitrogen levels are low in leaf litter. This element is often complexed as proteins with polyphenols and lignins, which presumably reduces its availability (Suberkropp et al. 1976; Odum et al. 1979). The maintenance of highly alkaline midguts could be a mechanism to obtain protein from polyphenol diets (Feeny 1970; Berenbaum 1980), and both alkaline midguts and alkaline-stable pro-teinases have been documented in aquatic detritivores (Dadd 1975; Martin et al. 1980). However, not all aquatic detritivores have alkaline midguts (Martin et al. 1981a, 1981b), and some that do may digest polyphenolic-bound protein in other ways.

Phosphorus levels are also low in many lake and stream habitats. Increased inputs of phosphorus may significantly increase the rate of detrital processing, and presumably increase turnover of detritus. Phosphorus stimulation of CPOM processing has been documented in streams (Howarth and Fisher 1976; Elwood et al. 1981) and littoral zones of lakes (Wissmar et al. 1977; Barnes et al. 1978; Richey and Wissmar 1979).

Role of Aquatic Insects in Sediment-Water Exchange of Nutrients

The turnover of nutrients by insects in depositional zones of lakes and streams may have both qualitative and quantitative significance. For example, aquatic insect feces may comprise a large portion of fine particulates in lakes (e.g. Brundin 1949; Davis 1974) and streams (e.g. Grafius and Anderson 1979). Therefore, species that ingest fecal matter from surficial sediments of lakes (gatherers-collectors) may play a major role in the continued mobilization of deposited organic matter (MacFadyen 1961). This mobilization by insects may be particularly important in profundal regions of lakes where low oxygen tensions prevail at least seasonally. As organic matter is deposited in the sediments, one or two species of chironomid midges may use this material and recycle as much as 20 percent of the gross primary production annually (Jónasson 1972).

Aquatic insects also move phosphorus and other nutrients between lake sediments and the overlying waters. They are the most abundant bottom-dwelling organisms in many lentic habitats, and they influence the concentration and form of phosphorus by (1) bioturbation and mineralization of sediments, and (2) uptake and excretion (Granéli 1979a; Nalepa et al. 1980).

Bioturbation and Mineralization of Sediments.   Bioturbation, the physical translocation of sediments, is accomplished through burrowing and feeding activity. The main aquatic insect groups involved in bioturbation are chironomids (Walshe 1950; Brundin 1951; Jónasson 1972; Gallepp et al. 1978; Nalepa et al. 1980) and burrowing mayflies (Hunt 1953). Due to insect activities, enriched interstitial water is pumped out of sediments, which allows oxygen-rich water to enter (Usinger and Kellen 1955; Edwards 1958; Edwards and Rolley 1965; Kimerle and Enns 1968; Hargrave 1975; Granéli 1979b). These insects also influence the release of phosphorus from the sediments to the overlying water by the (1) uptake and excretion of phosphorus (Gallepp et al. 1978; Granéli 1979a; Holdren and Armstrong 1980; Gardner et al. 1981); (2) active transport of particulate material to the sediment surface and into deeper sediment layers (Tessenow 1964; Davis 1974); and (3) egestion of fecal material at the water-sediment interface (McLachlan and McLachlan 1976; Petr 1977).

The tube-building activities of *Chironomus* larvae significantly influence nutrient exchange across the water-sediment interface. Granéli (1979a, 1979b) showed that densities of 250 *C. plumosus* larvae/m^2 increased the effective surface area for nutrient exchange by 50 percent. Their respiratory movements also accelerated the release of nutrients for phytoplankton, such as phosphorus, silica, iron, manganese, and, to a lesser extent, organic nitrogen, from the sediments (Tessenow 1964; Granéli 1979a). Bioturbation due to the activities of *Chironomus riparius* larvae increased the concentration of NH_4^+ in the water overlying sand filters and activated sludge (Ganapatí 1949; Edwards 1958). Andersen (1977) reported similar results when *C. plumosus* larvae were introduced into lake sediments.

Larval chironomids also increase the supply of dissolved oxygen to the mud and increase redox potentials in the mud surface layers, thereby facilitating nutrient transport across the sediment-water interface (Edwards 1958; Hargrave 1975). In the absence of these macroinvertebrates, the vertical transport of dissolved substances through the sediment occurs mainly via diffusion, which is a much slower process (Granéli 1979a). In some streams, insects occur deep in hyporheic sediments (25–30 cm), which could affect the translocation of materials (Hynes et al. 1976; Williams 1981 and in this volume, Chapter 14). Also, sediments containing burrowing aquatic insects are possibly more mineralized than sediments without such animals (e.g. Petr 1977; Gallepp et al. 1978; Granéli 1979a, 1979b).

Nutrient Uptake and Excretion/Egestion. The activities of chironomid larvae increase the release of both soluble reactive phosphorus and total phosphorus from sediments into overlying waters (Gallepp et al. 1978; Gallepp 1979; Granéli 1979a; Holdren and Armstrong 1980; Nalepa et al. 1980). However, it is not known whether the observed phosphorus release is the result of mixing, changes in sediment redox potential, or mobilization associated with larval digestion and excretion processes (Holdren and Armstrong 1980). Gallepp (1979) reported that chironomids could excrete their body equivalent of phosphorus in three days, which, based on the density of midges in his laboratory microcosms, could have easily accounted for the amount of phosphorus released during his experiment. Gardner et al. (1981) studied phosphorus release in Lake Michigan sediments and calculated that phosphorus excretion by tubificids and chironomids represented 13 to 20 percent of the total phosphorus released from the sediments, based on biomass measurements and release rates from these organisms. If other benthic invertebrates released phosphorus at similar rates, they estimated that invertebrate excretion could account for virtually all of the phosphorus released from aerobic sediments. Thus, the activities of these organisms must be an important mechanism for lake nutrient regeneration.

Recent studies by Gardner et al. (1983) have shown that chironomids and oligochaetes release nutrients differently, thereby affecting their relative importance in mobilizing nutrients from sediments. Tubificid worms release nitrogen continuously, whereas chironomid larvae release it in pulses several times per hour. Chironomids release o-phthalaldehyde reactive nitrogen (PRN) from their guts as discrete events, probably during defecation. Oligochaetes excrete PRN from nephridial openings located on each individual segment, and excretion products are probably released apart from the gut and in a continuous pattern. Because of the filter-feeding and respiratory habits of chironomids, nutrients appear to be transferred more efficiently to overlying water through the passage of large volumes of water in their larval burrows. In contrast, the physical movement of water around tubificids is minimal; therefore, mineralized nutrients accumulate in the pore water (and sediments) and are not released at the sediment-water interface.

The role of benthic insects in the sediment-water exchange of nutrients in lotic environments has received little attention when compared to the extensive studies of lentic environments. However, Chatarpaul et al. (1979, 1980) reported that tubificid worms significantly enchanced denitrification and nitrification processes in stream sediments. They also suggested that tubificids increased the removal of nitrate-N from the aerated water overlying stream sediments. Since midges and tubificids account for significant nutrient mobilization through sediment profiles and across substrate-water interfaces in lentic environments

(Wavre and Brinkhurst 1971), they may perform the same role for depositional zones of lotic environments. However, little information on this subject is available.

Role of Aquatic Insects in Translocation of Nutrients

Exports via Emergence. Emerging adult aquatic insects, especially midges, have been reported to remove energy from aquatic systems (Weston and Turner 1917; Kellen 1953; Tubb and Dorris 1965). However, Vallentyne (1952) calculated that insects emerging from an Indiana lake removed less than 1 percent of the organic matter that annually sedimented. Kimerle and Anderson (1971) investigated the bioenergetic role of the chironomid *Glyptotendipes barbipes* in an Oregon waste stabilization lagoon, and determined that larvae occupied approximately 30 percent of the lagoon bottom and removed about 7 percent of total net production. Emergence took 3 percent of total net production; respiration and mortality used 4 percent. The bioenergetic role of chironomids might be greater in shallow lagoons located in regions where the growing season of the midge would last most of the year (Kimerle and Anderson 1971). It is apparent though, that aquatic insects in lentic waters are responsible for only minimal energy (and carbon) losses from lakes.

The effect of insect emergence on phosphorus and nitrogen export was calculated for three North American lakes (Likens and Loucks 1978). Insect emergence accounted for 1 percent, 5 percent, and 14 percent of total annual phosphorus losses from Lake Wingra (Wisconsin), Findley Lake (Washington), and Mirror Lake (New Hampshire), respectively. Total nitrogen losses from each of the three lakes was 1 percent, 1 percent, and 10 percent. Based on this information, it appears that insect emergence accounts for no more than 10 to 15 percent of annual nutrient losses from lakes, and often represents only 1 to 5 percent of the annual export.

Little comparable data exists for streams, but exports due to insect emergence are probably small compared to the large amounts of nutrients transported downstream. For example, Meyer and Likens (1979) reported that geological export (e.g. runoff) in stream water was the only export vector of consequence for phosphorus losses from Bear Brook, New Hampshire.

Intrasystem Movements of Nutrients. With the exception of bioturbation and the associated movements of nutrients between sediment and water, few specific studies have dealt with how aquatic insects affect intrasystem movement of nutrients. Such movements could occur via (1) internal migrations of aquatic insects in lakes and streams, including stream drift; and (2) nutrient removal from water by filter-feeding organisms, as discussed previously.

The internal migrations of stream benthos (e.g. hyporheic movements, stream drift) have recently been reviewed by Williams (1981). Any of these migrations will move the nutrients that are stored in insect biomass from one part of a stream to another. However, such movements probably have little impact on stream nutrient movements, because the amount of nutrients stored in insect biomass is usually very small compared to the dissolved or particulate amounts that are transported in water and stored in sediments (discussed later in the chapter). Kitchell et al. (1979) suggested that zooplankton represent an upward vector of nutrient movement via their daily vertical migrations, which could be highly significant in dense zooplankton populations. Vertically migrating insects, such as larvae of the phantom midge *Chaoborus* that undergo nocturnal migrations and often attain high densities in lakes (Kajak et al. 1978), could have a similar impact.

Role of Aquatic Insects in Nutrient Storage

Nutrient storage in aquatic insect biomass also influences nutrient cycling in lakes and streams. For example, uptake and long-term storage (weeks to months) in insect biomass would reduce the downstream movement of nutrients in streams. Newbold et al. (1981) have suggested, however, that consumers such as insects have only minor direct effects on nutrient spiralling in streams, with most effects being indirect through their influence on particle size and on the aufwuchs community.

A few estimates of standing crops (i.e. concentration-biomass) of nutrients do exist. Webster and Patten (1979) measured standing crops of Ca and K for insects and crayfish in three North Carolina montane streams. Standing crops of K varied from 2.7 to 4.7 mg/m^2 for insects and 3.0 to 7.4 mg/m^2 for crayfish. Standing crops of Ca were 3.6 to 5.3 mg/m^2 for insects, but were much higher for crayfish (161 to 392 mg/m^2). In all cases, insect standing crops were relatively small compared to nutrients in the water, and to amounts ingested (0.15 to 0.27%) and egested (0.21 to 0.34%) by both insects and crayfish. This indicated that most nutrients passed through detritivore guts with little assimilation. Woodall (1972) reported similar findings from his studies in a North Carolina stream.

Although data are not extensive, aquatic insects probably are not a significant sink for nutrients in most aquatic systems. Other aquatic invertebrates, however, can constitute a significant portion of inorganic ion budgets for certain lakes; for example, calcium content of clams in Shell Lake, Canada, represented 50 percent of the Ca dissolved in lake water (Green 1980).

Indirect Effects of Aquatic Insects on Nutrient Cycling

Aquatic insects probably play a significant role in nutrient cycling by influencing community structure and turnover rates of the algal and microbial populations that constitute their food. Scrapers (periphyton grazers) in streams, particularly at high densities, may shift the structure of the affected periphyton communities from large, slow-turnover species to small, rapid-turnover species (Eichenberger and Schlatter 1978; Gregory 1980; Sumner and McIntire 1982; Lamberti and Moore, Chapter 7). A similar shift from slow- to rapid-turnover species may result from the feeding of shredders and collectors on fungal and bacterial communities (e.g. Bärlocher 1980). Digestion of microbial cells from particles is known to increase surfaces available for colonization (e.g. Cummins and Klug 1979), suggesting that increased turnover of these populations, and increased turnover times for nutrient resources, may be occurring.

SUMMARY AND CONCLUSION

Although the specific roles of aquatic insects in nutrient cycling in streams and lakes have not been adequately elucidated and quantified to date, their two primary roles appear to be the enhancement of microbial activities associated with particulate organic matter, and the translocation of nutrients at sediment-water interfaces.

By feeding on detritus, aquatic insects change the particle size of particulate organic matter (POM), which alters the surface-to-volume ratio of substrates and allows increased microbial utilization. In addition, biochemical changes are induced in POM during gut passage. Direct conversion of POM to CO_2 and the associated mineralization of inorganic nutrients is in the order of 10 to 20 percent (e.g. Petersen and Cummins 1974). Increases in particle surface area (as high as 50%) and the associated biochemical changes and microbial metabolism undoubtedly accelerate nutrient turnover, but the extent of such enhancement has yet to be quantified.

Aquatic insects can play a significant role in nutrient exchanges at the sediment-water interface (bioturbation). Some researchers have attributed virtually all phosphorus release from aerobic lake sediments to invertebrate activity.

Insect biomass constitutes only a minor nutrient storage component in streams and lakes. Thus, inputs and outputs of nutrients to and from aquatic ecosystems via insect movements, such as emergence, are of minor significance in total nutrient budgets. Intrasystem translocation of nutrients by insect migrations is also likely to be small compared to the physical processes of stream flow and turbulence, and lake turnover.

Clearly, a great deal of research will be required to elucidate the specific roles of aquatic insects in nutrient cycling in running and standing waters. Undoubtedly, a key feature of such investigations will be further detailed analysis of animal-microbial interactions.

Acknowledgments

We thank D. Lawson and M. Klug for their helpful comments and suggestions during the preparation of this manuscript. This work was supported by the National Science Foundation (Grant DEB-78-11145 and DEB-80-22634) and the Department of Energy, Ecological Sciences Division, Contract DE-AT06-79E1004.

References

Andersen, J. M. 1977. Importance of the denitrification process for the rate of degradation of organic matter in lake sediments, pp. 357–62. *In:* H. L. Golterman (ed.). Interactions between sediments and freshwater. Dr. W. Junk B. V. Publishers, The Hague, The Netherlands. 473 pp.

Anderson N. H. and J. R. Sedell. 1979. Detritus processing by macroinvertebrates in stream ecosystems. Annual Review of Entomology 24:351–77.

Anderson, N. H., J. R. Sedell, L. M. Roberts, and F. J. Triska. 1978. The role of aquatic invertebrates in processing of wood debris in coniferous forest streams. The American Midland Naturalist 100:64–82.

Baker, J. H. and L. A. Bradnam. 1976. The role of bacteria in the nutrition of aquatic detritivores. Oecologia 24:95–104.

Bärlocher, F. 1980. Leaf-eating invertebrates as competitors of aquatic hyphomycetes. Oecologia 47:303–6.

Bärlocher, F. and B. Kendrick. 1973a. Fungi and food preferences of *Gammarus pseudolimnaeus.* Archiv für Hydrobiologie 72:501–16.

Bärlocher, F. and B. Kendrick. 1973b. Fungi in the diet of *Gammarus pseudolimnaeus* (Amphipoda). Oikos 24:295–300.

Bärlocher, F. and B. Kendrick. 1981. The role of aquatic hyphomycetes in the trophic structure of streams, pp. 743–60. *In*: G. C. Carroll and D. T. Wicklow (eds.). The fungal community: its organisation and role in the ecosystem. Marcel Dekker, Inc., New York, NY. 855 pp.

Barnes, J. R., R. Ovink, and K. W. Cummins. 1978. Leaf litter processing in Gull Lake, Michigan, U.S.A. Internationale Vereinigung für Theoretische und Angewandte Limnologie Verhandlungen 20:475–79.

Beaulac, M. N. 1980. Nutrient export coefficients: an examination of sampling design and natural variability within differing land uses. Unpublished M.S. thesis. Michigan State University, East Lansing, MI. 243 pp.

Benke, A. C. and J. B. Wallace. 1980. Trophic basis of production among net-spinning caddisflies in a southern Appalachian stream. Ecology 61:108–18.

Berenbaum, M. 1980. Adaptive significance of midgut pH in larval Lepidoptera. The American Naturalist 115:138–46.

Berrie, A. D. 1976. Detritus, microorganisms and animals in fresh water, pp. 323–38. *In:* J. M. Andersen and A. MacFadyen (eds.). The role of terrestrial and aquatic organisms in decomposition processes. Blackwell Scientific Publications, Oxford, England. 474 pp.

Bjarnov, N. 1972. Carbohydrases in *Chironomus, Gammarus* and some Trichoptera larvae. Oikos 23:261–63.

Boling, R. H., Jr., E. D. Goodman, J. A. VanSickle, J. O. Zimmer, K. W. Cummins, R. C. Petersen, and J. R. Reice. 1975. Toward a model of detritus processing in a woodland stream. Ecology 56:141–51.

Brundin, L. 1949. Chironomiden und andere Bodentiere der sudschwedischen Urgebirgsseen. Drottningholm Institute of Freshwater Research, Report 30:1–914.

Brundin, L. 1951. The relation of O_2-microstratification at the mud surface to the ecology of the profundal bottom fauna. Drottningholm Institute of Freshwater Research, Report 32:32–42.

Burton, T. M. and J. E. Hook. 1979. Non-point source pollution from abandoned agricultural land in the Great Lakes basin. Journal of Great Lakes Research 5:99–104.

Burton, T. M., R. R. Turner, and R. C. Harriss. 1978. Response of a soft water lake in North Florida, U.S.A., to increased chemical and particulate loading from its watershed. Internationale Vereinigung für Theoretische und Angewandte Limnologie Verhandlungen 20:546–49.

Cammen, L. M. 1980. The significance of microbial carbon in the nutrition of the deposit feeding polychaete, *Nereis succinea.* Marine Biology 61:9–20.

Chatarpaul, L., J. B. Robinson, and N. K. Kaushik. 1979. Role of tubificid worms on nitrogen transformations in stream sediment. Journal of the Fisheries Research Board of Canada 36:673–78.

Chatarpaul, L., J. B. Robinson, and N. K. Kaushik. 1980. Effects of tubificid worms on denitrification and nitrification in stream sediment. Canadian Journal of Fisheries and Aquatic Sciences 37:656–63.

Cheng, L. (ed.). 1976. Marine insects. North-Holland Publishing Co., Amsterdam, The Netherlands. 581 pp.

Coffman, W. P. 1973. Energy flow in a woodland stream ecosystem. II. The taxonomic composition and phenology of the Chironomidae as determined by the collection of pupal exuviae. Archiv für Hydrobiologie 71:281–322.

Coffman, W. P., K. W. Cummins, and J. C. Wuycheck. 1971. Energy flow in a woodland stream ecosystem. I. Tissue support trophic structure of the autumnal community. Archiv für Hydrobiologie 68:232–76.

Cudney, M. D. and J. B. Wallace. 1980. Life cycles, microdistribution and production dynamics of six species of net-spinning caddisflies in a large southeastern (U.S.A.) river. Holarctic Ecology 3:169–82.

Cummins, K. W. 1974. Structure and function of stream ecosystems. BioScience 24:631–41.

Cummins, K. W. 1977. From headwater streams to rivers. American Biology Teacher 39:305–12.

Cummins, K. W. 1980a. The natural stream ecosystem, pp. 7–24. *In:* J. V. Ward and J. A. Stanford (eds.). The ecology of regulated streams. Plenum Press, New York, NY. 398 pp.

Cummins, K. W. 1980b. The multiple linkages of forests to streams, pp. 191–198. *In:* R. H. Waring (ed.). Forests: fresh perspectives from ecosystem analysis. Proceedings of the 40th Annual Biology Colloquium, Oregon State University, Corvallis, OR. 198 pp.

Cummins, K. W. and M. J. Klug. 1979. Feeding ecology of stream invertebrates. Annual Review of Ecology and Systematics 10:147–72.

Cummins, K. W., M. J. Klug, R. G. Wetzel, R. C. Petersen, K. F. Suberkropp, B. A. Manny, J. C. Wuycheck, and F. O. Howard. 1972. Organic enrichment with leaf leachate in experimental lotic ecosystems. BioScience 22:719–22.

Cummins, K. W., R. C. Petersen, F. O. Howard, J. C. Wuycheck, and V. I. Holt. 1973. The utilization of leaf litter by stream detritivores. Ecology 54:336–45.

Cummins, K. W., J. R. Sedell, F. J. Swanson, G. W. Minshall, S. G. Fisher, C. E. Cushing, R. C. Petersen and R. L. Vannote. 1983. Organic matter budgets for stream ecosystems: problems in their evaluation, pp. 299–353. *In*: J. R. Barnes and G. W. Minshall (eds.). Stream ecology: application and testing of general ecological theory. Plenum Press, New York, NY. 399 pp.

Cushing, C. E., Jr. 1963. Filter-feeding insect distribution and planktonic food in the Montreal River. Transactions of the American Fisheries Society 92:216–19.

Dadd, R. H. 1975. Alkalinity within the midgut of mosquito larvae with alkaline-active digestive enzymes. Journal of Insect Physiology 21:1847–53.

Dahm, C. N. 1981. Pathways and mechanisms for removal of dissolved organic carbon from leaf leachate in streams. Canadian Journal of Fisheries and Aquatic Sciences 38:68–76.

Davis, R. B. 1974. Stratigraphic effects of tubificids in profundal lake sediments. Limnology and Oceanography 19:466–88.

Edwards, R. W. 1958. The effect of *Chironomus riparius* Meigen on the redox potentials of settled activated sludge. Annals of Applied Biology 46:457–64.

Edwards, R. W. and H. L. J. Rolley. 1965. Oxygen consumption of river muds. Journal of Ecology 53:1–19.

Egglishaw, H. J. 1964. The distributional relationship between the bottom fauna and plant detritus in streams. Journal of Animal Ecology 33:463–76.

Eichenberger, E. and A. Schlatter. 1978. Effect of herbivorous insects on the production of benthic algal vegetation in outdoor channels. Internationale Vereinigung für Theoretische und Angewandte Limnologie Verhandlungen 20:1806–10.

Elwood, J. W., J. D. Newbold, R. V. O'Neill, and W. Van Winkle. 1983. Resource spiralling: an operational paradigm for analyzing lotic ecosystems, pp. 3–27. *In*: T. D. Fontaine, III and S. M. Bartell (eds.). Dynamics of lotic ecosystems. Ann Arbor Science Publishers, Ann Arbor, MI. 494 pp.

Elwood, J. W., J. D. Newbold, A. F. Trimble, and R. W. Stark. 1981. The limiting role of phosphorus in woodland stream ecosystem: effects of P enrichment on leaf decomposition and primary producers. Ecology 62:146–58.

Evans, F. C. and W. W. Murdoch. 1968. Taxonomic composition, trophic structure and seasonal occurrence in a grassland insect community. Journal of Animal Ecology 37:259–73.

Feeny, P. P. 1970. Seasonal changes in oak leaf tannins and nutrients as a cause of spring feeding by winter moth caterpillars. Ecology 51:565–81.

Fenchel, T. 1970. Studies on the decomposition of organic detritus derived from the turtle grass *Thalassia testudinum*. Limnology and Oceanography 15:14–20.

Findlay, S. and K. Tenore. 1982. Nitrogen source for a detritivore: detritus substrate versus associated microbes. Science 21:371–73.

Fisher, S. G. and G. E. Likens. 1973. Energy flow in Bear Brook, New Hampshire: an integrative approach to stream ecosystem metabolism. Ecological Monographs 43: 421–39.

Gallepp, G. W. 1979. Chironomid influence on phosphorous release in sediment-water microcosms. Ecology 60:547–56.

Gallepp, G. W., J. F. Kitchell, and S. M. Bartell. 1978. Phosphorous release from lake sediments as affected by chironomids. Internationale Vereinigung für Theoretische und Angewandte Limnologie Verhandlungen 20:458–65.

Ganapatí, S. V. 1949. The role of the blood worm, *Chironomus plumosus*, in accounting for the presence of phosphates and excessive free ammonia in the filtrates from the slow sand filters of the Madras Water Works. Journal of the Zoological Society of India 6:41–43.

Gardner, W. S., T. F. Nalepa, M. A. Quigley, and J. M. Malczyk. 1981. Release of phosphorus by certain benthic invertebrates. Canadian Journal of Fisheries and Aquatic Sciences 38:978–81.

Gardner, W. S., T. F. Nalepa, D. R. Slavens, and G. A. Laird. 1983. Patterns and rates of nitrogen release by benthic Chironomidae and Oligochaeta. Canadian Journal of Fisheries and Aquatic Sciences 40:259–66.

Georgian, T. J., Jr., and J. B. Wallace. 1981. A model of seston capture by net-spinning caddisflies. Oikos 36:147–57.

Gibbs, R. J. 1970. Mechanisms controlling world water chemistry. Science 170:1088–90.

Goldman, C. R. and B. L. Kimmel. 1978. Biological processes associated with suspended sediment and detritus in lakes and reservoirs, pp. 19–44. *In*: J. Cairns Jr., E. F. Benfield, and J. R. Webster (eds.). Current perspectives on river-reservoir ecosystems. Proceedings of the 1977 Symposium of the North American Benthological Society, Roanoke, VA. North American Benthological Society. 85 pp.

Grafius, E. and N. H. Anderson. 1979. Population dynamics, bioenergetics, and role of *Lepidostoma quercina* Ross (Trichoptera: Lepidostomatidae) in an Oregon woodland stream. Ecology 60:433–41.

Granéli, W. 1979a. The influence of *Chironomus plumosus* larvae on the exchange of dissolved substances between sediment and water. Hydrobiologia 66:149–59.

Granéli, W. 1979b. The influence of *Chironomus plumosus* larvae on the oxygen uptake of sediment. Archiv für Hydrobiologie 87:385–403.

Green, R. H. 1980. Role of an unionid clam population in the calcium budget of a small arctic lake. Canadian Journal of Fisheries and Aquatic Sciences 37:219–24.

Gregory, S. V. 1980. Effects of light, nutrients, and grazing on periphyton communities in streams. Unpublished Ph.D. thesis. Oregon State University, Corvallis, OR. 151 pp.

Haefner, J. D. and J. B. Wallace. 1981. Production and potential seston utilization by *Parapsyche cardis* and *Diplectrona modesta* (Trichoptera: Hydropsychidae) in two streams draining contrasting southern Appalachian watersheds. Environmental Entomology 10:433–41.

Hargrave, B. T. 1970. The utilization of benthic microflora by *Hyalella azteca* (Amphipoda). Journal of Animal Ecology 39:427–37.

Hargrave, B. T. 1975. Stability in structure and function of the mud-water interface. Internationale Vereinigung für Theoretische und Angewandte Limnologie Verhandlungen 19:1073–79.

Hargrave, B. T. 1976. The central role of invertebrate faeces in sediment decomposition, pp. 301–21. *In*: J. M. Andersen and A. MacFadyen (eds.). The role of terrestrial and aquatic organisms in decomposition processes. Blackwell Scientific Publications, Oxford, England. 474 pp.

Hilsenhoff, W. L., J. L. Longridge, R. P. Narf, K. J. Tennessen, and C. P. Walton. 1972. Aquatic insects of Pine-Popple River, Wisconsin. Wisconsin Department of Natural Resources Technical Bulletin 54:1–44.

Holdren, G. C., Jr., and D. E. Armstrong. 1980. Factors affecting phosphorus release from intact lake sediment cores. Environmental Science and Technology 14:79–87.

Howarth, R. W. and S. G. Fisher. 1976. Carbon, nitrogen and phosphorus dynamics during leaf decay in nutrient-enriched stream microecosystems. Freshwater Biology 6:221–28.

Hunt, B. P. 1953. The lift history and economic importance of a burrowing mayfly, *Hexagenia limbata*, in southern Michigan lakes. Michigan Department of Conservation, Institute of Fisheries Research Bulletin 4:1–151.

Hutchinson, G. E. 1957. A treatise on limnology. Vol. 1. Geography, physics and chemistry. John Wiley and Sons Inc., New York, NY. 1015 pp.

Hynes, H. B. N. 1961. The invertebrate fauna of a Welsh mountain stream. Archiv für Hydrobiologie 57:344–88.

Hynes, H. B. N. 1963. Imported organic matter and secondary productivity in streams. Proceedings of the International Congress of Zoology 16:324–29.

Hynes, H. B. N. 1970a. The ecology of running waters. University of Toronto Press, Toronto. 555 pp.

Hynes, H. B. N. 1970b. The ecology of stream insects. Annual Review of Entomology 15:25–42.

Hynes, H. B. N. 1975. The stream and its valley. Internationale Vereinigung für Theoretische und Angewandte Limnologie Verhandlungen 19:1–15.

Hynes, H. B. N., D. D. Williams, and N. E. Williams. 1976. Distribution of the benthos within the substratum of a Welsh mountain stream. Oikos 27:307–10.

Iversen, T. M. 1973. Decomposition of autumn-shed beech leaves in a springbrook and its significance for the fauna. Archiv für Hydrobiologie 72:305–12.

Janzen, D. H. 1973. Sweep samples of tropical foliage insects: effects of seasons, vegetation types, elevation, time of day, and insularity. Ecology 54:687–708.

Janzen, D. H. and T. W. Schoener. 1968. Differences in insect abundance and diversity between wetter and drier sites during a tropical dry season. Ecology 49:96–110.

Jónasson, P. M. 1972. Ecology and production of the profundal benthos in relation to phytoplankton in Lake Esrom. Oikos Supplementum 14:1–148.

Kajak, Z., J. Rybak, and B. Ranke-Rybicka. 1978. Fluctuations in numbers and changes in the distribution of *Chaoborus flavicans* (Meigen) (Diptera: Chaoboridae) in the eutrophic Mikolajskie Lake and dystrophic Lake Flosek. Ekologia Polska 26:259–72.

Kaushik, N. K. and H. B. N. Hynes. 1968. Experimental study on the role of autumn-shed leaves in aquatic environments. Journal of Ecology 56:229–43.

Kaushik, N. K. and H. B. N. Hynes. 1971. The fate of the dead leaves that fall into streams. Archiv für Hydrobiologie 68:465–515.

Kellen, W. R. 1953. Laboratory experiments on the role of insects in sewage oxidation ponds. Journal of Economic Entomology 46:1041–48.

Killingbeck, K. T., D. L. Smith, and G. R. Marzolf. 1982. Chemical changes in tree leaves during decomposition in a tallgrass prairie stream. Ecology 63:585–89.

Kimerle, R. A. and N. H. Anderson. 1971. Production and bioenergetic role of the midge *Glyptotendipes barbipes* (Staeger) in a waste stabilization lagoon. Limnology and Oceanography 16:646–59.

Kimerle, R. A. and W. R. Enns. 1968. Aquatic insects associated with midwestern waste stabilization lagoons. Journal of the Water Pollution Control Federation Research Supplement (Part 2) 40:31–41.

Kitchell, J. F., R. V. O'Neill, D. Webb, G. W. Gallepp, S. M. Bartell, J. F. Koonce, and B. S. Ausmus. 1979. Consumer regulation of nutrient cycling. BioScience 29:28–34.

Ladle, M. 1972. Larval Simuliidae as detritus feeders in chalk streams. Memorie dell'Instituto Italiano di Idrobiologia Supplement 29:429–39.

Likens, G. E. and O. L. Loucks. 1978. Analysis of five North American lake ecosystems. III. Sources, loading and fate of nitrogen and phosphorus. Internationale Vereinigung für Theoretische und Angewandte Limnologie Verhandlungen 20:568–73.

Lopez, G. R., J. S. Levinton, and L. B. Slobodkin. 1977. The effect of grazing by the detritivore *Orchestia grillus* on *Spartina* litter and its associated microbial community. Oecologia (Berlin) 30:111–27.

Loucks, O. L. and W. E. Odum. 1978. Analysis of five North American lake ecosystems. I. A strategy for comparison. Internationale Vereinigung für Theoretische und Angewandte Limnologie Verhandlungen 20:556–61.

Lush, D. L. and H. B. N. Hynes. 1978. Particulate and dissolved organic matter in a small partly forested Ontario stream. Hydrobiologia 60:177–85.

MacFadyen, A. 1961. Metabolism of soil invertebrates in relation to soil fertility. Annals of Applied Biology 49:215–18.

Maciolek, J. A. 1966. Abundance and character of microseston in a California mountain stream. Internationale Vereinigung für Theoretische und Angewandte Limnologie Verhandlungen 16:639–45.

Maciolek, J. A. and M. G. Tunzi. 1968. Microseston dynamics in a simple Sierra Nevada lake-stream system. Ecology 49:60–75.

Mahan, D. C. 1980. Land-cover and coarse particulate organic influxes to a small stream. Unpublished Ph.D. thesis. Michigan State University, East Lansing, MI. 121 pp.

Malmqvist, B., L. M. Nilsson, and B. S. Svensson. 1978. Dynamics of detritus in a small stream in southern Sweden and its influence on the distribution of the bottom animal communities. Oikos 31:3–16.

Martin, M. M., J. S. Martin, J. J. Kukor, and R. W. Merritt. 1980. The digestion of protein and carbohydrate by the stream detritivore, *Tipula abdominalis* (Diptera: Tipulidae). Oecologia (Berlin) 46:360–64.

Martin, M. M., J. S. Martin, J. J. Kukor, and R. W. Merritt. 1981a. The digestive enzymes of detritus-feeding stonefly nymphs (Plecoptera: Pteronarcyidae). Canadian Journal of Zoology 59:1947–51.

Martin, M. M., J. J. Kukor, J. S. Martin, D. L. Lawson, and R. W. Merritt. 1981b. Digestive enzymes of larvae of three species of caddisflies (Trichoptera). Insect Biochemistry 11:501–6.

McAlpine, J. F., B. V. Peterson, G. E. Shewell, H. J. Teskey, J. R. Vockeroth, and D. M. Wood (coordinators). 1981. Manual of Nearctic Diptera. Volume 1. Monograph 27, Research Branch, Agriculture Canada, Ottawa, Canada. 673 pp.

McCullough, D. A., G. W. Minshall, and C. E. Cushing. 1979a. Bioenergetics of lotic filter-feeding insects *Simulium* spp. (Diptera) and *Hydropsyche occidentalis* (Trichoptera) and their function in controlling organic transport in streams. Ecology 60: 585–96.

McCullough, D. A., G. W. Minshall, and C. E. Cushing. 1979b. Bioenergetics of a stream "collector" organism, *Tricorythodes minutus* (Insecta: Ephemeroptera). Limnology and Oceanography 24:45–58.

McDiffett, W. F. 1970. The transformation of energy by a stream detritivore, *Pteronarcys scotti* (Plecoptera). Ecology 51:975–88.

McDiffett, W. F. and T. E. Jordan. 1978. The effects of an aquatic detritivore on the release of inorganic N and P from decomposing leaf litter. The American Midland Naturalist 99:36–44.

McLachlan, A. J. and S. M. McLachlan. 1976. Development of the mud habitat during the filling of two new lakes. Freshwater Biology 6:59–67.

Merritt, R. W. and K. W. Cummins (eds.). 1978. An introduction to the aquatic insects of North America. Kendall/Hunt Publishing Company, Dubuque, IA. 441 pp.

Merritt, R. W. and D. L. Lawson. 1979. Leaf litter processing in floodplain and stream communities, pp. 93–105. *In*: R. R. Johnson and J. F. McCormick (technical coordinators). Strategies for protection and management of floodplain wetlands and other riparian ecosystems. Forest Service, United States Department of Agriculture General Technical Report WO-12, Washington DC. 410 pp.

Merritt, R. W. and J. B. Wallace. 1981. Filter-feeding insects. Scientific American 244: 132–44.

Meyer, J. L. and G. E. Likens. 1979. Transport and transformation of phosphorus in a forest stream ecosystem. Ecology 60:1255–69.

Minshall, G. W. 1967. Role of allochthonous detritus in the trophic structure of a woodland springbrook community. Ecology 48:139–49.

Minshall, G. W., R. C. Petersen, K. W. Cummins, T. L. Bott, J. R. Sedell, C. E. Cushing, and R. L. Vannote. 1983. Interbiome comparison of stream ecosystem dynamics. Ecological Monographs 53:1–25.

Moon, H. P. 1939. Aspects of the ecology of aquatic insects. Transactions of the British Entomological Society 6:39–49.

Moore, J. W. 1980. Composition of benthic invertebrate communities in relation to phytoplankton populations in five subarctic lakes. Internationale Revue der gesamten Hydrobiologie 65:657–71.

Nalepa, T. F., D. S. White, C. M. Pringle, and M. A. Quigley. 1980. The biological component of phosphorus exchange and cycling in lake sediments, pp. 93–109. *In*: D. Scavia and R. Moll (eds.). Nutrient cycling in the Great Lakes: a summarization of factors regulating the cycle of phosphorus. Special Report 83, Great Lakes Research Division, University of Michigan, Ann Arbor, MI. 140 pp.

Nelson, D. J. 1969. The stream ecosystem: terrestrial-lotic community interactions, pp. 14–19. *In*: K. W. Cummins (ed.). The stream ecosystem. Michigan State University Institute of Water Research Technical Report 7. 42 pp.

Newbold, J. D., J. W. Elwood, R. V. O'Neill, and W. Van Winkle. 1981. Measuring nutrient spiralling in streams. Canadian Journal of Fisheries and Aquatic Sciences 38:860–63.

Newbold, J. D., P. J. Mulholland, J. W. Elwood, and R. V. O'Neill. 1982. Organic carbon spiralling in stream ecosystems. Oikos 38:266–72.

Nielsen, C. O. 1962. Carbohydrases in soil and litter invertebrates. Oikos 13:200–15.

Nielsen, C. O. 1963. Laminarinases in soil and litter invertebrates. Nature (London) 199: 1001.

Odum, E. P. 1971. Fundamentals of ecology. W. B. Saunders Co., Philadelphia, PA. 574 pp.

Odum, W. E., P. W. Kirk, and J. C. Zieman. 1979. Non-protein nitrogen compounds associated with particles of vascular plant detritus. Oikos 32:363-67.

Odum, W. E. and R. T. Prentki. 1978. Analysis of five North American lake ecosystems. IV. Allochthonous carbon inputs. Internationale Vereinigung für Theoretische und Angewandte Limnologie Verhandlungen 20:574-80.

Omernik, J. M. 1976. The influence of land use on stream nutrient levels. EPA-600/3-76-014, Corvallis Environmental Research Laboratory, U.S. Environmental Protection Agency, Corvallis, OR. 106 pp.

Omernik, J. M. 1977. Nonpoint source-stream nutrient level relationships: a nationwide study. EPA-600/3-77-105, Corvallis Environmental Research Laboratory, U.S. Environmental Protection Agency, Corvallis, OR. 151 pp.

Oswood, M. W. 1979. Abundance patterns of filter-feeding caddisflies (Trichoptera: Hydropsychidae) and seston in a Montana (U.S.A.) lake outlet. Hydrobiologia 63: 177-83.

Petersen, R. C. and K. W. Cummins. 1974. Leaf processing in a woodland stream. Freshwater Biology 4:343-68.

Petr, T. 1977. Bioturbation and exchange of chemicals in the mud-water interface, pp. 216-26. In: H. L. Golterman (ed.). Interactions between sediments and fresh water. Dr. W. Junk B. V. Publishers, The Hague, The Netherlands. 473 pp.

Rice, D. L. 1982. The detritus nitrogen problem: new observations and perspectives from organic geochemistry. Marine Ecology Progress Series 9:153-62.

Richey, J. E. and R. C. Wissmar. 1979. Sources and influences of allochthonous inputs on the productivity of a subalpine lake. Ecology 60:318-28.

Ross, H. H. 1963. Stream communities and terrestrial biomes. Archiv für Hydrobiologie 59:235-42.

Sedell, J. R., F. J. Triska, F. S. Hall, N. H. Anderson, and J. H. Lyford. 1974. Sources and fates of organic inputs in coniferous forest streams, pp. 57-69. In: R. H. Waring and R. L. Edmonds (eds.). Integrated research in the coniferous forest biome. Coniferous Forest Biome Ecosystem Analytical Study, US/IBP, Bulletin Number 5. University of Washington, Seattle, WA. 148 pp.

Sedell, J. R., R. J. Naiman, K. W. Cummins, G. W. Minshall, and R. L. Vannote. 1978. Transport of particulate organic material in streams as a function of physical processes. Internationale Vereinigung für Theoretische und Angewandte Limnologie Verhandlungen 20:1366-75.

Short, R. A. and P. E. Maslin. 1977. Processing of leaf litter by a stream detritivore: effect on nutrient availability to collectors. Ecology 58:935-38.

Stewart, B. A., D. A. Woolhiser, W. H. Wischmeier, J. H. Caro, and M. H. Frere. 1975. Control of water pollution from cropland. Vol. I. A manual for guideline development. EPA-600/2-75-026A, U.S. Environmental Protection Agency, Washington DC. 111 pp.

Stewart, B. A., D. A. Woolhiser, W. H. Wischmeier, J. H. Caro, and M. H. Frere. 1976. Control of water pollution from cropland. Vol. II. An overview. EPA-600/2-75-026B, U.S. Environmental Protection Agency, Washington DC. 187 pp.

Suberkropp, K., G. L. Godshalk, and M. J. Klug. 1976. Changes in the chemical composition of leaves during processing in a woodland stream. Ecology 57:720–27.

Sumner, W. T. and C. D. McIntire. 1982. Grazer-periphyton interactions in laboratory streams. Archiv für Hydrobiologie 93:135–57.

Teskey, H. J. 1976. Diptera larvae associated with trees in North America. Memoirs of the Entomological Society of Canada 100:1–53.

Tessenow, U. 1964. Experimentaluntersuchungen zur Kieselsäurerückführung aus dem Schlamm der Seen durch Chironomidenlarven (Plumosus-Gruppe). Archiv für Hydrobiologie 60:497–504.

Triska, F. J. 1970. Seasonal distribution of aquatic hyphomycetes in relation to the disappearance of leaf litter from a woodland stream. Unpublished Ph.D. thesis. University of Pittsburgh, Pittsburgh, PA. 189 pp.

Tubb, R. A. and T. C. Dorris. 1965. Herbivorous insect populations in oil refinery effluent holding pond series. Limnology and Oceanography 10:121–34.

Usinger, R. L. and W. R. Kellen. 1955. The role of insects in sewage disposal beds. Hilgardia 23:262–321.

Vallentyne, J. R. 1952. Insect removal of nitrogen and phosphorus compounds from lakes. Ecology 33:573–77.

Vannote, R. L., G. W. Minshall, K. W. Cummins, J. R. Sedell, and C. E. Cushing. 1980. The river continuum concept. Canadian Journal of Fisheries and Aquatic Sciences 37:130–37.

Wallace, J. B., J. R. Webster, and W. R. Woodall. 1977. The role of filter feeders in flowing waters. Archiv für Hydrobiologie 79:506–32.

Wallace, J. B. and R. W. Merritt. 1980. Filter-feeding ecology of aquatic insects. Annual Review of Entomology 25:103–32.

Walshe, B. M. 1950. Observations on the biology and behavior of larvae of the midge Rheotanytarsus. Journal of the Quekett Microscopical Club 3:171–78.

Ward, J. V. and R. G. Dufford. 1979. Longitudinal and seasonal distribution of macroinvertebrates and epilithic algae in a Colorado springbrook-pond system. Archiv für Hydrobiologie 86:284–321.

Wavre, M. and R. O. Brinkhurst. 1971. Interactions between some tubificid oligochaetes and bacteria found in the sediments of Toronto Harbour, Ontario. Journal of the Fisheries Research Board of Canada 28:335–41,

Webster, J. R. 1975. Analysis of potassium and calcium dynamics in stream ecosystems on three southern Appalachian watersheds of contrasting vegetation. Unpublished Ph.D. thesis. University of Georgia, Athens, GA. 232 pp.

Webster, J. R. and B. C. Patten. 1979. Effects of watershed perturbation on stream potassium and calcium dynamics. Ecological Monographs 49:51–72.

Welton, J. S. 1980. Dynamics of sediment and organic detritus in a small chalk stream. Archiv für Hydrobiologie 90:162–81.

Weston, R. S. and C. E. Turner. 1917. Studies on the digestion of a sewage-filter effluent by a small and otherwise unpolluted stream. Contributions of the Sanitary Research Laboratory and Sewage Experiment Station, Massachusetts Institute of Technology. 10:1–96.

Wetzel, R. G. 1975. Limnology. W. B. Saunders Company, Philadelphia, PA. 743 pp.

Wetzel, R. G. and P. H. Rich. 1973. Carbon in freshwater systems, pp 241–63. In: G. M. Woodwell and E. V. Pecan (eds.). Carbon and the biosphere. Proceedings of the 24th

Brookhaven Symposium of Biology. USAEC CONF-720510, National Technical Information Service, Springfield, VA. 392 pp.

Wiggins, G. B., R. J. Mackay, and I. M. Smith. 1980. Evolutionary and ecological strategies of animals in annual temporary pools. Archiv für Hydrobiologie Supplement 58:97–206.

Williams, D. D. 1981. Migrations and distributions of stream benthos, pp. 155–207. *In*: M. A. Lock and D. D. Williams (eds.). Perspectives in running water ecology. Plenum Press, New York, NY. 430 pp.

Williams, D. D. 1983. The natural history of a Nearctic temporary pond in Ontario with remarks on continental variation in such habitats. Internationale Revue der gesamten Hydrobiologie 68:239–53.

Williams, D. D. and H. B. N. Hynes. 1976. The ecology of temporary streams. I. The faunas of two Canadian streams. Internationale Revue der gesamten Hydrobiologie 61:761–87.

Winterbourn, M. J., J. S. Rounick, and B. Cowrie. 1981. Are New Zealand stream ecosystems really different? New Zealand Journal of Marine and Freshwater Research 15:321–28.

Wissmar, R. C., J. E. Richey, and D. E. Spyridakis. 1977. The importance of allochthonous particulate carbon pathways in a subalpine lake. Journal of the Fisheries Research Board of Canada 34:1410–18.

Woodall, W. R., Jr. 1972. Nutrient pathways in small mountain streams. Unpublished Ph.D. thesis. University of Georgia, Athens, GA. 118 pp.

Woodall, W. R., Jr., and J. B. Wallace. 1975. Mineral pathways in small Appalachian streams, pp. 408–22. *In*: F. G. Howell, J. B. Gentry, and M. H. Smith (eds.). Mineral cycling in southeastern ecosystems. ERDA Symposium Series (Conf-740513). 898 pp.

Wotton, R. S. 1980. Coprophagy as an economic feeding tactic in blackfly larvae. Oikos 34:282–86.

Zimmerman, M. C. and T. E. Wissing. 1980. The nutritional dynamics of the burrowing mayfly, *Hexagenia limbata*, pp. 231–57. *In*: J. F. Flannagan and K. E. Marshall (eds.). Advances in Ephemeroptera biology. Plenum Press, New York, NY. 552 pp.

chapter 7

AQUATIC INSECTS AS PRIMARY CONSUMERS

Gary A. Lamberti
James W. Moore

INTRODUCTION

Most aquatic insects are opportunistic feeders that consume a wide range of the food items available in their environments (Hynes 1970; Cummins 1973; Hutchinson 1981). Both life-history features and habitat type influence the nature and extent of consumption patterns in aquatic insects. For example, with increasing size or age, some species may switch from herbivory to predation; conversely, under adverse environmental conditions, some predatory species may revert to other modes of consumption such as filter-feeding or deposit-collecting. In turn, interactions between herbivorous aquatic insects and their algal food resources can have significant implications for the functioning of aquatic ecosystems.

We define a *primary consumer* in a broad sense; that is, as an animal that feeds on primary producers and/or decomposers. The latter food source includes living or dead plant matter and associated microorganisms, and also inorganic material that may serve as substrate for attached microflora. We consider a *herbivore* to be a special category of primary consumer that utilizes living plant material, such as algae or vascular hydrophytes, as its principal food source.

The objectives of this chapter are (1) to describe the generalized feeding mechanisms of primary consumers, and the factors that influence the composition of materials ingested by different insect groups; (2) to relate feeding and digestion rates of primary consumers to the variation in the chemical composition and nutritive value of their food; (3) to evaluate the responses of benthic algae and other microorganisms to grazing by herbivores; (4) to identify the principal deficiencies in our knowledge of the feeding ecology of primary consumers; and (5) to propose directions for future research.

FEEDING MECHANISMS OF PRIMARY CONSUMERS

Despite the diversity of aquatic insects, the feeding mechanisms used by primary consumers can be grouped into four general categories: *filter-feeding, deposit-collecting, scraping,* and *shredding* (Cummins 1973, 1978; Cummins and Klug 1979). Filter-feeders use various structures (e.g. silken nets or modified body parts) to filter organic material that is suspended in the water column. Deposit-collectors gather finely divided organic material that has settled on the sediments in aquatic depositional zones. Scrapers graze the organic film on submerged surfaces, and thus subsist mainly on benthic algae and other attached material. Shredders include insects that consume the living tissue of vascular hydrophytes and those that chew or bore into decomposing vascular plant tissue. Merritt et al. (Chapter 6) provide further definitions of these categories. Most species of aquatic insects are apparently committed to one of these feeding mechanisms throughout their aquatic life, although there are exceptions to this rule.

Filter-feeders

Caddisfly larvae of the families Hydropsychidae, Philopotamidae, and Polycentropodidae, along with black fly (Simuliidae) larvae, are the most common filter-feeding insects in streams and rivers (Wallace and Merritt 1980). Filter-feeding caddisfly larvae use salivary secretions to spin nets on any solid object that is located in a suitable water current regime (Wallace and Sherberger 1975). Except in the Philopotamidae, the mesh size of trichopteran nets is relatively coarse, generally exceeding 5×40 μm in late instars and often being much larger (Table 7.1). Thus, bacteria, finely divided detritus, and many species of algae are not fully exploited as food sources by these filter-feeding taxa. In addition, the mesh size of nets increases with increasing larval age or size, thereby further changing entrapment efficiency and diet (Wallace et al. 1977).

In contrast to trichopteran capture nets, the cephalic fans of larval black flies provide a much finer filtering apparatus ($\leq 0.1 \times 4$ μm between setae; Table 7.1). When coated with a mucouslike substance that is secreted by the larva (Ross and Craig 1980), these fans may capture bacteria and colloidal material (Fredeen 1964; Wotton 1976; Kurtak 1978). Comparable filtration mechanisms also occur in many mosquito larvae (e.g. Dadd 1971, 1975; Ameen and Iversen 1978).

Larvae of midges in the subfamily Chironominae are probably the most common filtering insects in lentic habitats. Such species construct tubes on plant stems or in bottom sediments where they spin coarse, irregular nets either on or below the substrate surface (Oliver 1971; McLachlan 1977). Planktonic algae and detritus are apparently their main food sources, and it is unlikely that these larvae subsist on bacteria alone (Rodina 1949; Margolina 1961; Izvekova and Sorokin 1969). Some midges, such as *Glyptotendipes*, can revert to deposit-collecting or shredding if sufficient plankton is unavailable (Walshe 1951).

Deposit-collectors

Deposit-collectors are commonly found on the sediments of lakes, ponds, and slow-moving streams where current does not interfere with their movements and food acquisition. Several families of mayflies feed primarily on algae,

TABLE 7.1: **Mesh Size (μm) of Capture Nets in Selected Species of Trichoptera, and Intersetal Distance (μm) of Cephalic Fans in Selected Species of Simuliidae**

SPECIES	INSTAR*	MESH SIZE (μm); INTERSETAL DISTANCE (μm)
Hydropsychidae		
Arctopsyche irrorota[1]	5	403 × 534
Diplectrona modesta[1]	4	125 × 195
	5	184 × 238
Hydropsyche betteni[2]	5	148 × 250
Hydropsyche incommoda[1]	5	150 × 260
Hydropsyche macleodi[1]	5	145 × 250
Hydropsyche orris[1]	5	63 × 137
Hydropsyche slossonae[2]	5	176 × 298
Hydropsyche sparna[1]	4	110 × 160
	5	180 × 270
Macronema sp.[3,4]	5	5 × 40
Parapsyche cardis[1]	4	176 × 291
	5	240 × 392
Philopotamidae		
Chimarra aterrima[5]	5	6 × 70
Chimarra socia[6]	5	1 × 9
Dolophilodes distinctus[1]	5	1 × 6
Wormaldia occipitalis[7]	NR	6 to 7.5 × 8.5 to 45
Simuliidae		
Prosimulium mixtum[8]	"late"	0.1 × 2 to 4
Simulium vittatum[8]	7	0.1 × 4
Stegopterna mutata[8]	6	0.1 × 2 to 4

*NR = not reported

Data from [1]Wallace et al. 1977; [2]Fuller and Mackay 1980; [3]Wallace and Sherberger 1974; [4]Wallace and Sherberger 1975; [5]Williams and Hynes 1973; [6]Wallace and Malas 1976; [7]Nielsen 1942; [8]Ross and Craig 1980

detritus, and bacteria that have been deposited on sediments (Moon 1938; Brook 1955; Brown 1960, 1961). Since such species do not scrape attached material, they generally have reduced mandibles and maxillae (Hynes 1970). Some insects can apparently switch between deposit-collecting and another feeding mechanism. For example, species in the chironomid subfamily Orthocladiinae are primarily deposit-feeders, but they may temporarily revert to passive filtration (Monakov

1972). Conversely, larvae of another midge subfamily, the Tanypodinae, possess stout mandibles and are considered to be predaceous (Kajak and Dusoge 1970), although many forms also consume detritus through deposit-collecting (Oliver 1971; Monakov 1972; Baker and McLachlan 1979).

Scrapers

Scrapers are most abundant in flowing waters, and they often possess elaborate mechanisms to remove algae and detritus from the surfaces of rocks and other solid objects. Many species of mayflies and some taeniopterygid stonefly nymphs bear stout bristles on the labial or maxillary palps. These bristles are effective in removing attached material from cracks and crevices (Brook 1955). Since these bristles usually increase in length with age, older animals scrape relatively large-sized particles from solid objects and leave smaller particles behind. Thus, food may be partitioned among the different instars in a population. Similarly, the mouthparts of the shredder *Tipula sacra* (Tipulidae) have many unarticulated setae, particularly on the inner surface of the mandibles (Hall and Pritchard 1975), which serve to dislodge attached material, especially diatoms, and allow the insect to function as a facultative scraper (Byers 1978; Pritchard 1983).

Other scrapers show different morphological adaptations. For example, larvae of the caddisfly *Glossosoma* possess scoop-shaped mandibles, and some mayflies (e.g. *Stenonema*) have modified grinding edges on the inner margins of their mandibles. Most *Orthocladius* species (Chironomidae) bear stout mandibles, each of which has four or five strong teeth that gradually wear down with age (Soponis 1977).

Shredders

Only a few species of shredders consume large amounts of living higher plants. Those insects that do (e.g. the chironomids *Paraponyx* and *Cricotopus*) possess small cutting mandibles and usually live in cases that they have constructed (Pennak 1978). Far more species, especially among the Chironomidae, mine directly into higher plants, after which they construct a net at the front of the tube to filter plankton, the main food source. However, because these mining insects eat only enough material to make space to build a silk-lined tube, they play only a limited role as shredders.

A major group of shredders, which includes members of the caddisfly families Limnephilidae and Lepidostomatidae, the tipulid subfamilies Limoniinae and Tipulinae, and a few species of Chironomidae, gouge fallen wood or consume large detrital particles such as autumn-shed leaves (Cummins 1973, 1974; Pritchard 1983). These species have cutting mandibles, and may become increasingly carnivorous with age.

HABITAT FEATURES AND FOOD COMPOSITION

Characteristics of a given habitat may greatly influence the composition of food that is available to each feeding group. For example, in different locations, or at the same location in different seasons, filter-feeding insects may ingest primarily algae, detritus, or animal tissue, depending on the relative abundance of these foods in the environment. In some instances, the role of the habitat may even overshadow the importance of the feeding mechanism in determining the choice of food. For example, the seasonal availability of algal food resources largely determines the consumption patterns of some lentic Chironomidae (Margolina 1961; Kajak and Warda 1968). As a result, both habitat and feeding mechanism can interact to exert primary control over trophic relations among insects and, consequently, the energy flow in aquatic systems.

Filter-feeders

Larval Simuliidae and Hydropsychidae probably exploit a wider food base than other filter-feeding organisms such as filter-feeding species of mayflies and mosquito larvae (Wallace and Merritt 1980). For example, in the laboratory, black fly larvae can be reared to maturity on a diet of bacteria (Fredeen 1960), and there may be enough suspended bacteria in some eutrophic habitats to support large larval populations (Fredeen 1964; Wotton 1978a). However, stream-inhabiting black flies may also feed on material as large as filamentous green algae (Davies and Syme 1958; Burton 1973). Similarly, middle- to late-instar hydropsychid caddisflies may consume particles ranging in size from 1 to 500 μm, including detritus and small animals (Benke and Wallace 1980). Consequently, algae, detritus, associated microflora, and animal material play a relatively large role in the nutrition of these organisms.

There can be substantial intraspecific variation in the consumption patterns of filter-feeding species, and this is at least partly related to location and season (Table 7.2). For example, at different sites and seasons, *Hydropsyche slossonae* may ingest 41 to 75 percent detritus or 23 to 46 percent algal material. Even greater variation is found in the midge larva *Chironomus plumosus*, which may ingest 0.1 to 72 percent algae. Maximum algal consumption by this species is generally found in eutrophic lakes during the summer when algal productivity is high. In addition, larvae inhabiting the littoral areas of these lakes consume more algae than those in profundal regions, which suggests a direct correspondence with the availability of algal food resources. However, in other seasons and collection sites, the detrital fraction may account for up to 99 percent of the gut contents of *C. plumosus* (Table 7.2).

TABLE 7.2: **Relative Abundance by Volume of the Gut Contents of the Caddisfly** *Hydropsyche slossonae,* **the Midge** *Chironomus plumosus,* **and the Mayfly** *Leptophlebia* **in Different Aquatic Habitats and Seasons**

SPECIES	RELATIVE ABUNDANCE (%)			SEASON
	Algae	*Detritus*	*Animals*	
Hydropsyche slossonae				
Farmland stream (Ontario)[1]	30–35	50–55	10	Summer
	35	60	<5	Fall
	45	50	<5	Winter
	40	50	5	Spring
Woodland stream (Pennsylvania)[2]	25	75	0	Fall
Woodland and farmland streams (Wisconsin)[3]	23	70	7	Early spring
	46	41	13	Late spring
	23	57	20	Fall
Chironomus plumosus				
Eutrophic lakes (Poland)[4]	66–72	25–32	1–2	Summer
Great Slave Lake (Canada)[5]	1.0–1.5	98–99	0	Summer
	0.5–1.0	99	0	Winter
Lady Burn Lough (England)[6]	42–61	39–58	0	Summer
	37–55	45–63	0	Winter
Rybinsk Reservoir (USSR)[7]	30–50	25–44	0	Summer
	5–20	58–87	0	Winter
Uchinskoye Reservoir (USSR)[8]	0.1–20	40–60	0	Summer
Leptophlebia spp.				
Woodland stream (Alberta)[9]	4–7	93–96	0	Summer
	3–6	94–97	0	Fall
	2–6	94–98	0	Spring
Subarctic stream (Canada)[10]	45–64	36–55	0	Summer
	10–20	80–90	0	Winter
Farmland stream (Wisconsin)[3]	26	74	0	Early spring
Marsh (Alberta)[9]	5	95	0	Spring

Data from [1]Fuller and Mackay 1980; [2]Cummins et al. 1966; [3]Shapas and Hilsenhoff 1976; [4]Kajak and Warda 1968; [5]Moore 1979a; [6]McLachlan 1977; [7]Margolina 1961; [8]Izvekova 1971; [9]Clifford et al. 1979; [10]Moore 1977

Deposit-Collectors and Scrapers

Habitat can have a significant impact on consumption patterns of deposit-collectors and scrapers as well. The diet of *Leptophlebia* mayflies apparently consists almost entirely of detritus in a Canadian woodland stream, regardless of season, and also in a Wisconsin farmland stream and an Alberta marsh (Table 7.2). However, in areas where autotrophic growth is abundant (e.g. in a subarctic stream; Moore 1977), algae are seasonally dominant dietary items. Many other species of collectors and scrapers that inhabit lotic habitats are similarly opportunistic in their food choice (e.g. Chapman and Demory 1963; Cummins 1973; Sain et al. 1977; Hart and Resh 1980).

The bottom of deep lakes provides a relatively stable environment for many collector species of Chironomidae, particularly those within the subfamily Orthocladiinae (Oliver 1971). In deep northern lakes, bottom temperatures remain fairly constant at about 4 to 5°C, thereby reducing thermal effects on consumption patterns. Algal growth on profundal sediments in deep lakes is generally minimal, regardless of season, and deposit-collectors and filterers must therefore depend on the sinking and settling of dead plankton and other detritus for food. Due to these factors, the gross composition of ingested material remains relatively constant throughout the year. For example, in Great Slave Lake, Canada, the midge larva *Heterotrissocladius changi* ingests 0 to 5 percent algae regardless of season (Moore 1979b). Although the importance of bacteria in the nutrition of these larvae has not been established, detritus and mineral particles appear to make up the remainder of the ingested material.

Information on the feeding activities of other invertebrates from deep-water habitats is limited. However, many species of oligochaetes rely for food almost exclusively on the bacterial film that is associated with detritus (Ivlev 1939; Sorokin 1966; Brinkhurst and Chua 1969; Brinkhurst et al. 1972). Similar consumption patterns have been reported for profundal populations of crustaceans and for some marine invertebrates (Larkin 1948; Green 1968; Hargrave 1971; Rich and Wetzel 1978; Whitlatch 1980). In contrast, some species of benthic filter-feeders, such as *Chironomus anthracinus*, feed heavily on the phytoplankton that has dropped from the photic zone and settled on the sediments (Jónasson and Kristiansen 1967).

Shredders

The importance of shredders for the processing of autumn-shed leaves has been extensively examined (see review by Anderson and Sedell 1979). This research has shown the following:

1. The major period of growth for many shredders is in the autumn and early winter.
2. Shredders function to break down fallen leaves into fine detritus which, in turn,

becomes available for consumption by filter-feeders and deposit-collectors (e.g. Grafius and Anderson 1980).

3. Seasonal differences in the movement rates of some species are influenced by the availability of detritus (e.g. Hart and Resh 1980).

4. Microdistribution patterns are positively correlated with the location of detritus (e.g. Nelson and Scott 1962; Egglishaw 1964; Ulfstrand 1967).

5. Shredders may preferentially select leaves that have a dense covering of aquatic hyphomycetes (e.g. Bärlocher 1980).

It remains unclear, however, whether the autumn peak in the growth of some insect species that inhabit temperate waters is solely dependent on detrital inputs or is also influenced by water temperature. Unlike many other primary consumers whose feeding rates decline in winter, some shredders are apparently adapted to take advantage of food supplies despite cold temperatures. For example, Short et al. (1980) reported that detritus consumption by invertebrates in a Colorado mountain stream was greatest during mid-winter when water temperature was near $0°C$, suggesting that this process was independent of temperature. However, other evidence suggests that peaks in invertebrate growth coincide with temperature maxima (Minshall 1968; Sweeney and Vannote 1978).

Many shredder species become increasingly more predaceous as they mature, and as the size of their mouthparts increases (Kubicek 1970; Mecom 1972; Cummins and Klug 1979). Such changes have been reported for species of both Trichoptera and Plecoptera, and generally serve to increase the growth rates of those species (Iversen 1974; Otto 1974). In some instances, supplemental animal matter may be necessary for pupation and normal adult development (Winterbourn 1971; Anderson 1976; Anderson and Cummins 1979).

A unique form of food acquisition is displayed by the limnephilid caddisfly *Desmona bethula*, a shredder that inhabits small mountain streams in California (Erman 1981). At dusk on certain nights, *D. bethula* larvae leave the water to climb and feed on riparian vegetation. Physical factors, including temperature, light, and humidity, control the onset and duration of this feeding migration, which may last until dawn. In this way, *D. bethula* exploits a food source unavailable to other aquatic herbivores, and also transports terrestrial material into the aquatic environment. The extent of this behavior within the Trichoptera is largely unknown, although Wiggins (1977) has collected larvae of another limnephilid caddisfly, *Cryptochia* sp., from moist leaves on a stream bank.

NUTRITIONAL VALUE OF FOODS

The nutritional value of the foods that aquatic insects consume varies widely, and may have a significant impact on population phenology (see also Sweeney, Chapter 4). For example, the growth rates of many lentic Chironomidae peak during the spring maximum of phytoplankton production, thereafter declining when live algal food is no longer available (Jónasson and Kristiansen 1967; Jónasson 1969; Ward and Cummins 1979). To ensure successful pupation, final instar larvae of the limnephilid caddisfly *Clistoronia magnifica* must eat food that has a high protein content, such as animal tissue (Anderson and Cummins 1979). In other instances, aquatic insects and other invertebrates increase their feeding rates to compensate for low nutrient or low caloric intake (House 1965; Cummins 1973; Cammen 1980).

Living food sources generally have a higher nutritive and caloric content than the same materials have when they are dead. However, the nutritional value of dead material is greatly enhanced when a surface microflora, composed of fungi and bacteria, develops (Booth and Anderson 1979).

Protein is the principle source of amino acids for insects, and it frequently forms a major portion of their diets (Chapman 1982). Carbohydrates are commonly used as the major source of energy, and may also be converted into fats for energy storage or into amino acids; however, in certain instances, carbohydrates can be replaced in the diet by proteins or lipids. Among the lipids ingested by insects, fats are usually not essential dietary items because they can be synthesized from carbohydrates, but all insects require a source of sterol in the diet for normal growth. Cellulose and lignin components of ingested material generally have no nutritive value, although a few aquatic insects may be able to hydrolyze cellulose (see Digestion and Assimilation in this chapter). Although the inorganic or ash component of insect diets serves as a source of essential salts and trace elements, most of this material provides no nutritional benefit.

Bacteria and microscopic algae have protein contents of up to 67 percent of their dry weight (Table 7.3) and probably offer the best nutrition to insects. In some species of microscopic algae, carbohydrates may exceed 50 percent of the dry weight and lipids may reach 40 percent, but amounts are usually much less (Table 7.3). Cellulose and lignin, both of which are difficult to digest, seldom exceed 20 percent in either bacteria or microscopic algae. Similarly, ash levels in bacteria and algae are usually less than 20 percent. However, silica in the cell wall of diatoms (Bacillariophyceae) greatly elevates the ash content in that group (e.g. ranging up to 57%; Table 7.3). Although protein, lipid, and carbohydrate levels are relatively low in diatoms, the presence of puncta in the cell wall permits many insects to rapidly digest the cytoplasm, and this makes diatoms a good food source despite their high ash levels.

Filamentous forms of macroscopic algae, such as *Cladophora* and *Spirogyra* (Chlorophyta), generally incorporate relatively large amounts of cellulose and lignin, which provide structural support for the plants, but also result in high ash levels (Table 7.3). Carbohydrate levels of filamentous algae are comparable to those in microscopic algae. However, since the cell wall of filamentous forms is generally thick, it accounts for much of the dry weight, and thus total protein and lipid levels are relatively low compared to microscopic species. In general, due to their thick cell walls or mucous coating, chlorophyte and cyanophyte algae are apparently less readily digested than diatoms.

Fine detritus, alone, is an extremely poor food source. In most instances, the combination of largely undigestible cellulose, lignin, and ash constitutes greater than 90 percent of the dry weight of both suspended and deposited fine detritus (Table 7.3). Protein and lipids are correspondingly low (usually $< 5\%$ each), reflecting their rapid rate of hydrolysis during decay (de Leeuw et al. 1977). Consequently, the bacteria, fungi, and other microorganisms associated with fine detritus may be an important food source for detritivores. However, because such microorganisms constitute only a small portion of the total weight of detritus (Paerl 1977), detritivores probably need rapid ingestion-egestion rates to obtain enough nutriment from detritus.

Nutritional quality not only varies considerably among potential food sources (e.g. bacteria cf. fine detritus), but also within many specific foods (e.g. microscopic algae; see ranges in Table 7.3). This, at least in part, reflects the influence of seasonal variability and environmental conditions on plant metabolism. For example, Spoehr and Milner (1949) were among the first to demonstrate that the protein content of *Chlorella* (Chlorophyta) and other algae could vary considerably (e.g. from 10–86%), depending on growth conditions. Seasonal variation in nutritional composition has also been recorded in a number of natural algal populations (Moore 1975).

In waters of the temperate zone, the highest organic levels often occur in the autumn, coincident with leaf deposition. Some detrital components such as sterols break down rapidly in water (Gaskell and Eglinton 1975), so the nutritional content of detritus is likely to show even greater seasonal variability. These leaf inputs likely promote the growth of heterotrophs through the release of organic compounds that are available for microbial growth and, as such, for the growth of aquatic insects. The organic content of detritus may also show marked differences within a particular water body. For example, Calow (1975) reported that lotic detritus on the surfaces of rocks had an organic content of 63 percent dry weight, whereas the organic content of detritus between stones was 44 percent. Similarly, in lakes where the bottom slope is steep, organic material accumulates in deep depositional basins (Hargrave and Nielsen 1977).

DIGESTION AND ASSIMILATION IN PRIMARY CONSUMERS

The gut of many primary consumers is apparently adapted to hydrolyze cellulose and other carbohydrates, but it is not known if cellulose is digested under natural conditions (Halliwell 1959; Dash et al. 1981). The enzyme cellulase is present in a number of consumer groups, including the Ephemeroptera (Ephemerellidae, Ecdyonuridae, Caenidae), Trichoptera (Hydropsychidae, Limnephilidae, Polycentropodidae), Diptera (Chironomidae), Megaloptera, Coleoptera, and Hemiptera (Bjarnov 1972; Monk 1976, 1977). In most insects, it is not clear if these enzymes are endogenous or of microbial origin. Although some invertebrates such as the marine wood-boring isopod *Limnoria lignorum*

TABLE 7.3: Nutritional Composition (% dry weight) of Bacteria, Microscopic and Macroscopic Algae, Vascular Plants, and Fine Detritus*

TAXON	PROTEIN	LIPID	CARBOHYDRATE	CELLULOSE AND LIGNIN	ASH
Bacteria					
Escherichia coli[1]	63–67	20–22	5–6	0	8–9
Microscopic algae					
Cosmarium laeve[2]	11.5–20.1	15.0–33.3	32.0–60.0	4.5–13.0	4.8–12.5
Scenedesmus dimorphus[2]	8.0–17.5	16.4–40.3	21.0–51.1	6.5–20.0	6.2–14.8
Freshwater algae					
mixed species[1]	12–14	21–23	19–21	3–5	39–41
mixed species[3]	41–54	20–27	20–38	NR	NR
Bacillariophyceae[4]	NR	NR	NR	NR	27.2–55.0
Chlorophyceae[4]	NR	NR	NR	NR	5.3–19.9
Myxophyceae[4]	NR	NR	NR	NR	18.6
Marine algae					
Bacillariophyceae[5]	17–37	1.8–6.9	4.1–24.0	NR	7.6–57.0
Chrysophyceae[5]	49–56	4.6–11.6	17.8–31.4	NR	6.4–36.5
Chlorophyceae[5]	52–57	2.9–6.4	15.0–31.6	NR	7.6–23.8
Dinophyceae[5]	28–31	15–18	30.5–37.0	NR	8.3–14.1
Myxophyceae[5]	36	12.8	31.5	NR	10.7
mixed species[6]	13.6–25.1	3.5–19.0	14.1–33.3	NR	33.8–50.5

Macroscopic algae				
Cladophora glomerata[2]	9.8–15.0	7.0–25.2	20–51	13.0–23.9	13.3–32.5
Palmaria palmata[7]	8–35	0.3–3.8	38–74	1.5–3.5	12–37
Spirogyra sp.[2]	6.1–20.2	10.8–20.5	33–64	4.6–16.5	10.0–27.5
Ulothrix zonata[2]	3.5–12.5	4.2–19.3	22–66	11.5–20.5	12.5–35.8
Vascular plants				
Hydrilla verticillata[8]	17.5	1.9	40.1	15.2	17.6
Lemna gibbs[8]	20.3	6.7	34.1	8.6	23.1
Pistia stratiotes[8]	30.5	9.5	24.9	9.1	18.9
mixed species[9]	4.0–21.6	NR	NR	NR	NR
Fine detritus				
Bottom sediment[1,10–12]	1.5–2.0	2–3	28–33	7–9	55–65
	3.9	3.2	NR	11.1	82.6
	0.4–3.0	0–1.5	0–4	8–55	37–91
Suspended solids[10,13]	NR	NR	NR	NR	87.5–98.4
	10.6	6.5	NR	17.3	67.2
	NR	NR	NR	NR	86–92

*NR = not reported

Data from [1]Moore and Potter 1976; [2]Moore 1975; [3]Ketchum and Redfield 1949; [4]Nalewajko 1966; [5]Parsons et al. 1961; [6]Platt and Irwin 1973; [7]Morgan et al. 1980; [8]Tan 1970; [9]Boyd 1970; [10]Baker and Bradnam 1976; [11]Sain et al. 1977; [12]Cranwell 1977; [13]Serruya 1977 digest cellulose by endogenous enzymes (Ray and Julian 1952), most species that rely on cellulose for food also possess a well-developed enteric microflora. The mechanisms by which such bacteria aid digestion in aquatic insects are incompletely understood at present.

The midgut pH of invertebrate consumers is generally either neutral or alkaline; these nonacidic conditions promote the dissociation of ingested protein, such as that in protein-tannin complexes (Berenbaum 1980). Casein hydrolysis by the amphipod *Gammarus pulex* is also highest at neutral to alkaline conditions, and includes peaks at pH 7.0 to 7.5 and, possibly, at pH >8.75 (Monk 1977). The midgut of the detritivorous cranefly larva *Tipula abdominalis* has an alkaline pH that is suitable for dissociation of protein-tannin complexes (Martin et al. 1980). In addition, the high proteolytic activity in *T. abdominalis* enables larvae to obtain nitrogen, perhaps from ingested microorganisms.

Primary consumers often have rapid ingestion-egestion rates, and the total amount of food eaten per day may exceed the body weight (Ladle et al. 1972; Mulla and Lacey 1976; Wotton 1978a). However, considerable variation in the intake rate exists. For example, the stonefly *Pteronarcys scotti*, with a mean dry weight equal to 8.4 mg, consumed only 1.5 mg of detritus per day (Cammen 1980), whereas the amphipod *Hyallela azteca*, with a mean dry weight equal to 0.32 mg, consumed 0.4 mg of detritus daily (Hargrave 1972). The gut-clearing time of deposit-feeding crustaceans may vary from 0.5 to 12 hours, depending on the species, temperature, and food type (Martin 1966; Hargrave 1970; Cammen 1980). Food retention times vary from 0.3 to 2 hours for larval Simuliidae (Ladle et al. 1972; Mulla and Lacey 1976; Wotton 1978a, 1978b; McCullough et al. 1979), and from 1.5 to 5.5 hours in the mayfly *Hexagenia* and the caddisfly *Hydropsyche* (Zimmerman et al. 1975; Zimmerman and Wissing 1978; McCullough et al. 1979; Wallace and Merritt 1980). At the extreme, actively feeding chironomid larvae may have a retention time of less than 10 minutes (Walshe 1951). In some cases, aquatic insects may use rapid ingestion-egestion rates to compensate for the low nutritive quality of specific foods. For example, *Tipula abdominalis* larvae grow about twice as fast on basswood than on hickory, but feed nearly two times faster on hickory (Cummins and Klug 1979).

Assimilation efficiencies in primary consumers are probably low, but little specific information on this is available (Benke and Wallace 1980; Giguère 1981). Depending on food type, assimilation efficiencies of different species of Simuliidae vary from 2 to 57 percent (Wotton 1978b; McCullough et al. 1979), whereas those for Hydropsychidae and *Hexagenia* range from 40 to 80 percent (Zimmerman et al. 1975; McCullough et al. 1979). A similar range of variation (5–50% efficiency) has been reported for freshwater and marine crustaceans, molluscs, and annelids (Welch 1968; Prus 1971; Cammen et al. 1980). Benke and Wallace (1980) reported that the assimilation efficiency of the caddisfly *Arctopsyche irrorata* is only about 10 percent for fine detritus and vascular plant detritus, but increases to about 30 percent for algae. However, in many consumer species, large quantities of viable algae may be found within the gut contents and feces, indicating that little digestion of these items has occurred (Chapman and Demory 1963; Moore 1977; Clifford et al. 1979).

ALGAL RESPONSES TO HERBIVORY

In the previous sections, we dealt with general patterns of primary consumption. In this section, we will specifically discuss herbivory, which is the ingestion of living plant material that is produced within the water body. Primary, or *autochthonous,* production within aquatic habitats involves vascular hydrophytes, phytoplankton (i.e. algae suspended in the water column), and periphyton or Aufwuchs (i.e. algae attached to a substrate). Periphyton, in a broad sense, includes benthic algae, bacteria, and their secretions, along with associated fine detritus and various species of microscopic invertebrates.

The importance of primary production to consumers in running waters has been reviewed by Minshall (1978), who found that autochthonous production often exceeded the contribution of *allochthonous* (terrestrially derived) inputs to aquatic energy budgets. Since periphyton is a dominant component of lotic autochthonous production, it is of major importance to organisms that use living plant material for growth and reproduction. Freshwater ecologists have identified a considerable number of aquatic insect species within the scraper and shredder groups that graze periphyton or consume aquatic macrophytes (Cummins 1973, 1978), but comparatively little is known about the interactions of consumers and periphyton food resources in freshwater ecosystems. McIntire (1973) concluded that this was due to both a historical research emphasis based on allochthonous control of stream energy budgets (e.g. Fisher and Likens 1972; Cummins 1974), and the less apparent nature of periphyton to observers when compared to highly visible allochthonous inputs such as logs and leaf accumulations.

Although there have been few experimental investigations of herbivore-periphyton interactions in freshwater habitats, a number of studies have documented correlations between herbivores and food resources. For example, Douglas (1958) noted an inverse relationship between the density of the diatom *Achnanthes* sp. and the abundance of the grazing caddisfly *Agapetus fuscipes.* In contrast, a large number of in situ experimental studies have been conducted in the marine rocky intertidal habitat, and these studies have shown that invertebrate herbivores can substantially alter the standing crop, species diversity, successional rates, distributional limits, and life histories of marine algae (see review by Lubchenco and Gaines 1981). These marine studies can be instructive in designing freshwater investigations. For example, the experimental animals used in marine studies, especially gastropods such as snails and limpets, generally have ecological analogs in freshwater habitats that can be easily manipulated. In addition, many of these studies have used herbivore-exclusion methods, such as simple cages, barriers, or chemicals, that are easily adapted for use in fresh waters. The remainder of this chapter will emphasize the effects of grazing by macroinvertebrates on the standing crop, productivity, species composition, and diversity of freshwater periphyton, as determined by experimental investigations.

Grazer Effects on Standing Crop

Experimental investigations of grazing effects on freshwater periphyton have primarily used gastropods. For example, of the 12 experimental studies summarized in Table 7.4, seven used snails, two used insects, and one used a crustacean; two other studies used tadpoles and fish.

In one of the first in situ experimental studies of herbivory, Dickman (1968) demonstrated that grazing by *Rana aurora* tadpoles reduced the standing crop of periphyton (in particular the amount of filamentous green algae) in Marion Lake, British Columbia. Apparently, the annual spring blooms of filamentous algae in the littoral regions of the lake were controlled by the feeding of newly born tadpoles. Both laboratory (Doremus and Harman 1977) and in situ (Hunter 1980; Cuker 1983a, 1983b) experiments have shown that freshwater snails may also greatly reduce the standing crop of algae. In fact, Hunter (1980) reported that combinations of the snails *Lymnaea, Physa,* and *Helisoma* reduced virtually all measured features of pond periphyton, including dry weight, chlorophyll *a*, algal cell density, and carbon content. Cattaneo (1983) further demonstrated that grazer assemblages, which included chironomid larvae, could significantly reduce the biomass of lentic epiphytic algae. This relationship may partly control the seasonal cycling in epiphyte abundance in certain lakes (Mason and Bryant 1975; Cattaneo 1983). Grazing may also improve the nutritional quality of algae by increasing the relative proportions of nitrogen to total dry weight (Hunter 1980), or of active chlorophyll *a* to its degradation products (Cuker 1983a).

Some investigators, however, have reported that grazers have little or no effect on standing crop. For example, Kesler (1981) found that the snail *Amnicola limosa* had little overall effect on the standing crop of pond periphyton, although the densities of some small diatoms were reduced. In a study using laboratory streams, Kehde and Wilhm (1972) reported that the snail *Physa gyrina* also had no effect on the dry weight of periphyton and, in fact, grazing increased the amount of chlorophyll *a*. Sumner and McIntire (1982) showed that grazing by the snail *Juga plicifera* in laboratory streams reduced periphyton biomass by only up to 30 percent, and the amount was frequently much less.

Two experimental studies of grazing by insects used markedly different approaches for assessing the responses of periphyton. Eichenberger and Schlatter (1978) excluded chironomid larvae from outdoor artificial channels by using periodic insecticide treatments. They observed a higher standing crop of periphyton in the treated, ungrazed channels, but only after a 50-day growth period. In contrast, Lamberti and Resh (1983) excluded larvae of the caddisfly *Helicopsyche borealis* from in situ lotic plots by elevating substrate 15 cm above the stream bottom. In three replicate experiments, grazing reduced the amount of benthic chlorophyll *a*, numbers of bacteria, and total organic matter to less than 10 percent of the ungrazed levels after 4 weeks.

Grazer Effects on Primary Productivity

In both aquatic and terrestrial systems, primary productivity, or the *rate* of tissue elaboration, appears to be maximized at relatively immature (low biomass) states (e.g. Marker 1976; McNaughton 1976; Sumner and McIntire 1982). By using radioisotopes to quantify primary production and grazing rates in a stream, Elwood and Nelson (1972) suggested that stream grazers influenced productivity by maintaining periphyton at a low standing crop.

McIntire (1973) provided the impetus for further studies of this relationship by using a simulation model to show that a small, but highly productive, amount of periphyton could support a relatively large standing crop of consumers. Empirical confirmation of this relationship was provided soon afterward when Cooper (1973) showed that low densities of grazers (the phycophagous minnow *Notropis spilopterus*) stimulated primary productivity compared with an ungrazed control, whereas higher densities of grazers reduced productivity. He suggested that the increased productivity of grazed algae was due to their higher turnover rate. Flint and Goldman (1975), using carbon-14 techniques, found a similar enhancement of productivity at low densities of the grazing crayfish *Pacifastacus leniusculus* in laboratory aquaria and in Lake Tahoe, California. They concluded that grazing maintained periphyton in a vigorous growth phase.

Other researchers have evaluated periphyton productivity by using chlorophyll *a* estimates as part of various ratios, although such ratios are indirect measurements of productivity. For example, the increased proportion of chlorophyll *a* per unit total biomass (Hunter 1980) or per unit phaeophytin (Cuker 1983a) may indicate a higher turnover rate in grazed lentic periphyton. Lamberti and Resh (1983) attempted to quantify the relationship between productivity and biomass of grazed lotic algae in two ways, by using an O_2-evolved/chlorophyll *a* ratio and by estimating the periphyton turnover time with the $P/\overline{B}$ ratio. The first ratio showed that algae grazed by *Helicopsyche* caddisflies had a much higher turnover rate (O_2 evolved per unit chlorophyll $a = 34$ $\mu g \cdot \mu g^{-1} \cdot h^{-1}$) than ungrazed algae ($O_2$ evolved per unit chlorophyll $a = 7$ $\mu g \cdot \mu g^{-1} \cdot h^{-1}$). Using the second ratio, they calculated a turnover time of 8 days for grazed algae compared with 16 days for ungrazed algae. Thus, both ratios indicated that the turnover of periphyton was markedly accelerated by grazing.

The increased productivity or turnover rate of periphyton at certain grazer densities may be explained by one or a combination of several mechanisms

TABLE 7.4: Summary of Selected Experimental Studies of Grazer Effects on Various Features of Freshwater Periphyton*

STUDY	HABITAT	GRAZER	EFFECT ON PRODUCTION	EFFECT ON OTHER FEATURES
Beyers 1963	Laboratory aquaria	Snail: *Marisa* sp.	(−) 1° productivity	Not reported
Dickman 1968	Lake enclosures	Tadpole: *Rana aurora*	(−) algal standing crop (dry weight)	(−) density of filamentous green algae (−) density of desmid diatoms
Kehde and Wilhm 1972	Laboratory streams	Snail: *Physa gyrina*	(o) algal standing crop (dry weight, AFDW)	(o) algal pigment diversity (o) algal species diversity
Cooper 1973	Laboratory aquaria	Cyprinid fish: *Notropis spilopterus*	(+, −) 1° productivity	(+?) algal turnover rate
Flint and Goldman 1975	Lake enclosures, laboratory aquaria	Crayfish: *Pacifastacus leniusculus*	(+, −) 1° productivity	(+?) algal turnover rate
Doremus and Harman 1977	Laboratory aquaria	Snails: *Physa heterostropha, Promenetus exacuous*	(−) algal standing crop (chlorophyll *a*)	Not reported
Eichenberger and Schlatter 1978	Outdoor artificial streams	Chironomid midge larvae (Diptera)	(−) algal standing crop (AFDW)	(−) density of blue-green algae (o) algal export

Reference	System	Grazer	Effects	Effects
Hunter 1980	Pond enclosures	Snails: *Helisoma trivolvis, Lymnaea elodes, Physa gyrina*	(−) algal standing crop (dry weight, carbon, cell density, chlorophyll *a*); (−) 1° productivity	(+) chlorophyll/dry weight of algae; (+) nitrogen/dry weight of algae; (−) algal species diversity; (+?) algal nutritional quality
Kesler 1981	Pond enclosures	Snail: *Amnicola limosa*	(?) algal standing crop (AFDW); (+,o) organic content of algae (AFDW/dry weight)	(o) density of *Cocconeis* diatoms; (−) density of other small diatoms
Sumner and McIntire 1982	Laboratory streams	Snail: *Juga plicifera*	(o) 1° productivity; (−,o) algal standing crop (AFDW)	(+) algal export; (o) chlorophyll/AFDW of algae; (o) algal species diversity
Lamberti and Resh 1983	Stream in situ exclosures	Caddisfly: *Helicopsyche borealis*	(−) algal standing crop (chlorophyll *a*, phaeophytin, AFDW); (−) bacterial density; (−) 1° productivity	(+) algal turnover rate (chlorophyll/O_2, $P/\overline{B}$); (−) algal turnover time; (−) density of filamentous green algae
Cuker 1983a, 1983b	Lake in situ enclosures	Snail: *Lymnaea elodes*	(−) algal standing crop (chlorophyll *a*, dry weight); (+,−) 1° productivity	(+) chlorophyll/phaeophytin of algae; (+) density of small green and blue-green algae

* (−) = decrease in periphyton feature (per unit habitat area)
  (+) = increase
  (o) = no change
  AFDW = ash-free dry weight

(Hunter 1980; Sumner and McIntire 1982; Cuker 1983b; Lamberti and Resh 1983), as follows:

1. Grazing generally reduces the number of cells and thickness of the algal layer, and may thus prevent nutrient and light limitations from occurring.
2. Grazers may ingest or dislodge, and thereby remove, dead and senescent algal cells that reduce the overall productivity of the population.
3. Waste products of the grazers may provide nutrients for algal growth.
4. Mechanical disruption of some algal cells by grazers may regenerate nutrients for other cells.
5. Grazing may prevent the accumulation of algal wastes that are detrimental to normal growth.
6. Grazers may selectively remove less productive species, while leaving fast-growing species.

It is difficult to evaluate the relative importance of these mechanisms because detailed information is lacking. The first two mechanisms, however, may have more universal importance than the others (Sumner and McIntire 1982; Lamberti and Resh 1983). For example, in the lower layers of thick periphyton mats, self-shading and nutrient depletion may reduce productivity in much the same way as they do in dense phytoplankton blooms. Grazers thin the periphyton mat while possibly removing older, less productive cells; in this way, productivity remains high throughout the grazed algal mat. Such mechanisms may also be important in lentic systems since there is some evidence that moderate grazing by zooplankton stimulates phytoplankton productivity (Porter 1977 and references therein).

Selective Feeding

Freshwater grazers appear to display a limited degree of selectivity in the ingestion of various algal species, but the selectivity that does exist may be based more on a combination of algal features and mechanical limitations of the grazers' scraping mouthparts than on active choice *per se*. For example, the cranefly larva *Tipula sacra* uses setal fringes on the mouthparts to consume large diatoms preferentially, but this is probably a case of passive (rather than active) selection that is mediated by the intersetal distance (Pritchard 1983). The inability of the oligochaete *Nais elinguis* to ingest colonial or filamentous algae is probably due to the morphology of that species' pharynx (Bowker et al. 1983).

Some microbial and algal forms appear to be relatively resistant to grazing. For example, Dickman (1968) reported that certain tightly adherent bacteria are usually missed by grazers. There is also evidence that small, closely adherent diatoms, such as *Achnanthes* and *Navicula*, are also only lightly grazed (Hunter 1980; Sumner and McIntire 1982). Patrick (1970) found that *Physa* snails did not feed on the small, adherent diatom *Cocconeis placentula*, and experimental studies have also shown that other gastropods have little effect on *Cocconeis* densities (Hunter 1980; Kesler 1981). Although some species of chironomid larvae may preferentially ingest and digest certain algal taxa (Kajak and Warda 1968; Izvekova 1971; Davies 1975), the basis for such selectivity is not known.

Some forms of algae seem to be susceptible to grazing, and these include filamentous algae (Dickman 1968; Kesler 1981) and other large, overstory algae such as stalked diatoms (Sumner and McIntire 1982). For example, freshwater snails feed on most species of large-sized algae (Hunter 1980) and, similarly, medium-sized to large-sized pennate diatoms are readily grazed (Calow 1970; Hunter 1980; Bowker et al. 1983).

In addition to growth form, the reproductive ability of algae may explain observed consumption patterns. For example, resistant species may be readily consumed, but the remaining cells may reproduce quickly enough to compensate for consumption. Conversely, susceptible species may grow too slowly to compensate for grazing losses (Dickman 1968; Sumner and McIntire 1982). This may also explain why grazed algae have higher turnover rates than ungrazed algae.

Certain freshwater algae may gain some degree of resistance to grazing through chemical defenses. For example, the filamentous green alga *Cladophora* may produce chemical acids that inhibit grazing by benthic invertebrates (Hutchinson 1981), just as some species of phytoplankton may produce toxins that inhibit feeding by zooplankton (Porter 1977). Unfortunately, insufficient information is available to assess the importance of antiherbivore, or allelo-chemical, compounds in freshwater plants.

Algal Succession and Diversity

Most marine studies have shown that invertebrate grazers accelerate algal succession by cropping the first algal colonists, with the result that late-succession species can become established (e.g. Sousa 1979; Robles and Cubit 1981). However, in freshwater habitats, most information suggests that grazing prevents successional changes. For example, Eichenberger and Schlatter (1978) reported that in ungrazed stream channels periphyton changed successively from fila-mentous green algae to blue-green algae, whereas, in grazed stream channels, periphyton remained a mixture of filamentous green algae and diatoms. Lamberti and Resh (1983) observed that while grazed periphyton remained a predominantly diatom monolayer, ungrazed periphyton changed from a diatom film to a dense turf of filamentous green algae. However, in at least one instance, grazers may have accelerated successional changes; for example, Dickman and Gochnauer (1978) reported that stream grazers removed an early colonizing diatom species, which allowed other algal species to become established.

The apparent tendency of freshwater grazers to halt succession may be related to their low degree of food selectivity. In freshwater algal mats, late-succession species are typically large, overstory algae, such as stalked or filamentous types that are highly susceptible to grazing, and are apparently consumed in approximately equal proportions to their abundance. Consequently, fast-growing, closely adhering species such as diatoms predominate.

Algal species diversity may be altered by mechanisms similar to those that limit succession, by size-based selective feeding, or by a combination of these factors. Freshwater grazers have been shown to increase (Dickman and Gochnauer 1978), decrease (Hunter 1980), or have no effect on algal diversity (Kehde and Wilhm 1972; Sumner and McIntire 1982). Even if the number of species remains unchanged, grazers frequently alter the relative abundances of different algal species (e.g. Eichenberger and Schlatter 1978; Sumner and McIntire 1982; Lamberti and Resh 1983). For example, Dickman (1968) observed that filamentous green algae declined in abundance and desmid diatoms were completely eliminated within grazed pond enclosures.

Importance of Grazing: Synthesis

The effects of grazing on the productivity and biomass of periphyton are summarized by a general model (Fig. 7.1) in which grazers reduce the standing crop of periphyton as their own density and, thus, grazing pressure increases. At very low densities of grazers (i.e. undergrazing; Zone A), the periphyton is only

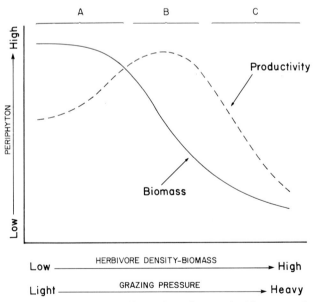

Figure 7.1 Empirical model of the effects of grazing on the biomass and productivity of freshwater periphyton per unit habitat area. Zone A = undergrazing; B = optimal grazing level; C = overgrazing.

TABLE 7.5: Effects of Different Invertebrate Densities or Biomasses on Periphyton Standing Crop and/or Primary Productivity (per unit habitat area) from Selected Experimental Studies*

GRAZER	DENSITY (number/m²)	BIOMASS (g/m²)	EFFECT ON PERIPHYTON
Snail[1]	45–107	NR	(o,−) standing crop
Snail[2]	120	15.6	(o) standing crop
Snail[3]	125	1.3	(o,−) standing crop
			(o) productivity
Snail[4]	8–192	4–90	(−) standing crop
			(+,−) productivity
Snail[5]	216	NR	(−) standing crop
			(−) productivity
Snail[3]	500	5.3	(o,−) standing crop
			(o) productivity
Snail[6]	NR	2.6–7.7	(−) standing crop
Crayfish[7]	NR	68–131	(+) productivity
Crayfish[7]	NR	203–295	(−) productivity
Caddisfly[8]	4,167	NR	(−) standing crop
			(−) productivity
Snail[9]	NR	575	(−) productivity

* (−) = decrease in periphyton feature
   (+) = increase
   (o) = no change
   NR = not reported

Data from [1]Kesler 1981; [2]Kehde and Wilhm 1972; [3]Sumner and McIntire 1982; [4]Cuker 1983a, 1983b; [5]Hunter 1980; [6]Doremus and Harman 1977; [7]Flint and Goldman 1975; [8]Lamberti and Resh 1983; [9]Beyers 1963

moderately productive even though its biomass is high. However, productivity increases as the density of grazers increases and the biomass of periphyton declines; productivity is highest at some intermediate grazer density (i.e. optimal grazing level; Zone B). Further increases in grazing pressure decrease both periphyton productivity and biomass because the resource is overutilized (i.e. overgrazing; Zone C).

Although few studies have experimentally varied the density of grazers, the existing data (Table 7.5) generally support the above model. For example, Flint and Goldman (1975) reported that, at grazer densities less than 131 g/m², primary productivity was greater than in ungrazed controls; conversely, at grazer densities greater than 203 g/m^2, productivity was less than in the controls. Cooper (1973) reported a similar trend for two densities of grazing fish, and he suggested that herbivore densities in natural systems form a continuum from undergrazing to overgrazing. Lamberti and Resh (1983) showed that productivity declined, although turnover rate increased, at natural grazer densities of 4,167 individuals/m^2, in what is probably an example of severe overgrazing. In contrast, Sumner and McIntire (1982) reported that algal productivity at grazer densities of 1.3 and 5.3 g/m^2 was the same as that of the ungrazed control; thus, both densities were probably examples of undergrazing (cf. grazer densities of Flint and Goldman 1975). In general, the investigations in which grazers have had little or no effect on periphyton standing crop or productivity (e.g. Kesler 1981; Kehde and Wilhm 1972 in Table 7.5) probably documented undergrazing.

In recent studies of grazing by snails in an arctic lake, Cuker (1983a, 1983b) experimentally varied grazing pressure in littoral-zone enclosures. All grazing levels tested reduced the biomass of periphyton, but intermediate densities increased the ratio of chlorophyll a to phaeophytin (Cuker 1983a), an indicator of algal activity. Although Cuker (1983b) claimed that primary productivity was also consistently reduced by grazing, his data indicate that low grazer densities (e.g. 4 mg/m^2; Cuker 1983b, Fig. 3) consistently resulted in higher productivity than in the ungrazed controls. Since that low density of snails roughly corresponded to the natural lake density, periphyton production may have been optimized for the natural snail population as well.

Overgrazing is the most frequently encountered condition in experimental studies, as demonstrated by the findings of Flint and Goldman (1975) in a lake littoral zone, Hunter (1980) in a pond, and Lamberti and Resh (1983) in a stream. Hart (1981) suggested that many natural systems may in fact be overgrazed, and his experiments with the stream-dwelling caddisfly *Dicosmoecus gilvipes* indicated that individuals competed exploitatively for limited periphyton resources. He further reasoned that since *D. gilvipes* was ony one member of a diverse grazer guild, overexploitation of the periphyton resource was probably even more severe. In fact, it is important to consider the influence of the entire grazer assemblage when extrapolating experimental evidence to natural systems. Therefore, although primary production may be enhanced by low grazer densities, grazing pressure in natural systems is usually intense enough to result in a net decline in primary production.

CONCLUSIONS AND FUTURE RESEARCH NEEDS

The characterization of feeding mechanisms by function, and the subsequent use of such functional groups in aquatic insect studies, has been a popular approach to the analysis of primary consumption patterns. However, since such functional groups are frequently generalized to the generic or family level, caution must be exercised in projecting those food habits to individual species (Resh 1976). Likewise, since substantial variation in feeding habits may occur within individual species as well, trophic studies should not be based on *anticipated* food habits.

It is unfortunate that so little information is available on the processes of digestion and assimilation in aquatic insects. Certain types of ingested matter are probably of little or no benefit to consumers, whereas small amounts of other materials may be critical. As our knowledge of these relationships expands, many of our theories on trophic relations among aquatic insects may prove to be inaccurate or at least more complex than we previously suspected. For example, although current information suggests that collectors directly depend on shredders for some of their detrital food, the low digestibility of fine detritus may minimize the importance of that interaction. In a more fundamental sense, current research on aquatic insect foods has not firmly established the relative importances of the microflora and the substrate on which it exists in insect nutrition.

Although there have been many advances in our knowledge of the feeding ecology of aquatic insects in the filterer and shredder groups, little information exists on the trophic importance of scrapers and deposit-collectors in aquatic systems. In particular, our knowledge of grazer-periphyton interactions in freshwater ecosystems is extremely limited. Most of the relationships covered in this chapter, such as grazing effects on primary production, are incompletely understood, and features such as plant defenses have yet to be examined in detail. The following key questions concerning the interactions of grazers and periphyton in freshwater systems deserve special attention in future experimental investigations:

1. What are the relative importances of grazers and physical-chemical factors in determining the structure of freshwater periphyton communities? For example, does the importance of grazing and physical features (such as ice scour during the winter) alternate seasonally?

2. To what extent do freshwater grazers alter the distribution patterns, morphologies, and phenologies of freshwater algal species?

3. Do herbivorous aquatic insects prefer certain foods or are there few specialist grazers of freshwater periphyton? How widespread is the use of chemical defenses by freshwater plants and what is their effectiveness against grazers?

4. Is freshwater periphyton typically overgrazed and, if so, to what extent does competition among grazers for resources influence grazer-periphyton interactions?

The template for grazer-algal interactions is the set of abiotic environmental features, such as light, temperature, and nutrient levels, that regulate the potential rate of increase in algal populations. Grazers operate on this template to influence algal community development or to modify certain features of established communities. Such interactions undoubtedly have significant implications for the structuring and operation of aquatic ecosystems.

References

Ameen, M. and T. M. Iversen. 1978. Food of *Aëdes* larvae (Diptera, Culicidae) in a temporary forest pool. Archiv für Hydrobiologie 83:552–64.

Anderson, N. H. 1976. Carnivory by an aquatic detritivore, *Clistoronia magnifica* (Trichoptera: Limnephilidae). Ecology 57:1081–85.

Anderson, N. H. and K. W. Cummins. 1979. Influences of diet on the life histories of aquatic insects. Journal of the Fisheries Research Board of Canada 36:335–42.

Anderson, N. H. and J. R. Sedell. 1979. Detritus processing by macroinvertebrates in stream ecosystems. Annual Review of Entomology 24:351–77.

Baker, A. S. and A. J. McLachlan. 1979. Food preferences of Tanypodinae larvae (Diptera: Chironomidae). Hydrobiologia 62:283–88.

Baker, J. H. and L. A. Bradnam. 1976. The role of bacteria in the nutrition of aquatic detritivores. Oecologia (Berlin) 24:95–104.

Bärlocher, F. 1980. Leaf-eating invertebrates as competitors of aquatic hyphomycetes. Oecologia (Berlin) 47:303–6.

Benke, A. C. and J. B. Wallace. 1980. Trophic basis of production among net-spinning caddisflies in a southern Appalachian stream. Ecology 61:108–18.

Berenbaum, M. 1980. Adaptive significance of midgut pH in larval Lepidoptera. The American Naturalist 115:138–46.

Beyers, R. J. 1963. The metabolism of twelve aquatic laboratory microecosystems. Ecological Monographs 33:281–306.

Bjarnov, N. 1972. Carbohydrases in *Chironomus, Gammarus* and some Trichoptera larvae. Oikos 23:261–63.

Booth, R. G. and J. M. Anderson. 1979. The influence of fungal food quality on the growth and fecundity of *Folsomia candida* (Collembola: Isotomidae). Oecologia (Berlin) 38:317–23.

Bowker, D. W., M. T. Wareham, and M. A. Learner. 1983. The selection and ingestion of epilithic algae by *Nais elinguis* (Oligochaeta: Naididae). Hydrobiologia 98:171–78.

Boyd, C. E. 1970. Amino acid, protein, and caloric content of vascular aquatic macrophytes. Ecology 51:902–6.

Brinkhurst, R. O. and K. E. Chua. 1969. Preliminary investigation of the exploitation of some potential nutritional resources by three sympatric tubificid oligochaetes. Journal of the Fisheries Research Board of Canada 26:2659–68.

Brinkhurst, R. O., K. E. Chua, and N. K. Kaushik. 1972. Interspecific interactions and selective feeding by tubificid oligochaetes. Limnology and Oceanography 17:122–33.

Brook, A. J. 1955. The aquatic fauna as an ecological factor in studies of the occurrence of freshwater algae. Revue Algologique 1:142–45.

Brown, D. S. 1960. The ingestion and digestion of algae by *Chloëon dipterum* L. (Ephemeroptera). Hydrobiologia 16:81–96.

Brown, D. S. 1961. The food of the larvae of *Chloëon dipterum* L. and *Baëtis rhodani* (Pictet) (Insecta, Ephemeroptera). Journal of Animal Ecology 30:55–75.

Burton, G. J. 1973. Feeding of *Simulium hargreavesi* Gibbons larvae on *Oedegonium* algal filaments in Ghana. Journal of Medical Entomology 10:101–6.

Byers, G. W. 1978. Tipulidae, pp. 285–310. *In:* R. W. Merritt and K. W. Cummins (eds.)

An introduction to the aquatic insects of North America. Kendall/Hunt Publishing Co., Dubuque, IA. 441 pp.

Calow, P. 1970. Studies on the natural diet of *Lymnaea peregor obtusa* (Kobelt) and its possible ecological implications. Proceedings of the Malacological Society of London 39:203–15.

Calow, P. 1975. On the nature and possible utility of epilithic detritus. Hydrobiologia 46:181–89.

Cammen, L. M. 1980. Ingestion rate: an empirical model for aquatic deposit feeders and detritivores. Oecologia (Berlin) 44: 303–10.

Cammen, L. M., E. D. Seneca, and L. M. Stroud. 1980. Energy flow through the fiddler crabs *Uca pugnax* and *U. minax* and the marsh periwinkle *Littorina irrorata* in a North Carolina salt marsh. The American Midland Naturalist 103:238–50.

Cattaneo, A. 1983. Grazing on epiphytes. Limnology and Oceanography 28:124–32.

Chapman, D. W. and R. L. Demory. 1963. Seasonal changes in the food ingested by aquatic insect larvae and nymphs in two Oregon streams. Ecology 44:140–46.

Chapman, R. F. 1982. The insects. Structure and function. 3rd ed. Harvard University Press, Cambridge, MA. 919 pp.

Clifford, H. F., H. Hamilton, and B. A. Killins. 1979. Biology of the mayfly *Leptophlebia cupida* (Say) (Ephemeroptera: Leptophlebiidae). Canadian Journal of Zoology 57: 1026–45.

Cooper, D. C. 1973. Enhancement of net primary productivity by herbivore grazing in aquatic laboratory microcosms. Limnology and Oceanography 18:31–37.

Cranwell, P. A. 1977. Organic compounds as indicators of allochthonous and autochthonous input to lake sediments, pp. 133–40. *In*: H. L. Golterman (ed.). Interactions between sediments and fresh water. Dr. W. Junk B. V. Publishers, The Hague, The Netherlands. 473 pp.

Cuker, B. E. 1983a. Competition and coexistence among the grazing snail *Lymnaea*, Chironomidae, and microcrustacea in an arctic epilithic lacustrine community. Ecology 64:10–15.

Cuker, B. E. 1983b. Grazing and nutrient interactions in controlling the activity and composition of the epilithic algal community of an arctic lake. Limnology and Oceanography 28:133–41.

Cummins, K. W. 1973. Trophic relations of aquatic insects. Annual Review of Entomology 18:183–206.

Cummins, K. W. 1974. Structure and function of stream ecosystems. BioScience 24: 631–41.

Cummins K. W. 1978. Ecology and distribution of aquatic insects, pp. 29–31. *In*:R. W. Merritt and K. W. Cummins (eds.). An introduction to the aquatic insects of North America. Kendall/Hunt Publishing Co., Dubuque, IA. 441 pp.

Cummins, K. W. and M. J. Klug. 1979. Feeding ecology of stream invertebrates. Annual Review of Ecology and Systematics 10:147–72.

Cummins, K. W., W. P. Coffman, and P. A. Roff. 1966. Trophic relationships in a small woodland stream. Internationale Vereinigung für Theoretische und Angewandte Limnologie Verhandlungen 16:627–38.

Dadd, R. H. 1971. Effects of size and concentration of particles on rates of ingestion of latex particulates by mosquito larvae. Annals of the Entomological Society of America 64:687–92.

Dadd, R. H. 1975. Ingestion of colloid solutions by filter-feeding mosquito larvae: relationship to viscosity. Journal of Experimental Zoology 191:395–406.

Dash, M. C., B. Nanda, and P. C. Mishra. 1981. Digestive enzymes in three species of Enchytraeidae (Oligochaeta). Oikos 36:316–18.

Davies, D. M. and P. D. Syme. 1958. Three new Ontario black flies of the genus *Prosimulium* (Diptera: Simuliidae). Part II. Ecological observations and experiments. The Canadian Entomologist 90:744–59.

Davies, I. J. 1975. Selective feeding in some arctic Chironomidae. Internationale Vereinigung für Theoretische und Angewandte Limnologie Verhandlungen 19:3149–54.

Dickman, M. 1968. The effect of grazing by tadpoles on the structure of a periphyton community. Ecology 49:1188–90.

Dickman, M. D. and M. B. Gochnauer. 1978. Impact of sodium chloride on the microbiota of a small stream. Environmental Pollution 17:109–26.

Doremus, C. M. and W. N. Harman. 1977. The effects of grazing by physid and planorbid freshwater snails on periphyton. Nautilus 91:92–96.

Douglas, B. 1958. The ecology of the attached diatoms and other algae in a small stony stream. Journal of Ecology 46:295–322.

Egglishaw, H. J. 1964. The distributional relationship between the bottom fauna and plant detritus in streams. Journal of Animal Ecology 33:463–76.

Eichenberger, E. and A. Schlatter. 1978. The effect of herbivorous insects on the production of benthic algal vegetation in outdoor channels. Internationale Vereinigung für Theoretische und Angewandte Limnologie Verhandlungen 20:1806–10.

Elwood, J. W. and D. J. Nelson. 1972. Periphyton production and grazing rates in a stream measured with a ^{32}P material balance method. Oikos 23:295–303.

Erman, N. A. 1981. Terrestrial feeding migration and life history of the stream-dwelling caddisfly, *Desmona bethula* (Trichoptera: Limnephilidae). Canadian Journal of Zoology 59:1658–65.

Fisher, S. G. and G. E. Likens. 1972. Stream ecosystem: organic energy budget. BioScience 22:33–35.

Flint, R. W. and C. R. Goldman. 1975. The effects of a benthic grazer on the primary productivity of the littoral zone of Lake Tahoe. Limnology and Oceanography 20:935–44.

Fredeen, F. J. H. 1960. Bacteria as a source of food for black-fly larvae. Nature (London) 187:963.

Fredeen, F. J. H. 1964. Bacteria as food for blackfly larvae (Diptera: Simuliidae) in laboratory cultures and in natural streams. Canadian Journal of Zoology 42:527–48.

Fuller, R. L. and R. J. Mackay. 1980. Feeding ecology of three species of *Hydropsyche* (Trichoptera: Hydropsychidae) in southern Ontario. Canadian Journal of Zoology 58:2239–51.

Gaskell, S. J. and G. Eglinton. 1975. Rapid hydrogenation of sterols in a contemporary lacustrine sediment. Nature (London) 254:209–11.

Giguère, L. A. 1981. Food assimilation efficiency as a function of temperature and meal size in larvae of *Chaoborus trivittatus* (Diptera: Chaoboridae). Journal of Animal Ecology 50:103–9.

Grafius, E. and N. J. Anderson. 1980. Population dynamics and role of two species of *Lepidostoma* (Trichoptera: Lepidostomatidae) in an Oregon coniferous forest stream. Ecology 61:808–16.

Green, R. H. 1968. A summer-breeding population of the relict amphipod *Pontoporeia affinis* Lindström. Oikos 19:191-97.

Hall, H. A. and G. Pritchard. 1975. The food of larvae of *Tipula sacra* Alexander in a series of abandoned beaver ponds (Diptera: Tipulidae). Journal of Animal Ecology 44:55-66.

Halliwell, G. 1959. The enzymic decomposition of cellulose. Nutrition Abstracts and Reviews 29:747-59.

Hargrave, B. T. 1970. The utilization of benthic microflora by *Hyalella azteca* (Amphipoda). Journal of Animal Ecology 39:427-37.

Hargrave, B. T. 1971. An energy budget for a deposit-feeding amphipod. Limnology and Oceanography 16:99-103.

Hargrave, B. T. 1972. Prediction of egestion by the deposit-feeding amphipod *Hyalella azteca.* Oikos 23:116-24.

Hargrave, B. T. and L. K. Nielsen. 1977. Accumulation of sedimentary organic matter at the base of steep bottom gradients, pp. 168-73. *In:* H. L. Golterman (ed.). Interactions between sediments and fresh water. Dr. W. Junk B. V. Publishers, The Hague, The Netherlands. 473 pp.

Hart, D. D. 1981. Foraging and resource patchiness: field experiments with a grazing stream insect. Oikos 37:46-52.

Hart, D. D. and V. H. Resh. 1980. Movement patterns and foraging ecology of a stream caddisfly larva. Canadian Journal of Zoology 58:1174-85.

House, H. L. 1965. Effects of low levels of the nutrient content of a food and of nutrient imbalance on the feeding and nutrition of a phytophagous larva, *Celerio euphorbiae* (Linnaeus) (Lepidoptera: Sphingidae). The Canadian Entomologist 97:62-68.

Hunter, R. D. 1980. Effects of grazing on the quantity and quality of freshwater Aufwuchs. Hydrobiologia 69:251-59.

Hutchinson, G. E. 1981. Thoughts on aquatic insects. BioScience 31:495-500.

Hynes, H. B. N. 1970. The ecology of running waters. University of Toronto Press, Toronto. 555 pp.

Iversen, T. M. 1974. Ingestion and growth in *Sericostoma personatum* (Trichoptera) in relation to the nitrogen content of ingested leaves. Oikos 25:278-82.

Ivlev, V. S. 1939. Transformation of energy by aquatic animals. Coefficient of energy consumption by *Tubifex tubifex* (Oligochaeta). Internationale Revue der gesamten Hydrobiologie 38: 449-58.

Izvekova, E. I. 1971. On the feeding habits of chironomid larvae. Limnologica 8:201-2.

Izvekova, E. I. and Y. E. Sorokin. 1969. A study of feeding *Chironomus anthracinus* Zett. larvae with C[14]. Fisheries Research Board of Canada Translation Series No. 1661: 1-11.

Jónasson, P. M. 1969. Bottom fauna and eutrophication, pp. 274-305. *In:* Eutrophication: causes, consequences, correctives. National Academy of Sciences, Washington DC. 661 pp.

Jónasson, P. M. and J. Kristiansen. 1967. Primary and secondary production in Lake Esrom. Growth of *Chironomus anthracinus* in relation to seasonal cycles of phytoplankton and dissolved oxygen. Internationale Revue der gesamten Hydrobiologie 52:163-217.

Kajak, Z. and K. Dusoge. 1970. Production efficiency of *Procladius choreus* Mg (Chironomidae, Diptera) and its dependence on the trophic conditions. Polskie Archiwum Hydrobiologii 17:217-24.

Kajak, Z. and J. Warda. 1968. Feeding of benthic nonpredatory Chironomidae in lakes. Annales Zoologici Fennici 5:57–64.

Kehde, P. M. and J. L. Wilhm. 1972. The effects of grazing by snails on community structure of periphyton in laboratory streams. American Midland Naturalist 87: 8–24.

Kesler, D. H. 1981. Periphyton grazing by *Amnicola limosa:* an enclosure-exclosure experiment. Journal of Freshwater Ecology 1:51–59.

Ketchum, B. H. and A. C. Redfield. 1949. Some physical and chemical characteristics of algae growth in mass culture. Journal of Cellular and Comparative Physiology 33:281–99.

Kubicek, F. 1970. On the drift of a brook running through a pond. Vestnik Ceskoslovenske Spolecnosti Zoolgicke 34:219–26.

Kurtak, D. C. 1978. Efficiency of filter feeding of black fly larvae (Diptera: Simuliidae). Canadian Journal of Zoology 56:1608–23.

Ladle, M., J. A. B. Bass, and W. R. Jenkins. 1972. Studies on production and food consumption by the larval Simuliidae (Diptera) of a chalk stream. Hydrobiologia 39: 429–48.

Lamberti, G. A. and V. H. Resh. 1983. Stream periphyton and insect herbivores: an experimental study of grazing by a caddisfly population. Ecology 64:1124–35.

Larkin, P. A. 1948. *Pontoporeia* and *Mysis* in Athasbaska, Great Bear and Great Slave Lakes. Bulletin of the Fisheries Research Board of Canada 78:1–33.

de Leeuw, J. W., W. I. C. Rijpstra, J. J. Boon, F. de Lange, and P. A. Schenck. 1977. The relationship between lipids from *Fontinalis antipyretica,* its detritus and the underlying sediment: the fate of wax esters and sterolesters, pp. 141–47. *In:* H. L. Golterman (ed.). Interactions between sediments and fresh water. Dr. W. Junk B. V. Publishers, The Hague, The Netherlands. 473 pp.

Lubchenco, J. and S. D. Gaines. 1981. A unified approach to marine plant-herbivore interactions. I. Populations and communities. Annual Review of Ecology and Systematics 12:405–37.

Margolina, G. L. 1961. Contribution to the problem on feeding *Tendipes plumosus* in the Rybinsk water reservoir. Fisheries Research Board of Canada Translation Series No. 1798:1–13.

Marker, A. F. H. 1976. The benthic algae of some streams in southern England: II. The primary production of the epilithon in a small chalk-stream. Journal of Ecology 64: 359–73.

Martin, A. L. 1966. Feeding and digestion in two intertidal gammarids: *Marinogammarus obtusatus* and *M. pirloti.* Journal of Zoology (London) 148:515–25.

Martin, M. M., J. S. Martin, J. J. Kukor, and R. W. Merritt. 1980. The digestion of protein and carbohydrate by the stream detritivore, *Tipula abdominalis* (Diptera, Tipulidae). Oecologia (Berlin) 46:360–64.

Mason, C. F. and R. J. Bryant. 1975. Periphyton production and grazing by chironomids in Alderfen Broad, Norfolk. Freshwater Biology 5:271–77.

McCullough, D. A., G. W. Minshall, and C. E. Cushing. 1979. Bioenergetics of lotic filter-feeding insects *Simulium* spp. (Diptera) and *Hydropsyche occidentalis* (Trichoptera) and their function in controlling organic transport in streams. Ecology 60: 585–96.

McIntire, C. D. 1973. Periphyton dynamics in laboratory streams: a simulation model and its implications. Ecological Monographs 43:399–420.

McLachlan, A. J. 1977. Some effects of tube shape on the feeding of *Chironomus plumosus* L. (Diptera: Chironomidae). Journal of Animal Ecology 46:139-46.

McNaughton, S. J. 1976. Serengeti migratory wildebeest: facilitation of energy flow by grazing. Science 191:92-94.

Mecom, J. O. 1972. Feeding habits of Trichoptera in a mountain stream. Oikos 23:401-7.

Minshall, G. W. 1968. Community dynamics of the benthic fauna in a woodland spring-brook. Hydrobiologia 32:305-39.

Minshall, G. W. 1978. Autotrophy in stream ecosystems. BioScience 28:767-71.

Monakov, A. V. 1972. Review of studies on feeding of aquatic invertebrates conducted at the Institute of Biology of Inland Waters, Academy of Science, USSR. Journal of the Fisheries Research Board of Canada 29:363-83.

Monk, D. C. 1976. The distribution of cellulase in freshwater invertebrates of different feeding habits. Freshwater Biology 6:471-75.

Monk, D. C. 1977. The digestion of cellulose and other dietary components, and pH of the gut in the amphipod *Gammarus pulex* (L.). Freshwater Biology 7:431-40.

Moon, H. P. 1938. The growth of *Caenis horaria* (L.), *Leptophlebia vespertina* (L.), and *L. marginata* (L.) (Ephemeroptera). Proceedings of the Zoological Society of London 108:507-12.

Moore, J. W. 1975. Seasonal changes in the proximate and fatty acid composition of some naturally grown freshwater chlorophytes. Journal of Phycology 11:205-11.

Moore, J. W. 1977. Some factors affecting algal consumption in subarctic Ephemeroptera, Plecoptera and Simuliidae. Oecologia (Berlin) 27:261-73.

Moore, J. W. 1979a. Some factors influencing the distribution, seasonal abundance and feeding of subarctic Chironomidae (Diptera). Archiv für Hydrobiologie 85:302-25.

Moore, J. W. 1979b. Factors influencing algal consumption and feeding rate in *Heterotrissocladius changi* Saether and *Polypedilum nebeculosum* (Meigen) (Chironomidae: Diptera). Oecologia (Berlin) 40:219-27.

Moore, J. W. and I. C. Potter. 1976. A laboratory study on the feeding of larvae of the brook lamprey *Lampetra planeri* (Bloch). Journal of Animal Ecology 45:81-90.

Morgan, K. C., J. L. C. Wright, and F. J. Simpson. 1980. Review of chemical constituents of the red alga *Palmaria palmata* (Dulse). Economic Botany 34:27-50.

Mulla, M. S. and L. A. Lacey. 1976. Feeding rates of *Simulium* larvae on particulates in natural streams (Diptera: Simuliidae). Environmental Entomology 5:283-87.

Nalewajko, C. 1966. Dry weight, ash, and volume data for some freshwater planktonic algae. Journal of the Fisheries Research Board of Canada 23:1285-88.

Nelson, D. J. and D. C. Scott. 1962. Role of detritus in the productivity of a rock-outcrop community in a Piedmont stream. Limnology and Oceanography 7:396-413.

Nielsen, A. 1942. Über die Entwicklung und Biologie der Trichopteren mit besonderer Berücksichtigung der Quelltrichopteren Himmerlands. Archiv für Hydrobiologie Supplementband 17:255-631.

Oliver, D. R. 1971. Life history of the Chironomidae. Annual Review of Entomology 16: 211-30.

Otto, C. 1974. Growth and energetics in a larval population of *Potamophylax cingulatus* (Steph.) (Trichoptera) in a south Swedish stream. Journal of Animal Ecology 43: 339-61.

Paerl, H. W. 1977. Bacterial sediment formation in lakes: trophic implications, pp. 40-47. *In*: H. L. Golterman (ed.). Interactions between sediments and fresh water. Dr. W. Junk B. V. Publishers, The Hague, The Netherlands. 473 pp.

Parsons, T. R., K. Stephens, and J. D. H. Strickland. 1961. On the chemical composition of eleven species of marine phytoplankters. Journal of the Fisheries Research Board of Canada 18:1001–16.

Patrick, R. 1970. Stream benthic communities. American Scientist 58:546–49.

Pennak, R. W. 1978. Fresh-water invertebrates of the United States. John Wiley and Sons, New York, NY. 803 pp.

Platt, T. and B. Irwin. 1973. Caloric content of phytoplankton. Limnology and Oceanography 18:306–10.

Porter, K. G. 1977. The plant-animal interface in freshwater ecosystems. American Scientist 65:159–70.

Pritchard, G. 1983. Biology of Tipulidae. Annual Review of Entomology 28:1–22.

Prus, T. 1971. The assimilation efficiency of *Asellus aquaticus* L. (Crustacea, Isopoda). Freshwater Biology 1:287–305.

Ray, D. L. and J. R. Julian. 1952. Occurrence of cellulase in *Limnoria*. Nature (London) 169:32–33.

Resh, V. H. 1976. Life histories of coexisting species of *Ceraclea* caddisflies (Trichoptera: Leptoceridae): the operation of independent functional units in a stream ecosystem. The Canadian Entomologist 108:1303–18.

Rich, P. H. and R. G. Wetzel. 1978. Detritus in the lake ecosystem. The American Naturalist 112:57–71.

Robles, C. D. and J. Cubit. 1981. Influence of biotic factors in an upper intertidal community: dipteran larvae grazing on algae. Ecology 62:1536–47.

Rodina, A. G. 1949. The role of bacteria in the feeding of tendipedid larvae. Fisheries Research Board of Canada Translation Series No. 1848:1–6.

Ross, D. H. and D. A. Craig. 1980. Mechanisms of fine particle capture by larval black flies (Diptera: Simuliidae). Canadian Journal of Zoology 58:1186–92.

Sain, P., J. B. Robinson, W. N. Stammers, N. K. Kaushik, and H. R. Whiteley. 1977. A laboratory study on the role of stream sediment in nitrogen loss from water. Journal of Environmental Quality 6:274–78.

Serruya, C. 1977. Rates of sedimentation and resuspension in Lake Kinneret, pp. 48–56. *In*: H. L. Golterman (ed.). Interactions between sediments and fresh water. Dr. W. Junk B. V. Publishers, The Hague, The Netherlands. 473 pp.

Shapas, T. J. and W. L. Hilsenhoff. 1976. Feeding habits of Wisconsin's predominant lotic Plecoptera, Ephemeroptera, and Trichoptera. Great Lakes Entomologist 9:175–88.

Short, R. A., S. P. Canton, and J. V. Ward. 1980. Detrital processing and associated macroinvertebrates in a Colorado mountain stream. Ecology 61:727–32.

Soponis, A. R. 1977. A revision of the Nearctic species of *Orthocladius (Orthocladius)* van der Wulp (Diptera: Chironomidae). Memoirs of the Entomological Society of Canada 102:1–187.

Sorokin, J. I. 1966. Carbon-14 method in the study of the nutrition of aquatic animals. Internationale Revue der gesamten Hydrobiologie 51:209–24.

Sousa, W. P. 1979. Experimental investigations of disturbance and ecological succession in a rocky intertidal algal community. Ecological Monographs 49:227–54.

Spoehr, H. A. and H. W. Milner. 1949. The chemical composition of *Chlorella*: effect of environmental conditions. Plant Physiology (Bethesda) 24:120–49.

Sumner, W. T. and C. D. McIntire. 1982. Grazer-periphyton interactions in laboratory streams. Archiv für Hydrobiologie 93:135–57.

Sweeney, B. W. and R. L. Vannote. 1978. Size variation and the distribution of hemi-metabolous aquatic insects: two thermal equilibrium hypotheses. Science 200: 444–46.

Tan, Y. T. 1970. Composition and nutritive value of some grasses, plants and aquatic weeds tested as diets. Journal of Fish Biology 2:253–57.

Ulfstrand, S. 1967. Microdistribution of benthic species (Ephemeroptera, Plecoptera, Trichoptera, Diptera: Simuliidae) in Lapland streams. Oikos 18:293–310.

Wallace, J. B. and D. Malas. 1976. The significance of the elongate, rectangular mesh found in capture nets of fine particle filter feeding Trichoptera larvae. Archiv für Hydrobiologie 77:205–12.

Wallace, J. B. and R. W. Merritt. 1980. Filter-feeding ecology of aquatic insects. Annual Review of Entomology 25:103–32.

Wallace, J. B. and F. F. Sherberger. 1974. The larval retreat and feeding net of *Macronema carolina* Banks (Trichoptera: Hydropsychidae). Hydrobiologia 45: 177–84.

Wallace, J. B. and F. F. Sherberger. 1975. The larval dwelling and feeding structure of *Macronema transversum* (Walker) (Trichoptera: Hydropsychidae). Animal Behaviour 23:592–96.

Wallace, J. B., J. R. Webster, and W. R. Woodall. 1977. The role of filter feeders in flowing waters. Archiv für Hydrobiologie 79:506–32.

Walshe, B. M. 1951. The feeding habits of certain chironomid larvae (subfamily Tendipedinae). Proceedings of the Zoological Society of London 121:63–79.

Ward, G. M. and K. W. Cummins. 1979. Effects of food quality on growth of a stream detritivore, *Paratendipes albimanus* (Meigen) (Diptera: Chironomidae). Ecology 60: 57–64.

Welch, H. E. 1968. Relationships between assimilation efficiencies and growth efficiencies for aquatic consumers. Ecology 49:755–59.

Whitlatch, R. B. 1980. Patterns of resource utilization and coexistence in marine intertidal deposit-feeding communities. Journal of Marine Research 38:743–65.

Wiggins, G. B. 1977. Larvae of the North American caddisfly genera (Trichoptera). University of Toronto Press, Toronto. 401 pp.

Williams, N. E. and H. B. N. Hynes. 1973. Microdistribution and feeding of the net-spinning caddisflies (Trichoptera) of a Canadian stream. Oikos 24:73–84.

Winterbourn, M. J. 1971. The life histories and trophic relationships of the Trichoptera of Marion Lake, British Columbia. Canadian Journal of Zoology 49:623–35.

Wotton, R. S. 1976. Evidence that blackfly larvae can feed on particles of colloidal size. Nature (London) 261:697.

Wotton, R. S. 1978a. The feeding-rate of *Metacnephia tredecimatum* larvae (Diptera: Simuliidae) in a Swedish lake outlet. Oikos 30:121–25.

Wotton, R. S. 1978b. Growth, respiration, and assimilation of blackfly larvae (Diptera: Simuliidae) in a lake-outlet in Finland. Oecologia (Berlin) 33:279–90.

Zimmerman, M. C. and T. E. Wissing. 1978. Effects of temperature on gut-loading and gut-clearing times of the burrowing mayfly, *Hexagenia limbata*. Freshwater Biology 8:269–77.

Zimmerman, M. C., T. E. Wissing, and R. P. Rutter. 1975. Bioenergetics of the burrowing mayfly, *Hexagenia limbata*, in a pond ecosystem. Internationale Vereinigung für Theoretische und Angewandte Limnologie Verhandlungen 19:3039–49.

chapter 8

PREDATOR-PREY INTERACTIONS AMONG AQUATIC INSECTS

Barbara L. Peckarsky

Most insect orders with aquatic life stages contain species that are predaceous, and some of these orders are entirely so. The predatory activities of such aquatic insects, as well as those of their vertebrate and invertebrate predators, may have a strong influence in shaping the evolution of aquatic insect communities. Predator-prey relations also affect reproduction, feeding, adaptation to abiotic surroundings, and defense, which are the four basic requirements of animals for survival and perpetuation. For these reasons, the study of predation on or by aquatic insects is important.

Predator-prey interactions have received much attention in the last decade, and are the subject of some recent comprehensive reviews. Curio (1976) discussed predation from a behavioral perspective, Hassell (1978) used arthropod predator-prey interactions to develop a number of mathematical predator-prey models, and Edmunds (1974) provided a comprehensive survey of antipredator defenses in animals. Although each of these reviews make little reference to aquatic insects, they do provide an interesting conceptual framework within which any predator-prey interaction can be examined, regardless of the specific organisms or habitat involved.

Zaret (1980) synthesized much of the work of the last two decades in his book *Predation and Freshwater Communities*. However, this title is somewhat overstated, since the book neglects streams entirely and contains very little reference to insects, except to *Chaoborus* as an important predator of zooplankton. Three recent reviews have treated the topic of aquatic insect predator-prey relations specifically. Bay (1974) concentrated on predators of aquatic insect groups of nuisance importance to humans (i.e. Chironomidae, Culicidae, Simuliidae); Allan (1983a) reviewed predator-prey relationships in streams, including discussion of fish and invertebrate predators; Peckarsky (1982) summarized recent literature on aquatic insect predator-prey relations for audiences with general biological backgrounds.

The purpose of this chapter is two-fold. First, it provides a comprehensive review of the data available on aquatic insect predator-prey relations, with tabular summaries intended to serve as an annotated guide to the published literature. Second, it presents theoretical considerations that relate to various aspects of predator-prey interactions, and it discusses empirical studies in relation to theory where appropriate studies exist. The usefulness of existing theory in interpreting aquatic insect predator-prey systems and the need for additional data to provide answers to unanswered questions are discussed, and directions for future research efforts are proposed.

METHODS FOR STUDY OF PREDATOR-PREY RELATIONS

Scientific data are only as good as the methods used to obtain them, and the choice of effective methods should involve careful consideration of the specific questions being asked. In this chapter, I discuss five general problems that can be investigated as components of predator-prey relations and the classes of data that are required to test hypotheses relative to these problems (Table 8.1). The best studies are those that obtain measurements for all of the variables listed under "classes of data required"; these may be obtained using "alternative methods," each of which has strengths and weaknesses that should be clearly understood and incorporated into data interpretation.

Stomach content data, which are used to investigate predator responses to prey density, prey preferences, and community-level effects (Table 8.1), provide a record of undigested prey parts; but the gut clearance time, the potential overestimation of prey with heavily sclerotized parts compared to soft-bodied prey, regurgitation, the effect of preservatives on gut contents, and the feeding habits of predators must be considered in interpreting such data. I have observed that the perlodid stonefly *Kogotus modestus* will bite off the posterior end of the mayfly *Baetis bicaudatus* and discard the head and thorax. Thus, identification of gut contents may only show part of what has been eaten, not how much, and by counting only one body part (e.g. head capsules) rather than all recognizable parts (e.g. claws, mandibles, cerci, antennae), gross underestimates of the magnitude of prey consumption can result.

Prey preferences (Table 8.1) are often inferred from comparison of the proportion of a given prey item or prey size in the predator's stomach to the prey's proportion in the habitat, based on field samples of potential prey (Ivlev 1961; Jacobs 1974; Healey, Chapter 9). Chesson (1978), Johnson (1980), and Lechowicz (1982) have commented on problems associated with such electivity indexes, and have proposed appropriate modifications. Stomach content data cannot be used to show preference conclusively, because disproportionate numbers of prey in stomachs compared to those in habitat samples may reflect prey types available to predators, but not to biologists, or vice versa. Absence of a particular prey item from predator stomachs may indicate lack of microhabitat overlap, effective prey escape mechanisms, low availability, or any other aspect reducing predator-prey encounter probability, rather than reflecting predator choice.

Serological techniques (Table 8.1) may be used to identify the contents of the predator's stomach for sucking predators such as Hemiptera and some Coleoptera larvae (e.g. see reviews by Boreham 1979 and McIver 1981). Common

TABLE 8.1: **Summary of General Problems Concerning Predator-Prey Relations**

PROBLEM	CLASSES OF DATA REQUIRED	ALTERNATIVE METHODS
A. Hunting behavior	Search and attack modes	Behavioral observations
	Detection mode	Electrophysiological techniques, experiments eliminating sensory structures, behavioral observations
	Allocation of time (or energy) to feeding	Behavioral observations
B. Responses to prey density	Prey density	Field sampling, experimental manipulation (laboratory or field)
1. Numerical	Predator density	Field sampling (correlation to prey densities), comparison of densities in experimental treatments
2. Functional	Feeding rates	Stomach content analysis (?)*, serological techniques (?) (correlation to prey densities), rate of disappearance of prey from experimental enclosures
3. Developmental	Development rates	Measurements from laboratory experiments
4. Area-restricted search	Time spent feeding	Behavioral observations
C. Prey preferences or selectivity	Predator diet	Stomach analysis (?), serological techniques (?), behavioral observations, feeding experiments
	Prey availability (density, microhabitat, escape capabilities)	Field sampling, behavioral observations
	Prey profitability (energy gained/time spent, costs/benefits)	Behavioral experiments
D. Prey defenses	Sensory capabilities	Electrophysiology, behavioral observations, experiments eliminating sensory structures
	Vulnerability to predation	Behavioral observations

TABLE 8.1: Concluded

PROBLEM	CLASSES OF DATA REQUIRED	ALTERNATIVE METHODS
E. Community-level effects	Feeding rates	See B above
	Selectivity of predators	See C above
	Community composition (predator and prey densities)	See B, C above

*(?) = method often used but potentially invalid; see text for discussion.

precipitin tests operate on the same basic principle and approach. Potential prey material is injected into a mammal (as antigen) and antibodies specific to the antigen are produced. After about six weeks, antiserum is removed and tested against predator gut contents; a positive precipitin reaction indicates consumption. However, use of these data to infer preference in relation to samples of prey from the predator's habitat is subject to the same limitations as the stomach content analyses discussed above.

Correlations between natural populations of predators and prey, as used in examining predator responses to prey density (Table 8.1), are only suggestive of cause-effect relationships. Correlations based on accurately taken population data can form a descriptive basis for generating hypotheses on the responses of predators to prey density and on the effects of predators on prey community structure. However, an investigator may be unable to discern the relationships among a number of potential variables that are correlated with predator and prey densities.

Behavioral observations are used to investigate all five problems outlined in Table 8.1. Behavior of predators and prey can be observed in the laboratory or the field, but interpretation of these data must consider the conditions under which the observations were made. For example, removal of insects from their natural conditions and the presence of an observer can affect their performance. Wiley and Kohler (Chapter 5) and Peckarsky (1983) review some of the behavioral techniques that have been implemented in the study of predation by aquatic insects.

Experimental manipulations have been used extensively to study predator-prey relations among marine rocky intertidal invertebrates, between freshwater fish and zooplankton, and among zooplankton (e.g. see review by Connell 1975). These manipulations generally involve the use of enclosures to test responses to modified densities of predators and prey in the field or the laboratory. Recently, such techniques have been applied to the study of aquatic insect predator responses to prey densities and community-level effects of predators (Peckarsky and Dodson 1980a, 1980b; Benke et al. 1982). In contrast to other approaches, such experiments can provide data that are interpretable in examining cause-effect questions and, for that reason, experimental approaches are potentially more powerful than correlative data. However, care must be taken to ensure that the enclosures do not cause responses that are artifacts of the experiment rather than being indicative of natural responses (Hulberg and Oliver 1980).

In summary, investigators must carefully consider both the classes of data required to answer questions on relationships between aquatic insect predators and prey and the strengths and limitations of each method of testing hypotheses, before choosing the best method for answering a particular question. In addition, when reading the literature on predation, scrutiny of the authors' methods is critical in evaluating the strength of their conclusions.

AQUATIC INSECT PREDATOR BEHAVIOR

Aquatic Insect Predator Hunting Behavior

Visual, mechanical, and chemical cues are available to predators to detect prey. Some of the published data on sensory cues used by aquatic insect predators to detect prey are summarized in Table 8.2. The full range of sensory cues used by a predator can only be identified by experimental elimination of each potential cue and by subsequent tests of the predator's response to presentation of prey. Although this has been done in a few studies, most conclusions are based on less than convincing data.

From the limited data that have been reported, we can see that some aquatic insect predators use tactile or mechanical cues to detect prey (some Odonata, Plecoptera, Trichoptera, Diptera) and some use visual cues (some Odonata, Hemiptera, Coleoptera); only inferential data exist for the use of chemical cues (Lindstedt 1971) and many predators probably use a combination of cues (e.g. Hemiptera). In terrestrial insects, chemical cues may affect the frequency of turning of predators or increase the time spent in search behavior (Hassell 1978), but no such behavior has been observed for aquatic insects.

The use of visual cues is probably not more widespread for several reasons. First, many aquatic insects are negatively phototactic, residing under the surface of the substrate during periods of high light intensity (Hynes 1970). This prey behavior would render visual prey-detection ineffective. Second, the eyes of aquatic insects are generally not well developed, although there are exceptions in the Odonata (Pritchard 1966), adult Hemiptera (Cloarec 1969), and Coleoptera (Cloarec 1972).

Sjöström (1983) analyzed the compound eye of the perlid stonefly *Dinocras cephalotes* and concluded that the front part of the eye can resolve objects as small as 2 mm, but this requires high light intensity (15 lux). The top and back parts of the eye have poor resolution, but they can detect objects greater than 5 cm at very low light intensity (0.5 lux). Sjöström suggested that the top and back facets are used to detect predators, but that use of the front of the eyes in hunting is unlikely since *D. cephalotes* is extremely night active. Nighttime observations using an infrared sensitive camera, to which *D. cephalotes* is not sensitive, have suggested that this predator responds to tactile cues via its long antennae and the swimming hairs on its legs. Perhaps *D. cephalotes* does not hunt during the day, when its eyes are optimally adapted to perceive small objects, due to the high risk of fish predation.

Table 8.2 also shows that two basic search modes are used by aquatic insect predators: stalking, or active pursuit, and ambush, or passive "sit and wait" behavior. A pursuit predator expends most of its energy on active search, whereas ambush predators expend virtually no energy in prey search since the majority of their costs are in the capture and the handling of prey. The time-energy budgets of these two types of predators should differ considerably (Andersson 1981) and, although these variables have been measured for only a handful of pursuit and ambush predators (Table 8.2), the measurements are consistent with predicted differences.

According to Optimal Foraging Theory (Krebs 1978), with all other things being equal, evolution should favor predators that harvest prey efficiently or that maximize their net rate of prey intake, assuming that this condition confers maximum fitness. Among the predictions for optimal predator search behavior that have been generated by proponents of this theory are (1) that predators should search for prey in directional paths that maximize the number of prey encountered (Pyke 1978), and (2) that predators should use a low-cost, low-return method of foraging (e.g. ambush rather than search) when prey density is low, when the energy cost of moving between patches is high, and when there is little difference between good and poor foraging sites (Janetos 1982). Although no aquatic insect data have been reported on the first prediction, Akre and Johnson (1979) found that a damselfly predator shifted between pursuit and ambush search modes in response to changes in the relative abundance of stationary and mobile cladoceran prey. However, the authors did not vary absolute prey density or measure other variables that would have provided a test of the second prediction. In contrast, Formanowicz (1982) showed that dytiscid larvae moved less when offered higher densities of tadpole prey in aquaria, which is a behavior different from that predicted by Janetos (1982).

Responses of Aquatic Insect Predators to Prey Density

Two predictions from Optimal Foraging Theory relate predator response to variations in prey density. First, a higher number of predators should be found in prey-rich than in prey-poor patches; this has been termed the *aggregative* or

TABLE 8.2: Facets of Aquatic Insect Predator Hunting Behavior

PREDATOR TAXON	SEARCH (ATTACK) MODE	DETECTION MODE	TIME OR ENERGY BUDGETS	REFERENCES (nature of study)*
EPHEMEROPTERA (some species)	Stalk (engulf)			Edmunds (1957) (R); Merritt and Cummins (1978, Table 8A) (R)
ODONATA (all species)	(Engulf)		Corbet (1980)	Merritt and Cummins (1978, Table 9A) (R); Corbet (1980) (R)
Anisoptera				
Petaluridae	Stalk and ambush	Tactile (legs), visual		Pritchard (1965, 1966) (E)
	Ambush	Tactile and visual (mostly tactile)		Pritchard (1965, 1966) (E)
Cordulegastridae	Ambush	Tactile and visual (mostly tactile)		Pritchard (1965, 1966) (E); Johnson (1982) (O)
Gomphidae	Ambush	Tactile and visual (mostly tactile)		Pritchard (1965, 1966) (E); Johnson (1982) (O)
Aeshnidae				
Aeshna	Stalk, may ambush	Visual (no chemical)	Etienne (1972); Popham and Bevans (1979)	Pritchard (1965, 1966) (E); Oakley and Polka (1967) (E); Caillère (1972) (E); Etienne (1972) (E); Popham and Bevans (1979) (O)
Anax		Not visual		Caillère (1972) (E); Sievers and Haman (1972) (O)
Corduliidae	Ambush	Tactile and visual (No chemical)		Pritchard (1965) (E)
Libellulidae	Ambush	Tactile		Pritchard (1965) (E)
Leucorrhinia	Stalk	Tactile and visual		Oakley and Polka (1967) (E) Pritchard (1965) (E)

Taxon	Mode	Detection		Reference
Libellula	Ambush	Tactile		Pritchard (1965) (E)
Sympetrum	Stalk	Visual		Pritchard (1965) (E)
Macromiidae	Ambush			Merritt and Cummins (1978, Table 9A) (R)
Zygoptera				
Calopterygidae		Tactile (antennae)		Pritchard (1965) (E)
Calopteryx	Stalk	Tactile		Caillère (1972) (E)
Coenagrionidae				
Ischnura	Stalk, ambush	Visual		Crowley (1979) (O)
Anomalagrion	Stalk, ambush	Tactile	Johnson et al. (1975)	Akre and Johnson (1979) (E)
Pyrrhosoma	Ambush		Lawton (1970, 1971)	Johnson et al. (1975) (E)
PLECOPTERA				
(some species)	Stalk (engulf)	Tactile (antennae)		Merritt and Cummins (1978, Table 11B) (R)
Pteronarcidae				
Pteronarcella	Stalk	Tactile		Peckarsky (1980, 1983) (O)
Perlidae				
Acroneuria	Stalk	Tactile		Peckarsky (1979) (R), (1980) (O)
Phasganophora	Stalk	Tactile		Kovalak (1978) (O)
Dinocras	Stalk	Tactile	Malmqvist and Sjöström (1980)	Malmqvist and Sjostrom (1980) (O); Sjöström (1983) (O)
Perlodidae				
Megarcys	Stalk	Tactile		Peckarsky (1980, 1983) (O)
Kogotus	Stalk	Tactile		Peckarsky (1983) (O)
Chloroperlidae	Stalk			Merritt and Cummins (1978, Table 11B) (R)

TABLE 8.2: Concluded

PREDATOR TAXON	SEARCH (ATTACK) MODE	DETECTION MODE	TIME OR ENERGY BUDGETS	REFERENCES (nature of study)*
HEMIPTERA (all species)	(Pierce)			Merritt and Cummins (1978, Table 12A) (R)
Hydrometridae		Chemical (olfactory)		Maier (1977) (R), (O)
Veliidae		Tactile (surface waves), supplementary visual	Maier (1977)	Maier (1977) (R), (O)
Gerridae		Tactile (surface waves)		Maier (1977) (R), (O); Murphey (1971) (O)
Nepidae	Stalk	Visual		Jamieson and Scudder (1979) (E)
	Ambush	Visual, tactile		Cloarec (1969, 1975a, 1977, 1978, 1980a, 1980b, 1981) (E)
Notonectidae			Toth and Chew (1972), Gittleman (1974), Zalom (1978a), Cook and Cockrell (1978), Giller (1980, 1982), Sih (1980b), Streams (1982)	
Notonecta	Ambush	Tactile (surface waves)		Markl et al. (1973) (E); Gittleman (1974) (O); Lang (1980) (E)
		Visual and tactile (surface waves)		Giller and McNeill (1981) (O); Streams (1982) (O)
Buenoa		Tactile		Gittleman (1974) (O)
MEGALOPTERA	Stalk (engulf)			
Sialis	Stalk			Merritt and Cummins (1978, Table 13A) (R)
	Ambush			Townsend and Hildrew (1979a) (O); Pritchard and Leischner (1973) (O)
NEUROPTERA	(Pierce sponges)			Merritt and Cummins (1978, Table 13A) (R)

TRICHOPTERA (some species)				
Phyrganeidae				
Banksiola (Instar V)	Stalk and ambush (engulf)			Merritt and Cummins (1978, Table 14A) (R)
	Stalk	Tactile	Winterbourn (1971)	Winterbourn (1971) (O)
Polycentropodidae				
Plectrocnemia	Ambush	Tactile (net vibrations)	Townsend and Hildrew (1978); Hildrew and Townsend (1980)	Tachet (1977) (E); Townsend and Hildrew (1978, 1979a, 1979b) (O)
Rhyacophilidae	Stalk	Tactile		Wiley and Kohler (1981) (O)
COLEOPTERA (some) (adults and larvae)	Stalk, ambush (engulf, pierce)			Merritt and Cummins (1978, Table 16A) (R)
Gyrinidae (adults)	Stalk (engulf)	Visual and tactile (surface waves)		Heinrich and Vogt (1980) (O)
Dytiscidae	(Pierce)	Visual, chemical		
adults	Stalk			Cloarec (1972) (R)
larvae	Ambush			Young (1967) (O)
	Ambush or stalk			Formanowicz (1982) (E)
DIPTERA (some)	Stalk, ambush (engulf, pierce)	Tactile		Merritt and Cummins (1978, Tables 18B, 19A, 20A, 22A) (R)
Chaoboridae	Ambush	Tactile (acoustical stimuli)	Giguère (1981); Pastorok (1981); Swift and Forward (1981)	Swift and Fedorenko (1975) (E); Giguère and Dill (1979) (E); Pastorok (1979) (R); Swift and Forward (1981) (E)
Chironomidae	Ambush or stalk (Engulf)	Tactile		Swüste et al. (1973) (E)
Tanypodinae	Search			Loden (1974) (O)
Chironominae	Stalk, ambush			

*(R) = review article
(E) = experimental study
(O) = behavioral observation
blanks = lack of information

numerical response (Holling 1961; Crawley 1975). Second, individual predators should spend more time in prey-rich than in prey-poor patches; this has been termed *area-restricted search* (Hassell and May 1974). Two specific hypotheses have been constructed relative to the second prediction. First, predators should stay in a prey patch until the rate of food intake decreases to equal the average rate of intake for the entire habitat; this has been termed the Marginal Value Theorem (Charnov 1976). Second, predators should forage longer before changing prey patches in richer than in poorer prey patches; this has been termed *giving up times* (McNair 1982).

Few empirical studies of aquatic insect predators that test these predictions are available. Wiley and Kohler (Chapter 5) discuss in detail area-restricted search, the *Marginal Value Theorem,* and rules for giving up time, as they relate to aquatic insect feeding in general. The few available examples of published studies of area-restricted search by predators are summarized in Table 8.3.

Numerical Response. Most empirical studies of aggregative or numerical response of aquatic insect predators have failed to show predator aggregation in patches of high prey density (Table 8.3). This behavior may be due to the apparent conflict between the benefits of aggregation in patches of high prey density and the risks of incurring reduced efficiency of prey intake due to the presence of competitors (Murdoch and Sih 1978; Sih 1980a, 1982; Peckarsky and Dodson 1980b; Hildrew and Townsend 1980). In Sih's studies, adult backswimmers (*Notonecta hoffmanni*) consumed juveniles of the same species, while both age groups used mosquito larvae as prey. Juveniles were forced to retreat to less profitable prey patches to balance the need for optimal foraging with the need to avoid predation, since interference from adults in patches of high mosquito density reduced the efficiency with which juveniles foraged. Furthermore, the lack of aggregation of predators in regions of high prey density during some seasons, or in enclosures, may be due to mutual interference among these predators (Hildrew and Townsend 1980; Peckarsky and Dodson 1980b). Peckarsky (1983) presented behavioral data that corroborate the existence of interference competition among predatory stoneflies. The conflicting demands of optimal patch use and avoidance of interference from other predators should be incorporated into existing conceptual models of predator-prey interactions. The potential importance of mutual interference among predators in explaining observed spacing behavior, habitat selection, and possible territoriality is discussed by Wiley and Kohler (Chapter 5).

Developmental Response. Another possible predator response to variations in prey density that has been tested by few investigators is called developmental response theory (Murdoch 1971, 1973). This theory predicts that predator growth rate depends on feeding rate, which depends on size (developmental stage). Therefore, different life stages of predators should respond differently

TABLE 8.3: Responses of Aquatic Insect Predators to Prey Density*

PREDATOR TAXON	PREY TAXON	NUMERICAL RESPONSE	AREA-RESTRICTED SEARCH	DEVELOPMENTAL RESPONSE
ODONATA				
Ischnura verticalis	Cladocera			(+) Johnson (1973) (LE); (+) Johnson et al. (1975) (LE)
I. elegans	Cladocera			(+) Thompson (1975) (LE); (+) Lawton et al. (1980) (LE)
Coenagrion resolutum	Oligochaeta		(+,−) Baker (1980, 1981b) (LE), mutual interference	
Lestes disjunctus	Oligochaeta		(−) Baker (1981a) (LE)	
PLECOPTERA				
Calineuria californica, Hesperoperla pacifica Dinocras cephalotes	Benthos in artificial streams *Baetis*	(−) Brocksen et al. (1968) (LC)		(+) Malmqvist and Sjöström (1980) (LE)
Acroneuria lycorias, Megarcys signata	Ephemeroptera	(−) Peckarsky and Dodson (1980b) (FE); mutual interference, Peckarsky (1983) (FE)		
Kogotus modestus	Ephemeroptera	Mutual interference, Peckarsky (1983) (FE)		

TABLE 8.3: Concluded

PREDATOR TAXON	PREY TAXON	NUMERICAL RESPONSE	AREA-RESTRICTED SEARCH	DEVELOPMENTAL RESPONSE
HEMIPTERA *Notonecta hoffmanni*	*Drosophila, Culex*		(+ adults, − nymphs) (LE) Murdoch and Sih (1978); mutual interference, Sih (1980a, 1980b, 1981, 1982) (LE)	(+) Fox and Murdoch (1978) (LE)
TRICHOPTERA *Plectrocnemia conspersa*	Stream benthos	(+) Hildrew and Townsend (1976) (FC); (+) Townsend and Hildrew (1980) (FC)	(−) Townsend and Hildrew (1980) (LE); (−) Hildrew and Townsend (1980) (LE); mutual interference	

*(+) = positive response; (−) = no response; the nature of the study is indicated in parenthesis (L = laboratory, E = experiment, F = field, C = correlative); blanks = lack of information. Studies that documented mutual interference among predators are also shown.

to variations in prey density. Studies that provide data relating directly to this question support these predictions (i.e. the growth rate of aquatic insect predators increases with increased feeding rate as a function of prey density, and feeding rate per prey density changes with advancing instar development of the predators; see Table 8.3).

Functional Response. The relationship between feeding rate and prey density, termed the *functional response* (Holling 1959), is the most widely studied of the responses of aquatic insect predators to prey density. Three types of functional response curves have been generated from empirical data (Fig. 8.1), and these have undergone extensive theoretical discussion. Type I curves represent feeding rates of predators that increase prey consumption linearly as prey density increases, until some limit of feeding rate is reached. No aquatic insect predators have shown this type of curve. Type II functional responses are curvilinear, with slopes decelerating to an asymptote. A number of studies on aquatic insect predators have generated data conforming to this type of curve. Type III responses are sigmoidal, and also have been produced using data from aquatic insects (Table 8.4).

Figure 8.1 General shape of Type I, II, and III functional response curves.

TABLE 8.4: Experimental Studies on Functional Responses of Aquatic Insect Predators*

PREDATOR TAXON	PREY TAXON	TYPE OF RESPONSE (see Figure 8.1)	EXPERIMENTAL CONDITIONS	INCIDENCE OF SWITCHING	VARIATIONS IN RESPONSES	REFERENCES
ODONATA						
Anax imperator	Chironomidae (1 species)		Laboratory dishes		Predation peaks at dusk and dawn	Cloarec (1975b)
Coenagrion armatum	Cladocera, Chironomidae		Laboratory aquaria		Early instars consumed a higher daily ration	Sadyrin (1977)
Anomalagrion hastatum	Cladocera (1 or 2 species)	II	Laboratory aquaria, given one prey species		Increased attack rate with hunger	Akre and Johnson (1979)
		III	given two prey species	Yes		
Ischnura verticalis	Cladocera (2 species)	III	Heterogeneous habitat in laboratory	Yes	Increased attack rate with hunger	Crowley (1979)
I. elegans	Cladocera (1 species)	II	Laboratory aquaria, no refuges		Late instars ate more prey/day	Lawton et al. (1980) Lawton et al. (1974)
	Cladocera (2 species)		Given both prey species	Yes		
I. elegans	Cladocera (1 species)	II	Laboratory aquaria, no refuges		Increased attack rate with increased temperature	Thompson (1978a)

PLECOPTERA

Taxon		Prey	Setting		Notes	Reference
Dinocras cephalotes	II	*Baetis rhodani*	Laboratory aquaria with airstones, no refuges	No	Increased food intake after molt; large predators ate more prey than small predators	Malmqvist and Sjöström (1980)

HEMIPTERA

Taxon		Prey	Setting		Notes	Reference
Gerris (5 spp.)		*Drosophila*	Laboratory aquaria		Smaller instars had higher food intake than larger instars per unit weight, greater food intake with increased temperature, no effect of hunger	Jamieson and Scudder (1977)
Notonecta hoffmanni	II	*Culex*	Laboratory aquaria		Larger predators had higher feeding rate than smaller predators	Fox and Murdoch (1978)
N. glauca	III	*Asellus, Cloeon*	Plastic containers in the laboratory	Yes		Lawton et al. (1974)
Notonecta sp.	II	Culicidae larvae	Laboratory aquaria			Holling (1965, Table 3.1)
Corixa sp.	II	Culicidae larvae	Laboratory aquaria			
Lethocerus sp.	II	Tadpoles	Laboratory aquaria			

TABLE 8.4: Concluded

PREDATOR TAXON	PREY TAXON	TYPE OF RESPONSE (see Figure 8.1)	EXPERIMENTAL CONDITIONS	INCIDENCE OF SWITCHING	VARIATIONS IN RESPONSES	REFERENCES
TRICHOPTERA						
Plectrocnemia conspersa	Plecoptera (2 species)		Laboratory aquaria, no current,			Hildrew and Townsend (1977)
		II	no refuges			
		III	refuges			
		(II or III?)	Field correlation			Townsend and Hildrew (1978)
COLEOPTERA						
Acilius semisulcatus	Culicidae larvae	II	Laboratory aquaria			Holling (1965, Table 3.1)
DIPTERA						
Chaoborus americanus, C. trivittatus	Natural zooplankton assemblage	II	Field enclosures			Fedorenko (1975).
C. trivittatus	Copepoda		Laboratory aquaria		Feeding rates increased with temperature and predator size	Giguère (1981)
C. flavicans	Cladocera	II	Laboratory aquaria		Number caught depended on relative swimming speeds	Smyly (1980)
	Copepoda (3 species)	III	Laboratory aquaria			

*Blanks indicate lack of data. Additional information may be found in Hassell, et al. (1976) and Pastorak (1979).

The theoretical literature pertaining to functional responses relates primarily to considerations of the stability or persistence of a predator-prey link. Variation in prey harvesting rates are exemplified by the different curves. Empirical data and intuition indicate that predator-prey systems are stable and persist in nature. On the contrary, theoretical models consistently show that types I and II functional responses are not stable (i.e. they lead to extinction of the prey population or to escape of prey from predatory control); they also show that type III responses are only stable if there are no time delays between the predator rate of increase and the prey death rate (Holling 1961; Murdoch and Oaten 1975; Hassell et al. 1976; Hassell 1978).

Many authors have modified functional response stability theory to explain this contradiction between empirically measured functional-response data and observed natural population fluctuations by identifying realistic features that enhance the stability of predator-prey links (e.g. see Rosenzweig and MacArthur 1963; Hassell and May 1973; May 1974; Beddington et al. 1976). These authors have generated theoretical models showing that stability is enhanced by spatial heterogeneity of prey habitat, patchy or clumped distributions of prey, existence of refuges for prey, predator aggregation in areas of high prey density, and mutual interference among aggregating predators. Each of these conditions can promote stable density-dependent predation rates, in which predators consume prey at a faster rate when prey densities are high than when they are low. Therefore, the predominance of unstable type II functional responses reported for aquatic insect predators may be an artifact of the unnatural conditions under which they were measured (Table 8.4).

For example, most of the experiments on functional responses of aquatic insect predators have been conducted under laboratory conditions in which one species of predator is provided with one species of prey that is distributed homogeneously throughout the experimental habitat (Table 8.4). These unnatural conditions almost invariably have produced unstable type II functional response curves. However, data collected under laboratory conditions that provided heterogeneous habitats, prey refuges, or alternative prey species, conformed to potentially stable type III response curves. A type III functional response has also occurred when predators are given alternative prey in differing relative abundances. In these situations, predators will consume the most abundant prey species disproportionately; this has been termed *switching* (Murdoch 1969). These data are consistent with theoretical models that predict enhanced stability of the predator-prey link.

Variations in functional response curves are influenced by several factors (Table 8.4). Prey generally are consumed at a higher rate with increased predator hunger and increased temperature within a given range until an optimal temperature is reached at which the feeding rate is maximized. Adults or late-instar predators consume more prey biomass per prey density than younger predators, but on a per-weight basis, younger age classes probably eat more than older ones. In addition, many aquatic insects have been shown to cease feeding just before emergence or before a molt.

It is apparent that empirical evidence for predator responses to prey density is gravely lacking in the published literature. The few available published reports are neither sufficient to evaluate the validity of general models nor do they provide adequate support or refutation of alternative hypotheses to explain observed predator behavior. Some aquatic insect predators appear to conform to the predictions of optimal patch use under some circumstances, and some predators feed at rates that conform to the rules for persistence of the predator-prey link under some circumstances, but there are many exceptions to these observations. Clearly, more studies must be conducted with appropriate methods to test these hypotheses.

Prey Selectivity or Preference

The question of prey selectivity or preference has also been addressed theoretically. Optimal foraging theory (Krebs 1978) predicts that (all other things being equal) predators should only include the most profitable prey items in their diet (i.e. those that allow maximum total food intake over time). The energy content of the prey (calories), the rate at which the predator encounters prey (number of prey/time), which is a function of both predator and prey densities and swimming speeds, and the time spent searching for, pursuing, and handling prey are variables that must be considered in measuring prey profitability. The theory implies that a new prey item should be rejected or ignored if its profitability is lower than the average profitability of the prey items already in the predator's diet. However, such optimal behavior might not be expected when a predator is hungry. Also, since tactile predators can only recognize a prey item after the costs of search and pursuit have been spent, only handling costs would be important in a rejection decision. Therefore, the theory and its predictions must be modified for application to nonvisual predators, which include many aquatic insects (Elner and Hughes 1978).

Unfortunately, most investigations of aquatic insect predator preferences are based on stomach content analyses that are subject to the interpretation limits discussed earlier. Many studies are simply descriptions of the relative abundance of prey items in the stomachs of predators without comparison to estimates of prey available in the habitat. It is obvious from Table 8.5 that only a few studies of prey selection by aquatic insects have measured the variables necessary to test predictions of optimal diet theory. Table 8.5 also shows that many aquatic insect predators are opportunists, but some appear to select certain sizes or types of prey. In many cases, information on the same or related species is contradictory, largely due to differences in analytical techniques and possible experimental artifacts. Thus, limitations of each methodology should be considered when evaluating the data.

The best studies on feeding selectivity by aquatic insect predators have used larvae of the phantom midge *Chaoborus* (Table 8.5). An interesting conclusion of many of these studies is that although *Chaoborus* may appear to prefer Copepoda to Cladocera, and medium-size to large- or small-size zooplankton prey, these preferences may be due largely to probabilities of encountering prey, and prey escape capabilities rather than to actual prey choice or rejection by the predators. Pastorok (1981) showed that the profitability curve for *Chaoborus* feeding on different sizes of *Daphnia* coincided with the vulnerability curve for *Daphnia*. Vulnerability of a prey item is a function of prey size, susceptibility to encountering a predator, and ease of capture (Wilson 1975). Pastorok (1981) speculated that the variation in vulnerability of different sizes of *Daphnia* leads to different predation rates by *Chaoborus,* even though the predator does not actually select specific prey sizes.

Results of behavioral studies of stonefly predators feeding on mayfly prey (Peckarsky 1980; Molles and Pietruszka 1983) have seriously questioned the meaning of the terms *preference* or *prey selection* as they have been used in the ecological literature, particularly when they are based on indirect evidence such as stomach contents of predators. By presenting *Hesperoperla pacifica* and *Megarcys signata* with fresh and frozen mayflies (*Ephemerella altana* and *Baetis tricaudatus*), Molles and Pietruszka showed that the apparent preference for *E. altana* can be explained by greater effectiveness of escape by live *B. tricaudatus*. Also, frozen *B. tricaudatus* may have been selected over frozen *E. altana* because the lack of body armor of the former species allowed it to be swallowed very quickly. These authors concluded that stonefly predators choose mayfly prey only on the basis of handling costs that vary with prey mobility and morphology, and these data are consistent with modified predictions of optimal foraging theory for tactile predators. Peckarsky (1980), likewise, showed that different mayfly species have differing abilities to escape stonefly predation, and speculated that such differences may be responsible for the apparent preferences for prey types or prey sizes observed in the contents of predator stomachs.

Studies using *Chaoborus* (Pastorok 1980) and stoneflies (Molles and Pietruszka 1983) have produced apparently conflicting results on relative selectivity of hungry compared to satiated predators. *Chaoborus trivittatus* is selective only when not starved, which is in agreement with the predictions of optimal foraging theory (Charnov 1976). On the contrary, both stonefly species tested by Molles and Pietruszka consumed equal numbers of the two prey species when satiated, and disproportionately more *E. altana* when hungry. This observation could be erroneously interpreted as preference for *E. altana* without additional behavioral data presented by the authors. Both stonefly species attacked equal numbers of the mayfly species when hungry, but preferentially attacked *B. tricaudatus* when satiated. These results suggest actual preference for the latter prey species, which is obscured by different rates of capture success by predators due to varying prey escape effectiveness.

TABLE 8.5: Feeding Preferences of Aquatic Insect Predators, Based on Gut Analyses with (GE) or without (GO) Comparison to Prey Availability (electivity indices)*

PREDATOR TAXON	PREY TAXON	SELECTION OF PREY BY TAXON	SELECTION OF PREY BY SIZE	REFERENCES (type of data)
EPHEMEROPTERA				
Dolania americana	Chironomidae, microcrustacea, Nematoda, Tardigrada		No	Tsui and Hubbard (1979) (GE)
Ephemerella inermis	Chironomidae			Corkum (1980) (GO)
ODONATA	Pond benthos	No	No	Johnson and Crowley (1979) (review)
Anisoptera				
Aeshna, Libellula	Paper discs		3 mm² size preference	Pritchard (1965) (LEx)
Anax imperator	Pond fauna	(+) Trichoptera, Pleidae, Libellulidae, Lepidoptera, Chironomidae, (−) Mollusca, Hydracarina, Crustacea, Hemiptera, Coleoptera		Cloarec (1977) (fecal pellets, E)
Anax junius	Tadpoles		No	Caldwell et al. (1980) (BL)

Taxon	Habitat	Prey/selection	Notes	Reference
Many species (Anisoptera and Zygoptera)	Pond fauna	No	No	Pritchard (1964) (GO)
Cordulegaster	Stream benthos	(+) *Baetis*, (+) *Paraleptophlebia*, (−) *Litobrancha*		Johnson (1982) (GE)
Ophiogomphus	Stream benthos	*Hyalella azteca*, *Hydropsyche occidentalis*, *Simulium argus*		Koslucher and Minshall (1973) (GO)
Zygoptera				
Argia, *Enallagma*	Stream benthos	*Hyalella azteca*, *Hydropsyche occidentalis*, *Simulium argus*		Koslucher and Minshall (1973) (GO)
Ischnura elegans	Pond fauna	(+) Entomostraca, (+) Chironomidae	Larger predators ate larger prey, no selection within range of handling	Thompson (1978b) (fecal pellets, E)
Enallagma durum	Estuarine cove fauna	Chironomidae	Larger predators ate larger prey	Menzie (1981) (GO)
Coenagrion resolutum, *Lestes disjunctus*	Pond fauna	Chironomidae and Cladocera		Baker and Clifford (1981) (fecal pellets, GO)
PLECOPTERA				
Perla carlukiana, *Perlodes morioni*, *Isoperla grammatica*	Stream benthos	*Baetis*, *Simulium*, Chironomidae, not *Gammarus*		Peckarsky (1979); Malmqvist and Sjöström (1980) (reviews)
Isoperla clio, *Isogenus decicus*	Stream benthos	*Baetis* and Chironomidae, not *Gammarus*		Mackereth (1957) (GO)
Acroneuria (Calineuria) californica	Stream benthos	Diptera, when predators were small; Ephemeroptera, when predators were medium; Trichoptera, when predators were large	Larger predators ate larger prey	Minshall and Minshall (1966) (GO, BF); Sheldon (1969) (GO)

TABLE 8.5: Continued

PREDATOR TAXON	PREY TAXON	SELECTION OF PREY BY TAXON	SELECTION OF PREY BY SIZE	REFERENCES (type of data)
Perlodidae	Stream benthos	Chironomidae (in nearly same proportion as in benthos)		Thut (1969a) (GE)
Arcynopteryx (Skwala) curvata	Stream benthos	(+) Ephemeroptera		Sheldon (1972) (GO)
Isogenus nonus	Chironomidae		Larger predators ate larger prey	
Perlinodes aurea	Stream benthos	Diptera	Predators are listed in order of those con-	
Diura knowltoni	Stream benthos	Diptera and Trichoptera	suming smallest prey to those consuming	
Skwala curvata	Stream benthos	Ephemeroptera	largest prey	
Frisonia picticeps	Stream benthos	Diptera		
Stenoperla prasena	Stream benthos	Ephemeroptera, Chironomidae	Larger predators ate larger prey	Winterbourn (1974) (GO)
Megarcys signata	Stream benthos	Chironomidae, Baetis, and Chloroperlidae		Cather and Gaufin (1975) (GO)
Stenoperla prasena	Stream benthos	(+) Chironomidae, (+) Ephemeroptera	(+) middle and large size prey	Devonport and Winterbourn (1976) (GE)
Calineuria californica	Stream benthos	(+) Chironomidae, (+) Baetis, (−) others	Larger predators ate wider range of prey sizes	Siegfried and Knight (1976a) (GE)

Predator	Prey source	Prey taxa	Size selection	Reference
Pteronarcella badia	Stream benthos	Diptera		Fuller and Stewart (1977) (GO)
Cultus aestivalis	Stream benthos	Diptera		
Hesperoperla pacifica	Stream benthos	Chironomidae and Ephemeroptera		
Isoperla fulva	Stream benthos	Chironomidae		Fuller and Stewart (1979) (GO)
Claassenia sabulosa	Stream benthos	Ephemeroptera and Chironomidae		
Chloroperlidae	Stream benthos	Chironomidae		
Phasganophora capitata	Stream benthos	Ephemeroptera, caseless Trichoptera, and Diptera	Larger predators ate larger prey	Kovalak (1978) (GO)
Dinocras cephalotes	*Baetis*		No	Malmqvist and Sjöström (1980) (LEx)
Calineuria californica, *Hesperoperla pacifica*	Stream benthos		Larger predators ate larger prey	Sheldon (1980) (GO)
Arcynopteryx, Isoperla, Periodes, Perla, Dinocras	Stream benthos	Predator taxa are listed from most to least carnivorous	Larger predators ate larger *Baetis*	Berthélemy and Lahoud (1981) (GO)
Claassenia sabulosa	Stream benthos	Simuliidae	No size selection, larger predators ate wider range of prey sizes	Allan (1982a) (GO, GE)
Megarcys signata		Chironomidae, when predators were small, and *Baetis bicaudatus* when predators were large		

TABLE 8.5: Continued

PREDATOR TAXON	PREY TAXON	SELECTION OF PREY BY TAXON	SELECTION OF PREY BY SIZE	REFERENCES (type of data)
Kogotus modestus Megarcys signata, Hesperoperla pacifica	Baetis tricaudatus, Ephemerella altana	Chironomidae Ephemerella, when predator was starved; no preference, when predator was satiated	No	Allan (1982a) (GO, GE) Molles and Pietruszka (1983) (LEx)
HEMIPTERA				
Notonecta undulata	Pond fauna	(+) Culicidae, (−) Chironomidae (red)		Ellis and Borden (1970) (LEx)
N. hoffmanni	Stream pool benthos	Younger Notonecta ate aquatic, less mobile prey; older Notonecta ate surface prey, cannibalism		Fox (1975) (BF)
Buenoa	Pond fauna	Culicidae, Corixidae (mobile prey in water column)	Small prey	Gittleman (1975) (BF)
Martarega Neoplea	River fauna Pond fauna	Floating prey	Large prey Large prey relative to predator size	Gittleman (1977) (BF)
Buenoa	Prey from natural habitat (in laboratory)	Culicidae, Anostraca, Corixidae (ranked by capture success)		Zalom (1978b) (LEx)

	Habitat/food	Prey selectivity	Size selectivity	Reference
Notonecta lobata		Surface prey		
N. undulata		Surface prey		
N. kirbyi		Limnetic prey		
N. indica		Limnetic prey		
MEGALOPTERA				
Corydalus cornutus	Stream benthos	(+) Baetidae, Hydropsychidae, Simuliidae, opportunistic on other taxa	No	Stewart et al. (1973) (GE)
Sialis fuliginosa	Stream benthos	No	No	Townsend and Hildrew (1979a) (GO)
Sialis cornuta	Beaver pond fauna	(+) Chironomidae, Tubificidae, Ostracoda in proportion to availability; (−) Ephemeroptera, Trichoptera	Larger predators ate larger maximum size of prey	Pritchard and Leischner (1973) (fecal pellets, E; estimated field abundance)
TRICHOPTERA				
Rhyacophila sp.	Stream benthos	(+) Copepoda, Hydracarina, Chironomidae (eaten in proportion to availability); (−) Ephemeroptera, Plecoptera		Thut (1969b) (GE)
Clistoronia magnifica	Laboratory rearing, fed worm supplement	Opportunistic predator		Anderson (1976) (BL)

TABLE 8.5: Concluded

PREDATOR TAXON	PREY TAXON	SELECTION OF PREY BY TAXON	SELECTION OF PREY BY SIZE	REFERENCES (type of data)
Halesus digitalus, Potamophylax cingulatus	Bullhead eggs			Fox (1978) (BL)
Hydrobioses parumbry-sinnis	Stream benthos	Chironomidae, Ephemeroptera (opportunistic?)		Winterbourn (1978) (GO)
Psilochorema bidens	Stream benthos	Surface and hyporheic benthos		
Plectrocnemia conspersa	Stream benthos	Active prey	No	Townsend and Hildrew (1979a) (GO)
Rhyacophila acropedes	Stream benthos	Simuliidae		Wiley and Kohler (1981) (BF)
COLEOPTERA				
Dytiscus marginalis	Pond fauna	Small Dytiscus ate thin Lestes rectangularis, large Dytiscus ate stocky Pachydiplax longipennis, Eubranchipus vernalis, Bufo, and Pseudacris (selected prey by shape)		Young (1967) (BL)
Hydrophilus triangularis, Tropisternus lateralis	Rice field fauna	(+) Chironomidae		Zalom and Grigarick (1980) (GE)

DIPTERA
Chaoboridae

(Citations below are not included in the review by Pastorok 1979, Table 1)

Species	Prey	Notes	Citation
Chaoborus flavicans	Pond fauna	(+) *Diaptomus, Daphnia,* (−) Ostracoda, Oligochaeta	Swüste et al. (1973) (LEx)
C. flavicans	Cladocera Copepoda	*Diaptomus,* more active prey / Medium-sized Cladocera	Smyly (1980) (LEx)
C. punctipennis (instar IV)	Copepoda Cladocera	*Diaphanosoma*	Winner and Greber (1980) (LEx)
C. punctipennis	Reservoir fauna	(+) Rotifera, *Cyclops*	Chimney et al. (1981) (GE)
C. trivittatus	*Daphnia Diaptomus* (laboratory)	*Diaptomus,* when not starved, no preference when starved / (+) medium sizes of *Daphnia*	Pastorok (1980) (GE); Pastorok (1981) (LEx)
C. americanus, C. punctipennis, and *C. travittatus*	Cladocera, Copepoda, and Rotifera	Depended on prey speed, swimming pattern, and escape	Swift and Fedorenko (1975) (LEx); Swift and Forward (1981) (LEx)
Chironomidae *Procladius*	Pond fauna		Tarwid (1969) (GO)
Parachironomus forceps	Clear pool fauna	*Ambystoma* eggs	LeClaire and Bourassa (1981) (BF)
Culicidae *Toxorhynchites rutilis rutilis*	*Aedes aegypti*	Instar IV	Bay (1974) (review) Padgett and Focks (1981) (BL)

*(+) = positive electivity index, (−) = negative electivity index, in lab (L) or field (F). Experiments conducted in the laboratory (LEx) or the field (FEx) and behavioral observations in the laboratory (BL) or field (BF) are also indicated. E = electivity index. Blanks indicate no data available.

Another general result of studies on aquatic insect predator preferences is that most odonate, plecopteran, and megalopteran predators appear to feed on all sizes and types of prey within the range that they can handle (Pritchard and Leischner 1973; Thompson 1978b; Allan 1982a). Large size classes of these predators consume a wide range of size classes of prey, with the maximum prey size increasing with predator size. These observations may or may not contradict expectations of optimal diet theory, but to predict that smaller size classes should drop out of a predator's diet, data are needed on relative profitability of different size classes of prey.

Zaret (1980) applied a model of size-selective predation generated from data on planktivores to freshwater organisms in general. This model predicts that gape-limited predators (GLP) should select larger and larger prey until prey size exceeds maximum mouth gape (Fig. 8.2A). Zaret suggested that predators that swallow prey whole (e.g. Plecoptera) or those that use visual cues to locate prey (e.g. some Hemiptera) probably fall in this category. However, existing data on these taxa do not support this prediction (Table 8.5), and a semantic problem arises when sucking predators, such as Hemiptera, are called *gape-limited*. Size-dependent predators (SDP) should select prey that increase in size until difficulty in handling and increased probability of escape outweigh the returns gained by eating larger prey (Fig. 8.2B). Zaret predicted that *Chaoborus* and Odonata should show this response. He also stated that SDP curves should be applicable to predators with sucking mouthparts and predators that rely on tactile manipulation of prey, and that prey size should be more important in the capturing than in the handling of prey.

Table 8.5 indicates studies that have shown GLP or SDP electivity curves for aquatic insect predators. Data on sizes of prey selected by some Odonata and *Chaoborus* conform to the humped electivity curve for SDPs (Fig. 8.2B)

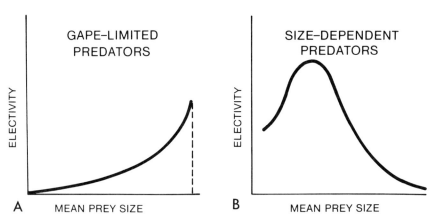

Figure 8.2 (A) Predicted electivity curve for gape-limited predators; (B) Predicted electivity curve for size-dependent predators (modified from Zaret 1980).

(Pritchard 1965; Pastorok 1981), but other studies of *Chaoborus* size-selectivity have contradicted this finding. For example, there may be an apparent maximum size of zooplankton prey consumed, indicating gape-limitation, but this maximum size is not related to the size of the predator (Swift and Forward 1981). *Chaoborus punctipennis* larvae eat the egg sacs of copepod prey that are larger than their gape limitation (Swift and Forward 1981). Thus, although Zaret's (1980) statements may have some applicability to fish-zooplankton systems, his model appears inappropriate for application to aquatic insect predator-prey relations.

In summary, few aquatic insect predators have been studied in enough detail to draw reliable generalizations on their conformity to expectations derived from existing predator-prey theory. Neither the cues by which most predators detect prey nor other aspects of their hunting behavior, such as energy budgets and search paths, have been determined. This limits our ability to evaluate the applicability of present theory. Limited data are available regarding aggregative responses, giving-up-time, area-restricted search, and functional and developmental responses of predators to patches of varying prey density. Some of these responses conform to expectations generated by existing theory, while others do not. In addition, few convincing data have been collected on prey size or species selectivity. Notable exceptions are the detailed studies on some Odonata (Aeshnidae, Coenagrionidae), Plecoptera (Perlidae and Perlodidae), Hemiptera (Notonectidae), and Diptera (Chaoboridae). However, these studies have indicated that apparent selectivity may be more a function of prey vulnerability to capture than of selection by the predator.

DEFENSES OF AQUATIC INSECT PREY

The previous discussion of predator preferences indicates that the apparent selection by predators of specific sizes or types of prey may, instead, be a reflection of effective or ineffective defenses by prey. For example, many authors have concluded that species of *Chaoborus* consume sizes and shapes of prey that are easiest to catch and subdue (Swift and Fedorenko 1975; Smyly 1980; Pastorok 1981). Therefore, a consideration of the defensive mechanisms of aquatic insect prey is essential to understand fully the relationships between aquatic insects and their predators.

A survey of the defenses used by aquatic insects reveals the evolution of a wide variety of mechanisms by which they avoid predation. Schall and Pianka (1980) predicted that when predation pressure is high, selection will favor a complex array of escape tactics. Aquatic insects are vulnerable to predation from various vertebrate as well as invertebrate predators. Many species of aquatic insects exhibit a complex array of morphological, chemical, and behavioral defenses, as predicted by Schall and Pianka (1980), while others rely on one simple escape response. Sih (1982) suggested that the degree to which predators are avoided should reflect the magnitude of the risk of predation and the cost of avoiding predators, but this hypothesis has not been tested elsewhere.

The antipredator defenses of animals have been classified as either *primary* or *secondary* (Edmunds 1974). Most aquatic insects possess both types. Primary defenses operate regardless of the predator's proximity, or before the predator initiates prey-catching behavior, and include most morphological and some chemical defenses. Secondary defenses operate during an encounter with the predator, thereby increasing the chances of the prey's survival. They may be relatively passive, such as some chemical defenses, requiring little or no energy consumption on the part of the prey, or they may be active, such as most behavioral responses elicited by predator presence. Some defenses are specific to a particular dominant predator, but a prey organism must evolve a whole system of defenses to protect it from an entire complement of predators.

Primary Defenses

Refuges.   *Anachoresis* is defined as the practice of living in crevices or holes, either continuously or leaving such surroundings only to feed or mate (Edmunds 1974). Such a practice decreases an organism's chances of encountering predators. Several types of refuges are defined by Woodin (1978) that broaden the realm of behaviors that could be classified in this particular primary defense category. For example, prey may find refuge from predators through spatial or temporal separation, as well as by residing in physically or biologically generated hiding places such as mineral or plant substrates, or even animals (e.g. sponges, bryozoans). Burrowing mayflies, such as *Hexagenia*, and the tube-dwelling larvae of chironomids are examples of prey that find refuge in substrates not likely to be disturbed by predators. Many aquatic insects reach high densities in the littoral zone of ponds, a region that presumably offers refuge from fish predation.

Few studies have been designed specifically to document the effectiveness of refuges in reducing predation, but a number of studies have suggested that prey populations may reach higher levels when refuges from predation are available. For example, Benke et al. (1982) suggested that the absence of vertical macrophytes that could be used as refuges inside their experimental field enclosures caused reduced prey densities compared to those outside the enclosures. Losses of two stonefly species to predatory caddisflies were reduced when the stoneflies occupied substrates that offered a spatial refuge from these predators (Hildrew and Townsend 1977). The occupation of suboptimal foraging areas by immature *Notonecta hoffmanni* (Sih 1980a, 1981) is an example of spatial refuge from predation by adult *N. hoffmanni*. Finally, Peckarsky and Dodson (1980b) showed that effects of competition between mayflies are only significant in enclosures that offer a spatial refuge from predators in a Wisconsin stream.

Many aquatic insects time their periods of highest activity to avoid overlap with those of their predators. For example, *Chaoborus* larvae show distinct diel vertical migration patterns in lakes. They forage in the hypolimnion during daylight and migrate to the epilimnion to forage on pelagic prey at night. Such a pattern allows their coexistence with fish (Swüste et al. 1973; von Ende 1979; Swift and Forward 1981; see review by Pastorok 1979). Some species of *Chaoborus* do not show vertical migration and are eliminated from lakes with fish (von Ende 1979). Also, Williams (1980) showed that predation on early instars of *Chaoborus americanus* by other invertebrate predators such as *Diaptomus shoshone* is reduced by their lack of temporal overlap.

Most stream biologists have casually observed that benthic invertebrates rarely reside on the substrate surface during daylight. Presumably, stream insects forage on the surface of the substrate at night primarily to avoid predation by visual feeders, but this explanation has not yet been tested. Such diel periodicity of activity has been suggested as one explanation for the nocturnal maximum in stream insect drift (Waters 1972), and recently has been linked to a predation-avoidance hypothesis. For example, Allan (1978) found that a higher proportion of large *Baetis bicaudatus* nymphs drifted at night during the day. The stomachs of trout predators contained more large than small *Baetis* prey, and Allan suggested that night drifting of large nymphs reduced their risk of being consumed by visual predators, thereby constituting a temporal refuge from trout feeding. Corkum and Pointing (1979) offered a similar explanation for the increased drift of large *Baetis vagans* nymphs at night. Edmunds and Edmunds (1979) speculated that differences in emergence and mating patterns between tropical and temperate mayflies may be explained by predator avoidance.

Crypsis. Crypsis is any adaptation that causes an animal to be less conspicuous with respect to visual, chemical, tactile, sound, or electrical perception (Edmunds 1974). Since predators have been shown to take conspicuous prey more often, structural or behavioral modifications that improve camouflage should be effective. "Aquatic insects are notoriously dingy in color" (Hutchinson 1981, p. 499), and few possess colors or shapes that contrast with their surroundings. Many adopt a resting position on appropriate background, and a few modify their environment to conceal their presence. A disadvantage of visual crypsis is that it only works if the insect does not move (Ware 1973). This conflicts with other important functions, such as feeding and reproduction, but aquatic insects often couple crypsis with temporal behavior patterns (described above) that improve their chances of avoiding encounters with predators.

Popham (1942, 1943) observed a close agreement between the coloration and habitat background of certain corixids, and he documented that corixids on contrasting backgrounds are much more vulnerable to predation by fish than those on like backgrounds. *Chaoborus* species with more pigmentation have higher susceptibility to predation by fish, and neither coexist with fish nor show diel patterns of vertical migration (Stenson 1979; see reviews by Kerfoot 1979 and Zaret 1980). Those species that do coexist with fish are inconspicuous due to their nearly complete transparency.

Zooplankton can attain low visibility through small size (Zaret 1980). This concept of a size refuge from predation is relevant to aquatic insects, but it has not been tested rigorously. A phenomenon like cyclomorphosis (Dodson 1974), which allows zooplankton prey to escape both vertebrate and invertebrate predators through cyclical alterations in their apparent size, has not yet been discovered in aquatic insects.

Caddisfly cases have several functions, including that of providing protection for larvae. This protective function may be crypsis or protection against predator bites. Otto and Svensson (1980) examined the types of case materials used by species of caddisflies in southern Sweden and North America and concluded that the energy expended in construction of cases is inversely related to case vulnerability from predation. For example, many Limnephilidae change from leaf disc cases to bark to mineral particle cases as the larva grows, thus increasing protection from biting predators. Cannibalism in Brachycentridae may have promoted the evolution of mineral pupal cases (Gallepp 1974), and the incorporation of snail shells into the cases of a glossosomatid species was attributed to defense against predators (Anderson and Vinikour 1980).

Aposematism. The advertisement of some dangerous or unpleasant attribute is a common characteristic of terrestrial insects (Edmunds 1974). Distasteful prey are often brightly colored or have other signals that facilitate learning by a predator to recognize and avoid them (Gittleman and Harvey 1980). Patterns of Batesian and Mullerian mimicry are based on this adaptation. Elton (1922) and Popham (1948) suggested that scarlet water mites are unpalatable, and this hypothesis was experimentally documented by Kerfoot (1982). Hutchinson (1981) cited Ellis and Borden (1970) as evidence that blood worms (Chironomidae) are not preferred prey of Notonectidae, and speculated that red Chironomidae are unpalatable, but there is no direct evidence of this. Hutchinson (1980) also suggested that the reddish brown or pinkish colors in the hemelytra of some *Notonecta* species may be aposematic, and that this deserves further study.

Secondary Defenses

Active predator-avoidance behavior requires effective predator-detection capabilities, so that a response can be made before a successful attack is completed. Aquatic insects such as some Odonata, Coleoptera, and Hemiptera probably detect predators visually. Adult whirligig beetles (Gyrinidae) have two pairs of eyes, one pair effective above and one pair effective below the water surface (Brown and Hatch 1929). Other aquatic insects, such as Ephemeroptera (Peckarsky 1980) and Diptera (Sih, unpublished), may use chemical cues to detect predators. For example, *Ephemerella* nymphs avoid regions of an observation box downstream of a stonefly predator confined in a screen tube (Peckarsky 1980). Most aquatic insects, however, probably use mechanical cues, either by predator contact or by long-range detection of mechanical waves. Gyrinidae, for example, echolocate. They possess a Johnston's organ at the base of their antennae, which detects surface waves from objects on the water surface and from which their own signals are rebounded (Tucker 1969). Notonectidae discriminate between prey and nonprey at the surface of ponds by detecting differences in the surface-wave characteristics of these two items (Lang 1980). I have observed that the mayfly *Baetis bicaudatus* responds to mechanical waves produced by foraging stonefly predators. The sensory cues available to an aquatic insect influence the evolution and effectiveness of its secondary defensive responses.

Withdrawal to a Prepared Retreat or Flight.   Although withdrawal to a retreat has been documented to occur in crayfish in response to predators (Stein and Magnuson 1976), and although it may be common for aquatic insects with prepared retreats such as burrowing mayflies, net-spinning caddisflies, and tube-dwelling Chironomidae, this phenomenon has not been studied specifically. More commonly, studies have shown that aquatic insects flee from their predators upon encounter, and that the path of such flight is generally straight rather than erratic, as in some terrestrial moths, birds, and mammals (cf. Edmunds 1974).

The drifting of mayflies and stoneflies increases in the presence of stonefly predators (Corkum and Pointing 1979; Corkum and Clifford 1980; Walton 1980), and the drifting of mayflies and Simuliidae is initiated by contact with stonefly and caddisfly predators (Peckarsky 1980; Wiley and Kohler 1981). Gerridae usually avoid cannibalism by other water striders by first moving at right angles to the direction of attack, and then circling behind their attackers (Jamieson and Scudder 1979). This defense is effective because once the predatory water strider has gained momentum, changing directions is difficult. Alternatively, prey may leap into the air allowing the predator to skim under them, or accelerate to foil the predator's line of interception.

Deimatic Behavior.   Many terrestrial insects show startle displays or adopt characteristic postures that intimidate, frighten, or warn a predator of their armor or unpalatability (Edmunds 1974). Such behavior is often a bluff. Adult *Tropisternus* beetles (Hydrophilidae) stridulate by moving a scraper located on the forefemora against striations on the head, a behavior that has been shown to deter predation by wolf spiders in an experimental situation (Masters 1979, 1980); acoustical signals in Dytiscidae might have a similar function (Smith 1973). Mayflies of the family Ephemerellidae adopt a characteristic scorpionlike posture, in which they raise their caudal filaments and posterior abdominal segments vertically through a 180° angle over their heads (Peckarsky 1980; Molles and Pietruszka 1983). This behavior apparently deters stonefly predators that respond by crawling over or around them in seeming disinterest. The posture may create an image of a spiny creature of larger size and different shape to a tactile predator, a behavioral analogue to cyclical development of spines by some zooplankton. Larger instars assume this posture more frequently than smaller nymphs, perhaps because small size is also an effective defense.

Thanatosis. Presumably, feigning death causes a predator to relax its attention or lose interest momentarily, allowing time for the prey to escape; it is also possible that a motionless animal does not elicit a killing strike (Edmunds 1974). Two species of stoneflies feign death and sink to the bottom of experimental chambers after being first attacked and captured, and then spit out by brown trout (Otto and Sjöström, in press), and early instar ephemerellid nymphs often freeze or stay motionless when encountered by a stonefly predator (Peckarsky 1980). Adult damselflies (Coenagrionidae) also drop from vegetation and feign death when encountered by predators (White 1980). Inactivity by *Drosophila* prey actually improves their chances of survival, compared to active *Culex* prey, when both are subjected to notonectid foraging (Sih 1979). If prey mobility enhances predator efficiency at high predator densities, this may lead to selection pressure for prey immobility.

Deflection of an Attack, or Retaliation. The cerci and antennae of stoneflies may be viewed as weapons of defense against trout predators (Otto and Sjöström, in press). The direction and success of attacks by brown trout depended on the presence of intact cerci. For example, brown trout preferred to attack toward the anterior of intact prey or prey without antennae, whereas trout preferred to attack stoneflies without cerci from the posterior. In addition, a high proportion of stoneflies were ejected when attack occurred from the less preferred direction. These data corroborate observations that related species of stoneflies turn their cerci toward other stoneflies upon encounter (Peckarsky, unpublished data). This behavior probably deters aggressive behavior of a competitor and serves as an effective defense against predation.

Protection against predation may also be achieved by loss of a body part. For example, crayfish autotomize their chelae in response to an attack from a perceived predator (Robinson et al. 1970). If the predator attacks a body part of the prey that is either conspicuous or oriented by the prey toward the predator, autotomy can be an effective defense. Mayflies and damselflies lose their caudal filaments readily, but it is not known whether these features are important in defense against predation.

Aggressive retaliation that is injurious to predators is little known for aquatic insect prey. However, some giant water bugs (*Lethocerus*) or predaceous beetles (*Dytiscus*) may attack vertebrate predators. Hungerford (1919) reported a giant water bug attacking a woodpecker, and C.O. Berg (personal communication) recorded an adult *Dytiscus* consuming a snapping turtle in an aquarium!

A number of studies have documented insect use of defensive glandular secretions that are poisonous or noxious to predators (Eisner and Meinwald 1966). Currently, there is controversy over whether the use of chemicals by aquatic insects is a primary defense (Eisner and Grant 1981 refer to this as *olfactory aposematism*) or a secondary defense that is elicited upon encounter with the predator (cf. Edmunds 1974). For example, beetles of the families Gyrinidae and Dytiscidae excrete substances from their pygidial and thoracic glands that have narcotic, anaesthetic, and toxic effects on fish (Benfield 1972; Miller and Mumma 1973; Miller et al. 1975; Miller and Mumma 1976a, 1976b; Newhart and Mumma 1979). Heinrich and Vogt (1980) observed a rock bass eat one gyrinid, but then regurgitate the second, and ignore the third one offered. Adult caddisflies also emit noxious secretions (Duffield et al. 1977; Duffield 1981). Noxious body chemistry, or bad taste, such as that of red water mites (Kerfoot 1982), may be an effective primary defense, but at least one prey organism must first fall victim to a predator to benefit the population (Edmunds 1974). Corixidae may taste bad to fish (Macan 1965), but this has not been tested rigorously.

Defensive Groups and Associations. Large groups of prey may reduce the chances of predation per individual (Edmunds 1974; Taylor 1976; Bertram 1978). The mechanism mediating this effect may be a confusion factor, group retaliation, or more effective detection of predators than occurs individually. For example, it has been suggested that insect swarms are an adaptation to confuse predators (Sullivan 1981). On the other hand, large aggregations of prey may attract predators or improve predator foraging efficiency at high prey densities (Holling 1961; Hassell 1978). Therefore, sparser spacing may be a better defense against predation in some instances (Tinbergen et al. 1967).

Little data exist to test these hypotheses. Diurnal rafts of whirligig beetles, which show rapid and confusing swarming behavior when disturbed, might be an effective defense against visual predators (Heinrich and Vogt 1980), and coordinated group movements of a species of surface dwelling Veliidae have been recorded in response to subsurface images resembling salmonid predators (Deshefy 1980). However, overhead models and water surface disturbances elicit immediate dispersal behavior of veliids. Flotillas of the marine water strider *Halobates robustus* are more effective than individuals in improving the distance at which aerial predators are detected (Treherne and Foster 1980, 1981; Foster and Treherne 1980). The predation rate by fish on *Halobates* decreases as flotilla size increases, and this has been termed the *dilution effect* (Treherne and Foster 1982).

In summary, aquatic insects show a wide array of primary and secondary defenses against predation by vertebrates as well as invertebrates. Further studies of these groups should provide increasingly interesting examples of the evolutionary responses of prey to predators.

COMMUNITY EFFECTS OF PREDATOR-PREY RELATIONS

Many variables influence the distribution and abundance of organisms in aquatic insect communities, including life histories, food, physical factors, competition, and dispersal, as well as predator-prey interactions. Each aspect of predator behavior discussed above (Tables 8.1 through 8.5), as well as prey defenses, can be important determinants of the role that predators play in structuring prey communities.

Predation may affect the structure of communities in several ways (e.g. see Hall et al. 1970, 1976; Connell 1975; Caswell 1978; Glasser 1979; Pastorok 1979; Sih 1982). Predators may reduce the density of individuals in all populations of a community. Indiscriminate predators that eat prey in the proportions in which the prey are found in the habitat should have this effect, given no compensatory increase in the production of prey. If predators are selective and remove some species of prey in greater numbers than others, a change in the relative species abundances in the prey community and elimination of certain vulnerable prey species should occur. Similarly, selective predation can alter the size structure or sex ratios of prey populations.

The effects of predation on prey-species diversity are complex and variable (Fox 1977; Friberg et al. 1977; Bärlocher 1980). If predators are not selective, prey diversity may still decrease in cases where a predator is extremely effective, and cause relatively rare species to become locally extinct. Such effects are more common when prey communities are composed of species with few interactions (e.g. no competition, short food chains), and show little response to removal of individuals by predation. In contrast, prey-species diversity may increase if predators selectively remove dominant competitors from the prey community, which allows inferior competitors to invade habitats where they were excluded previously; such organisms have been termed *keystone predators* by Paine (1966). Nonselective predators can also increase prey diversity by acting as a disturbance, thereby allowing new species to colonize or to increase population levels. Changes in species diversity, therefore, depend on the strength of interactions among prey species and the selectivity and effectiveness of the predator.

The available published experimental data on community-level effects of aquatic insect predators on their prey, and of other noninsect predators on aquatic insect prey communities are summarized in Table 8.6. Since such studies are extremely rare, Table 8.6 also includes experimentally documented predator-mediated decreases in prey population size. A number of correlative studies have estimated relationships among predator numbers and prey numbers or prey consumption rates. These studies have concluded that predation by Plecoptera (Wright 1975; Siegfried and Knight 1976b; Allan 1983b), Trichoptera (Wright 1975; Townsend and Hildrew 1978; Hildrew et al. 1980), *Chaoborus* (Lewis 1979), Chironomidae (Loden 1974; Dusoge 1980), and mixed species assemblages of predators (Davies and Reynoldson 1971; McDonald and Buchanan 1981; Zalom 1981) may numerically depress their prey populations or communities. Fish predation (Allen 1942; Arruda 1979; Eriksson et al. 1980) and triclad predation (Reynoldson and Bellamy 1975) have also been implicated, by correlation, as responsible for significantly reducing densities of aquatic insect prey populations.

According to Zaret (1980), the effects of selective fish predation are so clear that the size and species composition of zooplankton are reliable predictors of the presence or absence of fish predators in a pelagic community. However, the same generalization cannot be made for aquatic insect predators from an overview of the experimental data presented in Table 8.6. It may be safe to conclude from the published literature that many aquatic insect predators are highly efficient, causing potentially large reductions in some prey populations; however, feeding rates may be elevated artificially by experimental conditions. Although data on changes in relative species abundances or prey sizes as a result of predation are conclusive for a few species of aquatic insect predators, such as *Chaoborus* spp. preying on zooplankton communities, they are inconsistent or totally lacking for others. Reliable data on the effects of predators on aquatic insect community composition or diversity are virtually nonexistent.

CONCLUSIONS AND FUTURE NEEDS

The published data on aquatic insect predator-prey relations provide us with very little knowledge about each of the major topics outlined in Table 8.1 and discussed above. Studies designed to falsify hypotheses (Allan, Chapter 16) are necessary to quantify the search and detection modes of aquatic insect predators under realistic conditions. More data are required on responses of aquatic insect predators to variations in prey density, including improved estimates of prey density or manipulations of prey density in laboratory or field enclosures. Feeding trials conducted under more natural conditions and a divergence from reliance on stomach content analysis will improve estimates of functional responses. The relatively weak data available on prey selectivity or preferences, perhaps attributable to stomach content analyses, should be corroborated with data from experimental manipulations designed to measure all of the parameters necessary to assess prey availability, profitability, vulnerability, and predator diet

TABLE 8.6: Experimental Studies of Population- and Community-level Effects of Aquatic Insect and Non-insect Predators*

PREDATOR TAXON	PREY TAXON	EFFECTS	EXPERIMENTAL CONDITIONS	REFERENCES
ODONATA				
Ischnura verticalis	*Simocephalus serrulatus* (Cladocera)	(−) numbers	Plastic pools	Johnson (1973)
Enallagma durum	*Cricotopus sylvestris* (Chironomidae)	(−) numbers	Laboratory experiments and field enclosures	Menzie (1981)
Celithemis fasciata, Epitheca spp., *Ladona deplanata*	Pond benthos	(−) numbers of small Anisoptera	Large in situ enclosures (4 × 4 m)	Benke (1978)
Epitheca cynosura, Libellula incesta	Pond benthos	(0) numbers, biomass, diversity of total benthos (+) secondary effect on Zygoptera	Small in situ enclosures (20 l)	Benke et al. (1982)
Anax junius	Pond benthos	(−) *Chironomus tentans* (when nutrients low), (+) *Caenis simulans*, (0) biomass of total benthos	Experimental ponds	Hall et al. (1970)
Leucorrhinia sp.	*Chaoborus flavicans C. obscuripes*	(−) numbers (0) numbers	Laboratory experiment	Stenson (1981)

PLECOPTERA				
Dinocras cephalotes	Tricladida	(−) numbers	Laboratory observation	Davies and Reynoldson (1969)
Acroneuria lycorias, Megarcys signata	Stream benthos	(−) density of total benthos	Small in situ enclosures (40 × 30 × 10 cm)	Peckarsky and Dodson (1980a)
M. signata	*Zapada haysi*	(−) numbers	Leaf packs in in situ enclosures	Oberndorfer et al. (in press)
HEMIPTERA				
8 species of *Buenoa* and *Notonecta*	Pond fauna	(−) numbers	60 l containers with vegetation in laboratory	Zalom (1978a)
Anisops calcaratus	*Daphnia carinata*	Induced crest growth	Laboratory	Grant and Bayly (1981)
TRICHOPTERA				
Halesus digitatus, Potamophylax cingulatus	Bullhead eggs	(−) numbers	Laboratory	Fox (1978)
COLEOPTERA				
Acilius semisulcatus (larvae)	*Daphnia pulex*	(−) numbers	100 l field enclosures	Arts et al. (1981)
DIPTERA				
Chaoborus americanus	*Daphnia rosea*	(−) numbers (93% mortality)	Field enclosures	Dodson (1972)
C. americanus	*Diaptomus shoshone*	(−) numbers	Field enclosures	Sprules (1972)
C. punctipennis		(−) numbers	Field enclosures	von Ende (1979)
C. americanus	*Daphnia*	Induced spine formation (*D. pulex* → *D. minihaha*)	Laboratory experiment	Krueger and Dodson (1981)
C. americanus	*Daphnia, Diaptomus*	Sustained vertical segregation between predators and prey	Field enclosures	Melville and Maly (1981)

TABLE 8.6: Continued

PREDATOR TAXON	PREY TAXON	EFFECTS	EXPERIMENTAL CONDITIONS	REFERENCES
C. americanus	Bosmina longirostris	Eliminated prey in fish-less lakes	Field experiments	von Ende and Dempsey (1981)
C. americanus, C. trivittatus	Diaptomidae (Copepoda)	(−) numbers (5–25%/month)	Field enclosures	Fedorenko (1975)
C. americanus, C. flavicans (instar IV)	Zooplankton	Eliminated nearly all prey species when prey had high levels of food	17 m³ in situ enclosures	Neill and Peacock (1979)
C. flavicans	Mixed zooplankton	(−) numbers (3–13% biomass daily)	Small bottles in lakes	Kajak and Ranke-Rybicka (1970)
C. trivittatus	Daphnia pulex	(−) numbers	Laboratory experiments	Neill (1978)
C. trivittatus	Diaphanosoma	(−) numbers	Pond manipulation	Giguère (1979)
C. trivittatus	Zooplankton	Variable effects: (−) calanoid copepod numbers, when predators at very high densities (usually short-term); (0) Cladocera (rapid recovery from short-term effects), some changes in body size (varied with year);	20–27 m³ in situ enclosures	Neill (1981)

	Prey	Effect	Method	Reference
Chironomidae				
Allotrissocladius (instar IV)	*Paraborniella tonnoiri* (instars I, II) (Chironomidae)	(−) significant on rare species; (+) *Latona* (secondary effect); (−) numbers in shallow pools	Field observation	Jones (1974)
Procladius	Lake fauna	(−) Tubificidae biomass, (+) Tubificidae production, (−) Chironomidae	Field experiment	Kajak (1980)
Parachironomus forceps	*Ambystoma* eggs	(−) numbers (70–100% mortality)	Field observation	LeClair and Bourassa (1981)
Culicidae				
Culex, Anopheles	Protozoa, micrometazoa (in bracts of wild bananas)	Nearly eliminated	Laboratory experiment	Maguire et al. (1968)
Wyeomyia smithii	Protozoa (in pitcher plants)	(−) diversity (number of species)	Field experiment	Addicott (1974)
Toxorhynchites rutilis rutilis	*Aedes aegypti* (instar IV)	(−) numbers	Field experiment in automobile tires	Focks et al. (1980)
T. rutilis rutilis	*Aedes aegypti*	Ate 49.8 *Aedes*/predator/day	Laboratory containers (3.78 l)	Padgett and Focks (1981)
T. rutilis rutilis	*Aedes aegypti*	(−) numbers (instars III, IV of *T. rutilis* most effective)	Laboratory experiment	Bailey et al. (1981)

TABLE 8.6: Concluded

PREDATOR TAXON	PREY TAXON	EFFECTS	EXPERIMENTAL CONDITIONS	REFERENCES
OTHER INVERTEBRATE PREDATORS				
Flatworms				
Polycelis	Stream benthos	(−) Plecoptera, (−) Ephemeroptera	Artificial increase in predator density (septic tank overflow)	Macan (1966, 1977a)
Dugesia tigrina	Culex pipiens, C. restuans	(−) numbers	Laboratory experiment	George (1978)
D. tigrina	C. restuans, Chironomus sp.	(−) numbers	Laboratory experiment	Meyer and Learned (1981)
Mesostoma	Culex tarsalis, Anopheles freeborni (rice fields)	(−) numbers (4 times)	Field enclosures	Case and Washino (1979)
Leeches				
Helobdella stagnalis	Tendipes plumosus	(−) numbers	Laboratory experiment	Hilsenhoff (1963)
Helobdella nepheloidas	T. plumosus	(0) numbers		Hilsenhoff (1964)
FISH AND REPTILE PREDATORS				
Salmo trutta	Pond insects	(−) species diversity, restricted distributions of Corixidae, (−) eliminated Notonecta obliqua, Nymphula, Chaoborus, and (0) others	Artificial introduction of fish	Macan (1965, 1966, 1977a, 1977b)
Lepomis macrochirus	Insect benthos	(−) emergence, (−) body size,	Experimental ponds	Hall et al. (1970)

Predator	Prey	Response	Method	Reference
L. macrochirus	Insect benthos	(0) biomass of total benthos; (−) Chironomus tentans, (+) Caenis simulans, (−) Zygoptera	Experimental ponds	Crowder and Cooper (1982)
Fish	Lake benthos	(−) biomass, (−) mean prey size	Field experiment	Kajak (1977)
Fish	Glaenocorisa propinqua	(−) biomass (−) numbers	Field experiment	Henrikson and Oscarson (1978)
Trichogaster trichopterus	Chaoborus flavicans	(0) numbers	Laboratory experiment	Stenson (1981)
Perca fluviatilis	C. obscuripes	(−) numbers		
Umbra limi	Chaoborus sp.	(−) C. americanus, (0) C. punctipennis	Field enclosures	von Ende (1979)
Fish	Enallagma durum	(−) numbers	Field enclosures	Menzie (1981)
Fish, turtles	Pond benthos	(0) density, (0) diversity, (0) functional group abundances	4m² field exclosures	Thorp and Bergey (1981a)
Fish, turtles	Chironomidae	Same results	4m² field enclosures	Thorp and Bergey (1981b)
Trout	Stream benthos	(0) density, (0) drift density, some changes in relative species abundance, (0) diversity	Artificial removal of trout	Allan (1982b)
Fish	Stream benthos	(0) density, (0) diversity	Exclosures in field	Reice (1983)

*(+) = increased; (−) = decreased; (0) = no change in prey populations with predators.

(cf. Pastorok 1981). The available evidence of substantial effects of aquatic insect predators on local prey abundances and behavior needs further study to determine the importance of prey defense mechanisms in explaining apparent prey preferences and local population depression. Therefore, studies of community-level effects must not only accurately assess feeding rates of predators and their effects on community composition of prey, but such studies must also examine other possible explanations for such effects, such as predator preferences, prey defenses, and the potential of prey communities to respond to high predation pressure by high rates of turnover (Benke 1976, 1978).

In conclusion, the published data on aquatic insect predator-prey relations are not sufficient to evaluate the adequacy of existing predator-prey theory. For example, some models, such as that of Zaret (1980) on size-selective predation, appear inadequate to explain data on aquatic insect predators, and Optimal Foraging Theory must be modified in some cases for application to aquatic insect systems. Therefore, in the future, concepts of prey selection, prey-patch choice, and community-level effects should be developed to incorporate data on aquatic insects.

References

Addicott, J. F. 1974. Predation and prey community structure: an experimental study of the effect of mosquito larvae on the protozoan communities of pitcher plants. Ecology 55:475–92.

Akre, B. G. and D. M. Johnson. 1979. Switching and sigmoid functional response curves by damselfly naiads with alternative prey available. Journal of Animal Ecology 48: 703–20.

Allan, J. D. 1978. Trout predation and the size composition of stream drift. Limnology and Oceanography 23:1231–37.

Allan, J. D. 1982a. Feeding habits and prey consumption of three setipalpian stoneflies (Plecoptera) in a mountain stream. Ecology 63:26–34.

Allan, J. D. 1982b. The effects of reduction in trout density on the invertebrate community of a mountain stream. Ecology 63:1444–55.

Allan, J. D. 1983a. Predator-prey relationships in streams. In: J. R. Barnes and G. W. Minshall (eds.). Stream ecology: application and testing of general ecological theory. Plenum Press, New York, NY. 399 pp.

Allan, J. D. 1983b. Food consumption by trout and stoneflies in a Rocky Mountain stream, with comparisons to prey standing crop, pp 371–90. In: S. M. Bartell and T. D. Fontaine, III (eds.). Dynamics of lotic ecosystems. Ann Arbor Science Publishers, Inc., Ann Arbor, MI. 494 pp.

Allen, K. R. 1942. Comparison of bottom faunas as sources of available fish food. Transactions of the American Fisheries Society 71:275–83.

Anderson, N. H. 1976. Carnivory by an aquatic detritivore, Clistoronia magnifica (Trichoptera: Limnephilidae). Ecology 57:1081–85.

Anderson, R. V. and W. S. Vinikour. 1980. Shells of *Physa gyrina* (Gastropoda: Physidae) observed as substitute case-making material by *Glossosoma intermedium* (Trichoptera: Glossosomatidae). Entomological News 91:85–87.

Andersson, M. 1981. On optimal predator search. Theoretical Population Biology 19: 58–86.

Arruda, J. A. 1979. A consideration of trophic dynamics in some tallgrass prairie farm ponds. American Midland Naturalist 102:254–62.

Arts, M. T., E. J. Maly, and M. Pasitschniak. 1981. The influence of *Acilius* (Dytiscidae) predation on *Daphnia* in a small pond. Limnology and Oceanography 26:1172–75.

Bailey, D. L., D. A. Focks, and A. L. Cameron. 1981. Effects of *Toxorhynchites rutilus rutilus* (Coquillett) larvae on production of *Aedes aegypti* adults in laboratory tests. Mosquito News 41:522–27.

Baker, R. L. 1980. Use of space in relation to feeding areas by zygopteran nymphs in captivity. Canadian Journal of Zoology 58:1060–65.

Baker, R. L. 1981a. Use of space in relation to areas of food concentration by nymphs of *Lestes disjunctus* (Lestidae, Odonata) in captivity. Canadian Journal of Zoology 59: 134–35.

Baker, R. L. 1981b. Behavioural interactions and use of feeding areas by nymphs of *Coenagrion resolutum* (Coenagrionidae: Odonata). Oecologia (Berlin) 49:353–58.

Baker, R. L. and H. F. Clifford. 1981. Life cycles and food of *Coenagrion resolutum* (Coenagrionidae: Odonata) and *Lestes disjunctus disjunctus* (Lestidae: Odonata) populations from the boreal forest of Alberta, Canada. Aquatic Insects 3:179–91

Bärlocher, F. 1980. Leaf-eating invertebrates as competitors of aquatic hyphomycetes. Oecologia (Berlin) 47:303–6.

Bay, E. C. 1974. Predator-prey relationships among aquatic insects. Annual Review of Entomology 19:441–53.

Beddington, J. R., C. A. Free, and J. H. Lawton. 1976. Concepts of stability and resilience in predator-prey models. Journal of Animal Ecology 45:791–816.

Benfield, E. F. 1972. A defensive secretion of *Dineutes discolor* (Coleoptera: Gyrinidae). Annals of the Entomological Society of America 65:1324–27.

Benke, A. C. 1976. Dragonfly production and prey turnover. Ecology 57:915–27.

Benke, A. C. 1978. Interactions among coexisting predators—a field experiment with dragonfly larvae. Journal of Animal Ecology 47:335–50.

Benke, A. C., P. H. Crowley, and D. M. Johnson. 1982. Interactions among coexisting larval Odonata: an *in situ* experiment using small enclosures. Hydrobiologia 94: 121–30.

Berthélemy, C. and M. Lahoud. 1981. Regimes alimentaires et pieces buccales de quelques Perlodidae et Perlidae des Pyrenees (Plecoptera). Annales de Limnologie 17:1–24.

Bertram, B. C. 1978. Living in groups: predators and prey, pp 64–96. *In*: J. R. Krebs and N. B. Davies (eds.). Behavioural ecology: an evolutionary approach. Blackwell Scientific Publications, Oxford, England. 494 pp.

Boreham, P. F. L. 1979. Recent developments in serological methods for predator-prey studies. Miscellaneous Publications of the Entomological Society of America 11: 17–23.

Brocksen, R. W., G. E. Davis, and C. E. Warren. 1968. Competition, food consumption, and production of sculpins and trout in laboratory stream communities. Journal of Wildlife Management 32:51–75.

Brown, C. R. and M. H. Hatch. 1929. Orientation and "fright" reactions of whirligig beetles (Gyrinidae). Journal of Comparative Psychology 9:159–89.

Caillère, L. 1972. Dynamics of the strike in *Agrion* (syn. *Calopteryx*) *splendens* Harris, 1782 larvae (Odonata: Calopterygidae). Odonatologica (Utrecht) 1:11–19.

Caldwell, J. P., J. H. Thorp, and T. O. Jervey. 1980. Predator-prey relationships among larval dragonflies, salamanders, and frogs. Oecologia (Berlin) 46:285–89.

Case, T. J. and R. K. Washino. 1979. Flatworm control of mosquito larvae in rice fields. Science 206:1412–14.

Caswell, H. 1978. Predator-mediated coexistence: a nonequilibrium model. The American Naturalist 112:127–54.

Cather, M. R. and A. R. Gaufin. 1975. Life history and ecology of *Megarcys signata* (Plecoptera: Perlodidae), Mill Creek, Wasatch Mountains, Utah. Great Basin Naturalist 35;39–48.

Charnov, E. L. 1976. Optimal foraging, the marginal value theorem. Theoretical Population Biology 9:129–36.

Chesson, J. 1978. Measuring preference in selective predation. Ecology 59:211–15.

Chimney, M. J., R. W. Winner, and S. K. Seilkop. 1981. Prey utilization by *Chaoborus punctipennis* Say in a small, eutrophic reservoir. Hydrobiologia 85:193–99.

Cloarec, A. 1969. Étude descriptive et expérimentale du comportement de capture de *Ranatra linearis* au cours de son ontogenèse. Behaviour 35:84–113.

Cloarec, A. 1972. Revue générale de comportements alimentaires d'insectes prédateurs et leur régulation. Année Biologie 11:257–90.

Cloarec, A. 1975a. Variations quantitatives de la prise alimentaire chez *Ranatra linearis* L. (Hétéroptère Aquatique, carnivore). Annales de la Nutrition et de l'Alimentation 29: 245–57.

Cloarec, A. 1975b. Variations quantitatives circadiennes de la prise alimentaire des larves d'*Anax imperator* Leach (Anisoptera: Aeshnidae). Odonatologica (Utrecht) 4: 137–47.

Cloarec, A. 1977. Alimentation de larves d'*Anax imperator* Leach dans un milieu naturel (Anisoptera: Aeshnidae). Odonatologica (Utrecht) 6:227–43.

Cloarec, A. 1978. Estimation of hit distance by *Ranatra*. Biology of Behaviour 3:173–91.

Cloarec, A. 1980a. Ontogeny of hit distance estimation in *Ranatra linearis*. Biology of Behaviour 5:97–118.

Cloarec, A. 1980b. Post-moult behaviour in the water-stick insect *Ranatra linearis*. Behaviour 73:304–24.

Cloarec, A. 1981. Effect on predatory performance of the presence of prey after moulting in the water stick insect, *Ranatra linearis*. Physiological Entomology 6:241–49.

Connell, J. H. 1975. Some mechanisms producing structure in natural communities: a model and evidence from field experiments, pp. 460–90. *In*: M. L. Cody and J. M. Diamond (eds.). Ecology and evolution of communities. Belknap Press of Harvard University, Cambridge, MA. 545 pp.

Cook, R. M. and B. J. Cockrell. 1978. Predator ingestion rate and its bearing on feeding time and the theory of optimal diets. Journal of Animal Ecology 47:529–47.

Corbet, P. S. 1980. Biology of Odonata. Annual Review of Entomology 25:189–217.

Corkum, L. D. 1980. Carnivory in *Ephemerella inermis* nymphs (Ephemeroptera: Ephemerellidae). Entomological News 91:161–63.

Corkum, L. D. and H. F. Clifford. 1980. The importance of species associations and substrate types to behavioral drift, pp. 331–41. *In*: J. F. Flannagan and K. E. Marshall (eds.). Advances in Ephemeroptera biology. Plenum Publishing Co., NY. 552 pp.

Corkum, L. D. and P. J. Pointing. 1979. Nymphal development of *Baetis vagans* McDunnough (Ephemeroptera: Baetidae) and drift habits of large nymphs. Canadian Journal of Zoology 57:2348–54.

Crawley, M. J. 1975. The numerical responses of insect predators to changes in prey density. Journal of Animal Ecology 44:877–92.

Crowder, L. B. and W. E. Cooper. 1982. Habitat structural complexity and the interaction between bluegills and their prey. Ecology 63:1802–13.

Crowley, P. H. 1979. Behavior of zygopteran nymphs in a simulated weed bed. Odonatologica (Utrecht) 8:91–101.

Curio, E. 1976. Ethology of predation. Springer-Verlag, Berlin, Germany. 250 pp.

Davies, R. W. and T. B. Reynoldson. 1969. The incidence and intensity of predation on lake-dwelling triclads in the laboratory. Ecology 50:845–53.

Davies, R. W. and T. B. Reynoldson. 1971. The incidence and intensity of predation on lake-dwelling triclads in the field. Journal of Animal Ecology 40:191–214.

Deshefy, G. S. 1980. Anti-predator behavior in swarms of *Rhagovelia obesa* (Hemiptera: Veliidae). Pan-Pacific Entomologist 56:111–12.

Devonport, B. F. and M. J. Winterbourn. 1976. The feeding relationships of two invertebrate predators in a New Zealand river. Freshwater Biology 6:167–76.

Dodson, S. I. 1972. Mortality in a population of *Daphnia rosea*. Ecology 53:1011–23.

Dodson, S. I. 1974. Adaptive change in plankton morphology in response to size-selective predation: a new hypothesis of cyclomorphosis. Limnology and Oceanography 19:721–29.

Duffield, R. M. 1981. 2-nonanol in the exocrine secretion of the Nearctic caddisfly, *Rhyacophila fuscula* (Walker) (Rhyacophilidae: Trichoptera). Proceedings of the Entomological Society of Washington 83:60–63.

Duffield, R. M., M. S. Blum, J. B. Wallace, H. A. Lloyd, and F. E. Regnier. 1977. Chemistry of the defensive secretion of the caddisfly *Pycnopsyche scabripennis* (Trichoptera: Limnephilidae). Journal of Chemical Ecology 3:649–56.

Dusoge, K. 1980. The occurrence and role of the predatory larvae of *Procladius* Skuse (Chironomidae, Diptera) in the benthos of Lake Śniardwy. Ekologia Polska 28:155–86.

Edmunds, G. F., Jr. 1957. The predaceous mayfly nymphs of North America. Proceedings of the Utah Academy of Sciences 34:23–24.

Edmunds, G. F., Jr. and C. H. Edmunds. 1979. Predation, climate, and emergence and mating of mayflies, pp. 227–85. *In*: J. F. Flannagan and K. E. Marshall (eds.). Advances in Ephemeroptera biology. Plenum Publishing Co., New York. 552 pp.

Edmunds, M. 1974. Defense in animals. A survey of anti-predator defenses. Longman Group Ltd., New York, NY. 357 pp.

Eisner, T. and R. P. Grant. 1981. Toxicity, odor aversion, and "olfactory aposematism." Science 213:476.

Eisner, T. and J. Meinwald. 1966. Defensive secretions of arthropods. Science 153:1341–50.

Ellis, R. A. and J. H. Borden. 1970. Predation by *Notonecta undulata* (Heteroptera: Notonectidae) on larvae of the yellow-fever mosquito. Annals of the Entomological Society of America 63:963–73.

Elner, R. W. and R. N. Hughes. 1978. Energy maximization in the diet of the shore crab, *Carcinus maenas*. Journal of Animal Ecology 47:103–16.

Elton, C. S. 1922. On the colours of water mites. Proceedings of the Zoological Society of London 1922:1231–39.

Eriksson, M. O. G., L. Henrikson, and H. G. Oscarson. 1980. Predator-prey relationships among water-mites (Hydracarina) and other freshwater organisms. Archiv für Hydrobiologie 88:146–54.

Etienne, A. S. 1972. The behaviour of the dragonfly larva *Aeshna cyanea* M. after a short presentation of prey. Animal Behaviour 20:724–31.

Fedorenko, A. Y. 1975. Feeding characteristics and predation impact of *Chaoborus* (Diptera, Chaoboridae) larvae in a small lake. Limnology and Oceanography 20:250–58.

Focks, D. A., D. A. Dame, A. L. Cameron, and M. D. Boston. 1980. Predator-prey interaction between insular populations of *Toxorhynchites rutilus rutilus* and *Aedes aegypti*. Environmental Entomology 9:37–42.

Formanowicz, D. R., Jr. 1982. Foraging tactics of larvae of *Dytiscus verticalis* (Coleoptera: Dytiscidae): the assessment of prey density. Journal of Animal Ecology 51:757–67.

Foster, W. A. and J. E. Treherne. 1980. Feeding, predation and aggregation behaviour in a marine insect, *Halobates robustus* Barber (Hemiptera: Gerridae), in the Galapagos Islands. Proceedings of the Royal Society of London B. 209:539–53.

Fox, L. R. 1975. Some demographic consequences of food shortage for the predator, *Notonecta hoffmanni*. Ecology 56:868–80.

Fox, L. R. 1977. Species richness in streams: an alternative mechanism. The American Naturalist 111:1017–21.

Fox, L. R. and W. W. Murdoch. 1978. Effects of feeding history on short-term and long-term functional responses in *Notonecta hoffmanni*. Journal of Animal Ecology 47:945–59.

Fox, P. J. 1978. Caddis larvae (Trichoptera) as predators of fish eggs. Freshwater Biology 8:343–45.

Friberg, F., L. M. Nilsson, C. Otto, P. Sjöström, B. W. Svensson, Bj. Svensson, and S. Ulfstrand. 1977. Diversity and environments of benthic invertebrate communities in south Swedish streams. Archiv für Hydrobiologie 81:129–54.

Fuller, R. L. and K. W. Stewart. 1977. The food habits of stoneflies (Plecoptera) in the upper Gunnison River, Colorado. Environmental Entomology 6:293–302.

Fuller, R. L. and K. W. Stewart. 1979. Stonefly (Plecoptera) food habits and prey preference in the Dolores River, Colorado. American Midland Naturalist 101:170–81.

Gallepp, G. W. 1974. Behavioral ecology of *Brachycentrus occidentalis* Banks during the pupation period. Ecology 55:1283–94.

George, J. A. 1978. The potential of a local planarian, *Dugesia tigrina* (Tricladida, Turbellaria), for the control of mosquitoes in Ontario. Proceedings of the Entomological Society of Ontario 109:65–69.

Giguère, L. A. 1979. An experimental test of Dodson's hypothesis that *Ambystoma* (a salamander) and *Chaoborus* (a phantom midge) have complementary feeding niches. Canadian Journal of Zoology 57:1091–97.

Giguère, L. A. 1981. Food assimilation efficiency as a function of temperature and meal size in larvae of *Chaoborus trivittatus* (Diptera: Chaoboridae). Journal of Animal Ecology 50:103–9.

Giguère, L. A. and L. M. Dill. 1979. The predatory response of *Chaoborus* larvae to acoustic stimuli, and the acoustic characteristics of their prey. Zeitschrift für Tierpsychologie 50:113–23.

Giller, P. S. 1980. The control of handling time and its effects on the foraging strategy of a heteropteran predator, *Notonecta*. Journal of Animal Ecology 49:699–712.

Giller, P. S. 1982. Locomotory efficiency in the predation strategies of the British *Notonecta* (Hemiptera, Heteroptera). Oecologia (Berlin) 52:273–77.

Giller, P. S. and S. McNeill. 1981. Predation strategies, resource partitioning and habitat selection in *Notonecta* (Hemiptera/Heteroptera). Journal of Animal Ecology 50: 789–808.

Gittleman, J. L. and P. H. Harvey. 1980. Why are distasteful prey not cryptic? Nature (London) 286:149–50.

Gittleman, S. H. 1974. Locomotion and predatory strategy in backswimmers (Hemiptera: Notonectidae). American Midland Naturalist 92:496–500.

Gittleman, S. H. 1975. The ecology of some Costa Rican backswimmers (Hemiptera: Notonectidae). Annals of the Entomological Society of America 68:511–18.

Gittleman, S. H. 1977. Leg segment proportions, predatory strategy and growth in backswimmers (Hemiptera: Pleidae, Notonectidae). Journal of the Kansas Entomological Society 50:161–71.

Glasser, J. W. 1979. The role of predation in shaping and maintaining the structure of communities. The American Naturalist 113:631–41.

Grant, J. W. G. and I. A. E. Bayly. 1981. Predator induction of crests in morphs of the *Daphnia carinata* King complex. Limnology and Oceanography 26:201–18.

Hall, D. J., W. E. Cooper, and E. E. Werner. 1970. An experimental approach to the production dynamics and structure of freshwater animal communities. Limnology and Oceanography 15:839–928.

Hall, D. J., S. T. Threlkeld, C. W. Burns, and P. H. Crowley. 1976. The size-efficiency hypothesis and the size-structure of zooplankton communities. Annual Review of Ecology and Systematics 7:177–208.

Hassell, M. P. 1978. The dynamics of arthropod predator-prey systems. Monographs in Population Biology 13:1–237.

Hassell, M. P., J. H. Lawton, and J. R. Beddington. 1976. The components of arthropod predation. I. The prey death-rate. Journal of Animal Ecology 45:135–64.

Hassell, M. P. and R. M. May. 1973. Stability in insect host-parasite models. Journal of Animal Ecology 42:693–726.

Hassell, M. P. and R. M. May. 1974. Aggregation of predators and insect parasites and its effect on stability. Journal of Animal Ecology 43:567–94.

Heinrich, B. and F. D. Vogt. 1980. Aggregation and foraging behavior of whirligig beetles (Gyrinidae). Behavioral Ecology and Sociobiology 7:179–86.

Henrikson, L. and H. G. Oscarson. 1978. Fish predation limiting abundance and distribution of *Glaenocorisa p. propinqua* (Hemiptera). Oikos 31:102–5.

Hildrew, A. G. and C. R. Townsend. 1976. The distribution of two predators and their prey in an iron rich stream. Journal of Animal Ecology 45:41–57.

Hildrew, A. G. and C. R. Townsend. 1977. The influence of substrate on the functional response of *Plectrocnemia conspersa* (Curtis) larvae (Trichoptera: Polycentropodidae). Oecologia (Berlin) 31:21–26.

Hildrew, A. G. and C. R. Townsend. 1980. Aggregation, interference and foraging by larvae of *Plectrocnemia conspersa* (Trichoptera: Polycentropodidae). Animal Behaviour 28:553–60.

Hildrew, A. G., C. R. Townsend, and J. Henderson. 1980. Interactions between larval size, microdistribution and substrate in the stoneflies of an iron-rich stream. Oikos 35:387–96.

Hilsenhoff, W. L. 1963. Predation by the leech *Helobdella stagnalis* on *Tendipes plumosus* (Diptera: Tendipedidae) larvae. Annals of the Entomological Society of America 56:252.

Hilsenhoff, W. L. 1964. Predation by the leech *Helobdella nepheloidea* on larvae of *Tendipes plumosus* (Diptera: Tendipedidae). Annals of the Entomological Society of America 57:139.

Holling, C. S. 1959. Some characteristics of simple types of predation and parasitism. The Canadian Entomologist 91:385–98.

Holling, C. S. 1961. Principles of insect predation. Annual Review of Entomology 6: 163–82.

Holling, C. S. 1965. The functional response of predators to prey density and its role in mimicry and population regulation. Memoirs of the Entomological Society of Canada 45:3–60.

Hulberg, L. W. and J. S. Oliver. 1980. Caging manipulations in marine soft-bottom communities: importance of animal interactions or sedimentary habitat modifications. Canadian Journal of Fisheries and Aquatic Sciences 37:1130–39.

Hungerford, H. B. 1919. The biology and ecology of aquatic and semiaquatic Hemiptera. Kansas University Scientific Bulletin 11:3–265.

Hutchinson, G. E. 1980. Conjectures arising in a quiet museum. Bulletin of the Royal Entomological Society of London 4:92–98.

Hutchinson, G. E. 1981. Thoughts on aquatic insects. BioScience 31:495–500.

Hynes, H. B. N. 1970. The ecology of running waters. University of Toronto Press, Toronto. 555 pp.

Ivlev, V. S. 1961. Experimental ecology of the feeding of fishes. Yale University Press, New Haven, CT. 302 pp.

Jacobs, J. 1974. Quantitative measurement of food selection. A modification of the forage ratio and Ivlev's electivity index. Oecologia (Berlin) 14:413–17.

Jamieson, G. S. and G. G. E. Scudder. 1977. Food consumption in *Gerris* (Hemiptera). Oecologia (Berlin) 30:23–41.

Jamieson, G. S. and G. G. E. Scudder. 1979. Predation in *Gerris* (Hemiptera): reactive distances and locomotion rates. Oecologia (Berlin) 44:13–20.

Janetos, A. C. 1982. Active foragers vs. sit-and-wait predators: a simple model. Journal of Theoretical Biology 95:381–85.

Johnson, D. H. 1980. The comparison of usage and availability measurements for evaluating resource preference. Ecology 61:65–71.

Johnson, D. M. 1973. Predation by damselfly naiads on cladoceran populations: fluctuating intensity. Ecology 54:251–68.

Johnson, D. M., B. G. Akre, and P. H. Crowley. 1975. Modeling arthropod predation: wasteful killing by damselfly naiads. Ecology 56:1081–93.

Johnson, D. M. and P. H. Crowley. 1979. Odonate "hide-and-seek": habitat-specific rules?, pp. 569–79. *In*: W. C. Kerfoot (ed.). Evolution and ecology of zooplankton communities. University Press of New England, Hanover, NH. 793 pp.

Johnson, J. H. 1982. Diet composition and prey selection of *Cordulegaster maculata* Sel. larvae (Anisoptera: Cordulegastridae). Notulae Odonatologicae 1:151–53.

Jones, R. E. 1974. The effects of size-selective predation and environmental variation on the distribution and abundance of a chironomid, *Paraborniella tonnoiri* Freeman. Australian Journal of Zoology 22:71–89.

Kajak, Z. 1977. Factors influencing benthos biomass in shallow lake environments. Ekologia Polska 25:421–29.

Kajak, Z. 1980. Role of invertebrate predators (mainly *Procladius* sp.) in benthos, pp 339–48. *In*: D. A. Murray (ed.). Chironomidae ecology, systematics and physiology. Pergamon Press, NY. 354 pp.

Kajak, Z. and B. Ranke-Rybicka. 1970. Feeding and production efficiency of *Chaoborus flavicans* Meigen (Diptera, Culicidae) larvae in eutrophic and dystrophic lake. Polskie Archiwum Hydrobiologii 17:225–32.

Kerfoot, W. C. 1979. Commentary: transparency, body size, and prey conspicuousness, pp. 609–17. *In*: W. C. Kerfoot (ed.). Evolution and ecology of zooplankton communities. University Press of New England, Hanover, NH. 793 pp.

Kerfoot, W. C. 1982. A question of taste: crypsis and warning coloration in freshwater zooplankton communities. Ecology 63:538–54.

Koslucher, D. G. and G. W. Minshall. 1973. Food habits of some benthic invertebrates in a northern cool-desert stream (Deep Creek, Curlew Valley, Idaho-Utah). Transactions of the American Microscopical Society 92:441–52.

Kovalak, W. P. 1978. On the feeding habits of *Phasganophora capitata* (Plecoptera: Perlidae). Great Lakes Entomologist 11:45–49.

Krebs, J. R. 1978. Optimal foraging: decision rules for predators, pp 23–63. *In*: J. R. Krebs and N. B. Davies (eds.). Behavioral ecology: an evolutionary approach. Sinauer Associates Ltd., Sunderland, MA. 494 pp.

Krueger, D. A. and S. I. Dodson. 1981. Embryological induction and predation ecology in *Daphnia pulex*. Limnology and Oceanography 26:219–23.

Lang, H. H. 1980. Surface wave discrimination between prey and nonprey by the backswimmer *Notonecta glauca* L. (Hemiptera, Heteroptera). Behavioral Ecology and Sociobiology 6:233–46.

Lawton, J. H. 1970. Feeding and food energy assimilation in larvae of the damselfly *Pyrrhosoma nymphula* (Sulz.) (Odonata: Zygoptera). Journal of Animal Ecology 39:669–89.

Lawton, J. H. 1971. Ecological energetics studies on larvae of the damselfly *Pyrrhosoma nymphula* (Sulzer) (Odonata: Zygoptera). Journal of Animal Ecology 40:385–423.

Lawton, J. H., J. R. Beddington, and R. Bonser. 1974. Switching in invertebrate predators, pp. 141–58. *In*: M. B. Usher and M. H. Williamson (eds.). Ecological stability. Chapman and Hall Ltd., London. 196 pp.

Lawton, J. H., B. A. Thompson, and D. J. Thompson. 1980. The effects of prey density on survival and growth of damselfly larvae. Ecological Entomology 5:39–52.

Lechowicz, M. J. 1982. The sampling characteristics of electivity indices. Oecologia (Berlin) 52:22–30.

LeClaire, R., Jr. and J. P. Bourassa. 1981. Observation et analyse de la prédation des oeufs d'*Ambystoma maculatum* (Shaw) (Amphibia, Urodela) par des larves de

Diptères chironomidés, dans la région de Trois-Rivières (Québec). Canadian Journal of Zoology 59:1339–43.

Lewis, W. M., Jr. 1979. Evidence for stable zooplankton community structure gradients maintained by predation, pp. 625–34. *In*: W. C. Kerfoot (ed.). Evolution and ecology of zooplankton communities. University Press of New England, Hanover, NH. 793 pp.

Lindstedt, K. J. 1971. Chemical control of feeding behaviour. Comparative Biochemistry and Physiology 39A:553–81.

Loden, M. S. 1974. Predation by chironomid (Diptera) larvae on oligochaetes. Limnology and Oceanography 19:156–59.

Macan, T. T. 1965. Predation as a factor in the ecology of water bugs. Journal of Animal Ecology 34:691–98.

Macan, T. T. 1966. The influence of predation on the fauna of a moorland fishpond. Archiv für Hydrobiologie 61:432–52.

Macan, T. T. 1977a. The influence of predation on the composition of freshwater animal communities. Biological Reviews of the Cambridge Philosophical Society 52.45–70.

Macan, T. T. 1977b. A twenty-year study of the fauna in the vegetation of a moorland fishpond. Archiv für Hydrobiologie 81:1–24.

Mackereth, J. C. 1957. Notes on the Plecoptera from a stony stream. Journal of Animal Ecology 26:343–51.

Maguire, B., Jr., D. Belk, and G. Wells. 1968. Control of community structure by mosquito larvae. Ecology 49:207–10.

Maier, C. T. 1977. The behavior of *Hydrometra championana* (Hemiptera: Hydrometridae) and resource partitioning with *Tenagogonus quadrilineatus* (Hemiptera: Gerridae). Journal of the Kansas Entomological Society 50:263–71.

Malmqvist, B. and P. Sjöström. 1980. Prey size and feeding patterns in *Dinocras cephalotes* (Plecoptera). Oikos 35:311–16.

Markl, H., H. Lang, and K. Wiese. 1973. Die Genauigkeit der Ortungeines Wellenzentrums durch den Rückenschwimmer *Notonecta glauca* L. Journal of Comparative Physiology 86:359–64.

Masters, W. M. 1979. Insect disturbance stridulation: its defensive role. Behavioral Ecology and Sociobiology 5:187–200.

Masters, W. M. 1980. Insect disturbance stridulation: characterization of airborne and vibrational components of the sound. Journal of Comparative Physiology 135:259–68.

May, R. M. 1974. Stability and complexity in model ecosystems. 2nd ed. Princeton University Press, Princeton, NJ. 235 pp.

McDonald, G. and G. A. Buchanan. 1981. The mosquito and predatory insect fauna inhabiting fresh-water ponds, with particular reference to *Culex annulirostris* Skuse (Diptera: Culicidae). Australian Journal of Ecology 6:21–27.

McIver, J. D. 1981. An examination of the utility of the precipitin test for evaluation of arthropod predator-prey relationships. The Canadian Entomologist 113:213–22.

McNair, J. N. 1982. Optimal giving-up times and the marginal value theorem. The American Naturalist 119:511–29.

Melville, G. E. and E. J. Maly. 1981. Vertical distributions and zooplankton predation in a small temperate pond. Canadian Journal of Zoology 59:1720–25.

Menzie, C. A. 1981. Production ecology of *Cricotopus sylvestris* (Fabricius) (Diptera:

Chironomidae) in a shallow estuarine cove. Limnology and Oceanography 26: 467–81.

Merritt, R. W. and K. W. Cummins (eds.). 1978. An introduction to the aquatic insects of North America. Kendall/Hunt Publishing Co., Dubuque, IA. 441 pp.

Meyer, H. J. and L. W. Learned. 1981. Laboratory studies on the potential of *Dugesia tigrina* for mosquito predation. Mosquito News 41:760–64.

Miller, J. R., L. B. Hendry, and R. O. Mumma. 1975. Norsesquiterpenes as defensive toxins of whirligig beetles (Coleoptera: Gyrinidae). Journal of Chemical Ecology 1: 59–82.

Miller, J. R. and R. O. Mumma. 1973. Defensive agents of the American water beetles *Agabus seriatus* and *Graphoderus liberus*. Journal of Insect Physiology 19:917–25.

Miller, J. R. and R. O. Mumma. 1976a. Physiological activity of water beetle defensive agents. I. Toxicity and anesthetic activity of steroids and norsesquiterpenes administered in solution to the minnow *Pimephales promelas* Raf. Journal of Chemical Ecology 2:115–30.

Miller, J. R. and R. O. Mumma. 1976b. Physiological activity of water beetle defensive agents. II. Absorption of selected anesthetic steroids and norsesquiterpenes across gill membranes of the minnow. Journal of Chemical Ecology 2:131–46.

Minshall, G. W. and J. N. Minshall. 1966. Notes on the life history and ecology of *Isoperla clio* (Newman) and *Isogenus decisus* Walker (Plecoptera: Perlodidae). American Midland Naturalist 76:340–50.

Molles, M. C., Jr. and R. D. Pietruszka. 1983. Mechanisms of prey selection by predaceous stoneflies: roles of prey morphology, behavior and predator hunger. Oecologia (Berlin) 57:25–31.

Murdoch, W. W. 1969. Switching in general predators: experiments on predator specificity and stability of prey populations. Ecological Monographs 39:335–54.

Murdoch, W. W. 1971. The developmental response of predators to changes in prey density. Ecology 52:132–37.

Murdoch, W. W. 1973. The functional response of predators. Journal of Applied Ecology 10:335–42.

Murdoch, W. W. and A. Oaten. 1975. Predation and population stability. Advances in Ecological Research 9:1–131.

Murdoch, W. W. and A. Sih. 1978. Age-dependent interference in a predatory insect. Journal of Animal Ecology 47:581–92.

Murphey, R. K. 1971. Sensory aspects of the control of orientation to prey by the water-strider *Gerris remigis*. Zeitschrift für vergleichende Physiologie 72:168–85.

Neill, W. E. 1978. Experimental studies on factors limiting colonization by *Daphnia pulex* Leydig of coastal montane lakes in British Columbia. Canadian Journal of Zoology 56:2498–507.

Neill, W. E. 1981. Impact of *Chaoborus* predation upon the structure and dynamics of a crustacean zooplankton community. Oecologia (Berlin) 48:164–77.

Neill, W. E. and A. Peacock. 1979. Breaking the bottleneck: interactions of invertebrate predators and nutrients in oligotrophic lakes, pp. 715–24. *In*: W. C. Kerfoot (ed.). Evolution and ecology of zooplankton communities. University of New England Press, Hanover, NH. 793 pp.

Newhart, A. T. and R. O. Mumma. 1979. Seasonal quantification of the defensive secretions from Gyrinidae. Annals of the Entomological Society of America 72:427–29.

Oakley, B. and J. M. Polka. 1967. Prey capture by dragonfly larvae. American Zoologist 7:727–28. (Abstract only).

Oberndorfer, R. Y., J. V. McArthur, J. R. Barnes and J. Dixon. 1983. The effect of invertebrate predators on leaf litter decomposition in streams. Ecology. In press.

Otto, C. and P. Sjöström. 1983. Cerci as antipredatory attributes in stonefly nymphs. Oikos. In press.

Otto, C. and B. S. Svensson. 1980. The significance of case material selection for the survival of caddis larvae. Journal of Animal Ecology 49:855–65.

Padgett, P. D. and D. A. Focks. 1981. Prey stage preference of the predator, *Toxorhynchites rutilus rutilus* on *Aedes aegypti.* Mosquito News 41:67–70.

Paine, R. T. 1966. Food web complexity and species diversity. The American Naturalist 100:65–75.

Pastorok, R. A. 1979. Selection of prey by *Chaoborus* larvae: a review and new evidence for behavioral flexibility, pp. 538–54. *In*: W. C. Kerfoot (ed.). Evolution and ecology of zooplankton communities. University of New England Press, Hanover, NH. 793 pp.

Pastorok, R. A. 1980. The effects of predator hunger and food abundance on prey selection by *Chaoborus* larvae. Limnology and Oceanography 25:910–21.

Pastorok, R. A. 1981. Prey vulnerability and size selection by *Chaoborus* larvae. Ecology 62:1311–24.

Peckarsky, B. L. 1979. A review of the distribution, ecology, and evolution of the North American species of *Acroneuria* and six related genera (Plecoptera: Perlidae). Journal of the Kansas Entomological Society 52:787–809.

Peckarsky, B. L. 1980. Predator-prey interactions between stoneflies and mayflies: behavioral observations. Ecology 61:932–43.

Peckarsky, B. L. 1982. Aquatic insect predator-prey relations. BioScience 32:261–66.

Peckarsky, B. L. 1983. Use of behavioral experiments to test ecological theory in streams, pp. 79–97. *In*: J. R. Barnes and G. W. Minshall (eds.). Stream ecology: application and testing of general ecological theory. Plenum Press, New York, NY. 399 pp.

Peckarsky, B. L. and S. I. Dodson. 1980a. Do stonefly predators influence benthic distributions in streams? Ecology 61:1275–82.

Peckarsky, B. L. and S. I. Dodson. 1980b. An experimental analysis of biological factors contributing to stream community structure. Ecology 61:1283–90.

Popham, E. J. 1942. The variation in the colour of certain species of *Arctocorisa* (Hemiptera, Corixidae) and its significance. Proceedings of the Zoological Society of London 111:135–72.

Popham, E. J. 1943. Further experimental studies of the selective action of predators. Proceedings of the Zoological Society of London 112:105–17.

Popham, E. J. 1948. Experimental studies of the biological significance of non-cryptic pigmentation with special reference to insects. Proceedings of the Zoological Society of London 117:768–83.

Popham, E. J. and E. Bevans. 1979. Functional morphology of the feeding apparatus in larvae and adult *Aeshna juncea* (L.) (Anisoptera: Aeshnidae). Odonatologica (Utrecht) 8:301–18.

Pritchard, G. 1964. The prey of dragonfly larvae (Odonata: Anisoptera) in ponds in Northern Alberta. Canadian Journal of Zoology 42:785–800.

Pritchard, G. 1965. Prey capture by dragonfly larvae (Odonata: Anisoptera). Canadian Journal of Zoology 43:271–89.

Pritchard, G. 1966. On the morphology of the compound eyes of dragonflies (Odonata: Anisoptera) with special reference to their role in prey capture. Proceedings of the Royal Entomological Society of London 41:1–8.

Pritchard, G. and T. G. Leischner. 1973. The life history and feeding habits of *Sialis cornuta* Ross in a series of abandoned beaver ponds (Insecta; Megaloptera). Canadian Journal of Zoology 51:121–31.

Pyke, G. H. 1978. Optimal foraging: movement patterns of bumblebees between inflorescences. Theoretical Population Biology 13:72–98.

Reice, S. R. 1983. Interspecific associations in a woodland stream, pp 325–45. *In*: S. M. Bartell and I. D. Fontaine, III (eds.). Dynamics of lotic ecosystems. Ann Arbor Science Publishers, Inc. Ann Arbor, MI. 494 pp.

Reynoldson, T. B. and P. Bellamy. 1975. Triclads (Turbellaria: Tricladida) as predators of lake-dwelling stonefly and mayfly nymphs. Freshwater Biology 5:305–12.

Robinson, M. H., L. G. Abele, and B. Robinson. 1970. Attack autotomy: a defense against predators. Science 169:300–1.

Rosenzweig, M. L. and R. H. MacArthur. 1963. Graphical representation and stability conditions of predator-prey interactions. The American Naturalist 97:209–23.

Sadyrin, V. M. 1977. The daily ration of damselflies *Coenagrion armatum* (Odonata) under experimental conditions. Hydrobiological Journal 13:16–18.

Schall, J. J. and E. R. Pianka. 1980. Evolution of escape behavior diversity. The American Naturalist. 115:551–66.

Sheldon, A. L. 1969. Size relationships of *Acroneuria californica* (Perlidae, Plecoptera) and its prey. Hydrobiologia 34:85–94.

Sheldon, A. L. 1972. Comparative ecology of *Arcynopteryx* and *Diura* (Plecoptera) in a California stream. Archiv für Hydrobiologie 69:521–46.

Sheldon, A. L. 1980. Resource division by perlid stoneflies (Plecoptera) in a lake outlet ecosystem. Hydrobiologia 71:155–61.

Siegfried, C. A. and A. W. Knight. 1976a. Prey selection by a setipalpian stonefly nymph, *Acroneuria (Calineuria) californica* Banks (Plecoptera: Perlidae). Ecology 57:603–8.

Siegfried, C. A. and A. W. Knight. 1976b. Trophic relations of *Acroneuria (Calineuria) californica* (Plecoptera: Perlidae) in a Sierra foothill stream. Environmental Entomology 5:575–81.

Sievers, D. W. and A. C. Haman. 1972. Notes on snail feeding behavior of *Anax junius* (Drury): (Odonata). Proceedings of the Iowa Academy of Science 79:105–6.

Sih, A. 1979. Stability and prey behavioural responses to predator density. Journal of Animal Ecology 48:79–89.

Sih, A. 1980a. Optimal behavior: can foragers balance two conflicting demands? Science 210:1041–43.

Sih, A. 1980b. Optimal foraging: partial consumption of prey. The American Naturalist 116:281–90.

Sih, A. 1981. Stability, prey density and age/dependent interference in an aquatic insect predator, *Notonecta hoffmanni*. Journal of Animal Ecology 50:625–36.

Sih, A. 1982. Foraging strategies and the avoidance of predation by an aquatic insect, *Notonecta hoffmanni*. Ecology 63:786–96.

Sjöström, P. 1983. Hunting, spacing and antipredatory behaviour in nymphs of *Dinocras cephalotes* (Plecoptera). Ph.D. thesis, University of Lund, Lund, Sweden. 67 pp.

Smith, R. L. 1973. Aspects of the biology of three species of the genus *Rhantus* (Coleoptera: Dytiscidae) with special reference to the acoustical behavior of two. The Canadian Entomologist 105:909–19.

Smyly, W. J. P. 1980. Food and feeding of aquatic larvae of the midge *Chaoborus flavicans* (Meigen) (Diptera: Chaoboridae) in the laboratory. Hydrobiologia 70: 179–88.

Sprules, W. G. 1972. Effects of size-selective predation and food competition on high altitude zooplankton communities. Ecology 53:375–86.

Stein, R. A. and J. J. Magnuson. 1976. Behavioral response of crayfish to a fish predator. Ecology 57:751–61.

Stenson, J. A. E. 1979. Predation pressure from fish on two *Chaoborus* species as related to their visibility, pp. 618–22. *In*: W. C. Kerfoot (ed.). Evolution and ecology of zooplankton communities. University of New England Press, Hanover, NH. 793 pp.

Stenson, J. A. E. 1981. The role of predation in the evolution of morphology, behaviour and life history of two species of *Chaoborus*. Oikos 37:323–27.

Stewart, K. W., G. P. Friday, and R. E. Rhame. 1973. Food habits of hellgrammite larvae, *Corydalus cornutus* (Megaloptera: Corydalidae) in the Brazos River, Texas. Annals of the Entomological Society of America 66:959–63.

Streams, F. A. 1982. Diel foraging and reproductive periodicity in *Notonecta undulata* Say (Heteroptera). Aquatic Insects 4:111–19.

Sullivan, R. T. 1981. Insect swarming and mating. Florida Entomologist 64:44–65.

Swift, M. C. and A. Y. Fedorenko. 1975. Some aspects of prey capture by *Chaoborus* larvae. Limnology and Oceanography 20:418–25.

Swift, M. C. and R. B. Forward, Jr. 1981. *Chaoborus* prey capture efficiency in the light and dark. Limnology and Oceanography 26:461–66.

Swüste, H. F. J., R. Cremer, and S. Parma. 1973. Selective predation by larvae of *Chaoborus flavicans* (Diptera, Chaoboridae). Internationale Vereinigung für Theoretische und Angewandte Limnologie Verhandlungen 18:1559–63.

Tachet, H. 1977. Vibrations and predatory behaviour of *Plectrocnemia conspersa* larvae (Trichoptera). Zeitschrift für Tierpsychologie 45:61–74.

Tarwid, M. 1969. Analysis of the contents of the alimentary tract of predatory Pelopiinae larvae (Chironomidae). Ekologia Polska Series A 17:125–31.

Taylor, R. J. 1976. Value of clumping to prey and the evolutionary response of ambush predators. The American Naturalist 110:13–29.

Thompson, D. J. 1975. Towards a predator-prey model incorporating age structure: the effects of predator and prey size on the predation of *Daphnia magna* by *Ischnura elegans*. Journal of Animal Ecology 44:907–16.

Thompson, D. J. 1978a. Towards a realistic predator-prey model: the effect of temperature on the functional response and life history of larvae of the damselfly, *Ischnura elegans*. Journal of Animal Ecology 47:757–67.

Thompson, D. J. 1978b. Prey size selection by larvae of the damselfly, *Ischnura elegans* (Odonata). Journal of Animal Ecology 47:769–85.

Thorp, J. H. and E. A. Bergey. 1981a. Field experiments on responses of a freshwater, benthic macroinvertebrate community to vertebrate predators. Ecology 62:365–75.

Thorp, J. H. and E. A. Bergey. 1981b. Field experiments on interactions between vertebrate predators and larval midges (Diptera: Chironomidae) in the littoral zone of a reservoir. Oecologia (Berlin) 50:285–90.

Thut, R. N. 1969a. Feeding habits of stonefly nymphs of the suborder Setipalpia. Presented at the 1969 meeting of the Midwest Benthological Society.

Thut, R. N. 1969b. Feeding habits of larvae of seven *Rhyacophila* (Trichoptera: Rhyacophilidae) species with notes on other life-history features. Annals of the Entomological Society of America 62:894–98.

Tinbergen, N., M. Impekoven, and D. Franck. 1967. An experiment on spacing out as a defense against predation. Behaviour 28:307–21.

Toth, R. S. and R. M. Chew. 1972. Development and energetics of *Notonecta undulata* during predation on *Culex tarsalis*. Annals of the Entomological Society of America 65:1270–79.

Townsend, C. R. and A. G. Hildrew. 1978. Predation strategy and resource utilization by *Plectrocnemia conspersa* (Curtis) (Trichoptera: Polycentropodidae), pp. 283–91. *In*: M. I. Crichton (ed.). Proceedings of the Second International Symposium on Trichoptera. Dr. W. Junk B. V. Publishers, The Hague, The Netherlands. 359 pp.

Townsend, C. R. and A. G. Hildrew. 1979a. Foraging strategies and coexistence in a seasonal environment. Oecologia (Berlin) 38:231–34.

Townsend, C. R. and A. G. Hildrew. 1979b. Form and function of the prey catching net of *Plectrocnemia conspersa* larvae (Trichoptera). Oikos 33:412–18.

Townsend, C. R. and A. G. Hildrew. 1980. Foraging in a patchy environment by a predatory net-spinning caddis larva: a test of optimal foraging theory. Oecologia (Berlin) 47:219–21.

Treherne, J. E. and W. A. Foster. 1980. The effects of group size on predator avoidance in a marine insect. Animal Behaviour 28:1119–22.

Treherne, J. E. and W. A. Foster. 1981. Group transmission of predator avoidance behaviour in a marine insect: the Trafalgar effect. Animal Behaviour 29:911–17.

Treherne, J. E. and W. A. Foster. 1982. Group size and anti-predator strategies in a marine insect. Animal Behaviour 30:536–42.

Tsui, P. T. P. and M. D. Hubbard. 1979. Feeding habits of the predaceous nymphs of *Dolania americana* in northwestern Florida (Ephemeroptera: Behningiidae). Hydrobiologia 67:119–23.

Tucker, V. A. 1969. Wave making by whirligig beetles (Gyrinidae). Science 166:897–99.

von Ende, C. N. 1979. Fish predation, interspecific predation, and the distribution of two *Chaoborus* species. Ecology 60:119–28.

von Ende, C. N. and D. O. Dempsey. 1981. Apparent exclusion of the cladoceran *Bosmina longirostris* by invertebrate predator *Chaoborus americanus*. American Midland Naturalist 105:240–48.

Walton, O. E., Jr. 1980. Invertebrate drift from predator-prey associations. Ecology 61:1486–97.

Ware, D. M. 1973. Risk of epibenthic prey to predation by rainbow trout (*Salmo gairdneri*). Journal of the Fisheries Research Board of Canada 30:787–97.

Waters, T. F. 1972. The drift of stream insects. Annual Review of Entomology 17:253–72.

White, H. B. 1980. Avoidance behavior in Odonata. Notulae Odonatologicae 1:105–6.

Wiley, M. J. and S. L. Kohler. 1981. An assessment of biological interactions in an epilithic stream community using time-lapse cinematography. Hydrobiologia 78:183–88.

Williams, E. H. 1980. Disjunct distributions of two aquatic predators. Limnology and Oceanography 25:999–1006.

Wilson, D. S. 1975. The adequacy of body size as a niche difference. The American Naturalist 109:769–84.

Winner, R. W. and J. S. Greber. 1980. Prey selection by *Chaoborus punctipennis* under laboratory conditions. Hydrobiologia 68:231–33.

Winterbourn, M. J. 1971. An ecological study of *Banksiola crotchi* Banks (Trichoptera, Phryganeidae) in Marion Lake, British Columbia. Canadian Journal of Zoology 49:637–45.

Winterbourn, M. J. 1974. The life histories, trophic relations and production of *Stenoperla prasina* (Plecoptera) and *Deleatidium* sp. (Ephemeroptera) in a New Zealand river. Freshwater Biology 4:507–24.

Winterbourn, M. J. 1978. Food and occurrence of larval Rhyacophilidae and Polycentropodidae in two New Zealand rivers, pp. 55–66. *In*: M. J. Crichton (ed.). Proceedings of the Second International Symposium on Trichoptera. Dr. W. Junk B. V. Publishers, The Hague, The Netherlands. 359 pp.

Woodin, S. A. 1978. Refuges, disturbance and community structure: a marine soft-bottom example. Ecology 59:274–84.

Wright, J. F. 1975. Observations on some predators of stream-dwelling triclads. Freshwater Biology 5:41–50.

Young, A. M. 1967. Predation in the larvae of *Dytiscus marginalis* Linnaeus (Coleoptera: Dytiscidae). Pan-Pacific Entomologist 43:113–17.

Zalom, F. G. 1978a. A comparison of predation rates and prey handling times of adult *Notonecta* and *Buenoa* (Hemiptera: Notonectidae). Annals of the Entomological Society of America 71:143–48.

Zalom, F. G. 1978b. Backswimmer prey selection with observations on cannibalism (Hemiptera: Notonectidae). Southwest Naturalist 23:617–22.

Zalom, F. G. 1981. Interactions potentially affecting the seasonal abundance of selected aquatic invertebrates in a rice field habitat. Hydrobiologia 80:251–55.

Zalom, F. G. and A. A. Grigarick. 1980. Predation by *Hydrophilus triangularis* and *Tropisternus lateralis* in California rice fields. Annals of the Entomological Society of America 73:167–71.

Zaret, T. M. 1980. Predation and freshwater communities. Yale University Press, New Haven, CT. 187 pp.

chapter 9

FISH PREDATION ON AQUATIC INSECTS

Michael Healey

INTRODUCTION

The registered world catch of freshwater fishes is about 10.2 million metric tons (MMT). Fish ponds produce another 0.7 MMT. Subsistence fisheries (unregistered) may take as much as 6 MMT and world sport fisheries may take up to 3 MMT. Total world harvest of freshwater fishes is, therefore, on the order of 17 to 20 MMT. This harvest contributes significantly to the global supply of animal protein as human food (Borgstrom 1978).

The relationship between freshwater fishes and insects is an intimate one. Sterba (1962) commented on the food of 67 families of freshwater fishes; insects were among the food of 42 families and were the main food of 29 families. The food habits of temperate zone fishes have been well studied. Scott and Crossman (1973) listed 180 species of freshwater fishes in Canada and noted that 144 (80%) fed on insects. Ninety-seven of Canada's freshwater fish species are of direct commercial or recreational value and 85 (88%) of these species feed on insects. Maitland (1977) listed 215 species of freshwater fishes in Great Britain and Europe and noted that 143 (67%) fed on insects. One hundred fourteen of Europe's freshwater fish species are of direct commercial or recreational value and 73 (64%) of these species feed on insects. Species handbooks, like those by Scott and Crossman (1973) and Maitland (1977), probably list only the fish's most important foods; presumably an even higher percentage of species either feed on insects incidentally, or during life-history stages that are not yet well studied. The importance of insects in the diets of tropical fishes is less well documented but is probably comparable to temperate fishes (Fryer 1959; Corbet 1960; Greenwood 1974). Aquatic insects may be said, therefore, to contribute substantially to world fishery production by providing the forage base for many freshwater fish populations. Understanding the interactions between the fish and insect communities of freshwater habitats is fundamental both to the dynamics of natural ecosystems and to the management of aquatic resources for food production.

The purpose of this chapter is to explore three aspects of the interaction between fish and insects: functional adaptations of fish as predators, methods for calculating the number of insects eaten by fish in natural populations, and the impact of fish predation on the abundance and composition of the insect community. Space will not permit an exhaustive treatment of even this short list

255

of topics. I shall, therefore, concentrate on fundamental problems and short-comings in our current understanding in the hope of stimulating some creative research in this field.

The feeding biology of fishes has been reviewed several times in the past few years (Ricker 1967; Gerking 1978; Hoar et al. 1979). Much of the information presented in these reviews will also be touched on here. The major differences between this and existing reviews will be emphasis and point of view. I shall, whenever possible, emphasize information on interactions between fish and their insect prey with a view to predicting the consequences of predation on insect community structure.

FUNCTIONAL ADAPTATIONS FOR FEEDING ON INSECTS

Predation may be considered the summation of three subprocesses: prey detection, prey capture, and prey ingestion. Failure of any of the subprocesses effectively inhibits predation. Insects are so important in the diet of fish, and both insects and fish display such variation in physical and sensory structures, that considerable functional adaptation among fishes for feeding on insects should be evident. Surprisingly, however, there appear to be no studies dealing specifically with the mechanism of fish predation on insects, although a number of studies have dealt with morphological and physiological adaptations of fish for predation (Hasler 1957; Fryer 1959; Keast and Webb 1966; Fryer and Iles 1972; Greenwood 1974; Liem and Osse 1975; Liem 1978). Fryer (1959) and Greenwood (1974), in their studies of African great lakes, emphasized trophic specialization involving insect prey. Much of what I present in this section, therefore, is speculative and is based upon some data on fish predation, and considerable data on fish morphology and physiology.

Prey Detection

Prey detection, as used here, means not only making sensory contact with potential prey, but also that the contact stimulus is strong enough to trigger prey-capture behavior. Fishes make sensory contact with potential prey by means of all the traditional senses (sight, sound, taste, smell, touch) and possess a "sixth sense" involving electric field distortion. The sensory capabilities of fish are well studied but comparatively little is known about precisely how these senses are used in feeding or what are the characteristics of prey that trigger a feeding response.

Vision. Fish have well-developed eyes and in a well-lighted environment can discriminate rather small gradations in the size, shape, orientation, color, brightness, contrast, and movement of objects (e.g. see Hoar and Randall 1971).

There is little doubt that vision is the principal sensory mechanism of prey detection in many fishes (Woodhead 1966). All of the visible characteristics of prey mentioned above probably contribute to their detection and recognition as prey by fishes. Markl (1972), for example, demonstrated that piranhas (*Serrasalmus nattereri*) distinguish conspecifics from potential prey at least partly on the basis of body shape. De Groot (1969) showed that flatfishes displayed the strongest feeding reaction to models that mimicked the shape of their normal prey.

The characteristics of color, brightness, and contrast are often confounded in studies involving selection of natural or semi-natural foods, but all may be important. Ginetz and Larkin (1973) reported that rainbow trout (*Salmo gairdneri*) fed on trout eggs dyed blue, red, black, orange, brown, yellow, and green in descending order of preference when presented on a pale green background. This order of selection apparently reflected contrast with the background rather than color. Particular color combinations tended to be eaten more frequently than others, however, suggesting that color was also important to the choices being made by the fish. Ivlev (1961) and Ware (1973) also present evidence that food detection is related to contrast between the prey and the background. The transparency of many prey animals that spend part of their lives in the pelagic zone of lakes (e.g. *Chaoborus* larvae) may well be an adaptation to reduce detection by fish predators (Stenson 1978).

Moving prey are detected much more easily than stationary prey (De Groot 1969; Maly 1970; Ware 1973; Confer and Blades 1975). While movement of any sort is important, pattern and speed of movement are probably also important (Maly 1970; Ingle 1971; Confer and Blades 1975).

Like movement, prey size has a direct influence on probability of detection. Reactive distance to potential prey increases linearly over the normal range of prey sizes for a particular fish (Confer and Blades 1975; Vinyard and O'Brien 1976).

Olfaction and Taste. Olfaction and taste, although they both involve the detection of chemicals in dilute solution, are separate senses in fish. Both are important in food detection in some fishes. The olfactory sense is localized in a pair of pits situated bilaterally near the front of the head and structured so that a current of water is directed across the olfactory epithelium as the fish swims. Taste is mediated through taste buds that may be located in the mouth and pharynx, in the gill cavity, on the gill arches, on barbels and fins, and, in some fishes, on all external surfaces of the body. Fishes also have a general chemical sense mediated through free nerve endings on the surface of the body (Hara 1971). The general chemical sense is low in sensitivity and is probably unimportant in feeding.

The importance of olfaction in food detection among fishes has been known for a long time. Parker (1910) gave normal bullheads (*Ameiurus*) a choice between two wads of cheesecloth, one of which concealed earthworms. The fish repeatedly seized the wad containing earthworms yet ignored the other. Fish whose barbels (containing taste buds) were removed continued to attack the cheesecloth wad containing earthworms, but fish with their olfactory nerves severed showed no reaction to either wad. Hasler (1957) reported that the minnow, *Hyborhynchus*, could be trained to discriminate among the odors of the amphipods, *Gammarus* and *Hyalella*, and the caddisfly larva *Hesperophylax*. More recently, De Groot (1969) and Carr et al. (1976) demonstrated the importance of olfaction in feeding behavior of flatfishes and pinfishes respectively.

Although less well studied, the gustatory sense is also important in food detection for some species. For instance, taste buds on the barbels of catfish are probably important in detecting food buried in the mud (Olmsted 1918; Nikolsky 1963).

Touch and Sound.   Fish possess two systems for detecting underwater vibrations, the inner ear and the lateral line system. The inner ear is sensitive to sound waves propagated through water and is considered to be the principal organ of distant sound detection in fishes. The lateral line system consists of a series of receptors located in the skin of the head and along the lateral surface of the body and responds to low frequency vibrations close by (e.g. see Hoar and Randall 1971). The sense of touch is mediated by nerve endings over the whole body surface.

All movements of prey in water generate vibrations that may be detected by predatory fish. Banner (1972) has shown that sound is important in food detection among lemon sharks (*Negaprion brevirostris*) although not all sounds produced by prey were attractive to the sharks. More interesting from the point of view of fish feeding on insects, however, is the detection of surface waves by the lateral line system. Schwartz and Hasler (1966) demonstrated that the top minnow (*Fundulus notatus*), conditioned to associate surface waves with food, always reacted to small surface disturbances up to distances of 40 cm and still reacted over 50 percent of the time to disturbances 80 cm away. Blinded fish could orient toward, and precisely estimate the distance to, the source of the disturbance when it was within 15 cm of the fish.

The importance of the sense of touch in prey detection has not been systematically investigated, but touch may be important in separating edible from inedible particles among fishes that feed at night or that sift bottom sediments for food.

Electrodetection.   A few groups of fishes, including sharks, rays, catfish, gymnotids, mormyrids, and gymnarchids, comprising several hundred species, possess receptor organs that are specialized for detection of electric fields (Bennett 1971). These fish are extremely sensitive both to weak electric fields from other animals and to distortions in their own electric field caused by an object passing near them (Lissmann 1958; Lissmann and Machin 1958, 1963). Kalmijn (1971) elegantly demonstrated that sharks and rays were able to detect buried prey by the electric field around the prey. The electric sense probably greatly enhances the ability of fish to detect prey in turbid water or at night, or prey buried in the substrate, particularly if the sense can discriminate among types of objects, as recent evidence suggests (Bastian 1981).

Stimulus Summation and Response Modulation. While a single prey characteristic may often be sufficient to elicit a feeding response, several characteristics may act synergistically to increase the probability of response. De Groot (1969) found that shape and odor together were more likely to stimulate feeding behavior of flatfishes than either stimulus alone. Vinyard (1980) found that sunfish (*Lepomis* sp.) selected *Daphnia* preferentially over copepods, presumably because they had learned to associate the visual aspects of *Daphnia* with weak escape behavior. It is not surprising that individual fish should make use of a variety of sensory stimuli in prey detection, even though one sense may predominate. Anecdotal evidence suggests that sometimes very subtle characteristics of the prey stimulus are important in triggering feeding. Every angler knows, for example, that occasionally a seemingly insignificant difference in color or size makes the difference between acceptance or rejection of his or her lure. Thus, it is impossible to determine the precise cause of prey detection without carefully controlled experiments. With so many insect forms contributing to the diet of even a single species of fish, any prediction of their risk to predation based on an hypothesis of stimulus response is impractical, unless it can be ascertained that a few characteristics of the prey predominate in triggering a feeding response. For example, Griffiths (1975) found that in a number of instances, prey size, irrespective of other characteristics, was a sufficient predictor of risk to predation.

The physiological state of the predator affects prey detection by regulating the strength of the stimulus required to trigger feeding. Most work on this factor has been conducted with visually hunting predators, but similar consequences probably occur regardless of the sensory mechanism. Satiated predators show little response to any prey, although a fish that has been fed one type of prey until it ceases to respond may show a brief resurgence of interest if a new prey of a preferred type is introduced (E. Holst 1948, cited in Curio 1976). The distance at which prey will be detected appears not to relate to how hungry the predator is (Beukema 1968), although response distance does increase with increasing experience with novel prey, and then wanes again when the novel prey are removed (Ware 1971). Response to a particular prey may also depend on the presence and abundance of alternate prey (Murdoch et al. 1975; Ringler 1979).

The searching behavior of the predator influences the risk to predation of various potential foods. Stream-dwelling salmonids, for example, may prefer riffle or pool habitats depending on species and time of year (Hartman 1965). In riffles, the fish defend a feeding territory so that the density of potential predators is restricted by the behavior of the fish. Food is delivered to the fish mainly by the current and, on average, more potential food organisms pass through areas where the water velocity is high than where it is low. Salmonids take advantage of this fact by occupying the highest velocity water that their swimming ability and the availability of velocity shelters will allow (Chapman and Bjorn 1969; Lister and Genoe 1970). Thus, they are able to increase their opportunity for encounters with prey.

In slow or still waters, fish adopt other tactics for increasing their chance of encounter with prey. The particular behavior displayed is generally characteristic of the species (Keast and Webb 1966; Fryer 1959; Schutz and Northcote 1972; Greenwood 1974). Some predators wait in ambush and depend on the movements of their prey to bring them into detection distance. Others search actively for prey. Among those that search actively, some examine a small area thoroughly whereas others range more widely but may miss many prey that closer inspection would have revealed. For both types of search pattern there is an advantage in not re-searching areas that have already been covered. Beukema (1968) found that sticklebacks (*Gasterosteus acculeatus*) swam through a familiar maze along a path that increased their opportunity for locating prey over that of a purely random path. Some predators orient to the surface whereas others orient to the substrate. Dolly Varden (*Salvelinus malma*) normally swim close to the bottom and are more effective at searching the bottom for food than are cutthroat trout (*Salmo clarki*), which swim well off the bottom. Cutthroat trout, on the other hand, are more effective than Dolly Varden at searching the surface for food (Schutz and Northcote 1972). Kokanee (*Oncorhynchus nerka*) increase their chance of encountering small prey by swimming very close to the bottom, but by so doing, reduce the area they are able to search visually. Rainbow trout increase their visual field by staying 15 to 20 cm off the bottom, but thereby reduce their chance of detecting small prey (Hyatt 1979). Capture of a food item generally increases the intensity of search in the vicinity of the capture for a while, thus increasing the risk to other prey organisms in the immediate area (Curio 1976).

Prey Capture

Prey capture obviously also involves one or more of the six senses, since the fish must maintain sensory contact with its prey during the process of capture. Capture of prey after detection is by no means certain, even when the prey is several orders of magnitude smaller than the predator, as is often the case with fishes preying on insects. The fish may lose sensory contact with the prey for a variety of reasons or may simply be unable to capture it. Prey have well developed escape mechanisms and other defenses against predators (see Peckarsky, Chapter 8) that have presumably coevolved with the predator's mechanisms of prey detection and capture. A discussion of coevolution is outside the scope of this review, but the reader is referred to Schall and Pianka (1980) for an introduction to this important topic.

The capture process may itself be subdivided into two components: approach and strike. Prey are often detected at a distance too great for an effective feeding strike. Thus the approach is designed to bring the predator within striking distance. In many instances the approach is merely a dash toward the prey, which brings the predator within striking distance before the prey can escape or, if it initially escapes, the predator simply outswims the prey. Approaches of this nature are most common among open water visual predators. However, the approach may be more complicated. Sharks tracking prey by their odor search with side-to-side head movements or swim against the current to keep within the odor gradient. They also switch from odor to vision during the final stages of approach (Hodgson and Mathewson 1971). If prey are small with low escape velocity, the predator may approach slowly and inspect potential prey before striking. Ambush predators may conceal themselves and wait for prey to roam within striking distance even after they have responded to the presence of the prey. Body form appears adapted to increasing capture success in the habitat in which the fish normally feeds. Open water visual predators often have an elongate fusiform shape, while bottom feeders and those that forage in weed beds have shorter broader bodies (Keast and Webb 1966). Plankton feeders show adaptations for high maneuverability; nekton feeders show adaptations for speedy pursuit (Werner 1977). Body size and shape must, however, be affected by a wide variety of selection pressures in addition to the need to obtain food. Thus it is difficult to demonstrate unequivocally that a particular body form is an adaptation for feeding.

Once the fish is within striking distance a whole new set of adaptations takes over. Some fish pick up their prey, using the mandibles as forceps, but most suck in their prey. When a fish strikes, it opens its mouth and expands its buccal cavity in a way that creates strong negative pressure inside the mouth. Water is sucked into the mouth to equalize the pressure, and the food is swept in along with the water. Pressures from 80 to 400 cm of water (80–400 thousand dynes cm^{-2}) occur during the strikes of different fishes (Alexander 1970). The anatomy of the buccal apparatus is highly organized to perform this function, with muscles operating near their maximum recorded isometric tension during the strike (Alexander 1970; Lauder 1980). Potential prey vary in their vulnerability to buccal suction, however, and this influences their risk of capture (Confer and Blades 1975; Drenner et al. 1978).

Numerous other aspects of head anatomy influence capture success. Mouth size and shape differ widely among fishes (Hartman 1958; Nikolsky 1963; Keast and Webb 1966; Werner 1977; Hyatt 1979) and this obviously affects the size of prey that they can capture. Mouth position together with body form can enhance capture success in particular habitats. The brook silverside (*Labidesthes sicculus*) is an important insect predator in temperate waters. It is adapted to surface feeding in having an almost straight dorsal line, a supraterminal mouth, and a small dorsal fin. *Labidesthes* can swim just under the surface to feed on insects and other organisms caught in the surface film. The antithesis of *Labidesthes* is the log perch (*Percina caproides*), which is adapted for bottom feeding. *P. caproides* has a ventroterminal mouth, large pectoral fins, and is a poor swimmer. In searching for food it moves in short hops across the bottom and pounces on prey living in or on the bottom (Keast and Webb 1966).

Specific adaptations also assist in the capture of prey that hide in tubes or burrow into the sediment. Bottom feeders typically suck in a mouthful of sediment, trap the prey organisms that were concealed there, and expel the sediment through the opercula (Nikolsky 1963; Fryer 1959; Schutz and Northcote 1972). Ivlev (1961) and Nikolsky (1963) provide data on the depth to which various species of fish can bite into the substrate. According to Nikolsky (1963), carp (*Cyprinis carpio*) can penetrate almost 13 cm into soft substrate, followed, in descending order of penetration, by ruffe (*Acerina cernua*), tench (*Tinca tinca*), bream (*Abramis brama*), roach (*Rutilus rutilus*), and crucian carp (*Carassius carassius*). Least effective was the perch (*Perca fluviatilis*), which could penetrate less than 1 cm. The depth of penetration is, of course, dependent on substrate type; carp are able to penetrate only 6 cm into clay soils compared to 12 to 13 cm in silt. The cichlids, *Lethrinops brevis* and *L. furcifer*, both specialized predators of chironomid larvae in Lake Nyasa, Africa, have sharp recurved outer teeth and backwardly directed inner teeth, ideally suited for picking up and holding onto insects (Fryer 1959). The cichlid, *Haplochromis chilotes*, from Lake Victoria, Africa, has protruding medial teeth that form an effective forceps for extracting insect larvae from their tubes (Greenwood 1974). Finally there are more bizarre adaptations for capture of insects such as that shown by the archer fish (*Toxotes jaculator*), which spits a jet of water at insects sitting on vegetation above the surface, splashing them into the water; or the terrestrial excursions of mudskippers (*Periopthalmus* spp.) and the cyprinid, *Rivulus hartii*, to feed on ants and other terrestrial insects (Nikolsky 1963; Seghers 1978).

Ingestion

The final predatory process is ingestion. Organisms captured must be retained in the buccal cavity and passed to the esophagus. Again, a number of circumstances may protect potential prey from ingestion. Things taken into the mouth receive a final screening there before being passed to the esophagus. This final screening is particularly important when inedible objects are sucked into the mouth along with the prey, as when bottom feeders bite into the substrate. In some way the fish must separate food organisms from the debris and selectively retain them in the buccal cavity while the debris is expelled. Gill rakers are believed to play an important role in this process (Nikolsky 1963; Fryer 1959; Fryer and Iles 1972). The retention of small animals is supposed to be associated with the presence of long, closely spaced gill rakers that act as a kind of sieve. Long, closely spaced gill rakers are, however, by no means a necessary condition for the retention of small prey (Kliewer 1970; Seghers 1975). Even when prey are taken into the mouth individually, a screening process occurs, probably mediated through taste and texture (Sutterlin and Sutterlin 1970). Food particles are often taken into the mouth and then spat out. Some organisms appear to be unpalatable and are rejected for that reason (Ivlev 1961). Others may be palatable but, because they are novel, they initially may be rejected. Cases and tubes containing insects, such as caddisfly larvae, may have to be crushed before they can be swallowed. Some fishes have elaborately designed pharyngeal jaws for dealing with armored prey such as caddisfly larvae (Nikolsky 1963; Liem and Osse 1975), whereas others may not be able to ingest such prey at all. Ivlev (1961) investigated the importance of cases in protecting the trichopteran larva *Phryganea* from predation. Larvae without cases were ingested at a much greater frequency than those with cases and even a partial case afforded some protection. Perch were not able to eat cased *Phryganea* at all, but ate uncased larvae quite readily.

Fish predation of insects obviously is regulated by the complex of signals emanating from the prey and its defense mechanisms, and by the sensory detectors of the predator, its physiological state, and its mode of hunting, pursuit, and capture. We know a great deal about some of these characteristics in both predator and prey. The relative importance of the various signals transmitted by the prey and their relationship to risk of predation have not, however, been adequately investigated. It is time, in my view, for some rigorous experimentation to test hypotheses relating risk of predation to variation in physical and behavioral characteristics of insect prey.

MODELS FOR ESTIMATING PREDATION RATE

In taking the step from describing predation to estimating predation rate it is important to be aware of the multiplicity of sensory modalities and behavioral responses that can affect the interaction between predator and prey. Although predation rate may be estimated in specific situations without a knowledge of the mechanism of prey selection, such estimates cannot be generalized without this knowledge. Ultimately the prediction of predation rate in changing environments depends on understanding the mechanism of prey selection. All predation rate models seek, in some way, to incorporate the elements of predatory behavior discussed above. It is not practical, however, to include all the state variables in a model of predation. Much of the art of such modeling lies in selecting a set of variables that adequately describes the process without becoming unwieldy. The models described in this section represent successively simpler descriptions of the process of predation by using the fish as integrators of the process, and measuring parameters of the fish population alone as opposed to measuring parameters of both predator and prey. I shall discuss these models in terms of their applicability to predicting predation rate and the impact of predation in natural communities.

Functional Response Models

Perhaps the most analytic of the current predation rate models is that developed by Holling (1965, 1966) and validated through a series of experiments with deer mice and preying mantids as predators. Holling (1965) coined the term *functional response* to describe the individual components of predation from which he assembled his model. In many respects his model is directly applicable to fish that feed on insects. Holling's (1965, 1966) argument may be summarized as follows:

Predators feed during a certain period of time each day comprised of a number of bouts of predation

$$TI = TD + TS + TP + TE \qquad (1)$$

where:

TI = time between the capture of one prey and the capture of the next

TD = the digestive pause, or the length of time after eating one prey that the predator's hunger is below the threshold for demonstrating predatory behavior

TS = time spent searching for prey

TP = time spent pursuing prey

TE = time spent eating prey

The rate of predation (feeding bouts per day) is the number of TI that must be accumulated to equal the length of the daily feeding periods. The variables TD, TS, TP, and TE that determine the length of each successive TI encompass a number of the characteristics of predator-prey interaction that I have discussed earlier, i.e.:

1. TD depends on hunger, which in turn depends on the fullness of the predator's gut. Holling (1965) proposed that there was a threshold gut fullness, FT, above which the predator would not show a feeding response. While other factors undoubtedly affect hunger in fish, stomach fullness is probably a dominant one (Colgan 1973).
2. TS depends on the density and signal characteristics of prey and the effectiveness of the predator's prey detection senses and search behavior.
3. TP depends, in fishes, mainly on the pursuit speed of the predator relative to the prey and the reactive distance to the prey. Escape speeds of insect prey are

generally low relative to the burst speed of fish. In the case of ambush style predators, however, pursuit is more prolonged and dependent on the speed of the prey.

4. TE depends on the type of prey, how much it must be manipulated, whether it is in a protective case, or whether it must be sieved from the substrate. In many situations in which fish feed on insects, the prey are small relative to the predator and are soft bodied so that they are swallowed very quickly; thus, TE has an insignificant effect on the predation rate. When the prey are large, armored, or concealed, however, TE can be significant. Werner (1974) and Mittelbach (1981) showed large increases in time taken to capture and eat prey as prey size increased.

Holling's model can be used to describe predation rate over any desired time span but it is most practical to develop longer term rates of predation by accumulating successive estimates of the daily food intake of the predator. Model calculations for daily food intake proceed logically from the state of hunger (defined by F, the weight of food in the stomach) of the predator at the start of a feeding period (there may be more than one per day). Presumably $F < FT$ at this time so that the predator begins to feed at a rate determined by TS + TP + TE. Provided the weight of each prey eaten is greater than the loss of stomach contents due to digestion in the interval TS + TP + TE, then the weight of stomach contents will increase until $F \geq FT$. When this point is reached, the predation rate will drop to TD + TS + TP + TE and continue at this rate for the balance of the feeding period, unless prey become so scarce that digestion over the period TD + TS + TP + TE exceeds the average weight of prey. In some instances, however, filling the stomach is followed by a latent period during which the fish will not feed even though stomach volume is significantly reduced. For brown trout (*Salmo trutta*), Elliott (1975) found that feeding would not occur after a full meal until stomach volume had been reduced to 10 to 20 percent of the original meal size.

Holling's model is attractive because it appears to capture the essence of predatory behavior and is well supported by empirical observation. The model is, however, still a gross simplification of the predatory process. Numerous real problems faced by predators in nature are not addressed by the model. As constituted, it is a "one predator, one prey" model, although it can be modified to include several prey types and several predator types. Variation in the reactive distance of the predator, as determined by light intensity, turbidity, and habitat complexity, and the visible characteristics, size, and movement of the prey are not accounted for in the model. Likewise, the importance of prey type to FT is not accounted for. Only the average density of prey enters into model calculations, although Ivlev (1961) demonstrated that feeding rate increased with increasing contagion in the distribution of benthic prey. The competitive interactions among predators that may reduce their feeding rate are also not accounted for. Functions incorporating these and other variables can certainly be included in the model. Each time a new state variable is incorporated into the model, however, appropriate experiments must be done to determine the functional relationship, and the variable must be monitored in nature to determine the range of values that should be included in the model. The number of measurements required to simulate even a simple natural system by means of this kind of model is beyond the capacity of most investigations. The model is, therefore, impractical in most instances as a tool for predicting the impact of fish predation in natural systems.

Holling's approach has, however, had considerable influence on the study of predation in fishes and several authors have fruitfully explored aspects of prey risk, the predatory process, and optimal foraging by means of related models (Beukema 1968; Ware 1973; Werner 1974, 1977; Confer and Blades 1975; Eggers 1976; Vinyard and O'Brien 1976; Mittelbach 1981). These studies have focused on predation by hungry fish, however, and use only the search and attack portions of Holling's model. This is generally expressed as

$$C = SPT/(1 + SPTh) \qquad (2)$$

where:

C = the number of prey captured in time T

S = the area or volume searched by the predator per unit time

P = the density of prey

T = the length of the feeding period

Th = the "handling time" for each prey = TP + TE

One of the best examples of this approach to determining prey risk is Ware's (1971, 1972, 1973) investigation of rainbow trout feeding on epibenthic prey. Ware's (1973) model included variables for prey size, movement, contrast, and population density, and the effect of light intensity, time of day, population density, and temperature on the proportion of prey that were exposed and active at the mud-water interface. The model included variables for the probability that the predator would recognize an object as prey (different for moving and stationary prey), the swimming speed of the predator and its distance from the bottom (which affects the width of the search path), and the handling time for individual prey. Risk was defined as the ratio of model predictions for capture rate of a particular prey to total capture rate. Model predictions of seasonal variation in prey risk were compared with the observed proportions of four principal prey types (amphipods, caddisfly larvae, odonate larvae, planorbid snails) in the stomachs of trout from Marion Lake, British Columbia. The model explained 47 percent of the observed variation in prey risk. The model predicted seasonal variation in risk for caddisfly larvae very well ($r^2 = 0.9$) but predicted risk for odonates ($r^2 = 0.03$) and planorbids ($r^2 = 0.09$) very poorly. Ware (1973) concluded that, in the situation that he examined, prey activity, exposure, density, and size were most important in determining prey risk, but that it was unlikely that these variables would prove to be most important in another situation. The failure of even this ambitious research to account very well for prey risk in a relatively simple situation (one predator, four prey) emphasizes the difficulty of predicting predation rate in natural communities from such deductive models.

Other studies have concentrated on elements of the functional response and, in a few instances, have attempted to predict risk to predation from a more limited set of parameters. Beukema (1968), for example, investigated the effects of hunger and experience on predatory behavior of sticklebacks in a multicompartmented maze. Hunger level (defined as hours of food deprivation) correlated positively with swimming speed in the maze, frequency of bursts of searching activity, frequency of feeding reactions to inedible objects, and proportion of prey discovered that were grasped and eaten. All these relationships were asymptotic. The efficiency of searching for prey and the reactive distance to prey were unrelated to hunger, although searching efficiency increased dramatically as the fish gained experience with the maze. Ware (1972) and Werner (1974) noted positive correlations between hunger and frequency of attacks by rainbow trout and handling time in sunfish respectively, results that are consistent with Beukema's (1968) more analytical approach. It seems logical that intelligent animals would at first be wary of novel prey, but with growing familiarity would attack them more and more frequently. This response has been observed in several studies (Beukema 1968; Ringler 1979; Ware 1972) and the reactive distance to prey apparently increases asymptotically with increasing experience. The influence of prey size on reactive distance has also attracted considerable study (Ware 1971; Confer and Blades 1975; Vinyard and O'Brien 1976). Reactive distance increases linearly with increasing prey size within the range of sizes normally fed upon. However, reactive distance can differ between similar-sized prey species, presumably being related to other signal characteristics of the prey. As environmental conditions favoring visual recognition of prey deteriorate (e.g. reduced light, increased turibidity), both the reactive distance and the rate of increase in reactive distance with prey size decline. In situations where visibility is very poor, reactive distance is almost independent of prey size, suggesting that the fish are switching to nonvisual means of prey detection. Werner (1974, 1977) and Mittelbach (1981) present data showing that both handling time and pursuit time are increasing functions of prey size. Handling time tends to be relatively constant over a wide range of prey sizes, then increases very rapidly as the upper limit of prey size is reached. Pursuit time increases linearly with prey size, but the rate of increase is more rapid in planktivores than in piscivores. Presumably this reflects the pursuit speeds that these morphologically different predators can achieve.

These results are consistent with the known functional adaptations for feeding. Unfortunately they fall short of providing a practical basis for estimating feeding rate in natural populations. They have, however, provided a basis for predicting patterns of size selection and optimal foraging strategies in situations involving one or several predator species (Werner 1974, 1977; Werner and Hall 1974, 1979; Griffiths 1975; Mittelbach 1981; Werner et al. 1981). It is encouraging that, in a number of instances, prey size and density were virtually sufficient statistics on which to base a qualitative prediction of risk to predation. The very important relationship between predation rate and prey density is the basis of the following models.

Ivlev's Models

One of the most thorough investigations of predation by fishes is Ivlev's (1961) classic study on the feeding ecology of fishes. Ivlev's (1961) approach was to measure directly the feeding rate of fish under different experimental situations. This approach avoids some of the problems inherent in Holling's functional approach because the fish integrate many of the state variables that appear explicitly in the functional response model. Ivlev's approach was no less analytical, however, and, as we shall see, his model is no less difficult to apply to natural communities.

Ivlev (1961) investigated both the rate of feeding of hungry fish under different conditions and the process of selecting prey. He observed that, as prey density increased, the feeding rate of a hungry fish increased to some maximum value that was characteristic of both the predator and the prey species. The function describing this relationship was (Ivlev 1961)

$$C = C_{max} (1 - e^{-KP}) T \tag{3}$$

where:

C = total prey captured in time T

C_{max} = maximum possible rate of prey capture per unit of time

K = a constant (for a particular species of fish and species of prey) determining how fast the capture rate approaches C_{max} as prey density (P) increases

This equation describes what Holling (1965) later termed a "type 2" functional response of predators to prey density, a "type 1" response being a linear increase to C_{max} and a "type 3" response being a sigmoidal increase to C_{max}. Of particular importance to determining predation rate in natural communities is Ivlev's (1961) discovery that the rate at which C approaches C_{max} increases with increasing contagion in prey distribution. Thus, the use of average prey densities, as is normally done in predation rate models, may cause predation rate in natural communities to be underestimated.

From this simple description of the relationship between predation rate and prey density and dispersion, Ivlev (1961) went on to describe the characteristics of food selection by predators faced with a choice of foods. Predators seldom take prey strictly in proportion to their density. Ivlev termed the process of selecting certain prey types over others "electivity" and defined it quantitatively as

$$E = (r - p)/(r + p) \qquad (4)$$

where:

E = electivity $(-1 < E < + 1)$

r = proportion of the prey item in the diet

p = proportion of the prey item in the natural community

Electivity is clearly dependent on the sensory and morphological adaptations of the predator, and the defense mechanisms of the prey. It is also, however, a consequence of the ill-defined food preferences of predators. Ivlev (1961) investigated the effect on electivity of a number of factors including predator type, predator experience with the prey, predator satiation, the absolute and relative density of prey, prey dispersion, prey speed and size relative to the predator, the extent and toughness of the prey's protective covering, and the depth of the prey's burrowing. All of these factors, alone and in combination, have some influence on selective feeding by the predator. The trend in electivity with change in these factors depended on whether the prey were preferred or avoided. For example, the electivity of most preferred prey increased as total prey density increased, but that of avoided prey decreased. The electivity of moderately preferred prey at first increased, then decreased. These kinds of changes make sense in terms of what we know about predator behavior. An increase in total prey density permits the predator to exercise greater choice in the prey it eats because it encounters more preferred prey per unit of feeding time. The existence of such responses complicates the prediction of predation rate in natural communities, however, because it means that prey risk is a function of the whole community structure rather than just of the individual prey and predator species.

Ivlev (1961) extended his investigation to the impact of several predators and predator types feeding on individual species and mixtures of species of prey. He correctly observed that competition between predators for food would result in changes in the electivity values, since some predators would be forced by this competition to feed on less preferred prey. This situation has been elegantly demonstrated in natural populations by Nilsson (1967) and in experimental assemblages by Werner and Hall (1979). Ivlev (1961) proposed that the average shift in electivity values be used as an index of interaction between competitors. This is a rather cumbersome index, however, and its rigorous application requires determination of electivity indices in a situation where competition is absent. There is no straightforward way to set confidence limits on electivity.

To assess the impact of selective feeding on prey populations by means of Ivlev's (1961) models, the parameters for selective feeding must be combined with the estimation of feeding rate. Ivlev (1961) did not do this. Intuitively, however, it seems likely that the process of electivity that Ivlev investigated so thoroughly is reflected in variation in the parameters C_{max} and K in equation (3). This equation can be modified to permit calculation of daily food intake by including terms for satiation and digestion as in Holling's functional response model.

Ivlev's model is a "one predator, one prey" model. It can easily be modified to include several prey types, however, provided the separate values of K, P, and C_{max} are known for each prey. Inclusion of terms to modify feeding rate in the presence of intra- and interspecific competition is conceivable, but the model rather quickly becomes unwieldly when several predators and prey are involved. The problems of parameter estimation and monitoring with this model are no less formidable than they were with the functional response model.

As was the case with Holling's model, elements of Ivlev's model, in particular the relationship between feeding rate and prey density for hungry fish, have made an important contribution to recent studies of predation by fishes (Ware 1972; Mittelbach 1981; Werner et al. 1981). The principal advances in these recent studies have been in the evaluation of net energy gain from feeding on different sizes and densities of prey. A consideration of net energy gain often leads to the prediction of maximum predation rate on intermediate sized prey (Werner 1977) and such predictions may differ from those predictions based on type 2 functional responses alone.

In spite of their disadvantages as tools for predicting absolute predation rates in nature, the impact of Ivlev's and Holling's models cannot be overstated. These models have led to a much clearer understanding of the elements of prey risk and have largely explained why organisms appear in different proportions in fish guts than in nature. Furthermore, predictions of daily ration based on these or analagous models, even when only a few state variables are included, are often as correct as predictions by any other model. For the purpose of comparing predation rates against prey population statistics, however, these models have a significant confounding defect. They both depend on measures of prey density, so that the estimate of prey abundance and predation rate are both determined from the same sample values. Thus, any analysis of the impact of predation on prey abundance runs the risk of being tautological if such a model is used to generate the estimate of predation rate.

Bioenergetic, Gastric Evacuation, and Growth Efficiency Models

Several models are available for estimating feeding rate that depend only on vital statistics of the fish population. These fall into three general categories: bioenergetic, gastric evacuation, and growth efficiency models. The models are interrelated but they depend on different parameters and, therefore, have somewhat different strengths and weaknesses. They all involve relatively few parameters compared with the functional response and Ivlev-type models.

Bioenergetic Models. Bioenergetic models arise from the knowledge that living organisms are governed by the laws of thermodynamics and that the energy of food ingested must all be accounted for in excretory losses, metabolic losses, or growth, i.e.:

$$Er = Em + Eg + Ee \qquad (5)$$

where:

Er = energy in the daily ration

Em = energy of metabolism (including standard metabolism, active metabolism, and specific dynamic action)

Eg = energy of growth (including somatic and gonadal growth)

Ee = energy of excretory products (including feces and nitrogenous waste products)

The classic work on this approach is that of Winberg (1956), although Fry (1957) expressed similar conclusions independently. Winberg (1956) reviewed all the known data on fish metabolism, growth, and excretion and came to a series of important general conclusions:

1. The metabolic rate of fishes is a function of body weight and is adequately described by an equation of the form

$$M = aW^b \qquad (6)$$

where:

M = metabolic rate, usually measured as oxygen consumption

W = weight of the fish

a and b = regression coefficients

> The coefficient a is a variable, dependent on temperature and other environmental factors, but at 20°C is about 0.3. The exponent b is constant and is approximately 0.8 for all species of fish.

2. Metabolic rate increases as a power function of temperature. This increase is adequately described by the normal curve of Krogh (Winberg 1956).

3. Metabolic rate increases with activity, but, under ordinary circumstances, fish living free in nature will have a metabolic rate about twice the rate determined for resting fish in the laboratory. (The rate for resting fish is termed the *routine metabolic rate.*)

4. For all carnivorous fishes, approximately 15 percent of the energy of the daily ration is lost as feces and a further 5 percent is lost in nitrogen excretion. Thus, 80 percent of the ingested energy is available for metabolism and growth.

Winberg (1956) proposed, therefore, a set of standards, applicable to all fishes, from which daily food intake could be calculated provided the weight of the fish, its growth rate, and the water temperature were known. Subsequent work has, not surprisingly, revealed exceptions to Winberg's standards. The weight exponent, b, differs significantly among some species (Beamish 1964), among seasons of the year in some species (Wohlschlag and Juliano 1959; Moore and Wohlschlag 1971), and with swimming speed (Brett 1965). Winberg (1956) observed that b varied from 0.63 to 0.98 among species, but recent investigation has extended the range to 0.55 to 1.06 with good statistical evidence for significant departure from 0.8; variation within a species is generally less. Seasonal variation in b values is about 30 percent and variation with swimming speed is about 20 percent. Seasonal variation is at least partly due to different metabolic processes in maturing and immature fish (Wohlschlag and Juliano 1959; Brett 1965; Paloheimo and Dickie 1965a). Brett (1965) suggested that the increase in b values with increasing swimming speed was due to the dominance of muscle metabolism at high speeds. Increases in metabolic rate with increasing temperature do not always conform to Krogh's curve (Brett 1970; Moore and Wohlschlag 1971). Sometimes the curve is rotated around an intermediate temperature in the fish's environment, presumably reflecting temperature adaptation. Similarly, Winberg's standard assimilation efficiency of 85 percent has been challenged by a number of investigators who show that assimilation can range from 73 to 99 percent, depending on food type (Webb 1978).

Considerable attention has focused on the active metabolism of fish in nature and the energetic cost of foraging. In laboratory studies, where the cost of searching for food is insignificant, two processes influence feeding metabolic rates. First, there is increased swimming activity in fish excited by the presence of food. Second, there is specific dynamic action (SDA), or energy cost, of digesting, absorbing, transporting, and storing foods in the body. SDA varies with ration size and is usually in the range of 9 to 20 percent of ingested energy (Jobling 1981a). For fishes foraging in nature, however, the energy expended in locating and capturing prey is likely to be of greater consequence than the energy required to process it. Estimating the level of activity of fish in nature is accomplished largely by inference rather than by direct measurement except in a few instances where larger fish have been tracked by radio telemetry or sonar (Hergenrader and Hasler 1967; Diana 1980). It appears that the energy expended in foraging and other daily activities results in metabolic rates ranging from slightly greater than routine to just over double routine, although this can be greatly influenced by prey size and density (Kerr 1971b; Elliott 1973; Ware 1975).

Despite the existence of some obvious exceptions, Winberg's standards have stood the test of time well (Kerr 1971a; Mann 1978) and still provide a reasonably reliable way to estimate the daily food intake of fishes from very limited data. Many of the departures from Winberg's standards (e.g. low metabolic rates among sedentary species, low assimilation efficiency of foods with a high proportion of chitin) are logical and the standards can be adjusted appropriately to meet these circumstances when the natural history of the system is known.

Winberg's model, or some variant of it, has been used numerous times to estimate daily or annual rations of fish and to compare these with the standing crop or productivity of food resources (Mann 1965; Burbidge 1974; Miroschnichenko 1979).

Growth Efficiency Models. Growth efficiency models are intimately related to bioenergetic models since growth efficiency is the ratio of energy of growth (Eg) to total energy ingested (Er), i.e.:

$$g = Eg/Er \qquad (7)$$

where:

g = growth efficiency

Daily food intake can be predicted theoretically from knowledge of growth rate, fish size, and growth efficiency. Various authors have explored the interrelations among daily food intake, metabolism, and growth (Winberg 1956; Paloheimo and Dickie 1956a,b; Brett et al. 1969; Brett 1971; Kerr 1971a,b,c; Healey 1972). The results of these studies confirm the possibilities of a growth efficiency model, but they also show that the model must be more complicated than it at first appears. Growth efficiency is not constant but varies with ration size, temperature, and fish size (Paloheimo and Dickie 1965b; Brett et al. 1969; Brett 1971). Growth efficiency is also a function of the physiological state of the animal, being greater when the fish is undergoing a period of rapid somatic growth, and less when some other physiological function is dominant (Healey 1972). Kerr (1971b) developed a model to predict growth efficiency from metabolic coefficients, fish weight, cruising speed, and the density and average weight of food organisms. Kerr's model takes into account a number of the factors known to influence growth efficiency and predicts it independently of ration size. Unfortunately, it requires data on density of prey animals. If the object is to predict the consequences of predation for the prey, then this model suffers from the same problem of confounding as described for the functional response and Ivlev-type models.

Despite the apparent difficulty of applying the simple concept of growth efficiency to estimating daily food intake, the technique holds promise for certain situations. Gerking (1962) used it effectively in his thorough study of production and food utilization in bluegill sunfish. The clear relationships among growth rate, temperature, and ration size presented by Brett et al. (1969) suggest that a satisfactory model for growth efficiency might be developed from laboratory data.

Digestion Rate Models. Digestion rate models seek to estimate daily food intake through knowledge of gastric evacuation rates and stomach contents. Models of this sort, involving some intuitive assumptions about the rate of gastric evacuation and daily feeding patterns, were among the first models to be used in estimating the feeding rate of fish in nature (Allen 1951; Bajkov 1935). The pattern of gastric evacuation has now been investigated for numerous fishes (Windell 1978). In general, the weight of food in the stomach declines exponentially with time after feeding stops and the rate of this decline is a power function of temperature (Elliott and Persson 1978). Many factors besides temperature can influence the absolute rate of evacuation, however, including the species of fish and its size, and the meal size, meal frequency, food particle size, food digestibility, and fat content (Windell 1978). Thus, the digestion rate model to be used needs to be thoroughly worked out if accurate estimates of daily food intake are to be obtained.

Elliott and Persson (1978) reviewed the various models for calculating feeding rate from information on stomach contents and rate of gastric evacuation. They concluded that all the existing models suffered from errors in assumptions that would lead to underestimation of daily food intake. They proposed two new models for estimating daily food intake, the first applicable to situations where the feeding rate of the fish was constant, i.e.:

$$Ct = (Ft - Foe^{-Dt})Dt/(1 - e^{-Dt}) \tag{8}$$

where:

Ct = the weight eaten over the time period t

Fo and Ft = the weight of food in the stomach at time 0 and t, respectively

D = the instantaneous rate of gastric evacuation

The second model was applicable to situations where the feeding rate declined with time, i.e.:

$$Ct = C_{max} - (C_{max} - Fo)e^{-ct} \tag{9}$$

$$Ft = Foe^{-Dt} + (C_{max} - Fo)ce^{-Dt}(1 - e^{(D-c)t})/c - D) \tag{10}$$

where:

C_{max} = the maximum meal size for the species of fish and type of food

c = constant

The reader is referred to Elliott and Persson (1978) for the derivation of these equations.

Recently Jobling (1981b) suggested that, in many species, the rate of gastric emptying is better described by a square root model than by an exponential model, i.e.:

$$\sqrt{Ft} = \sqrt{Fo} - Dt \tag{11}$$

Jobling (1981b) develops equations to estimate daily ration based on this model that are analagous to those proposed by Elliott and Persson (1978) based on the exponential model.

Direct application of either model to a field situation requires rather frequent sampling of the predator population over 24 hours so that Ct can be estimated for relatively short periods of time and these values can be summed to get daily food intake. Laboratory estimates of D and C_{max} for various temperatures and types of food are also required. Frequent 24-hour sampling excursions are tedious and costly. If ration is to be estimated on a seasonal or annual basis, then it is probably worth developing a model for diel changes in stomach content so that stomach weights at several times of day can be predicted from a few samples.

Elliott and Persson (1978) compared the predictions of their models with known feeding rates in laboratory experiments. They found that predicted feeding rates were generally within 10 percent of known rates, even under situations of periodic feeding, provided the interval between samples was short. They also found that their models predicted daily food intake more accurately than a similar model developed by Thorpe (1977), which they considered to be the best of the models that predated theirs. Jobling (1981b) used the data in Elliott and Persson (1978) to test his model and found that it gave estimates of daily ration that were equal in accuracy to the estimates of Elliott and Persson (1978).

Various gastric evacuation models have been used in field studies to estimate the feeding rate of fishes (Elliott 1973; Staples 1975; Doble and Eggers 1978). In general, shortcomings in these models result in the feeding rate being underestimated (Elliott and Persson 1978). The tools are now available, however, to estimate feeding rate by this method with considerable accuracy.

Estimates of food consumption by either bioenergetic, growth efficiency, or gastric evacuation models must be coupled with taxonomic analysis of stomach contents so that food consumption may be partitioned among the various prey species, and with estimates of predator abundance so that total predation rate can be estimated. Such analyses are likely to be part of any investigation of predation in insect communities. An additional advantage to these models, therefore, is that many of the measurements necessary for their application will be made in the course of determining which are the predators of the insect community, and how abundant they are. Thus, additional sampling effort and laboratory analyses are kept to a minimum if one of these models is employed.

Although bioenergetic, growth efficiency, and gastric evacuation models are comparatively simple to use and require few parameter estimates, they still involve numerous simplifying assumptions when applied to natural communities and any of these models must be used with caution. Further caution must be exercised when apportioning the predation rate among prey. Individual fish are, characteristically, highly selective in what they eat and the probability distributions for diet items in fish stomachs are often other than normal (Sibert and Obrebski 1976). The sampling design for the predators must be carefully chosen to provide estimates of prey frequency that are representative of the population as a whole.

Whatever model is chosen as the principal means of estimating predation rates, it is always desirable to collect sufficient information so that one of the other models can be applied in a few instances, as a check on the principal method. Since all the models involve rather severe simplifying assumptions when applied to natural situations, checking the predictions of one model against another is, in my view, the most satisfactory way of validating estimates.

EMPIRICAL OBSERVATION OF THE EFFECTS OF PREDATION

The analysis of predatory behavior and estimation of predation rate suggest certain important consequences for the insect prey community provided the fish predators are sufficiently abundant. Without ever having looked in the stomach of a fish taken from a natural population, one could reasonably predict that predation pressure would be greatest on the larger components of the aquatic insect community, on those most conspicuous by virtue of their activity patterns or habitat, and on those least protected by cases or tubes. A further prediction would be that these components of the community would decline in abundance relative to other components under the influence of fish predation. Although these and more complicated predictions about prey risk can be tested in the laboratory, their final corroboration must come from observations of the consequences of predation in natural communities. Two types of studies provide evidence of the impact of fish predation on insect communities. These are studies in which natural variation in abundance of fish is compared with the occurrence of particular components of the insect community and studies in which the abundance of fish predators or their access to the insect community is artificially manipulated. A good example of the former type is the study by Pope et al. (1973) of *Chaoborus* abundance and species composition in 26 lakes in the Matamek River watershed, Quebec. *C. americanus* occurred only in fishless lakes while *C. punctatus* coexisted with fish but was absent from lakes having only one or no fish species. *C. trivittatus* and *C. flavicans* apparently occurred independently of the presence of fish. Both large and small larvae of these two species tended to be more abundant in fishless lakes but this was not universally so. The conclusion of this study was that *C. americanus* was strong in competition with other *Chaoborus* species, but was vulnerable to predation by fish. *C. punctatus,* on the other hand, was a weak competitor, but was relatively invulnerable to fish predation.

A number of studies (Ball and Hayne 1952; Lellak 1965; Welch and Ball 1966; Kajak 1972; Kajak et al. 1972; Andersson et al. 1978) in which fish populations were manipulated have demonstrated the general conclusion that benthic invertebrates in ponds and small lakes (which are usually dominated by insect larvae) increase and decrease in abundance in response to changes in the abundance of fish predators. The techniques used to manipulate the fish populations in these studies encompass poisoning and netting to remove existing fish, introductions of particular species of fish in known densities, and the use of wire mesh enclosures either to confine a known number of fish to a particular area or to protect an area of bottom from predation. These methods are all open to criticism regarding their influence on elements of the ecosystem other than the fish. The methods used in sampling the prey communities are also open to criticism. Nevertheless, these investigations provide exciting evidence of the important role of fish predation in structuring the insect community of lakes and ponds.

Not all such investigations have recorded a general change in abundance of benthos with changing fish populations. Macan (1977) observed that although rare and conspicuous forms were strongly influenced by trout predation in Hodson's Tarn, England, most of the common and abundant species were unaffected. The results of Macan's study are, nevertheless, consistent with predictions of the effects of moderate predation on the insect community. For example, Macan (1977) attributed the failure of some of the common insects in Hodson's Tarn to show a change in abundance with changes in predation rate not to a lack of response, but rather to well-developed self-regulatory mechanisms that removed any numerical response.

Thorp and Bergey (1981a, 1981b) were also unable to detect any overall changes in density or species composition of benthos in littoral areas of Par Pond, South Carolina, that were protected from foraging fish and turtles. The density of some benthic components, however, did change significantly. Thorp and Bergey gave no data on the abundance of vertebrate predators or on the proportion of their food that was taken from the littoral. Par Pond apparently has a very dense population of bass (Bennett and Gibbons 1972), which are chiefly predators of other fishes. Conceivably, the bass may have reduced the abundance of smaller benthic feeding fishes sufficiently to prevent the demonstration of any effect of enclosures.

Comparable studies in flowing waters are much rarer. Straskraba (1965) attributed an abrupt drop in the abundance of the amphipod, *Rivulogammarus*, within a section of stream to predation by brown trout and, by inference, explained other abrupt changes in faunal abundance in streams the same way. Zelinka (1974, quoted in Macan 1977) manipulated the abundance of trout and cottids in two zones of a stream and observed little change in the abundance or composition of Ephemeroptera. Allan (1982) has conducted the most thorough study to date of the effects of predator (trout) removal on the benthic and drift fauna of a small stream. Reduction of predator biomass to 10 to 25 percent of initial values had no measurable effect on either total abundance or on the abundance of the most important insect prey species during four years of investigation. Allan attributed his failure to detect a response to three causes.

First, high variance of sample means would have prevented detection of anything less than a two-fold change in abundance. Second, the fish may have been cropping only a small percentage of the standing crop of benthic and drifting animals. Third, the insect community may be so well adapted to fish predation that it is essentially unaffected by that predation. The low annual production that Allan (1982) estimated for trout in his experimental stream ($3.1g/m^2/yr$), and the relatively high standing crops of benthic and drifting fauna, suggest to me that the first two causes are probably a sufficient explanation of Allan's results. Insects are tremendously important in the food of stream-dwelling fishes, and there is good evidence for a relationship between food abundance and fish abundance in streams (Mason 1976). No doubt some significant consequences of fish predation to stream-dwelling insects will soon be demonstrated.

Not only is the general phenomenon of changes in abundance of benthos with changes in fish abundance reasonably well established, but changes in the composition of the benthos are evident as well. Ball and Hayne (1952), Macan (1965, 1966), Kajak et al. (1972), Henrikson and Oscarson (1978), Andersson et al. (1978), Stenson et al. (1978), and Stenson (1978, 1979) all provide information on compositional changes in the benthic insect community related to changes in the predatory fish population.

Ball and Hayne (1952) removed the fish (mainly bluegill, *Lepomis macrochirus*) from a 10-acre (~4-hectare) lake in Southeastern Michigan by poisoning with rotenone and observed the changes in benthic populations following the poisoning. The poisoning killed not only the fish but also *Chaoborus* and many of the dragonflies and leeches. The volume of benthic organisms per unit area increased significantly after fish removal. Insect components of the population either did not change in abundance or decreased in abundance. The overall increase in volume of benthic organisms was due to large increases in Amphipoda and Mollusca. Prior to the removal of fish the benthic community was dominated numerically by chironomid larvae, with Trichoptera and Amphipoda being about one-half as abundant, and with Mollusca, Anisoptera, Zygoptera, and Ephemeroptera being about one-quarter as abundant. The biomass was dominated by Anisoptera and Trichoptera, with Chironomidae and Mollusca being about one-half the trichopteran biomass. After removal of the fish, Chironomidae, Amphipoda, and Mollusca dominated the benthos numerically and were about equally abundant. Mollusca and Anisoptera dominated the biomass, with Chironomidae and Amphipoda being one-third to one-quarter the Anisopteran biomass.

Macan (1965, 1966) reported changes in the composition of various components of the insect community of Hodson's Tarn following introduction of brown trout (*Salmo trutta*). Nymphs of the dragonfly, *Lestes sponsa*, were common everywhere in the tarn before introduction of the fish. Afterward they were rare and restricted to small areas of the pond where thick vegetation protected them from predation. The beetle, *Rhantus exsoletus*, and other dytiscid larvae, which had been scarce but consistently encountered in samples before introduction of fish, completely disappeared. The same was true of nine casual species and three rare but consistently collected species of hemipterans. The common species of hemipterans (*Corixa castanea, C. scotti, Notonecta obliqua*) were all reduced in numbers and restricted to shallow nearshore areas with cover. Thus, the numbers of species encountered and their distribution were both reduced after the introduction of trout, with the rare and occasional species eliminated. Macan commented that it was also the most conspicuous species that were eliminated, as would be expected with a visual predator like trout. After the introduction of trout, the relative abundance and distribution of water bugs was like that commonly observed in other tarns with trout, and like the other tarns, the bugs were also rare in the stomachs of the fish. The investigation of tarns with trout, therefore, offered no clues as to the potential contribution of water bugs to the diet of trout, or to their rapid population expansion in the absence of predation.

Stenson (1978) eliminated the fish from a 1-hectare lake by poisoning with rotenone. Prior to poisoning, *Chaoborus flavicans* was the only chaoborid present in the lake. Following poisoning there was a marked increase in the numbers of all instars of *Chaoborus* and the appearance in abundance of a new species, *C. obscuripes,* normally found in fishless lakes. *C. obscuripes* differs from *C. flavicans* in having a larger eye and more pigmentation, and in remaining in the surface waters throughout the day. These factors probably make *C. obscuripes* highly vulnerable to predation by visually feeding fish and contribute to its inability to coexist with fish.

Henrikson and Oscarson (1978) investigated the occurrence and abundance of the hemipteran *Glaenocorisa propinqua* in three Swedish lakes. One lake, fishless due to acidification, was divided in two by a mesh curtain and perch were introduced to one half. Of two similar lakes containing fish, one was poisoned and the other was retained as a control. Comparison of the two halves of the divided lake revealed a dense population of *Glaenocorisa* in the fishless half and none in the half with fish, although a few were found in fish stomachs. Similarly *Glaenocorisa* was abundant in the poisoned lake and absent from the control. Thus the absence of *Glaenocorisa* was clearly a consequence of fish predation and could not be ascribed to any subtle differences in other biotic or abiotic features of the lakes.

Kajak and his co-workers (1972), in an ambitious program of research on Lake Warniak in Northern Poland, demonstrated the overwhelming influence of the fish population on the abundance, composition, and productivity of the lower trophic levels in the lake. A two-fold increase in the standing stock of fishes (mainly carp and tench) resulted in about a 50-percent reduction in the biomass of organisms living in the substrate and on aquatic vegetation. The reduction was almost certainly a consequence of fish predation as biomass in enclosures protected from predation did not decrease. Benthic fauna was dominated by chironomid larvae, the next most abundant groups being Ephemeroptera and Trichoptera; this composition did not change significantly with increasing predator density. However, the proportion of invertebrate predators in the benthic community was reduced by half at the higher fish density, and the individual chironomid larvae were smaller. As the density of fish increased, the proportion of Chironomidae and Gastropoda in their diet decreased whereas the proportion of Trichoptera, Ephemeroptera, Odonata, and Cladocera increased. Other changes in lake community dynamics associated with the increase in fish biomass included a reduction in primary production, an increase in bacterial decomposition rate, a small increase in the abundance and average size of crustacean zooplankton, and a shift in composition of the crustacean zooplankton community from dominance by large zooplankton to dominance by small zooplankton. Kajak et al. (1972) ascribed all these changes to the impact of foraging by the fish, both from the direct effects of selective feeding and from the indirect effects of stirring and sifting the bottom sediments during feeding, and from dislodging vegetation.

Similar effects of fish foraging on the community dynamics of lakes (Stenson et al. 1978) and enclosures (Andersson et al. 1978) have been observed in other studies. Stenson et al. (1978) poisoned the fish in Lake Lilla Stockelidsvatten, Sweden, and observed a dramatic reduction in primary production, an increase in biomass of net phytoplankton and dominance by large forms, an increase in biomass of zooplankton, a shift from small Cladocera to large Copepoda, lower pH, lower total phosphorous, and increased abundance of *Chaoborus* and *Glaenocorisa*. No comparable changes occurred in a nearby control lake with fish. Andersson et al. (1978) set up enclosures of plastic sheeting in the littoral zone of two eutrophic lakes in southern Sweden. One enclosure in each lake was densely stocked with fish and the other was fish-free. The biomass of phytoplankton was much higher and its composition was different in enclosures with fish compared with fishless enclosures. Zooplankton was denser in fishless enclosures and was dominated by large Cladocera, compared with dominance by Rotifera in enclosures with fish. Total phosphorous and pH were higher in enclosures with fish. Total benthic biomass was higher in enclosures without fish mainly due to increases in Oligochaeta and Chironomidae in one lake and Chaoboridae in the other.

These studies confirm some of the more obvious predictions from laboratory data on fish predation: that predation will change the composition of the insect community by reducing the relative abundance of larger and more conspicuous members. Less obvious consequences of predation in natural communities, but still consistent with known predatory behavior, are the elimination of rare and occasional forms and restriction of the distribution of some species to particular areas (Macan 1966; Stenson 1978, 1979). An important modulating influence of predatory fish is apparent, however, through their indirect effects on many aspects of aquatic community dynamics (Kajak et al.

1972; Andersson et al. 1978; Stenson et al. 1978). Species like carp that sift large quantities of sediment in their quest for food reduce the abundance and distribution of macrophytes through their feeding activity, and can affect primary productivity, detritus formation, and organic decomposition. Some of these indirect effects may stimulate benthic productivity and thereby offset the effects of predation (Lellak 1965; Kajak et al. 1972).

Much of the work that has been done in manipulating ponds and using enclosures is tantalizing rather than conclusive. Most of the studies were undertaken to describe consequences rather than to test specific hypotheses about predator-prey interactions. Unfortunately, sampling and analysis methods were sometimes inadequate and replication was usually insufficient. Nevertheless, it is apparent that these approaches offer exciting opportunities for directly testing hypotheses about the effects of fish predation on the aquatic insect community.

References

Alexander, R. McN. 1970. Mechanics of the feeding action of various teleost fishes. Journal of Zoology (London) 162:145–56.

Allan, J. D. 1982. The effects of reduction in trout density on the invertebrate community of a mountain stream. Ecology 63:1444–55.

Allen, K. R. 1951. The Horokiwi stream: a study of a trout population. New Zealand Marine Department Fisheries Bulletin 10:1–238.

Andersson, G., H. Berggren, G. Cronberg, and C. Gelin. 1978. Effects of planktivorous and benthivorous fish on organisms and water chemistry in eutrophic lakes. Hydrobiologia 59:9–15.

Bajkov, A. D. 1935. How to estimate the daily food consumption of fish under natural conditions. Transactions of the American Fisheries Society 65:288–89.

Ball, R. C. and D. W. Hayne. 1952. Effects of the removal of the fish population on the fish-food organisms of a lake. Ecology 33:41–48.

Banner, A. 1972. Use of sound in predation by young lemon sharks, *Negaprion brevirostris*. Bulletin of Marine Science 22:251–83.

Bastian, J. 1981. Electrolocation: 1. How the electroreceptors of *Ateronotus albifrons* code for moving objects and other electrical stimuli. Journal of Comparative Physiology 144:465–79.

Beamish, F. W. H. 1964. Respiration of fishes with special emphasis on standard oxygen consumption. II. Influence of weight and temperature on respiration of several species. Canadian Journal of Zoology 42:177–88.

Bennett, D. H. and J. W. Gibbons. 1972. Food of largemouth bass (*Micropterus salmoides*) from a South Carolina reservoir receiving heated effluent. Transactions of the American Fisheries Society 101:650–54.

Bennett, M. V. L. 1971. Electroreception, pp 493–574. *In*: W. S. Hoar and D. J. Randall (eds.). Fish physiology. Vol. 5: sensory systems and electric organs. Academic Press, Inc., New York, NY. 600 pp.

Beukema, J. J. 1968. Predation by the three-spined stickleback (*Gasterosteus acculeatus* L.): the influence of hunger and experience. Behaviour 31:1–126.

Borgstrom, G. 1978. The contribution of freshwater fish to human food, pp. 469–99. *In:* S. D. Gerking (ed.). Ecology of freshwater fish production. John Wiley and Sons, Inc., New York, NY. 520 pp.

Brett, J. R. 1965. The relation of size to rate of oxygen consumption and sustained swimming speed of sockeye salmon (*Oncorhynchus nerka*). Journal of the Fisheries Research Board of Canada 22:1491–501.

Brett, J. R. 1970. Temperature. Animals, fishes, pp. 513–60. *In:* O. Kinne (ed.). Marine ecology, Vol. 1, Environmental factors, Part 1. Wiley-Interscience, London. 682 pp.

Brett, J. R. 1971. Growth responses of young sockeye salmon (*Oncorhynchus nerka*) to different diets and planes of nutrition. Journal of the Fisheries Research Board of Canada 28:1635–43.

Brett, J. R., J. E. Shelbourn, and C. T. Shoop. 1969. Growth rate and body composition of fingerling sockeye salmon, *Oncorhynchus nerka*, in relation to temperature and ration size. Journal of the Fisheries Research Board of Canada 26:2363–94.

Burbidge, R. G. 1974. Distribution, growth, selective feeding, and energy transformations of young-of-the-year blueback herring, *Alosa aestivalis* (Mitchill), in the James River, Virginia. Transactions of the American Fisheries Society 103:297–311.

Carr, W. E. S., A. R. Gondeck, and R. L. Delanoy. 1976. Chemical stimulation of feeding behavior in the pinfish, *Lagodon rhomboides*: a new approach to an old problem. Comparative Biochemistry and Physiology 54A:161–66.

Chapman, D. W. and T. C. Bjornn. 1969. Distribution of salmonids in streams, with special reference to food and feeding, pp. 153–76. *In:* T. G. Northcote (ed.). Symposium on salmon and trout in streams. H. R. McMillan Lectures in Fisheries. University of British Columbia, Vancouver, Canada. 388 pp.

Colgan, P. 1973. Motivational analysis of fish feeding. Behaviour 45:38–66.

Confer, J. L. and P. I. Blades. 1975. Omnivorous zooplankton and planktivorous fish. Limnology and Oceanography 20:571–79.

Corbet, P. S. 1960. The food of non-cichlid fishes in the Lake Victoria basin, with remarks on their evolution and adaptation to lacustrine conditions. Proceedings of the Zoological Society of London 136:1–101.

Curio, E. 1976. The ethology of predation. Zoophysiology and ecology, Vol. 7. Julius Springer Verlag, Berlin, Germany. 250 pp.

De Groot, S. J. 1969. Digestive system and sensorial factors in relation to the feeding behavior of flatfish (*Pleuronectiformes*). Journal du Conseil permanent international pour l'Exploration de la Mer 32:385–95.

Diana, J. S. 1980. Diel activity pattern and swimming speeds of northern pike (*Esox lucius*) in Lac Ste. Anne, Alberta. Canadian Journal of Fisheries and Aquatic Sciences 37:1454–58.

Doble, B. D. and D. M. Eggers. 1978. Diel feeding chronology, rate of gastric evacuation, daily ration, and prey selectivity in Lake Washington juvenile sockeye salmon (*Oncorhynchus nerka*). Transactions of the American Fisheries Society 107:36–45.

Drenner, R. W., J. R. Strickler, and W. J. O'Brien. 1978. Capture probability: the role of zooplankter escape in the selective feeding of planktivorous fish. Journal of the Fisheries Research Board of Canada 35:1370–73.

Eggers, D. M. 1976. Theoretical effect of schooling by planktivorous fish predators on rate of prey consumption. Journal of the Fisheries Research Board of Canada 33: 1964–71.

Elliott, J. M. 1973. The food of brown and rainbow trout (*Salmo trutta* and *S. gairdneri*) in relation to the abundance of drifting invertebrates in a mountain stream. Oecologia 12:329–47.

Elliott, J. M. 1975. Number of meals in a day, maximum weight of food consumed in a day and maximum rate of feeding for brown trout, *Salmo trutta* L. Freshwater Biology 5:287–303.

Elliott, J. M. and L. Persson. 1978. The estimation of daily rates of food consumption for fish. Journal of Animal Ecology 47:977–91.

Fry, F. E. J. 1957. The aquatic respiration of fish, pp 1–63. *In*: M. E. Brown (ed.). Physiology of fishes, Vol. 1. Academic Press, Inc. New York, NY. 447 pp.

Fryer, G. 1959. The trophic interrelationships and ecology of some littoral communities of Lake Nyasa with especial reference to the fishes and a discussion of the evolution of a group of rock-frequenting Cichlidae. Proceedings of the Zoological Society of London 132:153–281.

Fryer, G. and T. D. Iles. 1972. The cichlid fishes of the great lakes of Africa. Their biology and evolution. Oliver and Boyd, Edinburgh, Scotland. 641 pp.

Gerking, S. D. 1962. Production and food utilization in a population of bluegill sunfish. Ecological Monographs 32:31–78.

Gerking, S. D. (ed.). 1978. Ecology of freshwater fish production. John Wiley and Sons, Inc., New York, NY. 520 pp.

Ginetz, R. M. and P. A. Larkin. 1973. Choice of colors of food items by rainbow trout (*Salmo gairdneri*). Journal of the Fisheries Research Board of Canada 30:229–34.

Greenwood, P. H. 1974. The cichlid fishes of Lake Victoria, east Africa: the biology and evolution of a species flock. Bulletin of the British Museum of Natural History (Zoology) Supplement 6:1–134.

Griffiths, D. 1975. Prey availability and the food of predators. Ecology 56:1209–14.

Hara, T. J. 1971. Chemoreception, pp. 79–120. *In*: W. S. Hoar and D. J. Randall (eds.). Fish physiology. Vol. 5. Sensory systems and electric organs. Academic Press, Inc., New York, NY. 600 pp.

Hartman, G. F. 1958. Mouth size and food size in young rainbow trout, *Salmo gairdneri*. Copeia 3:233–34.

Hartman, G. F. 1965. The role of behavior in the ecology and interaction of underyearling coho salmon (*Oncorhynchus kisutch*) and steelhead trout (*Salmo gairdneri*). Journal of the Fisheries Research Board of Canada 22:1035–81.

Hasler, A. D. 1957. The sense organs: olfactory and gustatory senses of fishes, pp. 187–209. *In*: M. E. Brown (ed.). Physiology of fishes. Vol. 2. Academic Press, Inc., New York, NY. 526 pp.

Healey, M. C. 1972. Bioenergetics of a sand goby (*Gobius minutus*) population. Journal of the Fisheries Research Board of Canada 29:187–94.

Henrikson, L. and H. G. Oscarson. 1978. Fish predation limiting abundance and distribution of *Glaenocorisa p. propinqua* (Hemiptera). Oikos 31:102–5.

Hergenrader, G. L. and A. D. Hasler. 1967. Seasonal changes in swimming rates of yellow perch in Lake Mendota as measured by sonar. Transactions of the American Fisheries Society 96:373–82.

Hoar, W. S. and D. J. Randall (eds.). 1971. Fish physiology. Vol. 5. Sensory systems and electric organs. Academic Press Inc., New York, NY. 600 pp.

Hoar, W. S., D. J. Randall, and J. R. Brett (eds.). 1979. Fish physiology. Vol. 8. Bioenergetics and growth. Academic Press Inc., New York, NY. 786 pp.

Hodgson, E. S. and R. F. Mathewson. 1971. Chemosensory orientation in sharks. Annals of the New York Academy of Sciences. 188:175–82.

Holling, C. S. 1965. The functional response of predators to prey density and its role in mimicry and population regulation. Memoirs of the Entomological Society of Canada 45:1–60.

Holling, C. S. 1966. The functional response of invertebrate predators to prey density. Memoirs of the Entomological Society of Canada. 48:1–86.

Holst, E. von 1948. Quantitative Untersuchungen über Umstimmungsvorgänge im Zentralnervensystem. 1. Der Einfluss des "Appetits" auf das Gleichgewichtsverhalten bei *Pterophyllum*. Zeitschrift für vergleichende Physiologie 31:134–48.

Hyatt, K. D. 1979. Feeding strategy, pp. 71–119. *In*: W. S. Hoar, D. J. Randall, and J. R. Brett (eds.). Fish physiology. Vol. 8. Bioenergetics and growth. Academic Press Inc., New York, NY. 786 pp.

Ingle, D. 1971. Vision: the experimental analysis of visual behavior, pp. 59–77. *In*: W. S. Hoar and D. J. Randall (eds.). Fish physiology. Vol. 5. Sensory systems and electric organs. Academic Press Inc., New York, NY. 600 pp.

Ivlev, V. S. 1961. Experimental ecology of the feeding of fishes. Yale University Press, New Haven, CT. 230 pp.

Jobling, M. 1981a. The influences of feeding on the metabolic rate of fishes: a short review. Journal of Fish Biology 18:385–400.

Jobling, M. 1981b. Mathematical models of gastric emptying and the estimation of daily rates of food consumption for fish. Journal of Fish Biology 19:245–58.

Kajak, Z. 1972. Analysis of the influence of fish on benthos by the method of enclosures, pp. 781–93. *In*: Z. Kajak and A. Hillbricht-Ilkowska (eds.). Productivity problems of freshwaters. International Biological Program P. F. Section. Polish Scientific Publishers, Warszawa, Poland. 918 pp.

Kajak, Z., K. Dusoge, A. Hillbricht-Ilkowska, E. Pieczynski, A Prejs, I. Spodniewska, and T. Weglenska. 1972. Influence of the artificially increased fish stock on the lake biocenosis. Internationale Vereinigung für Theoretische und Angewandte Limnologie Verhandlungen 18:228–35.

Kalmijn, A. J. 1971. The electric sense of sharks and rays. Journal of Experimental Biology 55:371–83.

Keast, A. and D. Webb. 1966. Mouth and body form relative to feeding ecology in the fish fauna of a small lake, Lake Opinicon, Ontario. Journal of the Fisheries Research Board of Canada 23:1845–74.

Kerr, S. R. 1971a. Analysis of laboratory experiments on growth efficiency of fishes. Journal of the Fisheries Research Board of Canada 28:801–8.

Kerr, S. R. 1971b. Prediction of fish growth efficiency in nature. Journal of the Fisheries Research Board of Canada 28:809–14.

Kerr, S. R. 1971c. A simulation model of lake trout growth. Journal of the Fisheries Research Board of Canada 28:815–19.

Kliewer, E. V. 1970. Gillraker variation and diet in lake whitefish *Coregonus clupeaformis* in northern Manitoba, pp. 147–65. *In*: C. C. Lindsey and C. S. Woods (eds.). Biology of coregonid fishes. University of Manitoba Press, Winnipeg, Canada. 560 pp.

Lauder, G. V., Jr. 1980. Evolution of the feeding mechanism in primitive actinopterygian fishes: a functional anatomical analysis of *Polypterus, Lepisosteus* and *Amia*. Journal of Morphology 163:283–317.

Lellak, J. 1965. The food supply as a factor regulating the population dynamics of bottom animals. Internationale Vereinigung für Theoretische und Angewandte Limnologie Mitteilungen 13:128–38.

Liem, K. F. 1978. Modulatory multiplicity in the functional repertoire of the feeding mechanism in cichlid fishes. 1. Piscivores. Journal of Morphology 158:323–60.

Liem, K. F. and J. W. M. Osse. 1975. Biological versatility, evolution, and food resource exploitation in African cichlid fishes. American Zoologist 15:427–54.

Lissmann, H. W. 1958. On the function and evolution of electric organs in fish. Journal of Experimental Biology 35:156–91.

Lissmann, H. W. and K. E. Machin. 1958. The mechanism of object location in *Gymnarchus niloticus* and similar fish. Journal of Experimental Biology 35:451–86.

Lissmann, H. W. and K. E. Machin. 1963. Electric receptors in a non-electric fish (*Clarias*). Nature (London) 199:88–89.

Lister, D. B. and H. S. Genoe. 1970. Stream habitat utilization by cohabiting underyearlings of chinook (*Oncorhynchus tshawytscha*) and coho (*O. kisutch*) salmon in the Big Qualicum River, British Columbia. Journal of the Fisheries Research Board of Canada 27:1215–24.

Macan, T. T. 1965. Predation as a factor in the ecology of water bugs. Journal of Animal Ecology 34:691–98.

Macan, T. T. 1966. Predation by *Salmo trutta* in a moorland fish pond. Internationale Vereinigung für Theoretische und Angewandte Limnologie Verhandlungen 16: 1081–87.

Macan, T. T. 1977. The influence of predation on the composition of fresh-water animal communities. Biological Reviews of the Cambridge Philosophical Society 52:45–70.

Maitland, P. S. 1977. The Hamlyn guide to freshwater fishes of Britain and Europe. Hamlyn Publishing Group Ltd., New York, NY. 256 pp.

Maly, E. J. 1970. The influence of predation on the adult sex ratios of two copepod species. Limnology and Oceanography 15:566–73.

Mann, K. H. 1978. Estimating the food consumption of fish in nature, pp. 250–73. *In*: S. D. Journal of Animal Ecology 34:253–75.

Mann, K. H. 1978. Estimating the food consumption of fish in nature, pp. 250–73. *In*: S. D. Gerking (ed.). Ecology of freshwater fish production. John Wiley and Sons Inc., New York, NY. 520 pp.

Markl, H. 1972. Aggression und Beuteverhalten bei Piranhas (Serrasalminae, Characidae). Zeitschrift für Tierpsychologie 30:190–216.

Mason, J. C. 1976. Response of underyearling coho salmon to supplemental feeding in a natural stream. Journal of Wildlife Management 40:775–88.

Miroschnichenko, M. P. 1979. The state and degree of utilization of bottom food resources by benthos-eating fish from Tsimlyansk reservoir. Journal of Ichthyology 19:70–78.

Mittelbach, G. G. 1981. Foraging efficiency and body size: a study of optimal diet and habitat use by bluegills. Ecology 62:1370–86.

Moore, R. H. and D. E. Wohlschlag. 1971. Seasonal variations in the metabolism of the

Atlantic midshipman, *Porichthys porosissimus* (Valenciennes). Journal of Experimental Marine Biology and Ecology 7:163–72.

Murdoch, W. W., S. Avery, and M. E. B. Smyth. 1975. Switching in predatory fish. Ecology 56:1094–105.

Nikolsky, G. V. 1963. The ecology of fishes. Academic Press, New York, NY. 352 pp.

Nilsson, N. A. 1967. Interactive segregation between fish species, pp. 295–313. *In:* S. D. Gerking (ed.). The biological basis of freshwater fish production. Blackwell Scientific Publications, Oxford, England. 495 pp.

Olmsted, J. M. D. 1918. Experiments on the nature of the sense of smell in the common catfish, *Ameiurus nebulosus* (LeSueur). American Journal of Physiology 46:443–55.

Paloheimo, J. E. and L. M. Dickie. 1965a. Food and growth of fishes. II. Effects of food and temperature on the relation between metabolism and body weight. Journal of the Fisheries Research Board of Canada 23:869–908.

Paloheimo, J. E. and L. M. Dickie. 1965b. Food and growth of fishes. III. Relations among food, body size, and growth efficiency. Journal of the Fisheries Research Board of Canada 23:1209–48.

Parker, G. H. 1910. Olfactory reactions in fishes. Journal of Experimental Zoology 8: 535–42.

Pope, G. F., J. C. H. Carter, and G. Power. 1973. The influence of fish on the distribution of *Chaoborus* spp. (Diptera) and density of larvae in the Matamek River system, Québec. Transactions of the American Fisheries Society 102:707–14.

Ricker, W. E. (ed.). 1967. Methods for the assessment of fish production in fresh waters. International Biological Program Handbook No. 3. Blackwell Scientific Publications, Oxford, England. 313 pp.

Ringler, N. H. 1979. Selective predation by drift-feeding brown trout (*Salmo trutta*). Journal of the Fisheries Research Board of Canada 36:392–403.

Schall, J. J. and E. R. Pianka. 1980. Evolution of escape behavior diversity. The American Naturalist 115:551–66.

Schutz, D. C. and T. G. Northcote. 1972. An experimental study of feeding behavior and interaction of coastal cutthroat trout (*Salmo clarki clarki*) and Dolly Varden (*Salvelinus malma*). Journal of the Fisheries Research Board of Canada 29:555–65.

Schwartz, E. and A. D. Hasler. 1966. Perception of surface waves by the blackstripe topminnow, *Fundulus notatus*. Journal of the Fisheries Research Board of Canada 23: 1331–52.

Scott, W. B. and E. J. Crossman. 1973. Freshwater fishes of Canada. Fisheries Research Board of Canada Bulletin 184:1–966.

Seghers, B. H. 1975. Role of gillrakers in size-selective predation by lake whitefish, *Coregonus clupeaformis* (Mitchell). Internationale Vereinigung für Theoretische und Angewandte Limnologie Verhandlungen 19:2401–05.

Seghers, B. H. 1978. Feeding behavior and terrestrial locomotion in the cyprinodontid fish, *Rivulus hartii* (Boulenger). Internationale Vereinigung für Theoretische und Angewandte Limnologie Verhandlungen 20:2055–59.

Sibert, J. and S. Obrebski. 1976. Frequency distributions of food item counts in individual fish stomachs, pp. 107–14. *In:* C. A. Simenstad and S. Lipovsky (eds.). Fish food habits studies. Washington Sea Grant. University of Washington, Seattle, WA. 193 pp.

Staples, D. J. 1975. Production biology of the upland bully, *Philypnodon breviceps* Stokell, in a small New Zealand lake. 1. Life history, food, feeding and activity rhythms. Journal of Fish Biology 7:1–24.

Stenson, J. A. E. 1978. Differential predation by fish on two species of *Chaoborus* (Diptera, Chaoboridae). Oikos 31:98–101.

Stenson, J. A. E. 1979. Predator-prey relations between fish and invertebrate prey in some forest lakes. Drottningholm Institute of Freshwater Research Report 58:166–83.

Stenson, J. A. E., T. Bohlin, L. Henrikson, B. I. Nilsson, H. G. Nyman, H. G. Oscarson, and P. Larsson. 1978. Effects of fish removal from a small lake. Internationale Vereinigung für Theoretische und Angewandte Limnologie Verhandlungen 20: 794–801.

Sterba, G. 1962. Freshwater fishes of the world. Studio Vista Books, London. 877 pp.

Straskraba, M. 1965. The effect of fish on the number of invertebrates in ponds and streams. Internationale Vereinigung für Theoretische und Angewandte Limnologie Mitteilungen 13:106–27.

Sutterlin, A. M. and N. Sutterlin. 1970. Taste responses in Atlantic salmon (*Salmo salar*) parr. Journal of the Fisheries Research Board of Canada 27:1927–42.

Thorp, J. H. and E. A. Bergey. 1981a. Field experiments on responses of a freshwater, benthic macroinvertebrate community to vertebrate predators. Ecology 62:365–75.

Thorp, J. H. and E. A. Bergey. 1981b. Field experiments on interactions between vertebrate predators and larval midges (Diptera: Chironomidae) in the littoral zone of a reservoir. Oecologia 50:285–90.

Thorpe, J. E. 1977. Daily ration of adult perch, *Perca fluviatilus* L. during summer in Loch Leven, Scotland. Journal of Fish Biology 11:55–68.

Vinyard, G. L. 1980. Differential prey vulnerability and predator selectivity: effects of evasive prey on bluegill (*Lepomis macrochirus*) and pumpkinseed (*L. gibbosus*) predation. Canadian Journal of Fisheries and Aquatic Sciences 37:2294–99.

Vinyard, G. L. and W. J. O'Brien. 1976. Effects of light and turbidity on the reactive distance of bluegill (*Lepomis macrochirus*). Journal of the Fisheries Research Board of Canada 33:2845–49.

Ware, D. M. 1971. Predation by rainbow trout (*Salmo gairdneri*): the effect of experience. Journal of the Fisheries Research Board of Canada 28:1847–52.

Ware, D. M. 1972. Predation by rainbow trout (*Salmo gairdneri*): the influence of hunger, prey density, and prey size. Journal of the Fisheries Research Board of Canada 29: 1193–201.

Ware, D. M. 1973. Risk of epibenthic prey to predation by rainbow trout (*Salmo gairdneri*). Journal of the Fisheries Research Board of Canada 30:787–97.

Ware, D. M. 1975. Growth, metabolism, and optimal swimming speed of a pelagic fish. Journal of the Fisheries Research Board of Canada 32:33–41.

Webb, P. W. 1978. Partitioning of energy into metabolism and growth, pp. 184–214. *In*: S. D. Gerking (ed.). Ecology of freshwater fish production. John Wiley and Sons, Inc., New York, NY. 520 pp.

Welch, E. B. and R. C. Ball. 1966. Food consumption and production of pond fish. Journal of Wildlife Management 30:527–36.

Werner, E. E. 1974. The fish size, prey size, handling time relation in several sunfishes and some implications. Journal of the Fisheries Research Board of Canada 31:1531–36.

Werner, E. E. 1977. Species packing and niche complementarity in three sunfishes. The American Naturalist 111:553-78.

Werner, E. E. and D. J. Hall. 1974. Optimal foraging and the size selection of prey by the bluegill sunfish (*Lepomis macrochirus*). Ecology 55:1042-52.

Werner, E. E. and D. J. Hall. 1979. Foraging efficiency and habitat switching in competing sunfishes. Ecology 60:256-64.

Werner, E. E., G. G. Mittelbach, and D. J. Hall. 1981. The role of foraging profitability and experience in habitat use by the bluegill sunfish. Ecology 62:116-25.

Winberg, G. G. 1956. Rate of metabolism and food requirements of fishes. Belorussian State University, Minsk, USSR. 251 pp. (Fisheries Research Board of Canada Translation Series No. 194).

Windell, J. T. 1978. Digestion and the daily ration of fishes, pp. 159-83. *In*: S. D. Gerking (ed.). Ecology of freshwater fish production. John Wiley and Sons, Inc., New York, NY. 520 pp.

Wohlschlag, D. E. and R. O. Juliano. 1959. Seasonal changes in bluegill metabolism. Limnology and Oceanography 4:195-209.

Woodhead, P. M. J. 1966. The behavior of fish in relation to light in the sea. Oceanography and Marine Biology. An Annual Review 4:337-403.

Zelinka, M. 1974. Die Eintagsfliegen (Ephemeroptera) in Forellenbächen der Beskiden. III. Der Einfluss des verschiedenen Fischbestandes. Vestnik Ceskoslovenske spolecnosti Zoologicke 38:76-80.

chapter 10

SECONDARY PRODUCTION OF AQUATIC INSECTS

Arthur C. Benke

Knowledge of the secondary production of aquatic insects is of considerable ecological importance from both a population and a community perspective. In terms of population dynamics, it combines in a single measurement two parameters (individual growth and population survivorship) that are generally considered to be of major ecological significance. In terms of communities, secondary production estimates are of critical importance when we attempt to quantify energy flow pathways in food webs. Furthermore, secondary production may often be the most appropriate response to consider in attempting to understand mechanisms of population or community regulation (e.g. see Hall et al. 1970; Hall 1971). There are applied aspects to secondary production information as well. Production of aquatic insects represents a large part of the food that is available to the fish species that are important for food and recreation. Indeed, aquatic insects typically comprise the major portion of what fisheries biologists sometimes demeaningly refer to as "fish food" (e.g. see Larkin 1979; see also Healey, Chapter 9).

Secondary production can be defined as the living organic matter, or biomass, that is created or produced by an animal population during an interval of time. It might be thought of as the ". . . net production of consumers above respiration and excretion" (Edmondson and Winberg 1971, p. xxi), although when referring to animals, the adjective "net" is superfluous (Winberg 1971). Secondary production is thus the flow rate of biomass produced, regardless of its fate (e.g. loss to predators, emergence), and its units are biomass or energy per unit area, per unit time. Production is distinct from standing-stock biomass, which is the amount of biomass or energy per unit area at a *given point in time*. However, the maximum standing-stock biomass is sometimes used as a minimum estimate of production for a growth interval. The relationship of production to mean biomass provides important insights into the production process and will be discussed in detail shortly.

Present-day methods of analyzing secondary production of aquatic invertebrates can be traced to Boysen-Jensen (1919), who studied production of marine benthic invertebrates in the Limfjord, Denmark. His general approach, one of summing production losses over a period of time, is now called the removal-summation method (Waters 1977). Much of the early production work was done by Russian scientists such as Borutski (1939), who studied the midge *Chironomus plumosus* in Lake Beloie. The highlights of this early Russian literature are covered nicely in Winberg (1971). The mathematical principles of what is now called the instantaneous growth method of estimating production were developed by Ricker (1946), Clarke et al. (1946), and Allen (1949), and the Allen curve method was presented shortly thereafter (Allen 1951).

In spite of these early methodological developments, relatively few estimates of aquatic insect production had been made until the last decade. For example, in Waters' (1977) review of the literature on aquatic secondary production, only four out of the 67 estimates he listed for aquatic insects were published prior to 1970. Much of the proliferation of production work in the early 1970s was due to the efforts of the International Biological Program (IBP), which was responsible for initiating numerous field studies throughout the world (see summary of IBP production literature in Morgan et al. 1980). Two excellent handbooks on the estimation of secondary production of aquatic invertebrates (Edmondson and Winberg 1971; Winberg 1971) were also published during this busy period of production studies. The recent upsurge in production studies also appears to be due in part to the development of a new method by Hynes (1961; Hynes and Coleman 1968) and Hamilton (1969), now called the size-frequency method (Hynes 1980; Waters and Hokenstrom 1980).

In addition to the two handbooks just mentioned, mathematical or graphical descriptions and comparisons of several methods can be found in many book chapters and theoretical articles (e.g. see Chapman 1971; Crisp 1971; Zaika 1973; Tanaka 1976, 1977; Gillespie and Benke 1979). Also, Waters (1977) provides an excellent summary of the basic methods in his comprehensive review of the freshwater production literature. Morgan et al. (1980) summarize much of the IBP secondary production studies, but, unfortunately, they only include literature through the early 1970s due to a long publication delay. Nonetheless, their thought-provoking synthesis of the IBP literature is reading that is highly recommended.

After little more than a decade of rapid growth in studies of aquatic insect secondary production, it seems appropriate to take stock of what we have learned and consider how this knowledge can help us guide our future efforts. Thus, this chapter has the following objectives:

1. To illustrate the theory and practice of estimating production of aquatic insects with different methods, emphasizing the importance of life history information
2. To evaluate the magnitude of secondary production and biomass turnover ($P/\bar{B}$ ratio) estimates in both single-species populations and multi-species communities, based upon direct estimates of production, theoretical possibilities, and indirect estimates of predator consumption
3. To explore some of the important ecological issues relating to secondary production such as sampling problems, factors that limit production, the influence of stress effects on production, and the role that secondary production analysis can play in elucidating the trophic dynamics of aquatic insects
4. To summarize the state of the art in aquatic secondary production studies and, by identifying gaps in our knowledge, to indicate some useful directions for future research.

METHODS OF MEASUREMENT

Various methods have been developed for estimating secondary production. Some use field data alone, whereas others use some combination of field and laboratory growth data. Emphasis in this chapter will be placed upon the basic field methods, since that is the approach most commonly used for aquatic insects. Most of the field methods are closely related to one another, and an understanding of key life-history features is essential to all field methods. Waters (1979) points out that voltinism and length of aquatic life are perhaps the two most critical life-history features necessary for obtaining good estimates of secondary production. The distinctiveness of individual cohorts, also termed developmental synchrony, is another important life-history feature that may dictate the most appropriate field method to use. The amount of sample replication and the temporal frequency of sampling in relation to length of aquatic life may also influence the investigator regarding choice of the most appropriate method.

Actual Cohort Methods

Many insect species develop in a reasonably synchronous pulse. In this life-cycle pattern, breeding adults are present over a short time span, and oviposition and hatching occur in sequence. Newly hatched aquatic larvae all commence growth at about the same rate, are all about the same size at any point in time, and upon reaching the final instar or pupa, they all emerge synchronously (Fig. 10.1). Such a sequence has been illustrated a number of times with both actual and hypothetical data (e.g. see Waters 1969; Winberg 1971; Waters and Crawford 1973; Cummins 1975).

Estimating secondary production of such a population is quite straightforward and several field methods have been proposed, most of which are closely related (see Gillespie and Benke 1979). These methods usually depend upon frequent quantitative sampling of the population throughout its life cycle. Typically, monthly samples are collected for species with a 1-year life span, with the idea being to describe the true growth and survivorship curves (Fig. 10.1), as accurately as possible from the sample data.

Four closely related methods have been widely used in estimating production from the kind of data that can be collected from an actual cohort: (1) Allen curve, (2) removal-summation, (3) increment-summation, and (4) instantaneous growth (Waters 1977). Rather than describing these methods separately, the relationships among them will be described graphically, and then with an example based on actual data.

The methods can be compared by reference to a plot of density against individual biomass (Fig. 10.2). Successive data points theoretically should represent the decline in numbers and the increase in individual biomass during a

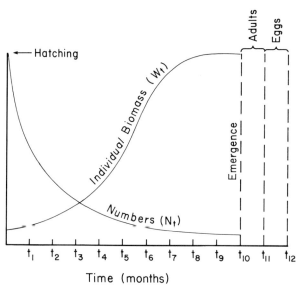

Figure 10.1 Individual growth and survivorship curves for a hypothetical aquatic insect population with synchronous cohort development.

sampling interval (Δt). For the moment, the reader should ignore reference in the figure to initial numbers (N_0, N_0') and initial biomass (W_0, W_0'), which relate to practical matters considered below. The standing stock biomass for any sampling date is equal to the area of the rectangle defined by that data point (or $N_t \times W_t$). Production for the entire cohort is equal to the total area under the curve, and the four methods mentioned above are simply different ways of calculating this area. Note that if there is no mortality, production is equal to $N_0 \times W_f$. Also note that the standing stock biomass of the final instars just prior to emergence is $N_f \times W_f$. This represents the portion of production that reached the final instar and potentially could emerge.

One way to approach calculating production of the entire cohort is to fit a smoothed Allen (1951) curve to the data of Figure 10.2 and calculate the area under the curve. Either an exponential curve can be fitted to the data as drawn in the figure, or individual data points can be connected with a smoothed curve (e.g. see Waters and Crawford 1973). Area can be determined with a planimeter or by some manual method such as counting the number of squares under a curve plotted on graph paper. Alternatively, production can be estimated algebraically using the exponential equation of Neess and Dugdale (1959).

Rather than drawing a continuous curve across all samples, successive N and W values can be used to calculate production (or loss) between sample dates. If time intervals are reasonably short, linear approximations are usually sufficient and easier to calculate. For the sake of brevity, only linear approximations will be presented here, but Gillespie and Benke (1979) discuss comparable forms of linear and exponential models.

defined by $Y + Z$ can be calculated as $\overline{N}\Delta W$. This is equal to production during Δt, and total cohort production can be determined by adding all these vertical trapezoids until the area under the entire curve is approximated. This approach has been referred to as the increment-summation method (Waters 1977).

Production lost during a sample interval, such as mortality from predation, is roughly equal to the trapezoid defined by area $X + Y$, which is calculated as $\overline{W}\Delta N$. Total cohort production can be determined by adding all these horizontal trapezoids until the area under the curve is approximated. This is the removal-summation method.

It can be verified that ΔB, the increase or decrease in standing stock biomass during Δt, is equal to the area $Z - X$. Thus, the actual production $(Y + Z)$ during Δt is equal to the production lost $(X + Y)$ plus ΔB $(= Z - X)$. Clearly, the increment-summation and removal-summation methods are just different ways of adding up areas under the curve and should always give exactly the same results.

The instantaneous growth method assumes an exponential increase in individual biomass. Production during the sampling interval is calculated as

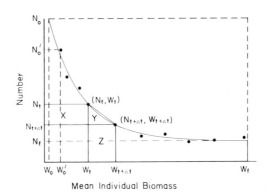

Figure 10.2 A hypothetical plot of number vs. mean individual biomass (i.e. an Allen curve). There are different ways of fitting a curve to these data points (see text), and the area under the curve equals cohort production.

$\overline{N} = (N_t + N_{t+\Delta t})/2 = $ mean number during Δt.

$\overline{W} = (W_{t+\Delta t} + W_t)/2 = $ mean individual biomass during Δt.

$N_o = $ total number of animals hatching.

$W_o = $ initial individual biomass of hatchlings.

$N_f = $ number of animals reaching largest size class.

$W_f = $ individual biomass of largest size class.

$B_o = N_o W_o = $ total biomass of all hatchlings.

(part of production from previous generation)

$N_o W_f = $ potential production of cohort (with no mortality)

$N_f W_f = $ potential emergent production (if all final instars emerge)

$\Delta N = N_t - N_{t+\Delta t} = $ decline in numbers during Δt.

$\Delta W = W_{t+\Delta t} - W_t = $ increase in individual biomass during Δt.

$P = g\Delta t\overline{B}$, where $g\Delta t$ (or G, Waters 1977) is calculated as $\ln(W_{t + \Delta t}/W_t)$ and $\overline{B}$ is usually calculated as $(B_t + B_{t + \Delta t})/2$. Thus, like the increment-summation method, the instantaneous growth method estimates production during a sampling interval. However, the two methods only provide exactly equal results if continuous exponential models are used for both (see Gillespie and Benke 1979). As before, total cohort production is the summation of production between all sampling intervals.

Before illustrating how to measure secondary production with an example, it is important to recognize an inherent problem associated with all the cohort methods discussed above. The problem is mentioned here since these actual cohort methods are often viewed as the ideal methods against which others might be compared (e.g. see Waters and Crawford 1973; Lapchin and Neveu 1980). For the cohort methods, we assume that sample data are collected from a population whose members are growing in perfect synchrony. That is, all individuals hatch at an instant in time, and grow at exactly the same rate. We assume that we are following the true survivorship curve and the true growth curve. In reality, even the population that appears most synchronous is not perfectly synchronous, and we are always following *apparent* rather than *true* survivorship and growth curves. A computer simulation of the growth and mortality of a univoltine population is illustrated in Figure 10.3. Hatching of first instars occurs as the positive half of a sinusoidal function over a 56-day period, and the relative numbers of each instar can be followed through time. Note that some animals are dying and some are growing into the next size class before all animals have hatched. Even if we census the entire population at daily intervals, the total number of hatchlings is never present on a given date. As one moves through time, and the mean individual weight and density are calculated, the values do not really equal the actual number of individuals that grew to that size. Thus, the solid upper curve is an apparent survivorship curve, and the dashed line (not observed, but determined from the simulation) is the true survivorship. The accuracy of any production calculation using actual cohort methods depends upon how closely the apparent survivorship approximates actual survivorship. Although these methods should usually provide good estimates for reasonably synchronous cohorts, they should not be considered standards against which other methods are judged, except in an approximate sense.

The data of Waters and Crawford (1973) for the stream mayfly *Ephemerella subvaria* is very useful to illustrate the practice of these methods. Waters and Crawford collected 20 quantitative samples from a small Minnesota stream on each of 11 dates throughout the year. Larvae were actually recovered on only 8 of the 11 dates since the population had all emerged by late June and hatchlings were not found until late August. The life history of this mayfly was fairly typical of species in which we follow cohort growth, in that perfect synchrony did not exist. Although a pulse was clearly discernible as the population was followed through time, it was common to have at least five out of ten size classes present on a

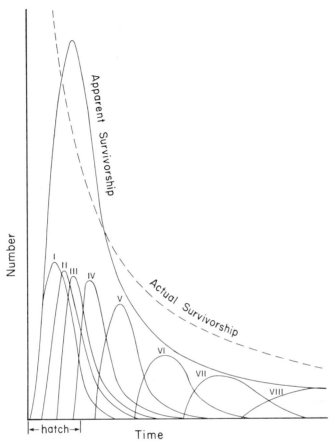

Figure 10.3 Computer simulations of the density of a hypothetical insect population with eight instars, in which hatching occurs over a 56-day period. The rise and fall in abundance of each instar is shown. The cumulative curve of numbers (unnumbered *solid curve*) is the apparent survivorship, but the *dashed curve* is the actual survivorship (i.e. the actual number of animals that lived until a given age). (Based upon simulations by A. C. Benke and J. B. Waide, unpublished data.)

sampling date (Fig. 5 of Waters and Crawford 1973). Nonetheless, Waters and Crawford were able to apply the removal-summation, instantaneous growth, Allen curve, and Hynes (size-frequency method; see following) methods in calculating production, and they achieved comparable results with all methods.

The basic data of Waters and Crawford are summarized in Tables 10.1 and 10.2 to show the relationships among the removal-summation, increment-summation, and instantaneous growth methods. The calculations are not exactly

TABLE 10.1: Production Statistics for *Ephemerella subvaria* Based upon Data of Waters and Crawford (1973)*

DATE	N (No./m^2)	$\bar{N}$	ΔN	W (mg)	$\bar{W}$	ΔW	B ($\mu g/m^2$)	$\bar{B}$	ΔB
26 Sept	6350			0.23					
		5391	1918		0.64	0.82	1.46	3.06	3.19
22 Oct	4432			1.05					
		4257	350		1.35	0.59	4.55	5.67	2.04
27 Nov	4082			1.64					
		4068	29		2.16	1.03	6.69	8.76	4.13
27 Dec	4053			2.67					
		3859	393		3.12	0.89	10.82	11.93	2.21
30 Jan	3660			3.56					
		2623.5	2073		4.97	2.82	13.C3	11.58	−2.90
12 Mar	1587			6.38					
		908.5	1357		9.0	5.23	10.13	6.40	−7.46
17 Apr	230			11.61					
		137	186		16.6	9.98	2.67	1.81	−1.72
06 June	44			21.59					
		22	44		25.3	7.41	0.95	0.48	− .95
01 July	0			29.0			0		

*All terms are defined in Figure 10.2 or in the text. All biomass values are wet weight.

TABLE 10.2: **Calculation of *Emphemerella subvaria* Production by Three Methods Using Production Statistics of Table 10.1***

DATE	REMOVAL-SUMMATION			INCREMENT-SUMMATION	INSTANTANEOUS GROWTH		
	$\overline{W}\Delta N$	ΔB	$\Delta B + \overline{W}\Delta N$	$\overline{N}\Delta W$	$G (=g\Delta t)$	$\overline{B}$	$g\Delta_t B$
26 Sept							
	1.23	3.19	4.42	4.42	1.52	3.06	4.65
22 Oct							
	0.47	2.04	2.51	2.51	0.44	5.67	2.49
27 Nov							
	0.06	4.13	4.19	4.19	0.49	8.76	4.29
27 Dec							
	1.22	2.21	3.43	3.43	0.29	11.93	3.46
30 Jan							
	10.30	−2.90	7.40	7.40	0.58	11.58	6.72
12 Mar							
	12.21	−7.46	4.75	4.75	0.60	6.40	3.84
17 Apr							
	3.09	−1.72	1.37	1.37	0.62	1.81	1.12
06 June							
	1.11	−0.95	0.16	0.16	0.30	0.48	0.14
01 July							
Σ	29.69	−1.46	28.23	28.23			26.71

*All units are g wet wt/m^2.

equivalent to those of Waters and Crawford because they used a slightly different method of calculating $\overline{W}$, and the data presented here are sometimes carried to an extra significant figure.

We can see from Table 10.2 that the removal-summation calculation ($\Sigma\overline{W}\Delta N = 29.69$ g wet wt/m^2) is larger than the increment-summation result ($\Sigma\overline{N}\Delta W = 28.23$). This is because the former includes the initial standing stock biomass ($B_0' = 1.46$) as part of the cohort's production, and the latter does not. If 1.46 is added to the $\overline{N}\Delta W$ column (or to the $\Delta B + \overline{W}\Delta N$ column), the result is exactly equal to the removal-summation calculation. Whether or not to include this initial biomass in the production of this new cohort is a debatable point since one can argue that the initial biomass of hatchlings ($N_0 \times W_0 = B_0$, Fig. 10.2) is production by the previous generation. However, by the time a population is actually sampled, a significant amount of growth has already occurred within the new cohort, and the initial standing stock biomass is $N_0' \times W_0' = B_0'$ (Fig. 10.2). In view of the fact that some unaccounted mortality also has occurred (N_0 to N_0'), it seems simplest and reasonably justifiable to include B_0' as part of the new production. Thus, the best estimate of annual production is probably 29.69 g/m^2.

The instantaneous growth calculation is also shown in Table 10.2, and the individual values between time intervals can be compared to the increment-summation column of production. As expected, these short-interval calculations are close, but not equal. By the same logic as above, the initial standing stock biomass ($B'_0 = 1.46$) should be added to the annual production column value ($\Sigma g \Delta t \bar{B} = 26.71$) to give 28.17 g/m^2.

Waters and Crawford's Allen curve was a smoothed curve connecting points and production was estimated to be 26.4 g/m^2. Again, if we add 1.46, we obtain 27.86. In summary, all four methods are very closely related and will usually provide very similar results, provided that the initial standing stock biomass is handled correctly.

The Concept of the $P/\bar{B}$ Ratio

Before discussing the size-frequency method, it will be useful to consider the relationship between production (P) and mean standing stock biomass ($\bar{B}$). The ratio of production to biomass, or $P/\bar{B}$ ratio, provides some particularly useful information regarding the production dynamics of animals.

We need to distinguish between the *cohort* $P/\bar{B}$ and the *annual* $P/\bar{B}$, since these provide very different kinds of information. The cohort $P/\bar{B}$ is the production of a real or hypothetical cohort divided by the mean biomass for the length of time it takes the cohort to complete development. This development time of the aquatic stage is called the cohort production interval or CPI by Benke (1979). For *Ephemerella subvaria*, the mean biomass for this interval of time is the sum of the B column in Table 10.1 (50.4) divided by the number of dates (nine) and equals 5.60 g/m^2. Cohort $P/\bar{B}$ for *E. subvaria* is calculated by taking any of the estimates for P above, and dividing by 5.6. The answer is very close to 5 in each case, and other field studies indicate that cohort $P/\bar{B}$ is often about 5 (Waters 1979). The cohort $P/\bar{B}$ is more related to the shapes of the growth and survivorship curves than it is to individual growth *per se*, and is independent of the actual amount of time it takes the cohort to grow from hatching to emergence. Using hypothetical populations, Waters (1969) varied the shapes of these curves, as well as the minimum and maximum values of individual size and population size, and he found that the cohort $P/\bar{B}$ can vary between about 2 and 8 under a variety of conditions.

The annual $P/\bar{B}$ is annual production divided by the mean biomass over the entire year. The annual $P/\bar{B}$ (or, for that matter, the weekly $P/\bar{B}$, daily $P/\bar{B}$, etc.) *is* directly related to individual growth rate, and is roughly equivalent to biomass turnover rate, or turnover ratio (Waters 1969). For *E. subvaria,* annual $\bar{B}$ is calculated by dividing 50.4 by the total number of sampling dates in the year (11) and equals 4.58. Thus, annual $P/\bar{B} = 29.69/4.58 = 6.48$ for the removal-summation method. The units of annual $P/\bar{B}$ are inverse time (yr^{-1}) and the reciprocal is turnover time; the latter is the amount of time it takes to replace the biomass of the population. Unlike the cohort $P/\overline{B}$, the annual $P/\overline{B}$ is subject to tremendous variation among insect populations—depending upon the species and its particular environmental conditions. The cohort $P/\overline{B}$ and the annual $P/\overline{B}$ are directly related by the ratio $365/CPI$ (see following).

Size-frequency Method

Unfortunately, for the aquatic ecologist interested in secondary production, most species do not have a neat synchronous development to which an actual cohort method can be applied, and many biologists are either ill-equipped for, uninterested in, or skeptical of laboratory approaches. Hynes (1961) recognized this dilemma and proposed what he considered to be an abbreviated approach, which he later realized needed modification (Hynes and Coleman 1968). Although his logic proved slightly flawed, Hynes' critical insight provided the stimulus for the correct application by Hamilton (1969) of what is now called the size-frequency method (Hynes 1980; Waters and Hokenstrom 1980). Hynes' original purpose was to provide order-of-magnitude approximations of production for an entire community; however, Hamilton's (1969) version of the method, when used with the CPI correction proposed by Benke (1979), should usually provide very good estimates when applied to individual species, or to groups of coexisting species with similar growth characteristics.

The basis of the size-frequency method is that an average size-frequency distribution calculated from samples taken over a period of a year will approximate the average survivorship of a hypothetical average cohort (Hamilton 1969; Benke and Waide 1977). The rationale for this is similar to that used in constructing a time-specific life table (Southwood 1978) in which survivorship of a population with overlapping generations is approximated from its age distribution at a single point in time. It is important to realize that the *average* cohort does not represent the actual number of individuals reaching each size class as if we were following an *actual* cohort (Table 10.1, second column). However, from this average cohort, if certain assumptions are made, it is possible to approximate the number of individuals that reach each size class over a period of a year. Suppose, for example, we are dealing with a univoltine insect that grows through N arbitrary size classes over a full year. If we assume linear growth, the insect will remain in each of N size classes for only $1/N$ part of the year. Since each size class lasts $1/N$ part of the year, the mean number of individuals in each size class of the average cohort must be multiplied by N to estimate the number of individuals that grew to that size class during the course of the year. This concept is somewhat difficult to visualize and the reader may wish to study Hamilton's (1969) detailed examples, or consider the alternative explanation by Menzie (1980). The assumption of linear growth is probably never true, and Hamilton (1969) showed how to correct for this (his n_g column). However, laboratory data on the relative amount of time spent in each size group are usually required.

Hamilton (1969) and Benke and Waide (1977) suggested that failure to adjust the average cohort for such nonlinear growth will not usually lead to large errors.

The production calculations for the size-frequency method is analogous to the removal-summation procedure except that $\overline{W}\Delta N$ values are based upon changes between size groups rather than between sampling dates. The calculations are illustrated for *Ephemerella subvaria* (Table 10.3). Using data from all 11 sampling dates, annual mean numerical density is calculated for each of the ten size groups. Notice that the $\overline{W}\Delta N$ values are multiplied by the number of size classes (ten) in the last column of the table. One of the problems with not adjusting for different times spent within size classes is that densities in the smaller size groups may be lower than in subsequent size classes (see first two size groups in Table 10.3). Obviously, these first size groups have been underrepresented, and Benke and Wallace (1980) suggest that the apparent negative production should not be included in the production summation (see last two columns of Table 10.3). The cohort $P/\overline{B}$ ratio can be determined directly from Table 10.3 by dividing the sum of the last column by the sum of the B column. In this case, the cohort $P/\overline{B} = 7.5$, which is somewhat larger than that obtained from actual cohort methods.

The size-frequency method as presented above requires the assumption that larval development takes a full year. Hamilton (1969) recognized that the cohort production value as calculated in Table 10.3 must be corrected if the population is not univoltine, and he suggested multiplying by the number of generations per year. However, it is the mean length of the aquatic stage (i.e. the CPI) relative to a full year, rather than voltinism *per se*, that is the critical factor in the correction. Benke (1979) showed that the size-frequency calculation must be multiplied by 365/CPI where 365 is the number of days per year and the CPI is given in days. This correction follows from the discussion above that an insect will remain in each size class for a period of time equal to $CPI/N = T$. Thus, the average number of individuals in each size class of the average cohort must be multiplied by 365/T to estimate the number of individuals that grew to each size class during the year. Obviously, other time units like weeks or months can be used in the correction. The CPI correction should always be used unless the CPI is very close to 1 year. Therefore, Benke et al. (in press) refer to the size-frequency calculation illustrated in Table 10.3 (sum of final column) as the uncorrected P.

When a population is univoltine, such as *E. subvaria* in our example, the CPI correction does not usually result in a large change. Since the CPI for *E. subvaria* is about 10 months, the uncorrected P should be multiplied by 12/10 or 1.2. Thus, annual production is actually $34.6 \times 1.2 = 41.5 \, g/m^2$. The cohort $P/\overline{B}$ also can be multiplied by the CPI correction to obtain annual $P/\overline{B}$ ($7.5 \times 1.2 = 9.0$). The difference between this corrected size-frequency production estimate and estimates from actual cohorts (which are about $30 \, g/m^2$) is larger than that originally determined by Waters and Crawford (1973), but because either approach involves several assumptions, it is currently impossible to say which is the more accurate.

TABLE 10.3: Calculation of *Ephemerella subvaria* Production by the Size-frequency Method Based upon Data of Waters and Crawford (1973)*

SIZE GROUP LENGTH (mm)	N (no./m²)	W (mg)	B (g/m²)	ΔN	WEIGHT AT LOSS $\bar{W}$	WEIGHT LOSS $\bar{W}\Delta N$	× 10 (g/m²)
0-1	217	0.06	0.01	-324	0.15	-0.05	(-0.5)
1-2	541	0.24	0.13	-147	0.62	-0.09	(-0.9)
2-3	688	1.0	0.69	216	1.95	0.42	4.2
3-4	472	2.9	1.37	286	4.25	1.22	12.2
4-5	186	5.6	1.04	108	7.35	0.79	7.9
5-6	78	9.1	0.71	61	11.10	0.68	6.8
6-7	17	13.1	0.22	-1	15.35	-0.02	-0.2
7-8	18	17.6	0.32	14	20.05	0.28	2.8
8-9	4	22.5	0.09	3	25.75	0.08	0.8
9-10	1	29.0	0.03	1	29.0	0.03	0.3
			$\bar{B} = 4.61$				34.6

Total production = 34.6×1.2 (CPI correction) = 41.5 g/m²

*Annual production is based on 11 sets of samples, including those with no nymphs. Values of last column in parentheses are not included in summation (see text). All biomass values are wet weight.

The CPI correction will be much more important for populations that are not univoltine than for a population like *Ephemerella subvaria*. If larval development time of a population is much less than a year, such as for many dipterans, the CPI correction will result in the annual production estimates being much higher than the initial uncorrected P, and the annual $P/\overline{B}$ will be much higher than the cohort $P/\overline{B}$. For example, Menzie (1981) used the size-frequency method in calculating the production of the midge *Cricotopus sylvestris* in a shallow cove of the Hudson River estuary. He estimated CPI in the two following ways: (1) from field data following peaks of abundance (CPI = 42 days for a 214-day growing season); and (2) from laboratory development times at different temperatures. Menzie's application of the CPI correction using laboratory data is particularly instructive because he partitioned the growing season and applied different temperature-dependent CPI values for different months (CPI = 28 days at 15°C, CPI − 10 days at $\geq 22^{\circ}$C). Annual production of *C. sylvestris* using the conservative field-derived CPI was 5.8 g/m^2 (dry wt) with an annual $P/\overline{B}$ of 21. Production using the laboratory-derived CPIs was 11.6 g/m^2 with an annual $P/\overline{B}$ of 42. Menzie felt that the true value probably fell somewhere in the middle. This lack of precision in estimating production of fast-growing species such as *C. sylvestris* may be unsettling to many readers, but it is a state-of-the-art approach and it illustrates the dilemma faced by the production biologist. However, consider the fact that failure to use any CPI correction would result in an underestimate by an order of magnitude! Waters (1979) has further discussed the topic of magnitude of errors.

If the development time of a population is much more than a year, the CPI correction will substantially decrease the uncorrected P value and the annual $P/\overline{B}$ will be less than the cohort $P/\overline{B}$, regardless of whether there are coexisting cohorts. Cushman et al. (1975) found that the stonefly *Alloperla mediana* had a CPI of 17.5 months. Thus, they had to multiply their uncorrected P estimate by $12/17.5 = 0.68$. If a species has a four- or five-year life cycle, such as the dragonfly *Cordulia aenea amurensis* (Ubukata 1980), the annual $P/\overline{B}$ should be about 1 if the cohort $P/\overline{B} = 5$. Life cycles, and thus CPIs, that are longer than a year seem most common in the orders Odonata (Corbet 1980), Plecoptera (Hynes 1976), and probably the Megaloptera (Evans 1978). If the environment is extremely cold or stressed in some way, species from other orders also may have long life cycles and low annual $P/\overline{B}$ values. For example, Butler (1982) found that a two-species population of *Chironomus* had a seven-year life cycle and an annual $P/\overline{B}<1$ in an arctic tundra pond.

There are a variety of life-history patterns that might cause one to question how to apply the CPI correction; three will be mentioned here. First, suppose that a univoltine species takes only six months to complete larval development, with adult and egg stages taking the other six. The only interval of importance as far as production is concerned is that CPI = 6 months. One can sample throughout the year, and include zero values for six months in constructing the size-frequency table. In this case, the uncorrected P must be multiplied by $12/6 = 2$. Alternatively, one can include only those samples that contain animals in constructing the table (for 6 months), but this estimate should not be multiplied by 2 to obtain annual production. Regardless of which way one makes the calculation, if the cohort $P/\bar{B}$ is 5, then the annual $P/\bar{B}$ is 10. Thus, one can use the size-frequency method for short growth intervals without a CPI correction, but the CPI correction must always be taken into account in calculating annual $P/\bar{B}$.

The second illustration is for a species whose life cycle lasts three years. Visualize the situation where there is only a single cohort and that this is sampled throughout the three-year period to construct the size-frequency table. The average annual production will only be one-third the production of the actual cohort. Imagine an alternative situation where there are three cohorts of different ages present simultaneously, and samples are collected throughout only one year in constructing a size-frequency table. Now the average annual production will be approximately equal to the production of a single cohort if all three cohorts have similar growth and survivorship statistics. How does one make a CPI correction in either situation? The answer is the same for both, and is consistent with the guidelines just given. Multiply the uncorrected P value by $12/36$ or $1/3$. If the cohort $P/\bar{B} = 5$, the annual $P/\bar{B}$ will be $5/3$ in both situations.

Third, what do you do about insects such as Coleoptera and Hemiptera that are present in the aquatic habitat as larvae and adults? Since the adults are nongrowing, they should simply be excluded from the production calculations, and the CPI should be defined as the time from hatching to the maximum size of the final size class (Hamilton 1969; Benke and Waide 1977). On the other hand, the adults should probably be included in calculation of the annual $P/\bar{B}$ ratio. This adds what may be considerable biomass, but no production, and would thus lower the final annual $P/\bar{B}$ over that calculated for the larvae alone.

Laboratory Methods

It should be obvious from the preceding section that field studies alone sometimes will be inadequate to estimate production of some aquatic species, and that laboratory growth studies may be necessary. One laboratory/field approach involves determination of size-specific growth rates at different temperatures in the laboratory, and determination of the population size structure from field data (e.g. see Cooper 1965; Mackey 1977a). While this approach can be a very good one, it has not been used as much as strictly field approaches, probably due to the detailed laboratory studies required. The use of such an approach for an entire benthic community would require enormous resources. Also, there is always some question as to how accurately laboratory measurements represent what is actually happening in nature. Growth in the laboratory may be too fast due to unnaturally favorable conditions, or too slow if laboratory conditions are not favorable. Thus, attempts should be made to simulate conditions, especially temperature, as close to those of the natural environment as possible.

Laboratory growth studies sometimes may be required to obtain an estimate of the CPI for use in the size-frequency method. However, the detailed size-specific growth rates as determined by Mackey (1977a) and others are not essential. Only the length of time from hatching to pupation is required, and this greatly reduces the amount of laboratory work involved. Growth chambers suspended in the natural environment or detailed observations of a subpopulation, such as on a rock surface, would likely provide an even better estimate of CPI for a population whose growth could not be easily followed from regular sampling. There are other, more conservative, means of estimating CPI from field data, including an analysis of the temporal patterns of larval size distributions and documentation of the first appearance of early instars, pupae, or adults (e.g. Neves 1979; Menzie 1981).

Some investigators have estimated metabolic rates as well as growth rates in the laboratory, and the use of such information in calculating production has come to be known as the physiological approach (Edmondson and Winberg 1971; Winberg 1971; Zaika 1973; Waters 1977). This approach has been used a great deal in zooplankton studies, but has not yet been as popular for estimating aquatic insect production.

Abbreviated Approaches

Several simplified methods have been proposed that enable secondary production estimates to be calculated more easily, but they require some major assumptions. Waters (1977) provides a more detailed summary of some of these methods which include the following:

1. Production is estimated from mean standing-stock biomass. Waters (1969, 1977, 1979) has suggested that since cohort $P/\overline{B}$ ratios are fairly constant (about 5), one can assume an annual $P/\overline{B}$ of 5 for univoltine species. In the absence of population data, this seems a reasonable assumption. Waters (1979) points out that a CPI correction would be necessary for species which are not univoltine.

2. Production is estimated from emergence data. Emergence of adults often represents a significant portion of aquatic insect production, and quantifying emergence rates can provide valuable information. In aquatic ecosystems in which benthic sampling is difficult, emergence sampling may be the only practical alternative (e.g. see Rosenberg et al. 1980). Speir and Anderson (1974) showed that emergence is a fairly constant proportion of black fly production and suggested that emergence data alone might provide an indirect approximation of production. A large effort in Germany has provided much detailed information on emergence productivity (e.g. see Castro 1975; Illies 1975). While emergence data are undoubtedly useful in making general comparisons among populations and systems, the validity of assuming a constant ratio of emergence to total production seems questionable. Furthermore, the practical aspects of

quantifying emergence in large heterogeneous aquatic systems could be quite formidable.

3. Production is estimated from the maximum standing-stock biomass. Several French researchers (e.g. see Lavandier 1975; Laville 1975) have reported that the ratio of production to maximum biomass is often close to 1.5. While this observation could have some utility in approximating production from biomass, it would probably be necessary to account for variation in the CPI.

4. P/B̄ ratio is estimated as $T^2/10$, where $T = °C$ (Johnson and Brinkhurst 1971). Although apparently of value for some aquatic animals, this method appears limited in its range of applications since different insect groups can require widely differing numbers of degree days to complete development.

5. Production is estimated as a function of maximum individual size by regression analysis (Banse and Mosher 1980). While the regression on which this method is based is probably accurate in general, the relationship should be largely due to a correlation between maximum size and CPI. Since many aquatic insect species can vary greatly in their CPI under differing environmental conditions, especially temperature (e.g. see Menzie 1981; Sweeney, Chapter 4), the use of such a regression would probably not provide the level of resolution usually desired for production estimates.

MAGNITUDE OF PRODUCTION AND TURNOVER

High Estimates

The major goals that can be achieved from studying the production of aquatic insects include an increased understanding about how aquatic insects fit into total energy budgets, the roles they play in complex food webs, and the factors that limit production dynamics for both individual populations and the benthic community as a whole. This brings us to the question of magnitude. How high are high production estimates for aquatic insects?

The magnitude of production depends upon only two things: (1) the standing-stock biomass; and (2) the rate of biomass turnover, or annual P/B̄ ratio. Since the cohort P/B̄ is relatively constant, the length of aquatic life, or CPI, is the primary determinant of annual P/B̄. Production can be high because of high biomass alone, high turnover (short CPI) alone, or some combination of the two.

Although the number of production studies in lotic systems have increased in recent years, probably more has been done in lentic systems. This is possibly due to the greater ease of sampling lentic systems and, also, to the greater emphasis that has been placed on lentic habitats in the IBP programs (Morgan et al. 1980). Since Chironomidae are a predominant component of most lake benthos, more studies have been done on them than on any other group. Comparisons of chironomid production values have been made by many authors (e.g. Potter and Learner 1974; Waters 1977; Morgan et al. 1980).

Apparently, the highest benthic annual production estimate in lentic systems is for a species of midge (*Glyptotendipes barbipes*, 162 g dry wt/m^2) living along the shore of a sewage lagoon (Kimerle and Anderson 1971). A number of researchers have recently found that annual production of Chironomidae can approach or be higher than 50 g dry wt/m^2 in temperate and even subarctic lakes (e.g. Maitland and Hudspith 1974; Lindegaard and Jónasson 1979; Morgan et al. 1980). Such high estimates seem largely due to high biomass, which is often greater than 10 g/m^2, rather than high biomass turnover. However, Sokalova (quoted by Winberg 1971) found annual P/$\overline{B}$ ratios for Chironomidae in a Russian reservoir to range from 2.9 to 36, and Zieba (1971) estimated summertime P/$\overline{B}$ ratios of up to 30 for Chironomidae in a Polish pond. The higher P/$\overline{B}$ ratios were due to short development times during warm months. Other investigators (e.g. see Kajak and Rybak 1966) have also reported fairly high P/$\overline{B}$ values in temperate lakes.

In tropical (e.g. see Lévêque et al. 1972) and semitropical (Cowell and Vodopich 1981) lakes, some species of Chironomidae seem to complete development in as little as 2 weeks, which is comparable to laboratory growth rates at warm temperatures. What does this indicate about potential P/$\overline{B}$ ratios in warm environments? If we consider *Corynoneura*, a small midge in the subfamily Orthocladiinae, we find that these midges can complete development in as little as five days at 15°C (Mackey 1977b). In a tropical or semitropical environment, temperatures would rarely, if ever, drop below 15°C. Let us assume a relatively high cohort P/$\overline{B}$ of 8, and further assume that the animals complete larval development within five days throughout the year. Thus, we can approximate annual P/$\overline{B}$ as $8 \times 365/CPI$ (where $CPI = 5$) = 584. This means the biomass turnover time is less than a day. Such a high P/$\overline{B}$ ratio has never been reported, but the exercise illustrates what is possible for a single short-lived species in a warm environment.

Estimates of secondary production in lotic systems have been accumulating at a rapid rate from throughout the world. In the Speed River, Ontario, total benthic annual production is about 200 g dry wt/m^2 (Hynes and Coleman 1968, with correction by Waters 1977). Such high production is due to an extremely high biomass distributed well into the hyporheic portion of the stream bottom (see Williams, Chapter 14). High values for benthic community production in lotic systems are more usually in the 50 to 100 g/m^2 range (e.g. see Nelson and Scott 1962; Tilly 1968; Mann 1975; Flössner 1976; Benke et al. in press). Again, these high community values have usually been due to a high biomass (around 10 g/m^2).

High annual production estimates for single species or for one particular group in lotic systems seem to be in the 25 g/m^2 range. Trichoptera (Hydropsychidae) production has been found to exceed 25 g/m^2 on solid substrates such as stones (Flössner 1976) or submerged wood (i.e. "snags": Cudney and Wallace, 1980; Benke et al. in press). This high caddisfly production is due primarily to high biomass, although some populations are bivoltine. In a few instances, dipterans have been found to have high production in streams, but usually it has been due to high $P/\bar{B}$ ratios rather than high biomass. Mackey's (1977a) estimates of high midge production in the Thames River (>25 g/m^2) were due to a seven-month $P/\bar{B}$ of 50 to 60. Benke et al. (in press) estimated black fly production in the Satilla, a subtropical blackwater river in Georgia, to be >25 g/m^2 of snag surface, with an annual $P/\bar{B}$ >70. Midge production on snags and in sand substrates was 10 to 25 g/m^2, with an annual $P/\bar{B} > 100$. The highest annual production found for a stream mayfly was 8.56 g/m^2, and that was largely due to an annual $P/\bar{B}$ ratio of 26 (Hall et al. 1980).

Thus, annual production of a benthic community in the 50 g/m^2 range would be considered high in both lentic and lotic systems. It could be due to high $P/\bar{B}$ ratios (primarily dipterans), or high biomass (any group); the latter is often enhanced by a high ratio of effective habitat to bottom area.

The preceding discussion on magnitude of production includes only detritivores, herbivores, and omnivores. There are a limited number of studies of large aquatic insects that are primarily predators. Large insect predators have reached annual production values of 4 to 8 g/m^2 in both lentic (e.g. dragonflies: Benke 1976) and lotic habitats (e.g. stoneflies, hellgrammites and dragonflies: Nelson and Scott 1962; Benke et al. in press). These large insects typically have relatively low annual $P/\bar{B}$ ratios, since they almost always have life cycles of a year or more. Smaller predators with shorter development times, such as midges of the subfamily Tanypodinae (e.g. *Procladius*), could have much higher $P/\bar{B}$ ratios (Mackey 1977a, 1977b; Benke et al. in press).

Predator Consumption—the Allen Paradox

Allen (1951) calculated that fishes in the Horokiwi Stream of New Zealand required about 100 times more benthic prey biomass in a year than was available at a single point in time. He concluded that the benthic animals must therefore have an annual $P/\bar{B}$ of at least 100. Even though Gerking (1962) later showed that Allen overestimated this value by a factor of two to three, Allen's benthic $P/\bar{B}$ was still much higher than obtained by many other investigators using direct methods of estimating secondary production. This discrepancy came to be known as the Allen paradox (Hynes 1970).

Since Allen's study, many others have determined that high $P/\bar{B}$ ratios of benthic insects are necessary to meet the food requirements of both fish and invertebrate predators. For example, Kajak and Kajak (1975) estimated that Chironomidae in Marion Lake, British Columbia, needed a weekly $P/\bar{B}$ of about 1 in order to match the consumption of invertebrate predators (primarily a crustacean, *Crangonyx richmondensis*). Benke (1976) estimated that benthic primary consumers, mostly Chironomidae, in a South Carolina pond needed weekly $P/\bar{B}$s close to 1 during summer months, and an annual $P/\bar{B}$ of at least 30, in order to provide the consumption required for dragonfly production. Coffman et al. (1971) reached a similar conclusion for insect herbivores and detritivores in a Pennsylvania stream. Morgan et al. (1980) summarize several other studies of predator consumption in which similar conclusions are reached.

As the Allen paradox persisted, a number of authors realized that at least part of the reason for the discrepancy was related to the existence of prey species with multiple generations per year. It seems that most direct estimates of production were for species that had a life cycle of at least a year and, thus, annual $P/\overline{B}$ ratios were always low. Now that it is recognized that the CPI is more directly related to annual $P/\overline{B}$ than is voltinism *per se* (Benke 1979; Waters 1979), it is easy to see that the high prey turnover sometimes required by predators (e.g. annual $P/\overline{B}$ of 30–40) is easily obtainable by Chironomidae and other fast-growing invertebrates in reasonably warm environments (e.g. see Mackey 1977a, 1977b; Menzie 1981).

Although there are undoubtedly other reasons for the Allen paradox, the recognition that some insect species really do have high annual $P/\overline{B}$ ratios is probably a large part of the answer. The fact that the paradox has been with us for some time illustrates the value of using independent approaches to find the answer to a question. Clearly, predator consumption by itself is not recommended as a method to measure prey production unless all sources of prey mortality are estimated. However, this nagging paradox has served as a stimulus for finding weaknesses in both approaches.

IMPORTANT ISSUES RELATING TO AQUATIC INSECT PRODUCTION

The Sampling Dilemma

The complex physical environment found in both lentic and lotic ecosystems creates various sampling problems for any investigator attempting to quantify benthic populations. Physical heterogeneity of aquatic systems results in patchy distributions of individuals since many species are strongly associated with a specific substrate type (see Minshall, Chapter 12). The investigator is interested in obtaining a reasonably accurate estimate of production for a given benthic population or community, but efficient utilization of time and energy is also an important consideration. Therefore, the sampling strategy (i.e. sampling device, sampling design, sorting procedure) is very important to the success of the study. The sampling strategy represents an enormous number of questions and problems, but only some of the issues that relate directly to production estimation will be mentioned here. For a more detailed study of sampling problems as they relate to production, see Resh (1977, 1979).

We have a reasonable understanding of the statistical considerations (e.g.

sample size, replication, degree of aggregation) that are important in estimating confidence intervals for population parameters on a single sample date (Elliott 1977), but production estimation introduces some additional problems (Resh 1979). For one thing, the estimation of secondary production requires repeated sampling through time. There are also varying degrees of error in the production methodology itself (see above) that are difficult or currently impossible to measure. All of these simultaneous sources of error create statistical questions that are largely unresolved, and attempts to put confidence intervals on production estimates are rare. A notable exception is Krueger and Martin's (1980) paper on the size-frequency method, but Hynes (1980) issues a warning that we still have to be concerned with other sources of error. While investigators should generally attempt to achieve a reasonably good degree of confidence in standing stock estimates on any given sampling date, samples taken across dates actually might be considered as replicates when using the size-frequency method, and to some extent the same is true for the actual cohort methods. At least in the latter case, overestimates and underestimates tend to be cancelled out from date to date (e.g. see Benke 1976).

Environmental heterogeneity of aquatic habitats presents a dual problem. On the one hand, we are concerned with the statistical problems of estimating production of a given species across a heterogeneous environment. On the other hand, we might be concerned with the differences in either population or community production among different habitat or substrate types in a given system (e.g. see Resh 1977). Typical comparisons of production among contrasting habitat types might be macrophytes vs. sediments, riffles vs. pools, snags vs. sediments, and deep vs. shallow sediments.

If distinct habitat (e.g. riffle) or substrate (e.g. stone or snag surface) types can be identified and sampled in aquatic systems, it seems best to use a stratified sampling strategy in which numbers, biomass, and production are expressed per area of habitat or substrate type (e.g. see Resh 1977, 1979). Resh (1979) refers to the specific substrate as the "effective habitat," and production can be expressed on this basis. For example, Neves (1979) expressed production per area of cobble surface; Cudney and Wallace (1980) and Benke et al. (in press) expressed production per area of snag surface; and Menzie (1981) expressed production per unit of macrophyte leaf dry weight. If production is calculated on this basis, it is also desirable to estimate the amount of habitat or substrate type per total area of stream or lake bottom, so that the relative contribution of different habitats to the entire system can be determined. For example, Mackey (1977a) found that in the Thames River, chironomid production was about 25 $g \cdot m^{-2} \cdot yr^{-1}$ in the littoral habitats (*Acorus* and *Nuphar* zones) compared to 2 $g \cdot m^{-2} \cdot yr^{-1}$ in the erosional main channel (flint zone). However, because of its much greater extent, the flint zone accounted for about 65 percent of total chironomid production. Fish, on the other hand, obtained about 75 percent of their chironomid prey from the *Nuphar* zone. Benke et al. (1979) found a similar phenomenon in the Satilla River, where many fish species consumed insects primarily from the highly productive snag habitat, even though the sand habitat in the main channel was much more extensive.

Another problem common in lotic systems is the dispersal of animals among habitat areas. This may be a relatively continuous process such as we find in drift (e.g. see Waters 1972; Wiley and Kohler, Chapter 5), or it may be a movement from one habitat type to another after a species attains a certain developmental stage (Minshall, Chapter 12). It is usually assumed that emigration equals immigration when calculating production, but obviously this is not always the case. Hall et al. (1980) discuss this problem in detail. They found that early instars of the stream mayfly *Tricorythodes atratus* occur primarily in riffles, but the later instars tend to move downstream into pools.

It has been common practice to sample benthic populations monthly when measuring production. This procedure has worked well for species with at least a one-year life cycle; however, recent research has indicated that CPIs as short as a week or two occur for some Chironomidae. In these cases, it would be impossible to follow a cohort unless sampling was much more frequent. Cowell and Vodopich (1981), upon finding midge life cycles of two to three weeks duration, suggested sampling no less than every three days in order to follow a cohort. However, if a species does not develop synchronously, and one can obtain an independent assessment of the CPI, it would not be necessary to sample so frequently if the size-frequency method is used. This appears to be a major advantage of the size-frequency method over actual cohort methods.

What Limits Secondary Production?

The limitation in the amount of secondary production in benthic communities is a very complex issue and seems far from being resolved. Food quantity and quality, temperature, habitat complexity, and biological interactions can all play a role in this issue, and their relative importance undoubtedly differs among different benthic environments. Furthermore, the factors that limit the production of one species in a given community may not limit the production of another species or the community as a whole. For example, increasing the food level may increase the production of species A and that of the entire community, but species B may be eliminated simultaneously through competition or some other environmental stress. Thus, the major factors limiting production of a species may often be the same as those limiting life-history patterns (Butler, Chapter 3).

Production is determined by both standing-stock biomass and biomass turnover, but the factors that limit biomass and turnover may be quite different, even for a single population. For example, a high ratio of effective habitat to bottom area could result in a high biomass per unit of bottom area. Thus, availability of appropriate substrate may limit the production of certain species or species groupings (see Minshall, Chapter 12). A deep hyporheic zone in streams (e.g. see Coleman and Hynes 1970; Williams, Chapter 14), the availability of snags in rivers (e.g. see Cudney and Wallace 1980; Benke et al. in press), and the abundance of macrophytes (e.g. see Mackey 1977a; Menzie 1981), all seem to greatly enhance the standing-stock biomass and production of certain insect communities.

Food quantity or quality may be a factor that limits either biomass or turnover rates. A number of experimental studies have been done, particularly in ponds, where addition of nutrients has resulted in increased biomass and production of benthic insects (e.g. see Hall et al. 1970). In streams, the biomass or production of insects is often highest under organically enriched conditions (e.g. see Flössner 1976; Hopkins 1976), or sometimes below lake outlets (e.g. see Cushing 1963; Ulfstrand 1968). High nutrient levels may change substrate conditions or habitat complexity (e.g. the growth of macrophytes); therefore, to some extent, increased insect biomass may be due to an increase in habitat space. Thus, sometimes it may be difficult to know for certain that increased biomass and production of insects is entirely due to increased food under enriched conditions. In large rivers, food limitation does not seem to be as important a limiting factor. Here, standing stock biomass is usually high and food sources seem plentiful (Mann 1975), although the amount of solid substrate may be limiting benthic production in many of these larger systems (Fremling 1960; Cudney and Wallace 1980; Benke et al. in press).

Temperature is a factor that does not seem to limit standing-stock biomass (Morgan et al. 1980). High biomass can be found in very cold (e.g. see Lindegaard and Jónasson 1979) or warm (e.g. see Cudney and Wallace 1980) environments. However, temperature can be a major factor in determining biomass turnover if food is not limiting and the environment is not too stressed. The potential for rapid growth, short CPI, and high annual $P/\overline{B}$ ratios increases greatly for some groups, especially midges, when temperatures rise from $<10°C$ to $>20°C$ (e.g. see Mackey 1977b; Menzie 1981; Sweeney, Chapter 4). Several investigators have shown that even in subarctic environments, some midges show rapid growth in the warm summer months, sometimes completing a generation or more, but cease growth completely during cold months (e.g. see Rosenberg et al. 1977). Whether rapid growth at warm temperatures is generally realized, or whether other factors limit growth, is one of the major questions confronting ecologists concerned with insect production.

Biological interactions, such as competition and predation (see Wiley and Kohler, Chapter 5; Peckarsky, Chapter 8) may also affect the production dynamics of some species (see Morgan et al. 1980 for a general discussion). In some cases, predation may limit production by cropping prey biomass to a low level (e.g. see Kajak et al. 1972). In many studies, manipulation of invertebrate or vertebrate predators has resulted in little change in benthic numbers or biomass (e.g. see Hall et al. 1970; Kajak and Dusoge 1973; Benke 1978; Thorp and Bergey 1981; Healey, Chapter 9). Several authors (e.g. see Winberg 1971; Benke 1978;

Wisniewski 1978; Kajak 1980) have suggested that predation may decrease competition for space among primary consumers and may actually stimulate their production, but such stimulation is often difficult to demonstrate from field studies. Predation might actually depress standing-stock biomass of prey, but an increase in prey growth rate and decreased CPI could cause an overall increase in production.

In summary, Morgan et al. (1980) concluded that quality and quantity of food are the main factors limiting benthic production, but most of their analysis was limited to lentic environments. The limitation of aquatic insect production is a complex issue and the factors discussed above probably interact in a variety of ways in different environments.

The Ecological Role of Insect Production in Freshwater Ecosystems

Aquatic ecologists have recently begun to understand the qualitative roles of individual species and functional feeding groups in ecosystem processes (see Merritt et al., Chapter 6, for nutrient processing); however, quantification of these roles remains a major challenge. Accurate measurement of secondary production often can provide very useful information in the quantification process. For example, an estimation of production can be useful in assessing the effects of competition or predation (e.g. see Hall et al. 1970; Benke 1976; Morgan et al. 1980), and the availability of food to higher trophic levels, especially fish; it can also provide key information in quantifying energy and nutrient pathways.

The following example from Benke and Wallace (1980) illustrates the usefulness of production data in assessing the ecological role that a group of stream insects perform. In this study, the production of six species of net-spinning caddisflies was measured in a Southern Appalachian stream in Georgia. Using quantitative feeding analysis and assuming variable food-specific assimilation efficiencies, Benke and Wallace calculated the relative amount of caddisfly production that could be attributed to each of five food types: diatoms, filamentous algae, fine detritus, vascular plant detritus, and animals. The two extremes of food utilization and feces generation are illustrated in Figure 10.4. The large larvae of the caddisfly *Arctopsyche irrorata* capture and consume high-quality animal food with their large coarse nets (mesh opening = $403 \times 534 \ \mu m$ for fifth instars). Small larvae of the caddisfly *Dolophilodes distinctus* primarily utilize low-quality fine detritus that is captured in their extremely fine nets (mesh opening = $1 \times 6 \ \mu m$). Due to different assimilation efficiencies for different food types, *Arctopsyche* larvae convert a much higher proportion of their consumption into production than do *Dolophilodes*. In spite of large differences between the two species in $\bar{B}$ and P, they produce similar amounts of feces that contribute to the seston. Considering the group of six species, 80 percent of caddisfly production was attributed to eating animal food, even though roughly equivalent volumes of animals and detritus were consumed. Benke and Wallace concluded

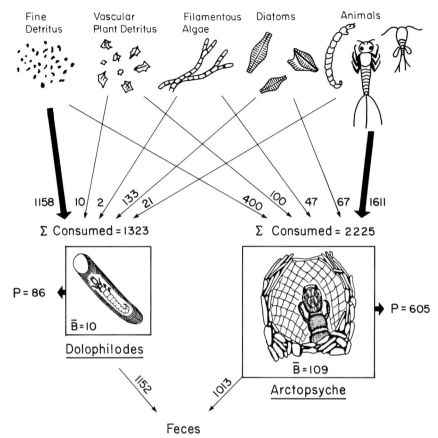

Figure 10.4 Differential utilization of five food types by two out of six coexisting species of net-spinning caddisflies (based on data from Benke and Wallace 1980). Units are mg ash-free dry wt/m^2 for $\bar{B}$, and mg·m^{-2}·yr^{-1} for P and all other pathways.

that the caddisflies, by filtering animals from the water and actually *egesting* more detritus than they ingest, appear to lower the food quality of the seston.

Haefner and Wallace (1981) used the same approach in another production study of two species of net-spinning caddisflies in two Southern Appalachian streams in North Carolina. They found that much of the production of *Parapsyche cardis* and *Diplectrona modesta* was attributed to eating animal food, just as Benke and Wallace (1980) found for the same two species. However, Haefner and Wallace also estimated food *availability* by analyzing organic seston. These investigators determined that the caddisflies remove a very small amount of total organic seston, but that they consume a substantial portion of the drifting benthic invertebrates. In fact, based upon the consumption of these two caddisflies alone, their invertebrate prey would need to have an annual $P/\bar{B}$ of at least 6 to 9.

In order to assess the ecological roles of all insect groups in an aquatic system, it is necessary to consider the entire benthic community simultaneously. The concept of functional feeding groups (e.g. see Merritt et al., Chapter 6; see also Cummins and Klug 1979, Lamberti and Moore, Chapter 7) has provided insight into the role of the various insect groups in benthic communities, but most studies thus far have only considered relative numbers or standing-stock biomass. The distribution of invertebrate production among the functional groups may be the most appropriate measure of their relative roles in ecosystems. Table 10.4 illustrates the distribution of mean annual density ($\bar{N}$), mean annual standing-stock biomass ($\bar{B}$), and annual production (P) among the major functional groups of insects found on the snag habitat in the Satilla River. The greatest variation among $\bar{N}$, $\bar{B}$, and P is obviously for the predators. The amount of variation for a given functional group that will be found in other aquatic systems remains to be seen. If finer groupings are used, greater discrepancies among $\bar{N}$, $\bar{B}$, and P are likely to be found. For example, black flies comprise 59 percent of $\bar{N}$, 18 percent of $\bar{B}$, and 60 percent of P on the snag habitat. This results in the counterintuitive observation that, in this instance, $\bar{N}$ is a better indicator of P than is $\bar{B}$. In future studies, if production is coupled with feeding analysis and bioenergetics data (i.e. production efficiencies) for all functional or finer groups, it should provide a more complete picture of the role of benthic insects in the functioning of lentic and lotic systems than is now generally available.

The Effects of Stress

Aquatic insects have been used as biotic indicators of stress in streams and lakes for many years (e.g. see Hart and Fuller 1974; Wiederholm, Chapter 17). Major emphasis in assessing the effects of water pollution has thus far been on examining alterations in biotic *structure*, especially aspects of species composition (e.g. changes in community diversity, loss of "sensitive" species, or increases in the number of tolerant species). However, Cairns (1977) and many others have stated that a better understanding of the consequences of stress in aquatic systems would result if emphasis were placed on examining alterations in ecosystem *function*. Such alterations could include variations in biogeochemical cycles, and changes in the magnitude and pathways of energy flow. As primary consumers, aquatic insects often play a central role in ecosystem function (Lamberti and Moore, Chapter 7). They are major assimilators of organic matter, and the relative abundance of the various functional feeding groups can tell us much about the balance between autotrophy and heterotrophy in particular aquatic habitats (e.g. see Cummins and Klug 1979; Merritt et al., Chapter 6). Furthermore, as an important link in the food web, aquatic insects are largely responsible for the composition and productivity of higher trophic levels. Since secondary

TABLE 10.4: **Percentages of Mean Annual Density ($\overline{N}$), Mean Annual Standing-Stock Biomass ($\overline{B}$), and Annual Production (P) of Major Functional Groups on the Snag Habitat in the Satilla River, Georgia, Upper Site***

FUNCTIONAL GROUP	$\overline{N}$	$\overline{B}$	P
Filtering collectors	79.9	64.6	79.3
Gathering collectors	16.4	6.6	9.7
Predators	3.7	28.4	10.9
	100%	100%	100%

*Based upon data from Benke et al. in press

production is a useful measure for assessing the function of aquatic insects, it can provide considerable insight into understanding the ecosystem-level consequences of various stresses. Unfortunately there have been very few studies that have examined the effects of stress on secondary production.

Different kinds of stress are capable of influencing secondary production of consumers in very different, and often very subtle, ways. For example, elevated levels of nontoxic organic pollutants can increase overall secondary production, as shown by the high estimate of secondary production for a chironomid midge from a sewage lagoon (Kimerle and Anderson 1971). However, Flössner's (1976) study on the Middle Saale River, Germany, indicated that the greatest effect of organic pollutants was not so much a change in total production, but rather a shift in the most productive species. Toxic stresses can be expected to reduce secondary production of most species, although this is not always true. For example, Uutala (1981) examined the composition and production of chironomid midges in two New York lakes, one with a fish population that had apparently declined as a result of acidification, and another with a near neutral pH and a thriving fish community. Although it was not possible to assess causation, it is interesting to note that although there were fewer species of midges in the acidified lake, their secondary production was actually higher than in the neutral lake.

Rodgers (1982) conducted some experimental studies of the effects of elevated temperatures ($+3$, $+6$, and $+9°$C above ambient) on production of the mayfly *Caenis*. Although increased temperature might be expected to increase production by causing a shorter CPI, production at all elevated temperatures tested in this study was less than half the production at ambient temperatures. This was due to a reduction in standing-stock biomass. Haefner and Wallace

(1981) estimated that caddisfly production was about three times higher in an Appalachian stream with a clearcut watershed than in one with an undisturbed watershed covered by hardwood forest. Benke et al. (in press) found that highest invertebrate production was on the snag habitat in the Satilla River and that to remove submerged trees and branches, as is commonly done to improve navigation and increase channel flow capacity, would substantially alter the food supply for many fish species. Clearly, these few studies indicate that secondary production information is potentially valuable in assessing the effects of stress, and that although these effects often may not be obvious, further research in this area would undoubtedly be justified.

CONCLUSION

Our knowledge of aquatic insect productivity has increased enormously in little more than a decade. This knowledge includes many field studies from all over the world and from many different habitat types. Advances also have been made in methodology, and there are a variety of field and laboratory approaches available to investigators. The approach selected often depends upon life-history type, but also the individual preference of the investigator. Along with this rapidly expanding body of knowledge should be a recognition of sampling and methodology errors that have been and will continue to be difficult to assess. New methods designed to measure and hopefully reduce such errors will be welcome, but because of the fact that there are so many sources of error, the problem will probably continue to haunt production biologists for many years.

The largest data base for production of aquatic insects is for species that grow synchronously and take at least a year to complete development. We will continue to collect valuable production data on such species using the methods described above without major difficulties. The biggest gap in production information is for fast-growing animals such as dipterans, especially those that develop asynchronously. Such rapid growth can be found wherever temperatures approach 15°C for a month or so, which is from the tropics to perhaps as far as the subarctic. Life-history information, particularly an estimation of the CPI, is essential for obtaining accurate production estimates, and it is especially important when considering fast-growing species. The extrapolation of laboratory growth data to field data has often proved to be necessary for such species and, as a result, new strategies for obtaining such information from field data would be especially welcome.

The factors that limit the production of benthic populations and communities are complex. Investigators have rightly utilized experimental field approaches in attempting to differentiate the factors involved. It has been possible to detect changes in numbers and biomass of organisms as a response to experimentation, but it is very difficult to detect subtle changes in biomass turnover rates. New approaches to measuring turnover rates would thus be invaluable for use in experimental studies on factors limiting production.

Production analysis can provide much insight into the role of insects in the functioning of aquatic ecosystems, including a better understanding of the effects of man-made stress, but most production studies have considered only a limited number of species with little knowledge of the rest of the community. Unless the entire benthic community is considered simultaneously, it is difficult to gain an appreciation for the relative significance of a single species or a particular group of species. To study the entire benthic community to a desired level of detail is an enormous task and obviously requires team efforts. Aquatic biologists should attempt to work at the species level whenever taxonomic considerations permit because production measurements are only as good as the life-history information used in the study design. Grouping certain parts of the community into higher taxonomic or functional groups frequently will be necessary, but reasonable approximations of their production should be possible if a group has members of similar maximum size and CPI.

References

Allen, K. R. 1949. Some aspects of the production and cropping of fresh waters. Report, Sixth Science Congress, 1947. Transactions of the Royal Society of New Zealand 77:222–28.

Allen, K. R. 1951. The Horokiwi Stream. A study of a trout population. New Zealand Marine Department Fisheries Bulletin 10:1–238.

Banse, K. and S. Mosher, 1980. Adult body mass and annual production/biomass relationships of field populations. Ecological Monographs 50:355–79.

Benke, A. C. 1976. Dragonfly production and prey turnover. Ecology 57:915–27.

Benke, A. C. 1978. Interactions among coexisting predators—a field experiment with dragonfly larvae. Journal of Animal Ecology 47:335–50.

Benke, A. C. 1979. A modification of the Hynes method for estimating secondary production with particular significance for multivoltine populations. Limnology and Oceanography 24:168–71.

Benke, A. C., D. M. Gillespie, F. K. Parrish, T. C. Van Arsdall, Jr., R. J. Hunter, and R. L. Henry, III. 1979. Biological basis for assessing impacts of channel modification: invertebrate production, drift, and fish feeding in a southeastern blackwater river. Environmental Resources Center Publication No. ERC 06-79, Georgia Institute of Technology, Atlanta, GA. 187 pp.

Benke, A. C., T. C. Van Arsdall, Jr., D. M. Gillespie, and F. K. Parrish. In press. Invertebrate productivity in a subtropical blackwater river: the importance of habitat and life history. Ecological Monographs.

Benke, A. C. and J. B. Waide. 1977. In defence of average cohorts. Freshwater Biology 7:61–63.

Benke, A. C. and J. B. Wallace. 1980. Trophic basis of production among net-spinning caddisflies in a southern Appalachian stream. Ecology 61:108–18.

Borutski, E. V. 1939. Dynamics of the total benthic biomass in the profundal of Lake Beloie. Proceedings of the Kossino Limnological Station of the Hydrometeorological Service of U.S.S.R. 22:196–218.

Boysen-Jensen, P. 1919. Valuation of the Limfjord. I. Studies on the fish-food in the Limfjord 1909–191⁻, its quantity, variation and annual production. Danish Biological Station Report No. 26:3–44.

Butler, M. G. 1982. Production dynamics of some arctic *Chironomus* larvae. Limnology and Oceanography 27:728–36.

Cairns, J., Jr. 1977. Quantification of biological integrity, pp. 171–87. *In*: R. K. Ballentine and L. J. Guarraia (eds.). The integrity of water. United States Environmental Protection Agency, Office of Water and Hazardous Materials, Washington, DC. 230 pp.

Castro, L. B. 1975. Ecology and production biology of *Agapetus fuscipes* Curt. at Brietenbach 1971–1972 (Schlitz studies on productivity, No. 11). Archiv für Hydrobiologie Supplement 145:305–75.

Chapman, D. W. 1971. Production, pp. 199–214. *In*: W. E. Ricker (ed.). Methods for the assessment of fish production in fresh waters. 2nd ed. IBP Handbook No. 3, Blackwell Scientific Publications, Oxford. 348 pp.

Clarke, G. L., W. T. Edmondson, and W. E. Ricker. 1946. Mathematical formulation of biological productivity. Ecological Monographs 16:336–37.

Coffman, W. P., K. W. Cummins, and J. C. Wuycheck. 1971. Energy flow in a woodland stream ecosystem: I. Tissue support trophic structure of the autumnal community. Archiv für Hydrobiologie 68:232–76.

Coleman, M. J. and H. B. N. Hynes. 1970. The vertical distribution of the invertebrate fauna in the bed of a stream. Limnology and Oceanography 15:31–40.

Cooper, W. E. 1965. Dynamics and production of a natural population of a fresh-water amphipod, *Hyalella azteca*. Ecological Monographs 35:377–94.

Corbet, P. S. 1980. Biology of Odonata. Annual Review of Entomology 25:189–217.

Cowell, B. C. and D. S. Vodopich. 1981. Distribution and seasonal abundance of benthic macroinvertebrates in a subtropical Florida lake. Hydrobiologia 78:97–105.

Crisp, D. J. 1971. Energy flow measurements, pp. 197–279. *In*: N. A. Holme and A. D. McIntyre (eds.). Methods for the study of marine benthos. IBP Handbook No. 16, Blackwell Scientific Publications, Oxford. 336 pp.

Cudney, M. D. and J. B. Wallace. 1980. Life cycles, microdistribution and production dynamics of six species of net-spinning caddisflies in a large southeastern (U.S.A.) river. Holarctic Ecology 3:169–82.

Cummins, K. W. 1975. Macroinvertebrates, pp. 170–98. *In*: B. A. Whitton (ed.). River ecology. Blackwell Scientific Publications, Oxford. 725 pp.

Cummins, K. W. and M. J. Klug. 1979. Feeding ecology of stream invertebrates. Annual Review of Ecology and Systematics 10:147–72.

Cushing, C. E., Jr. 1963. Filter-feeding insect distribution and planktonic food in the Montreal River. Transactions of the American Fisheries Society 92:216–19.

Cushman, R. M., J. W. Elwood, and S. G. Hildebrand. 1975. Production dynamics of *Alloperla mediana* Banks (Plecoptera: Chloroperlidae) and *Diplectrona modesta* Banks (Trichoptera: Hydropsychidae) in Walker Branch, Tennessee. Oak Ridge

National Laboratory Environmental Sciences Division Publication No. 785, Oak Ridge, TN. 66 pp.

Edmondson, W. T. and G. G. Winberg (eds.). 1971. A manual on methods for the assessment of secondary productivity in fresh waters. IBP Handbook No. 17, Blackwell Scientific Publications, Oxford. 358 pp.

Elliott, J. M. 1977. Some methods for the statistical analysis of samples of benthic invertebrates. 2nd ed. Freshwater Biological Association Scientific Publication No. 25. Ambleside, England. 160 pp.

Evans, E. D. 1978. Megaloptera and aquatic Neuroptera, pp. 133–45. *In*: R. W. Merritt and K. W. Cummins (eds.). An introduction to the aquatic insects of North America. Kendall/Hunt Publishing Company, Dubuque, IA. 441 pp.

Flössner, D. 1976. Biomasse und produktion des Makrobenthos der mittleren Saale. Limnologica (Berlin) 10:123–53.

Fremling, C. R. 1960. Biology and possible control of nuisance caddisflies of the upper Mississippi River. Iowa State University, Agricultural and Home Economics Experiment Station, Research Bulletin 483:853–79.

Gerking, S. D. 1962. Production and food utilization in a population of bluegill sunfish. Ecological Monographs 32:31–78.

Gillespie, D. M. and A. C. Benke. 1979. Methods of calculating cohort production from field data—some relationships. Limnology and Oceanography 24:171–76.

Haefner, J. D. and J. B. Wallace. 1981. Production and potential seston utilization by *Parapsyche cardis* and *Diplectrona modesta* (Trichoptera: Hydropsychidae) in two streams draining contrasting southern Appalachian watersheds. Environmental Entomology 10:433–41.

Hall, D. J. 1971. The experimental field approach to secondary production, pp. 210–21. *In*: W. T. Edmondson and G. G. Winberg (eds.). A manual on methods for the assessment of secondary productivity in fresh waters. IBP Handbook No. 17, Blackwell Scientific Publications, Oxford. 358 pp.

Hall, D. J., W. E. Cooper, and E. E. Werner. 1970. An experimental approach to the production dynamics and structure of freshwater animal communities. Limnology and Oceanography 15:839–928.

Hall, R. J., T. F. Waters, and E. F. Cook. 1980. The role of drift dispersal in production ecology of a stream mayfly. Ecology 61:37–43.

Hamilton, A. L. 1969. On estimating annual production. Limnology and Oceanography 14:771–82.

Hart, C. W., Jr. and S. L. H. Fuller (eds.). 1974. Pollution ecology of freshwater invertebrates. Academic Press, New York, NY. 389 pp.

Hopkins, C. L. 1976. Estimate of biological production in some stream invertebrates. New Zealand Journal of Marine and Freshwater Research 10:629–40.

Hynes, H. B. N. 1961. The invertebrate fauna of a Welsh mountain stream. Archiv für Hydrobiologie 57:344–88.

Hynes, H. B. N. 1970. The ecology of running waters. University of Toronto Press, Toronto. 555 pp.

Hynes, H. B. N. 1976. Biology of Plecoptera. Annual Review of Entomology 21:135–53.

Hynes, H. B. N. 1980. A name change in the secondary production business. Limnology and Oceanography 25:778.

Hynes, H. B. N. and M. J. Coleman. 1968. A simple method of assessing the annual production of stream benthos. Limnology and Oceanography 13:569–73.

Illies, J. 1975. A new attempt to estimate production in running waters (Schlitz studies on productivity, No. 12). Internationale Vereinigung für Theoretische und Angewandte Limnologie Verhandlungen 19:1705–11.

Johnson, M. G. and R. O. Brinkhurst. 1971. Production of benthic macroinvertebrates of Bay of Quinte and Lake Ontario. Journal of the Fisheries Research Board of Canada 28:1699–714.

Kajak, Z. 1980. Role of invertebrate predators (mainly *Procladius* sp.) in benthos, pp. 339–48. D. A. Murray (ed.). Chironomidae: ecology, systematics, cytology and physiology. Pergamon Press, Oxford. 354 pp.

Kajak, Z. and K. Dusoge. 1973. Experimentally increased fish stock in the pond type Lake Warniak. IX. Numbers and biomass of bottom fauna. Ekologia Polska 21:563–73.

Kajak, Z., K. Dusoge, A. Hillbricht-Ilkowska, E. Pieczynski, A. Prejs, I. Spodniewska, and T. Weglenska. 1972. Influence of the artificially increased fish stock on the lake biocenosis. Internationale Vereinigung für Theoretische und Angewandte Limnologie Verhandlungen 18:228–35.

Kajak, Z. and A. Kajak. 1975. Some trophic relations in the benthos of shallow parts of Marion Lake. Ekologia Polska 23:573–86.

Kajak, Z. and J. I. Rybak. 1966. Production and some trophic dependences in benthos against primary production and zooplankton production of several Masurian lakes. Internationale Vereinigung für Theoretische und Angewandte Limnologie Verhandlungen 16:441–51.

Kimerle, R. A. and N. H. Anderson. 1971. Production and bioenergetic role of the midge *Glyptotendipes barbipes* (Staeger) in a waste stabilization lagoon. Limnology and Oceanography 16:646–59.

Krueger, C. C. and F. B. Martin. 1980. Computation of confidence intervals for the size-frequency (Hynes) method of estimating secondary production. Limnology and Oceanography 25:773–77.

Lapchin, L. and A. Neveu. 1980. The production of benthic invertebrates: comparison of different methods. II. Application to the benthos of the Nivelle River (Pyrénées-Atlantiques, France). Acta OEcologia 1:359–72.

Larkin, P. A. 1979. Predator-prey relations in fishes: an overview of the theory, pp. 13–22. *In*: R. H. Stroud and H. Clepper (eds.). Predator-prey systems in fisheries management. Sport Fishing Institute, Washington, DC. 504 pp.

Lavandier, P. 1975. Cycle biologique et production de *Capnioneura brachyptera* D. (Plécoptères) dans un ruisseau d'altitude des Pyrénées centrales. Annales de Limnologie 11:145–56.

Laville, H. 1975. Production d'un Chironomide semivoltin (*Chironomus commutatus* Str.) dans le lac de Port-Bielh (Pyrénées Centrales). Annales de Limnologie 11:67–77.

Lévêque, C., J. P. Carmouze, C. Dejoux, J. R. Durand, R. Gras, A. Iltis, J. Lemoalle, G. Loubens, L. Lauzanne, and L. Saint-Jean. 1972. Recherches sur les biomasses et la productivité du Lac Tchad, pp. 165–81. *In*: Z. Kajak and A. Hillbricht-Ilkowska (eds.). Productivity problems of freshwaters. Polish Scientific Publishers, Warsaw, Poland. 918 pp.

Lindegaard, C. and P. M. Jónasson. 1979. Abundance, population dynamics and production of zoobenthos in Lake Mývatn, Iceland. Oikos 32:202–27.

Mackey, A. P. 1977a. Quantitative studies on the Chironomidae (Diptera) of the Rivers Thames and Kennet. IV. Production. Archiv für Hydrobiologie 80:327–48.

Mackey, A. P. 1977b. Growth and development of larval Chironomidae. Oikos 28: 270–75.

Maitland, P. S. and P. M. G. Hudspith. 1974. The zoobenthos of Loch Leven, Kinross, and estimates of its production in the sandy littoral area during 1970 and 1971. Proceedings of the Royal Society of Edinburgh 74:219–39.

Mann, K. H. 1975. Patterns of energy flow, pp. 248–63. In: B. A. Whitton (ed.). River ecology. Blackwell Scientific Publications, Oxford. 725 pp.

Menzie, C. A. 1980. A note on the Hynes method of estimating secondary production. Limnology and Oceanography 25:770–73.

Menzie, C. A. 1981. Production ecology of Cricotopus sylvestris (Fabricius) (Diptera: Chironomidae) in a shallow estuarine cove. Limnology and Oceanography 26: 467–81.

Morgan, N. C., T. Backiel, G. Bretschko, A. Duncan, A. Hillbricht-Ilkowska, Z. Kajak, J. F. Kitchell, P. Larsson, C. Lévêque, A. Nauwerck, F. Schiemer, and J. E. Thorpe. 1980. Secondary production, pp. 247–340. In: E. D. Le Cren and R. H. Lowe-McConnell (eds.). The functioning of freshwater ecosystems. Cambridge University Press, Cambridge, England. 588 pp.

Neess, J. and R. C. Dugdale. 1959. Computation of production for populations of aquatic midge larvae. Ecology 40:425–30.

Nelson, D. J. and D. C. Scott. 1962. Role of detritus in the productivity of a rock-outcrop community in a Piedmont stream. Limnology and Oceanography 7:396–413.

Neves, R. J. 1979. Secondary production of epilithic fauna in a woodland stream. The American Midland Naturalist 102:209–24.

Potter, D. W. B. and M. A. Learner. 1974. A study of the benthic macro-invertebrates of a shallow eutrophic reservoir in South Wales with emphasis on the Chironomidae (Diptera); their life-histories and production. Archiv für Hydrobiologie 74:186–226.

Resh, V. H. 1977. Habitat and substrate influences on population and production dynamics of a stream caddisfly, Ceraclea ancylus (Leptoceridae). Freshwater Biology 7:261–77.

Resh. V. H. 1979. Sampling variability and life history features: basic considerations in the design of aquatic insect studies. Journal of the Fisheries Research Board of Canada 36:290–311.

Ricker, W. E. 1946. Production and utilization of fish populations. Ecological Monographs 16:373–91.

Rodgers, E. B. 1982. Production of Caenis (Ephemeroptera: Caenidae) in elevated water temperatures. Freshwater Invertebrate Biology 1:2–16.

Rosenberg, D. M., A. P. Wiens, and B. Bilyj. 1980. Sampling emerging Chironomidae (Diptera) with submerged funnel traps in a new northern Canadian reservoir, Southern Indian Lake, Manitoba. Canadian Journal of Fisheries and Aquatic Sciences 37:927–36.

Rosenberg, D. M., A. P. Wiens, and O. A. Saether. 1977. Life histories of Cricotopus (Cricotopus) bicinctus and C. (C.) mackenziensis (Diptera: Chironomidae) in the Fort Simpson area, Northwest Territories. Journal of the Fisheries Research Board of Canada 34:247–53.

Southwood, T. R. E. 1978. Ecological methods. 2nd ed. Chapman and Hall, London. 524 pp.

Speir, J. A. and N. H. Anderson. 1974. Use of emergence data for estimating annual production of aquatic insects. Limnology and Oceanography 19:154–56.

Tanaka, M. 1976. Relations among elimination, production and biomass in secondary production system. (I) Theoretical consideration. Publications from the Amakusa Marine Biological Laboratory, Kyushu University 4:57–69.

Tanaka, M. 1977. Relations among elimination, production and biomass in secondary production system. (II) Numerical consideration. 1. Ratio of elimination or production to mean biomass, and errors for estimate due to calculation methods or formulae. Publications from the Amakusa Marine Biological Laboratory, Kyushu University 4:147–62.

Thorp, J. H. and E. A. Bergey. 1981. Field experiments on responses of a freshwater, benthic macroinvertebrate community to vertebrate predators. Ecology 62:365–75.

Tilly, L. J. 1968. The structure and dynamics of Cone Spring. Ecological Monographs 38:169–97.

Ubukata, H. 1980. Life history and behavior of a corduliid dragonfly, *Cordulia aenea amurensis* Selys. III. Aquatic period, with special reference to larval growth. Kontyu 48:414–27.

Ulfstrand, S. 1968. Benthic animal communities in Lapland streams. Oikos Supplementum 10:1–120.

Uutala, A. J. 1981. Composition and secondary production of the chironomid (Diptera) communities in two lakes in the Adirondack Mountain region, New York, pp. 139–54. *In*: R. Singer (ed.). Effects of acidic precipitation on benthos. North American Benthological Society, Springfield, IL. 154 pp.

Waters, T. F. 1969. The turnover ratio in production ecology of freshwater invertebrates. The American Naturalist 103:173–85.

Waters, T. F. 1972. The drift of stream insects. Annual Review of Entomology 17:253–72.

Waters, T. F. 1977. Secondary production in inland waters. Advances in Ecological Research 10:91–164.

Waters, T. F. 1979. Influence of benthos life history upon the estimation of secondary production. Journal of the Fisheries Research Board of Canada 36:1425–30.

Waters, T. F. and G. W. Crawford. 1973. Annual production of a stream mayfly population: a comparison of methods. Limnology and Oceanography 18:286–96.

Waters, T. F. and J. C. Hokenstrom. 1980. Annual production and drift of the stream amphipod *Gammarus pseudolimnaeus* in Valley Creek, Minnesota. Limnology and Oceanography 25:700–10.

Winberg, G. G. (ed.). 1971. Methods for the estimation of production of aquatic animals. Academic Press, London. 175 pp.

Wisniewski, R. J. 1978. Effect of predators on Tubificidae groupings and their production in lakes. Ekologia Polska 26:493–512.

Zaika, V. E. 1973. Specific production of aquatic invertebrates. John Wiley and Sons, New York, NY. 154 pp.

Zieba, J. 1971. Production of macrobenthos in fingerling ponds. Polskie Archiwum Hydrobiologii 18:235–46.

chapter 11

HYDROLOGIC DETERMINANTS OF AQUATIC INSECT HABITATS

Robert W. Newbury

The purpose of this chapter is to present some practical techniques for describing the variability and effects of hydrologic processes in common aquatic insect habitats. At the outset, it is important to distinguish between the hydrologic and the fluid determinants of these habitats. Hydrologic determinants, such as storm waves on a lakeshore or flood flows in a stream, establish the gross dimensions and geometry of the habitat (e.g. width, depth, and slope). Fluid determinants establish the microscale structure and characteristics of the environment immediately surrounding the insect, regardless of whether this organism is clinging to a rock outcrop on a wave-swept shoreline or to a cobble in a turbulent stream. Studies of the fluid determinants are part of the emerging field of biofluidmechanics (Vogel 1981), whereas studies of the hydrologic determinants are part of the more traditional disciplines of engineering, geology, and geography.

The opening sections of the chapter discuss ways of summarizing the variability of hydrologic processes when records of the hydrological factors that govern their intensity are available. The latter sections address the question of what parameters can be measured to describe the effects of hydrologic processes when there are no records available. The references cited are a small sample of an immense number that I have found useful for students with nonhydrologic backgrounds who, with me, are struggling to find order in the seemingly unorganized hydraulic landscape. General references include the comprehensive hydrology handbook of Chow (1964), the founding textbook in fluvial geomorphology of Leopold et al. (1964), and the definitive work on drainage basin processes of Gregory and Walling (1973). Many of the field techniques and analytical methods presented in this chapter are discussed in detail in Dunne and Leopold (1978). More detailed techniques and complex methods of instrumentation are available but, for preliminary stages of exploration, traditional techniques and simple methods of measuring and surveying are adequate. The field equipment required is inexpensive and easily carried in a small knapsack.

For readers with nontechnical backgrounds, Vogel (1981) clearly presents the properties of fluids, fluid dynamics, drag, and boundary layer phenomena as they affect the immediate habitat of aquatic plants and animals. For readers with technical backgrounds, Statzner (1981a, 1981b, 1981c) and Statzner and Holm (1982) explore the effects of hydraulic stress and boundary layer conditions on aquatic organisms; Cummins and Lauff (1969), Reice (1980), Tolkamp (1980),

323

and Minshall (Chapter 12) correlate gross hydraulic factors such as depth, slope, and bed materials with insect behavior and habitat preferences; and Townsend (1980), Vannote et al. (1980), and Vannote and Minshall (1982) analyze the hydrologic factors that determine the geographic distribution of insect communities in a drainage basin.

THE VARIABILITY OF HYDROLOGIC DETERMINANTS

In 1909, the geographer William Morris Davis presented a framework that was to become a basic tenet of geomorphology. Davis stated that the morphology of landforms such as mountains, plains, and river valleys was determined by (1) structural properties of the earth's materials of which the landform was composed; (?) geomorphic processes (such as glaciation, gravity, and flowing water) that now act or have acted on the landform; and (3) the length of time that the processes had acted, or,

$$\frac{\text{geologic}}{\text{structure}} + \frac{\text{geomorphic}}{\text{processes}} + \text{time} \rightarrow \text{landform}.$$

This framework is sufficiently general that it can be used to analyze the form of aquatic habitats as well. For example, the properties of the earth's materials in the upland surrounding a lake or in a river valley bottom determine the basic structure of the habitat. Superimposed on these, the geomorphic processes (such as waves, currents, and river flows) that act upon the structure establish—on a geological time scale—the dimensions and configuration of lakeshore and river channel that are the gross templates of aquatic insect habitats.

On a much shorter time scale, the intensity of hydrologic processes is highly periodic. Thus, their effectiveness in shaping the habitat must be described in terms of duration at different levels of intensity. In the two examples that follow, a method of combining intensity and duration is described that is commonly used in hydrologic studies to summarize the variability of hydrologic processes.

Example 1: The Duration of the Wind and Waves in Southern Indian Lake

In 1976, the mean level of Southern Indian Lake, a large lake in northern Manitoba, Canada, was raised 3 m to facilitate the diversion of water from the Churchill River to southern hydroelectric plants. The lake lies on the western arm of the rocky Precambrian Shield in an area covered extensively with lacustrine clays, glacial tills, and thick peat deposits. The regional climate is subarctic, and the upland surrounding the lake exists in the zone of discontinuous permafrost. As the rising lake waters flooded the uplands, a sequence of permafrost melting, bank failure, and erosion became widespread around the lake margin. Wind-generated waves from the 40-km-wide lake erode the cliff-like shoreline at rates of up to 20 m³ of bank material per m of shoreline length, significantly affecting the habitats of littoral insects (Crawford and Rosenberg, in press; Resh et al., in press). In this case, the principal factors affecting the form of the shoreline habitats, in Davis' framework, are,

before impoundment,

$$\frac{\text{granitic}}{\text{bedrock}} + \frac{\text{wave}}{\text{erosion}} + \text{time} \rightarrow \frac{\text{stable rocky}}{\text{foreshores}}$$

and after impoundment,

$$\frac{\text{frozen}}{\text{silty clays}} + \frac{\text{wave and}}{\text{thermal erosion}} + \text{time} \rightarrow \begin{array}{l}\text{unstable shorelines}\\ \text{with retreating and}\\ \text{aggrading foreshores.}\end{array}$$

The variability of waves acting on a shoreline is characteristic of most hydrologic events. At a typical Southern Indian Lake shoreline, in an average open-water season, there are 133 storms with onshore winds exceeding 10 km/hour, 24 storms with winds exceeding 18 km/hour, and one storm with winds exceeding 25 km/hour. The variability of the wave activity (Table 11.1) can be characterized using a cumulative frequency curve (Fig. 11.1), which is often referred to as a duration curve. The duration curve illustrates the difference be-

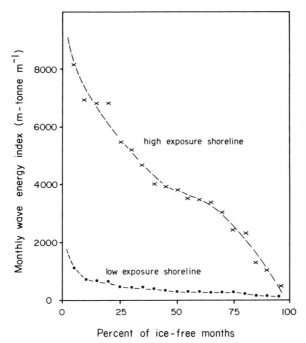

Figure 11.1 Duration curves of the monthly wave-energy index at high and low exposure shorelines on Southern Indian Lake, Manitoba, Canada (see Table 11.1 for data used in preparing curves).

TABLE 11.1: **Monthly Onshore Wave-Energy Index at High and Low Exposure Shorelines in Southern Indian Lake, Manitoba. The Wave-Energy Index, Expressed in Tonne-m per m Length of Shoreline, is Based on Hindcasting the Onshore Waves from Recorded Wind Velocities and Directions Using the Modified Sverdrup-Munk Method (U.S. Army Corps of Engineers 1966)**

| | | SHORELINES | | | | | |
| | | HIGH EXPOSURE | | | LOW EXPOSURE | | |
Year	Month	Wave Energy Index (tonne-m/m)	Rank	Plotting* Position (%)	Wave Energy Index (tonne-m/m)	Rank	Plotting* Position (%)
1977	June	2358	16	80	408	8	40
	July	3051	14	70	276	13	65
	Aug	5470	5	25	290	12	60
	Sept	1016	18	90	479	5	25
	Oct	6867	3	15	411	7	35
1978	June	3568	11	55	745	2	10
	July	4712	7	35	258	15	75
	Aug	6854	4	20	256	16	80
	Sept	483	19	95	626	4	20
	Oct	5187	6	30	680	3	15
1979	June	3957	9	45	272	14	70
	July	1269	17	85	333	10	50
	Aug	3390	13	65	166	18	90
	Sept	2427	15	75	295	11	55
	Oct	4037	8	40	354	9	45
1980	June	3539	12	60	1143	1	5
	July	3819	10	50	145	19	95
	Aug	6942	2	10	171	17	85
	Sept	8147	1	5	474	6	30

*Calculated using Weibull's empirical formula: $\dfrac{\text{rank}}{n+1}(100)$ (Chow 1964).

tween a protected shoreline site and a site exposed to the main body of the lake. At the low exposure site, wave energies ≥ 300 tonne-m per m length of shoreline will occur during 50 percent of the open-water months, while at the high exposure site, ≥ 3800 tonne-m/m will occur for the same duration. The amount of time that thresholds of turbulence, deposition, or foreshore instability significant to the benthic communities are exceeded can be predicted from the duration curve (Fig. 11.1) by relating these processes to the wave-energy index.

Example 2: The Duration of Wilson Creek Flows

Wilson Creek rises on the Manitoba escarpment, a 400-m-high bench of Cretaceous shale overlain by glacial till. In 1957, one of the two headwater branches of Wilson Creek was selected for investigating the feasibility of flood storage dams and sediment control works.The hydrological and meteorological monitoring program associated with the experimental project has provided a detailed record of the variability of the hydrologic processes in the basin. The Wilson Creek flows are estimated from continuous recordings of water depths in a concrete trapezoidal weir that was constructed across the channel at the bottom of the escarpment. The depths of flow are converted to discharge rates from a standard weir formula and adjusted by periodic streamflow measurements (Brater and King 1976).

Hydrologic behavior. Precipitation and consequent streamflow are highly variable in the Wilson Creek basin (Fig. 11.2). The sharply peaked periods of flow, called flood hydrographs, are typical of the rapid response of small headwater basins to snowmelt and rainstorm events. In most years, the flow increases abruptly in early April as the accumulated snowfall of the previous winter melts and collects in the stream channel to form a flood hydrograph. Rainstorms that occur during the snowmelt period produce high runoff-to-rainfall ratios (yields) because only a small portion of the rainfall infiltrates into the frozen ground and most of the meltwater flows directly into the stream. Similarly, early in the summer, rainstorms produce high yields and peak flows because the soil moisture condition is near saturation and there is limited storage available for the rainwater. However, later in the summer, the evapotranspiration demand of the mixed forest that covers the basin often produces soil moisture deficits that are equivalent to the amount of water provided by several cm of precipitation. Flood hydrographs then only occur when rainstorms follow in rapid succession, with the first storms restoring the soil moisture deficit. When the evapotranspiration demand decreases in the fall, high soil moisture conditions are restored and flood flows occur following every rainstorm.

Year-to-year variation in flow conditions is also evident in the three-year record presented in Fig. 11.2. In 1971, the normal sequence of spring snowmelt and early summer and late fall flood hydrographs occurred because precipitation was spread evenly throughout the year. In contrast, the dry summer of 1972 produced no summer and fall flood hydrographs. Dry conditions persisted throughout the winter, leading to a reduced spring meltwater hydrograph in 1973. However, precipitation in the following summer and fall produced normal flood hydrographs because the soil moisture conditions were restored in the basin.

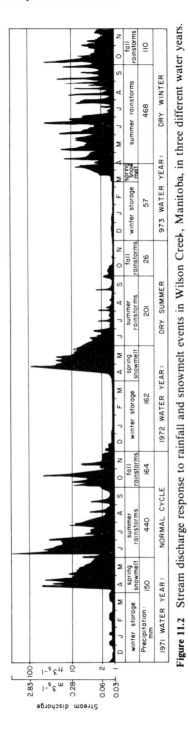

Figure 11.2 Stream discharge response to rainfall and snowmelt events in Wilson Creek, Manitoba, in three different water years.

The flood hydrographs of the open-water season are superimposed upon a steady year-round flow, called the baseflow, of the stream. The baseflow is supplied from water stored in small lakes on top of the escarpment and from the reservoir of groundwater that exists under the basin. Most of the groundwater is recharged by the infiltration of precipitation into the surface of the basin. In addition, a small component is derived from deeper strata that are recharged several thousand km to the west of the escarpment. The baseflow component of the discharge persists throughout the winter months when precipitation is stored on the ground surface as ice and snow because the groundwater reservoir remains unfrozen throughout the year. The zones of groundwater discharge into the stream can be observed readily in midwinter because the streambed and banks remain free of snow and ice, providing year-round open-water habitats for plant and insect communities (Fig. 11.3). During the summer, such reaches may be defined in mid-latitude streams by comparing the cooler water temperatures measured below the bed with those measured in the flow above the bed.

The variety of sources of streamflow contributes to the variability of the water quality of insect habitats. High concentrations of major ions in the baseflow are similar to those observed in water samples taken from the

Figure 11.3 The unique habitats of an open-water reach maintained throughout the winter by warm inflowing groundwater in Wilson Creek, Manitoba.

groundwater reservoir (Schwartz 1970). These high concentrations are abruptly diluted as surface flows and shallow transient groundwater flows enter the stream during snowmelt and rainstorm periods (Fig. 11.4). The effect of sudden changes in the quality of the streamflow on the productivity of the stream waters and, possibly, the habitat preferences of lotic insect communities may be an interesting area for future research.

Duration.   Duration of the various rates of Wilson Creek flows may be determined by ranking the flows and plotting them as a cumulative frequency curve similar to that produced for wave energy in the previous example (Fig. 11.5). The annual duration curve shows less variability than the daily duration curve because peak events are averaged over a longer period. A similar flattening of the daily duration curve can be artificially achieved if flood storage reservoirs that detain peak flows and release them slowly during subsequent low flow periods are constructed in the basin. The duration curves of a completely regulated stream or river are horizontal because the flow is constant.

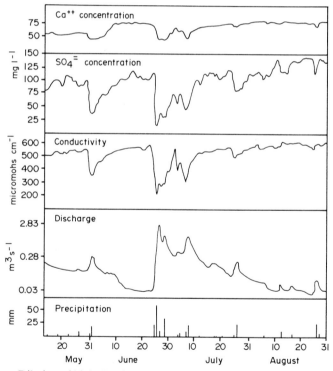

Figure 11.4   Dilution of high chemical concentrations in the stream baseflow component during peak streamflow periods caused by direct runoff from summer rainstorms in the Wilson Creek, Manitoba, basin in 1969 (Schwartz 1970).

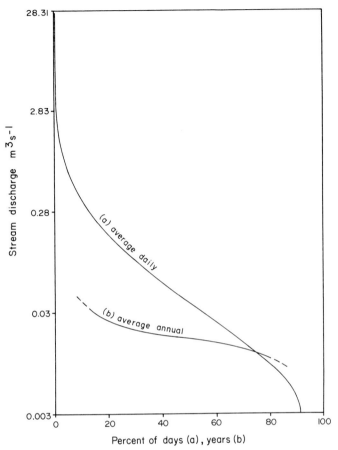

Figure 11.5 Duration curves of annual and daily stream discharges for 1959–1977 in Wilson Creek, Manitoba.

Periodicity.   Although the duration curve is a useful summary of flow rates, it does not reflect the periodicity of peak flow events. For example, on Wilson Creek the flow was ≥ 0.34 m³/s for 10 percent of all the days in the period of record. The duration curve will not distinguish between days that may have occurred in one or two wet summers, or days that were distributed evenly throughout the period of record. If the periodicity of high and low flow rates is an important habitat factor, then the flow record must be examined directly to determine the sequence of high and low flow periods.

Frequency.   Descriptions of the frequency of occurrence of flood flows are often based on a selected sample of the single maximum flow that occurred in each year of the period of record. Several theoretical distributions of the sample of peak flows have been proposed to characterize the frequency curve of annual flood peaks (Chow 1964). On Wilson Creek, for example, the logarithms of the annual flood peaks that occurred from 1959–1977 are normally distributed (Fig. 11.6). The annual peak discharge that is equalled or exceeded in two out of three years is 3.3 m^3/s. This flood condition can be related to the gross dimensions of the channel and flood plain, but it is not representative of the hydrologic conditions that occur in the channel most of the time because the duration of this and larger flows is less than two days per year.

Implications

The variability of the intensity of the hydrologic processes in the Southern Indian Lake and Wilson Creek examples is characteristic of the hydrologic determinants of aquatic habitats. Maximum rates occur for short periods of time,

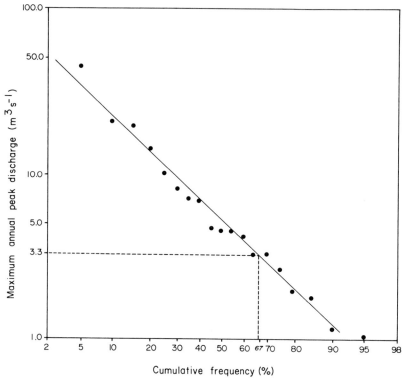

Figure 11.6 Cumulative frequency curve of the maximum annual instantaneous peak discharges that occurred in Wilson Creek, Manitoba, from 1959–1977. The *dashed line* indicates the discharge that is equalled or exceeded in two out of three years.

while intermediate and low rates occur over long periods of time. Thus, the majority of regular visits to a stream or shoreline will occur during the periods of relative quiescence. A stratified sampling program related to the intensity of the hydrologic processes is required to obtain a sample of high-intensity-habitat conditions. Duration, periodicity and, in some cases, the frequency of occurrence are three simple measures that can be applied to complete the time-related definition of the intensity of hydrologic determinants. In streams, the intensity and duration of the rates may be directly translated into insect habitat "usability" (Gore and Judy 1981) or "probability of use" curves for fish spawning and rearing (Bovee and Cochnauer 1977).

THE BEHAVIOR AND EFFECT OF HYDROLOGIC PROCESSES IN STREAM HABITATS

Two major factors of stream habitats, the flow characteristics and the ability of the flow to move the substrate downstream, are discussed in this section as examples of the variation in effect of hydrologic determinants.

Average Streamflow Characteristics

The velocity of flow across a section of a uniform, straight stream varies from zero at the channel boundaries to a maximum value just above the center of the flow. The velocity also varies along the channel as meanders, obstacles, and pools are encountered, and stream gradients, bed materials, and discharges change. In a uniform reach of channel that is several hundred meters long and has well-developed turbulent flow conditions, the average velocity is governed by the cross-sectional area of the flow, the longitudinal gradient of the channel, and the frictional resistance offered by the channel boundaries. The following flow velocity relationship based on these parameters was developed by Antoine Chezy in 1769 (Chow 1959):

$$V = C \ (RS)^{1/2}$$

where

C = Chezy's friction factor ($L^{1/2}/T$)
R = the hydraulic radius of the streamflow (L), which is the quotient of the wetted perimeter of the cross section divided by the cross-sectional area. This is approximately equal to the mean depth in wide, shallow streams.
S = the average longitudinal gradient of the water surface which, in the case of uniform flow, is equal to the gradient of the channel as well.
V = mean velocity of flow through the reach (L/T)

A commonly used form of the Chezy equation was introduced by Robert Manning in 1889 (Chow 1959):

$$V = \frac{R^{2/3}S^{1/2}}{n} \text{ (SI units)}$$

where

n = Manning's roughness coefficient observed for different channel types ($m^{1/6}$)

There are several difficulties encountered when the average velocity equation is applied to streams over a range of discharges. The shape of the cross-sectional area and the length of the wetted perimeter change as boulders and cobbles emerge at low flows, or when flood flows spill out of the central channel to the surrounding flood plains. Additionally, the gradient of the water surface changes as obstructions are submerged at higher flows. Usually, only the general cross section and channel gradient are known, and changes in the geometry of the flow are reflected in a wide range of Manning's roughness coefficients observed at different discharges. For example, in a carefully prepared study of roughness factors and errors in flow-prediction methods for a small Oregon stream, Bovee and Milhous (1978) found that Manning's roughness coefficient varied from 0.05 for flows >0.84 m^3/s to 0.35 for flows <0.03 m^3/s. To overcome the difficulty of estimating the roughness values, flow-rating curves should be based on at least three measurements of discharge covering low and high stages.

Where flow measurements are not available, as is often the case for infrequent high-flow stages, the flow equation must incorporate estimates of Manning's coefficient based on observations of the general configuration of the streambed and materials on it. Photographic keys for Manning's roughness coefficient, based on channels with known discharges and resistance factors, are presented in Barnes (1967) and Chow (1959). Relationships between bed materials and flow resistance similar to that proposed by Strickler (Chow 1959), where n = 0.041 $D_{50}^{1/6}$ and D_{50} = the median diameter of streambed materials (m), may be applied where the depths of flow are several times greater than the height of the materials on the streambed.

The durations of different velocities and depths taken from an observed rating curve or estimated by Chezy's equation can be determined from the stream discharge duration curve. The depths, roughnesses, and mean velocities in a typical reach of Wilson Creek (Fig. 11.7) are summarized in the first three columns of Table 11.2. The roughness coefficients at low flows are based on readily obtained observations, since the flow was <0.2 m^3/s for >80 percent of the time. The roughness coefficient for higher flows, up to but not exceeding the bankfull stage, is based on Strickler's relationship for a median bed material size equal to 5.5 cm. The discharge corresponding to each stage and velocity is calculated using the mean channel width of 4.3 m and an average slope of 0.008. The duration of each level of discharge, and the corresponding depth and velocity, are determined from the daily duration curve of flows (Fig. 11.5). In this example, the duration of various levels of velocity and depth are based on the

Figure 11.7   Wilson Creek, Manitoba, 300 m upstream from a permanent gauging station. The channel dimensions are summarized in Table 11.4.

total period of record, including both summer and winter flow conditions. Where these values correspond to thresholds that are significant to the behavior of the lotic insect community, the period of record for the duration curve may be shortened (e.g. to a particular season or segment of an insect's life cycle).

Changes in the turbulence of the flow, characterized by the Froude number (Fr), are summarized in column 4 of Table 11.2. The Froude number is a dimensionless parameter that combines both the depth of flow and velocity in one term where $Fr = V^2/gD$ and g = gravitational acceleration (L/T^2). Correlations between macroinvertebrate populations, substrate characteristics, and the Froude number have been proposed by Gore (1978) and Statzner (1981b, 1981c).

Local Velocities in Irregular Stream Channels

To the disappointment of many stream investigators, Chezy's flow equation, which he used so successfully to describe conditions in the uniform canals of Paris, can be applied only to generalized reaches of the stream channel where frictional resistance governs the average depth and velocity of the flow, and the mean slope of the water surface and stream bed are parallel. In steep sections of boulder-filled channels or in the pool and riffle sequences of mobile-bed streams, the water surface profile is often broken into many small, flat steps and steep

TABLE 11.2: Flow Characteristics and Duration in a Typical Stream Reach, Station WCW, Wilson Creek, Manitoba*

D DEPTH OF FLOW (m)	n MANNING'S ROUGHNESS CO-EFFICIENT (m^{1/6})	V MEAN VELOCITY (m/s)	Fr FROUDE NUMBER	Q DIS-CHARGE (m^3/s)	t DAILY DURATION (% ≥)
0.03	0.05	0.17	0.31	0.02	65
0.06	0,05	0.27	0.35	0.07	36
0.10	0.04	0.48	0.48	0.21	17
0.15	0.03	0.84	0.69	0.54	7
0.20	0.025	1.23	0.87	1.06	3
0.25	0.025	1.42	0.91	1.52	1.5
0.30	0.025	1.61	0.94	2.07	1.0
0.35	0.025	1.77	0.96	2.67	0.5

*The duration is based on the daily discharges recorded continuously in the 1959–1977 period. The bankfull stage is 0.35 m.

chutes as the bed varies and obstacles to the flow are encountered. Large boulders that have tumbled into the streams from the valley walls, irregular bedrock outcrops, buried logs, tree trunks that have fallen across the stream, beaver dams, and human-made structures all cause local variations in the velocity, depth of flow, and the slope of the water surface.

The factors governing the rapid variation of local flow conditions can be illustrated using the general energy relationships for water flowing in an open channel. The total energy of the flow at any section in the reach is the sum of the potential energy measured above a horizontal datum (Z), the depth of flow (D), and the velocity head or kinetic energy of the flow ($V^2/2g$) (Brater and King 1976). If these three components are plotted along the profile of a uniformly flowing channel, the lines joining the channel bottom, water surface, and total energy points will be parallel (Fig. 11.8). The total energy lost in the reach through turbulence and frictional heating at the channel boundaries is equal to the difference in elevations of the energy gradient from the top of the reach to the bottom of the reach.

The sum $D + (V^2/2g)$, called the specific energy of the flow, is measured with the sloping stream bed as datum. The specific energy will be constant between sections under uniform flow conditions. When a minor obstacle or

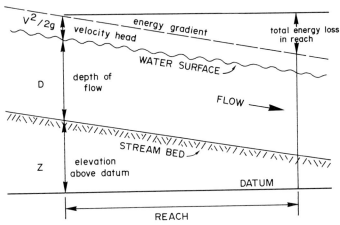

Figure 11.8 Distribution of potential and kinetic energy shown schematically in a uniformly flowing stream.

obstruction is encountered (Fig. 11.9), the flow velocity and depth change to accommodate the constant discharge without altering the position of the energy gradient. For example, if the stream bed is elevated by boulders (sections 2 and 3), the velocity head increases locally as the depth of flow decreases. For small obstructions, the specific energy of the flow is not changed significantly. However, if the depth of flow over the obstruction at section 3 decreases below a minimum value, called the critical depth (D_c), the entire water surface upstream will be raised and the specific energy of the flow will no longer be constant. The

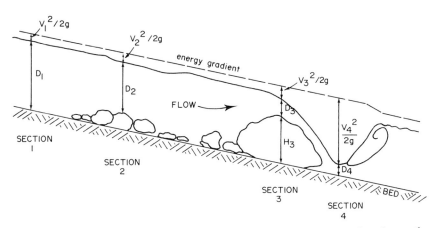

Figure 11.9 Local variations in flow conditions caused by submerged obstacles on the channel bed.

critical depth is the minimum depth that can occur in an open channel flowing over the top of an obstruction such as a weir, spillway crest, boulder, or log.

Beyond the large boulder (section 3), the flow accelerates down the steep lower face of the obstruction until the deeper water in the next segment of the profile is encountered. The adjustment from the shallow depth of the rapid flow zone that is just below the boulder to the deeper, more tranquil flow in the channel downstream is accomplished below section 4 in a steep transition zone called the hydraulic jump. Air bubbles entrained by the rapid flow at the bottom of the jump escape in the more tranquil zone, producing the familiar murmur of small streams or the roar of river rapids. The energy gradient drops abruptly in the hydraulic jump as energy is rapidly dissipated in this short zone of intense turbulence.

If the obstruction only partially fills the channel and extends above the water surface (e.g. a large isolated boulder), the water must accelerate to a higher velocity at the sides of the obstruction. For a constant specific energy, the increased velocity requires the flow depth to decrease alongside the boulder (Fig. 11.10A). Near the middle of the upstream face of the boulder, a stagnation point is created where the downstream velocity is reduced to zero (Fig. 11.10A, B). At this point, there is no velocity head, hence, the depth of flow rises to the level of the energy gradient. The approximate velocity head of the oncoming flow can be determined in the field by measuring the increase in water level relative to the approaching water surface.

The range of possible depths and velocities that can occur at a constant discharge for a range of obstruction heights can be summarized by comparing the depth and specific energy of the flow (Fig. 11.11). As the height of the obstruction increases and the depth of flow decreases, the depth and velocity combinations on the upper, subcritical limb of the curve decrease to a minimum value at which the critical depth and critical velocity (V_c) occur. In the example shown, the critical depth is 0.18 m and the critical velocity head is 0.09 m.

The critical velocity is used to predict the discharge over simple weirs placed across the stream bed where the depth of flow at the weir crest is monitored and the width of the weir section is known. The single critical depth value occurs at the point where two-thirds of the specific energy is composed of the depth of flow and one-third is composed of the critical velocity head, or $D_c = 2(V_c^2/2g)$. Thus, if the critical depth of flow is measured, the velocity head, velocity, and discharge at the weir section can be calculated as the product of the mean velocity and cross-sectional area of the flow.

The velocity of the flow may be accelerated beyond the critical value in steeply sloping segments of the channel that occur below sections of critical depth (i.e. between sections 3 and 4 in Fig. 11.9). The combinations of depth and velocity then occur on the lower, supercritical limb of the curve (Fig. 11.11). At the critical depth, the values of V_c^2 and gD_c are equal, producing a Froude number of 1. Thus, supercritical flow velocities occur when the Froude number is >1 and subcritical flow velocities occur when the Froude number is <1.

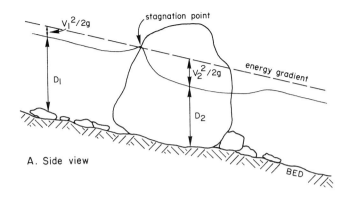

A. Side view

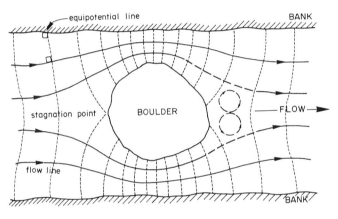

B. Plan view

Figure 11.10 Local variations in flow conditions caused by an object extending above the water surface. The "flow net" shown in *B* is formed by the trial and error sketching of mutually perpendicular flowlines and equipotential lines past the obstruction. Velocities in the plane of the flow net are inversely proportional to the lengths of the sides of the squares. The flow lines downstream from the obstruction are distorted in the zone of deceleration (*dashed lines*) and cannot be accurately predicted using the flow-net technique (Chow 1964; Davis and De Weist 1966).

A simple test of the local flow condition can indicate whether the velocity is subcritical or supercritical. The translational velocity or rate of travel of a wave in shallow still water (V_{sw}) varies with the depth where $V_{sw} = (gD)^{1/2}$. If the Froude number of the flow is 1, the critical velocity of the flow and the translational wave velocity are equal ($V_c = V_{sw}$). Waves propogated in a stream by obstructions or disturbances in the subcritical sections of the flow can move upstream until they are dissipated or until a critical depth section is encountered. At the critical depth section, the upstream progress halts and a standing wave is formed. If the flow is

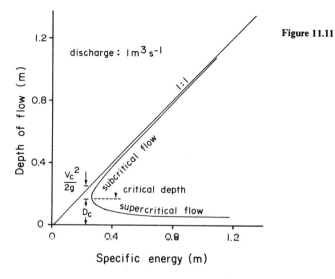

Figure 11.11 The relationship between specific energy and depth in Wilson Creek, Manitoba (see Fig. 11.7), for a discharge of 1 m³s⁻¹. Depending on the local configuration of the channel, the flow may be either subcritical (high depth, low velocity), or supercritical (low depth, high velocity). The two flow conditions are equivalent at the critical depth.

supercritical, the wave caused by the disturbance will be swept downstream. When exploring small streams, the direction of travel of a small wave formed with a stick or a boot can demonstrate where subcritical and supercritical segments of the flow occur.

An understanding of the variations that can occur in local flow conditions may be important in interpreting the results obtained using sampling devices that provide obstructions to the flow and that are subject to consequent changes in velocity distributions and turbulence (e.g. Surber and other netted sampler; Resh 1979). A few simple experiments using boulders or concrete blocks can create the full range of easily observed flow phenomena in a small stream channel. For the investigator wishing to pursue patterns and conditions of flow in more detail, reference should be made to comprehensive texts, such as Henderson (1966), that deal with the hydraulics of river channels.

Flow conditions also vary near the boundaries of the wetted surfaces within the stream channel. As the boundary is approached, the flow decelerates and shifts from the turbulent state, which exists in most of the channel section, to the laminar state, which exists only in a thin boundary layer on the wetted perimeter of the flow. In the laminar state, the viscous forces of the fluid overcome the turbulent momentum of the flow and smooth, independent streamlines of flow are formed. The specific energy relationships of the larger local flow variations are not applicable in the layer of flow adjacent to the boundary because the kinetic energy is not recovered as depth, but rather is lost through frictional heating. Observations of boundary layer phenomena in hydraulic flumes with isolated substrate conditions suggest that many of the wetted surfaces in the stream would provide opportunities for aquatic organisms to adapt feeding and survival strategies to the varying flow conditions (Vogel 1981; Statzner 1981c). The extension of these microscale phenomena to natural stream channels, as a further refinement of local flow variations, is an interesting challenge for benthic researchers.

Substrate Movement

Benthic insects, and the materials that form the stream bed on which they live, are subject to the same shear stresses as are exerted by the flow on the channel boundaries. If the stress is large enough, the substrate materials or organisms will be set in motion and begin rolling, bouncing, or being carried in suspension along the stream. The magnitude of the shear stress varies with the depth of flow, resulting in the movement of larger particles in the deep central thread of the stream, and finer materials along banks and flood plains. The intensity of the shear stress, and hence the stability of the substrate, varies in time as the stream discharge changes.

Under uniform flow conditions, the total shear stress acting on the stream bed is equal to the total frictional resistance to the flow, as proposed by Chezy. For a column of flowing water, the component of the gravitational force acting parallel to the sloping channel is exactly balanced by the shear stress exerted on the bottom of the column by the stream bed. The shear stress acting on a column of unit area is commonly called the tractive force, τ. By equating the gravitational component and the bed shear stress per unit area, the tractive force can be related to the specific weight of water ($1000 \, \mathrm{kg/m^3}$), depth of flow D (m), and the slope of the energy gradient S, in the relationship $\tau = 1000 \, DS \, (\mathrm{kg/m^2})$.

If the unit area is replaced by the area covered by an individual particle, or by an aquatic organism, that is of a representative diameter M and that is resting on the stream bed, the tractive force may be considered as acting directly on the particle. At the point of incipient motion, the drag force or critical tractive force (τ_c) will be just equal to the forces of static friction plus the forces exerted by surrounding particles that hold the individual particle in place. These forces are partially dependent on the submerged weight and hence, the size of the particle. Thus, the critical tractive force can be related to the diameter of the particle on the bed at incipient motion in the relationship $\tau_c = \theta \, (M)$, in which $\theta =$ an empirical coefficient based on the density and shape of the particle, the effect of surrounding particles in which the individual particle may be packed, and the friction factor between the particle and the underlying bed (Henderson 1966; Hendricks 1977). A wide range of observed values of the critical tractive force is apparent in Fig. 11.12, which is based on observations compiled by Lane (1955) for stable canals. Lane found that the critical tractive force ($\mathrm{kg/m^2}$) for rounded noncohesive particles >0.5 cm in diameter is approximately equal to the

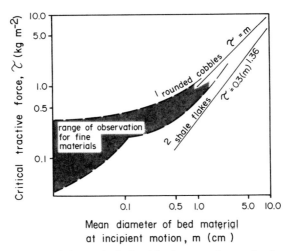

Figure 11.12 The range of observed relationships between the tractive force and the mean diameter of particles on the channel bottom at incipient motion ([1]data from Lane 1955; [2]data from Magalhaes and Chau, in press).

diameter of the particle (cm) at incipient motion. For flakelike shale particles similar to those found in the lower reaches of Wilson Creek, approximately one-half of the tractive force is required for an equivalent mean diameter at incipient motion (Magalhaes and Chau, in press).

The tractive force exerted on a reach of Wilson Creek is compiled for a range of discharges in Table 11.3. The duration at different levels of tractive force was obtained from the discharge duration curve (Fig. 11.5). The percentage of the substrate materials in motion at varying discharges compiled in the last column of Table 11.3 was determined from the tractive force relationship for shale particles (Fig. 11.12). Where the size of substrate materials is an important habitat factor, the fraction of time that a particular size will be unstable may be estimated from Table 11.3. For example, shale particles with a median diameter ≥4 cm, which represent 84 percent of the bed-paving materials, will be unstable for < 1.5 percent of the time, or < 5.5 days per year. This may be useful information in studies of substrate suitability or preference among stream benthic insects. It should be noted, however, that the substrate behavior is idealized in this analysis and the data can be interpreted only as a general indicator of the changes in habitat conditions caused by the variability of the hydrologic determinants. Just as uniform flow relationships are severely modified by local obstructions, the transport of stream-bed materials is altered by local variations in shear stress and the interaction of particles once they are in motion. Beyond the simplified tractive force approach, the complex mechanisms of sediment transport are not yet fully understood (Wolman 1977).

SOME PRACTICAL APPROACHES TO FIELD OBSERVATIONS

The characteristic geometry of a stream channel or a lakeshore can be established by comparing measurements taken at sites or reaches that are selected to represent a variety of hydrologic conditions. The choice of sampling sites may be restricted to the habitat of a particular species, or it may be extended to include all of the combinations of geological and hydrologic conditions that can occur. For example, in the study of wave erosion on Southern Indian Lake, all combinations of shoreline types and wave exposures were required to predict a lake sediment budget. Exploration of the shoreline was undertaken in the following stages: (1) a general reconnaissance of the shoreline was made using aerial photographs and a float-equipped aircraft; (2) sample sites were selected that represented the variety of shoreline configurations that occurred around the lake; and (3) the geology, topography, and vegetation features at each site were mapped in detail by site visits. When all of the site-specific data were compiled, a shoreline classification system was devised based on the bedrock and surficial geology, the configuration of the foreshore, beach, and backshore zones, and the

TABLE 11.3: Shear Stress and Bed Stability Duration at Station WCW, Wilson Creek, Manitoba

DISCHARGE (m^3/s)	UNIT SHEAR STRESS (kg/m^2)	TOTAL SHEAR STRESS PER 100 m (kg)	DAILY DURATION $(\%)$	MEAN DIAMETER OF SHALE PARTICLES AT INCIPIENT MOTION, A* (cm)	FRACTION OF BED MATERIALS $<A$ $(\%)$
0.02	0.24	103.2	65	0.8	0
0.07	0.48	206.4	36	1.4	0
0.21	0.80	344	17	1.0	1
0.54	1.2	516	7	2.8	3
1.06	1.6	688	3	3.4	7
1.52	2.0	860	1.5	4.0	16
2.07	2.4	1032	1.0	4.6	30
2.67	2.8	1204	0.5	5.1	41
4.84	4.0	1720	0.1	6.7	75

*Estimated from Figure 11.12

characteristics of the vegetation communities on the shoreline (McCullough 1978). The classification system was then applied to the entire shoreline and representative sites were selected to monitor shoreline erosion for each shoreline type. The classification system was also used to define biological sampling programs. For example, shoreline types with gravel or rocky foreshores were selected for studies of fish spawning (Fudge and Bodaly, in press), and similar shorelines in high and low exposure sites were selected to determine the effects of shoreline erosion on benthic macroinvertebrates (Crawford and Rosenberg, in press; Resh et al., in press).

The characteristics of stream channels may be established by measuring and mapping sample reaches as well. However, in contrast to the previous example of isolated shoreline segments, stream reaches are not independent samples of the habitat. Each reach is affected by inflows of water and detritus from upstream reaches, and the slope and energy dissipated in the reach depends, in part, on conditions in adjacent reaches. The hydrologic characteristics of stream habitats must, therefore, be based on a set of relationships that describe the changes in geology and channel geometry that occur along the stream as the drainage area and discharge change. In Canada, particularly beyond the agricultural zone, hydrometric surveys of small streams are seldom undertaken because of their limited potential for water supply or hydroelectric development. Methods for gathering the necessary hydrometric data in conjunction with biological sampling programs in a small river or stream where surveys and flow records are limited or nonexistent are discussed in this section.

Watershed Reconnaissance and the Selection of Sample Reaches

Drainage-basin boundaries can be sketched by interpolating between the highest contours of the surrounding landscape that slope toward and away from the tributary channels of the stream. This can be done using topographic maps and aerial photographs. The general profile of the main stem and major tributaries can be determined by scaling the distances between contour lines crossing the channels. If no maps are available, a geodetic survey of the major channels may be feasible for small basins (i.e. those less than a few hundred km^2 in area). Triangulation surveys or barometric leveling may be required in larger basins to determine point elevations that can be joined to produce approximate profiles of selected channels (Breed and Hosmer 1953).

Surficial and bedrock geology maps of the basin may be used to predict the regional pattern of groundwater movement. Where there are surficial deposits that are permeable and relatively homogeneous, groundwater moves downward from recharge areas in the upper reaches of a drainage basin to discharge zones in and beyond the lower basin. Vegetation patterns, springs, and channel-seepage zones in downstream reaches, where the elevation of the stream bed is lower than the adjacent groundwater table, are useful indicators of the pattern and direction of groundwater movement (Davis and De Wiest 1966).

The branching characteristics of the stream network can be established by numbering the tributaries and subsequent segments of the stream channel in a hierarchical sequence (Leopold et al. 1964). One of the simpler systems numbers all headwater channels with no tributaries (at that particular scale of mapping) as order 1 streams (Fig. 11.13). Two order 1 streams combine to form an order 2 segment of the stream; two order 2 segments combine to form an order 3 segment; and so on, until the main stem and highest order segment of the channel is reached at the bottom of the basin.

Several useful relationships may be developed between stream order numbers and channel characteristics in a particular basin. The tributary drainage area, length, and the number of streams in each order, for example, are often geometrically related to the order number (Fig. 11.14). The relationships developed in one portion of the basin are transferable to other portions of the basin if the geological and topographical characteristics are similar. In such instances, the characterization of aquatic habitats derived from sample reaches may be considered representative of similar order number segments throughout the

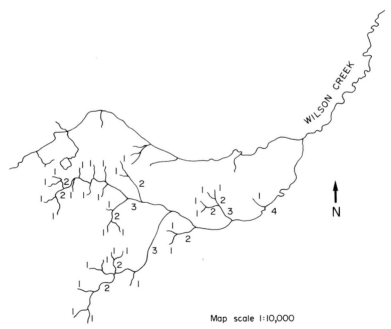

Map scale 1:10,000

Figure 11.13 Branching characteristics of the south branch of Wilson Creek, Manitoba, based on topographic mapping at a scale of 1:10,000.

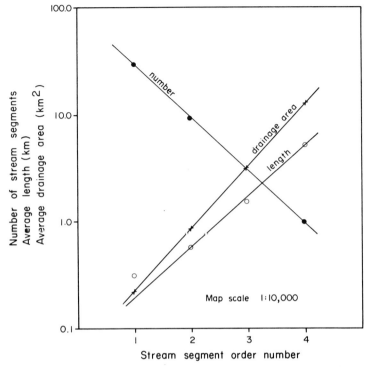

Figure 11.14 Stream order number relationships for the south branch of Wilson Creek, Manitoba. The number of stream segments, the average length, and the average drainage area for the tributaries and main channel are based on a 1:10,000, 6-m contour interval map. Stream length values for order 1 segments may be overestimated because the channels are obscured by beaver ponds.

basin. It is important to emphasize that the order numbers applied to the channels of one basin are unique to the mapping scale and mapping criteria used. Other channels in different geological and hydrological settings mapped at different scales may have the same order number, but completely different characteristics.

The selection of sample reaches should extend over a range of stream orders to include changes in the hydrologic behavior and channel geometry that occur as the tributaries of the stream unite, and the drainage area and stream discharge increase. The length of a sample reach should be 20 to 30 times the channel width so that measurements of channel geometry can be taken at several locations to determine the average characteristics in the reach. Isolated disturbances such as log jams, beaver dams, bridge piers, landfills, or snow-dumping sites should be avoided since they may alter the local stream gradient and dimensions. Duplicate reaches of the same order number may be chosen in adjacent branches of the stream or at different locations along the stream segment to ensure that the average characteristics of the stream segment order have been determined.

If flow records are available at a gauging station in the stream network, sample reaches in the vicinity but beyond the influence of the station should be included in the survey. High flow conditions predicted from channel characteristics can then be compared to the actual peak flows on record. For environmental monitoring studies, reaches of the same order number may be selected on other branches of the stream network, or in similar nearby basins, to distinguish between natural variation and artificial disturbances of the habitat.

Observations and Measurements in a Sample Reach

The hydrological characteristics of a sample reach can be determined with unsophisticated equipment from the observations presented in the following schedule:

1. Measure the present discharge.
2. Using several sample sections, determine the average cross section of the present flow in the reach.
3. Determine the average gradient of the stream bed in the reach.
4. Estimate the average cross section of the channel at the bankfull stage of the flow.
5. Determine the mean diameter of a randomly selected sample of the stream bed surface materials (50 to 100 samples).
6. Prepare a sketch of the sample reach showing the pattern of the bankfull channel, flood plains, and terraces; include notes on the geological materials that form the stream banks and valley walls. The sketch may be supplemented with photographs and other observations pertinent to the habitat.

The following notes on equipment and techniques have been prepared for streams similar in size to Wilson Creek (Figs. 11.3, 11.7). The techniques may be modified for rivers by using hydrometric procedures similar to those presented by Dunne and Leopold (1978). If the sample reaches are to be used as biological sampling sites as well, hydrologic measurements should be carried out after the sampling program is completed to avoid disturbing the substrate and associated benthic communities.

Observation 1: Present discharge. The velocity and discharge can be measured using standard metering techniques (Breed and Hosmer 1953; Dunne and Leopold 1978). In small, rock-filled streams, measurements with miniature current meters can be simplified by preparing a short section of channel with a uniform cross section by rearranging the bed materials. An estimate of the average flow velocity can be obtained by timing the passage of twigs or other floating debris through the prepared section to verify the proper operation of the current meter.

Observation 2: Present cross section. In small streams, the average cross section of the flow can be measured using a flexible tape stretched across the stream above the water surface to establish the horizontal distance to each depth measurement. Depths may be determined with a meter stick or a survey rod. If the channel in the sample reach is uniform, only three or four cross sections may be required to obtain an accurate average. However, additional cross sections will be required if the channel is tortuous and rock-filled, with frequent breaks in the water surface. The cross sections should be chosen to represent the variety of pools, riffles, chutes, and small ponds that may occur along the reach.

Observation 3: Average gradient. The average gradient of the channel may be determined by measuring the vertical drop along a measured length of stream bed at several locations in the sample reach. If the stream profile consists of a series of pools and riffles or short ponds and steep chutes, the measurements should extend over a sufficient length of channel to obtain the average gradient. If the slope of the sample reach is mild (i.e. <2%), a telescopic level and survey rod may be required to obtain an accurate measurement of the change in elevation over a measured length. For steeper reaches, the drop in elevation can be measured over 50 to 100 m lengths using a hand-held level or inclinometer.

Observation 4: Bankfull channel dimensions. Most stream surveys are conducted during low discharge periods, so the width and depth of the flow at the bankfull stage must be measured from an imaginary line that extends across the channel at the elevation of the flood plain (Fig. 11.15). Considerable judgment may be required to predict the bankfull stage. The entire sample reach should be examined before any sections are measured to determine the existence or the extent of flood plains, the general course of the central channel, the lengths of channel affected by local obstructions, and the textures of the stream bed and flood plain deposits. Cross sections may then be selected where the central channel boundaries are best defined.

Observation 5: Stream bed materials. Inorganic stream bed surface materials that pave the bed and affect the hydraulic roughness of the channel may be randomly sampled along the reach using several techniques. For example, wire grids may be laid on the stream bed and the mean diameter of the bed materials

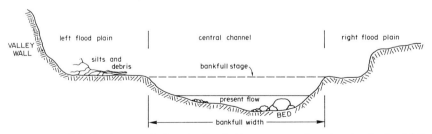

Figure 11.15   Idealized stream bed and valley bottom cross section showing the bankfull stage and width of the central channel.

can be measured at intersection points; or, randomly generated lengths may be measured from a central point to locate the sample sites. A less rigorous technique consists of walking through the reach and, without consciously preselecting the materials, stopping every few steps to measure the mean diameter of the bed materials underfoot. If 50 to 100 measurements are obtained and the sizes are arranged in a cumulative frequency curve, estimates of the median size of bed materials will not vary widely between techniques. Studies of aquatic insect microhabitats may require the use of more detailed sampling techniques, including the textural analysis of fine-grained deposits that occur in the spaces between larger cobbles and boulders. At the other extreme, only the largest sizes of the stream bed materials may be required to investigate stability of the stream bed under peak flow conditions.

Observation 6: Sketch and general observations. A sketch at a scale of approximately 1:1000 provides a useful reference to the general character and habitats of the sample reach (Tolkamp 1980; Oswood and Barber 1982; Minshall et al. 1982). Changes in the channel pattern, the distribution of vegetation, the valley bottom dimensions, and groundwater conditions may become apparent after the survey is completed when the sketch maps and observations for all of the sample reaches are compared.

For stream channels in unconsolidated materials rather than in bedrock, groundwater conditions in the stream bed may be determined qualitatively by comparing the temperature of the stream water with the temperature of the water below the stream bed. If the reach is in a zone of groundwater discharge, the stream water temperature during the summer months in mid-latitude streams will be greater than the temperature of the inflowing groundwater below the stream bed. The subsurface temperatures may be measured to a depth of 0.5 m using a metal thermistor probe and a portable telethermometer. Samples of the subsurface water may be obtained using seepage meters or miniature piezometers (Lee and Cherry 1978).

Characterization of Channels in the Drainage Network

The dominant or channel-forming discharge that occurs at the bankfull stage is the common link used to characterize the sample reaches. However, as this flood level is seldom observed, considerable interpretation of the sample reach measurements must be undertaken before a consistent reach-to-reach estimate of the flow can be made.

The observations and bankfull discharge estimates made by aquatic insect ecologists who attended a stream behavior workshop at Wilson Creek in 1981 are summarized in Table 11.4. The unadjusted bankfull velocity and discharge predictions were made using the measured bankfull channel dimensions and the Chezy equation with estimated Manning's roughness coefficients.

The relationship between bankfull discharge and drainage area for Wilson

TABLE 11.4: Stream Behavior Reconnaissance and Summary of Predicted Bankfull Conditions for Wilson Creek, Manitoba, September, 1981

SITE	ORDER NUMBER[a]	DRAINAGE AREA (km²)	PRESENT DISCHARGE (m³/s)	SLOPE	MEDIAN BED MATERIAL SIZE D_{50} (m)
Conway Creek	1	0.7	0.0	0.06	0.45
South Branch	2	3.0	0.002	0.06	0.36
Bald Hill Creek 1	3	8.6	0.011	0.036	0.34
Bald Hill Creek 2	3	10.6	0.012	0.037	0.23
Bald Hill Creek 3	3	11.1	0.010	0.027	0.22
Bald Hill Creek 4	3	12.7	0.015	0.015	0.16
Wilson Creek Weir	4	22.1	0.02	0.008	0.06

[a]1:50,000 map scale
[b]Estimated from observations on South Branch
[c]Estimated using photographic roughness guide (Barnes 1967)
[d]Estimated from median diameter of bed material, Strickler's equation (Henderson 1966)

Creek is similar to ones developed for streams in Wyoming and Idaho (Fig. 11.16), suggesting that simple observations, even when made by nonspecialists, can produce consistent estimates of bankfull flow conditions. The predicted bankfull flow at the sample reach immediately upstream from the Wilson Creek weir is 2.7 m³/s. Based upon a 19-year record for the Wilson Creek weir, the predicted discharge corresponds to an annual frequency of occurrence of 72 percent (Fig. 11.6), which does not vary widely from the 67 percent frequency of bankfull flows commonly observed in streams and rivers (Leopold et al. 1964).

The width, depth, and velocity of the sample reaches may be correlated individually with the bankfull discharge to discover anomalies in the observations taken at different sites (Fig. 11.17). The relationships can be written as three equations: $w = aQ^b$, $d = cQ^f$, and $v = kQ^m$, where w = average width, d = average depth, v = average velocity, and Q = the bankfull discharge. Since the product of w·d·v equals the bankfull discharge, the sum of the exponents b, f, and m and the product of the coefficients a, c, and k must both equal 1, or the observations are not consistent throughout the sample reaches. For the Wilson Creek data, $a \times c \times k = 3.02 \times 1.12 \times 0.30 = 1.02$ and $b + f + m = 0.31 + 0.09 + 0.61 = 1.01$.

Characterization of the Sediment Regime in the Basin

Once a consistent set of bankfull discharge estimates has been obtained, they may be combined with elevation changes on the longitudinal profile of the stream to indicate if, in general, reaches of net erosion or net deposition occur in

| BANKFULL | | | MANNING'S RESISTANCE "n" | | BANKFULL | | |
WIDTH (m)	DEPTH (m)	DEPTH D_{50}	PRESENT $(m^{1/6})$	PREDICTED BANKFULL $(m^{1/6})$	VELOCITY (m/s)	DISCHARGE (m^3/s)	TRACTIVE FORCE (kg/m^2)
2.0	0.29	0.6	—	0.20^b	0.5	0.3	17.4
3.0	0.24	0.7	0.23	0.08^c	1.2	0.9	14.4
3.8	0.30	0.9	0.38	0.06^c	1.4	1.6	10.8
3.5	0.33	1.4	0.24	0.05^c	1.8	2.1	12.2
3.5	0.31	1.4	0.66	0.05^c	1.5	1.6	8.4
3.6	0.32	2.0	0.20	0.03^d	1.9	2.2	4.8
4.3	0.35	6.4	0.05	0.025^d	1.8	2.7	2.8

the drainage basin. Natural streams cannot exist forever without net erosion occurring in some reaches unless other geomorphic processes restore sediments to the drainage basin. However, J. H. Mackin (1948) proposed that for short periods of geological time (i.e. less than a few thousand years) the main channel of a stream may achieve a "graded" state in which the slope of the stream profile is adjusted to provide, with the available discharge and prevailing channel characteristics, the velocity required to transport the sediment load delivered to the channel from the rest of the basin. However, the graded state is difficult to

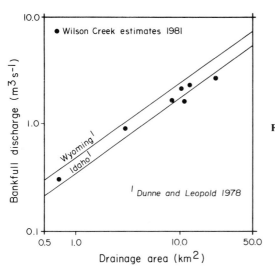

Figure 11.16 Average bankfull discharge-drainage area relationships for streams in two areas of the United States (summarized by Dunne and Leopold 1978) and the relationship (*dots*) for the Wilson Creek basin in Manitoba.

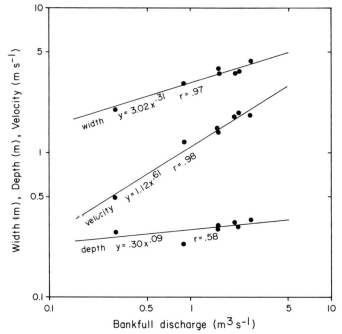

Figure 11.17 Channel geometry relationships for the south branch of Wilson Creek, Manitoba, using bankfull discharge estimates based on field surveys undertaken in September, 1981.

measure since most streamflow and sediment transport records have been gathered for only a few decades.

Leopold and Langbein (1962) proposed that the graded state exists when the powers of the bankfull discharge expended per unit length of a stream channel are equal. Power P (kw), is the product of the mass of the flow and the total drop in the reach, or P = 10Qh where Q = discharge (m³/s) and h = total drop in the reach (m). Equal power distribution per unit length of channel requires that the major portion of the stream profile be concave in a basin with uniform runoff characteristics. The small drainage areas and discharges of headwater channels are compensated for by steep gradients. At the mouth of the stream, the entire drainage area provides larger discharges that are compensated for by shallower gradients. The transition from steep to shallow gradients as the discharge increases produces the familiar concave profiles of many streams and rivers.

The distribution of power at bankfull stages in five equal reaches of channel along the Wilson Creek profile is summarized in Fig. 11.18. The upper two reaches exhibit approximately equal amounts of power at the bankfull stage, which suggests that the headwater channel in the escarpment exists in the graded state. The power decreases dramatically in reach 3, which is located on the flat-lying land at the bottom of the escarpment. Consequently, the coarse sediment load from the upper reaches is deposited, creating the apex of an alluvial fan. A slight increase in power occurs in reach 4, due to the lowering of the stream profile as it enters an excavated drainage canal in the agricultural zone below the escarpment. The excess energy in this reach has caused the stream to erode through 10 m of fan deposits since 1929. In the lowest reach of the stream, the power exerted is not sufficient to transport the sediment load from the upstream reaches and the drainage canal must be regularly excavated. The principal factors affecting the three major channel habitats, in Davis' framework, are,

in the upper reaches

$$\frac{\text{glacial}}{\text{till}} + \frac{\text{stream}}{\text{erosion}} + \text{time} \rightarrow \frac{\text{boulder-paved}}{\text{stable channels}}$$

in the middle reaches

$$\frac{\text{shale}}{\text{bedrock}} + \frac{\text{rapid weathering}}{\text{and stream erosion}} + \text{time} \rightarrow \frac{\text{mobile shale-}}{\text{filled streambeds}}$$

in the lower reaches

$$\frac{\text{shale}}{\text{deposits}} + \frac{\text{stream}}{\text{deposition}} + \text{time} \rightarrow \frac{\text{aggrading alluvial}}{\text{fan channels}}$$

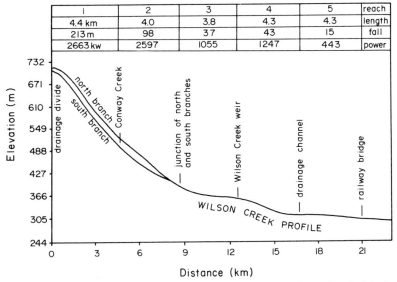

	1	2	3	4	5	reach
	4.4 km	4.0	3.8	4.3	4.3	length
	213 m	98	37	43	15	fall
	2663 kw	2597	1055	1247	443	power

Figure 11.18 Longitudinal profiles of the main branches of Wilson Creek, Manitoba, showing the distribution of hydraulic power at the bankfull discharge in five approximately equal reaches of the stream.

CONCLUSION

Most of this chapter has dealt with the concepts and empirical methods of hydrology that, if coupled to biological studies, may lead to a better understanding of aquatic insect habitats. Unfortunately, the determination of hydrological characteristics of aquatic insect habitats using field observations is like gathering the pieces of an ill-fitting jigsaw puzzle; some may be fitted to the puzzle immediately, while others require that the pattern and regime be resolved before a location can be found. In the field exploration example used in this chapter, the summary effect of highly variable stream processes, when combined with the geology of the Manitoba escarpment, produce the striking pattern and regime of Wilson Creek.

There are many combinations of hydrological processes and geological settings whose patterns can be discovered with a few simple field measurements within Davis' framework of structure, process, and time. Discovering the points of fitness (Henderson 1913) between the patterns of hydraulically determined habitats and biological systems is an absorbing challenge for all aquatic researchers. Several avenues of inquiry may lead to new discoveries. For example, what is the correlation between the spatial distribution or complexity of aquatic habitats and the ordered branching characteristics of stream networks? Can habitat suitability be related to thresholds of hydrologic behavior based on frequency or duration, and intensity? Can changes in the distribution and composition of biological communities be related to the downstream changes in the hydraulic geometry of stream channels?

The river continuum concept proposed by Vannote et al. (1980) suggests that the distribution of insect communities is analogous to the distribution of power and corresponding changes in the channel geometry and discharge in the basin. However, just as the graded state is an idealized concept of equilibrium conditions that river systems seek but seldom achieve without interference from other geomorphic factors, the structured distribution of stream communities of the continuum ideal may be disturbed by the ". . . continual, but not necessarily predictable recruitment of materials and organisms from outside" (Winterbourn et al. 1981, p. 326). The use of some of the techniques for the measurement and characterization of stream habitats presented in the previous section, in conjunction with biological sampling, may resolve the continuum debate.

References

Barnes, H. H., Jr. 1967. Roughness characteristics of natural channels. U. S. Geological Survey Water-Supply Paper 1849. U. S. Department of the Interior, Washington, DC. 213 pp.

Bovee, K. D. and T. Cochnauer. 1977. Development and evaluation of weighed criteria, probability-of-use curves for instream flow assessments: fisheries. Instream Flow Information Paper No. 3. U. S. Fish and Wildlife Service, Ft. Collins, CO. 39 pp.

Bovee, K. D. and R. T. Milhous. 1978. Hydraulic simulation in instream flow studies: theory and techniques. Instream Flow Information Paper No. 5. U. S. Fish and Wildlife Service, Ft. Collins, CO. 147 pp.

Brater, E. F. and H. W. King. 1976. Handbook of hydraulics. 6th ed. McGraw-Hill Book Co., New York, NY. 573 pp.

Breed, C. B. and G. L. Hosmer. 1953. The principles and practice of surveying. Vol. II. Higher surveying. 7th ed. John Wiley and Sons Inc., New York, NY. 675 pp.

Chow, V. T. 1959. Open channel hydraulics. McGraw-Hill Book Co., New York, NY. 680 pp.

Chow, V. T. (ed.). 1964. Handbook of applied hydrology. McGraw-Hill Book Co., New York, NY. 1200 pp.

Crawford, P. J. and D. M. Rosenberg. In press. The breakdown of conifer needle debris in a new northern reservoir, Southern Indian Lake, Manitoba. Canadian Journal of Fisheries and Aquatic Sciences.

Cummins, K. W. and G. H. Lauff. 1969. The influence of substrate particle size on the microdistribution of stream macrobenthos. Hydrobiologia 34:145–81.

Davis, S. N. and R. J. M. De Wiest. 1966. Hydrogeology. John Wiley and Sons Inc., New York, NY. 463 pp.

Davis, W. M. 1909. Geographical essays. Ginn, New York, NY. 777 pp.

Dunne, T. and L. B. Leopold. 1978. Water in environmental planning. W. H. Freeman and Co., San Francisco. 818 pp.

Fudge, R. J. P. and R. A. Bodaly. In press. Post-impoundment winter sedimentation and survival of lake whitefish (*Coregonus clupeaformis* (Mitchill)) eggs in Southern Indian Lake, Manitoba. Canadian Journal of Fisheries and Aquatic Sciences.

Gore, J. A. 1978. A technique for predicting in-stream flow requirements of benthic macroinvertebrates. Freshwater Biology 8:141–51.

Gore, J. A. and R. D. Judy, Jr. 1981. Predictive models of benthic macroinvertebrate density for use in instream flow studies and regulated flow management. Canadian Journal of Fisheries and Aquatic Sciences 38:1363–70.

Gregory, K. J. and D. E. Walling. 1973. Drainage basin form and process: a geomorphological approach. Edward Arnold (Publishers) Ltd., London. 458 pp.

Henderson, F. M. 1966. Open channel flow. MacMillan Publishing Co. Inc., New York, NY. 522 pp.

Henderson, L. J. 1913. The fitness of the environment, an inquiry into the biological significance of the properties of matter. Beacon Press, Boston. 317 pp.

Hendricks, T. 1977. Current velocities required to move sediments, pp. 71–76. *In*: The Southern California Coastal Water Research Project Annual Report for the Year Ended 30 June 1976. Southern California Coastal Water Research Project, El Segundo, CA. 268 pp.

Lane, E. W. 1955. Design of stable channels. Transactions of the American Society of Civil Engineers 120:1234–60.

Lee, D. R. and J. A. Cherry. 1978. A field exercise on groundwater flow using seepage meters and mini-piezometers. Journal of Geological Education 27:6–10.

Leopold, L. B. and W. B. Langbein. 1962. The concept of entropy in landscape evolution.

U. S. Geological Survey Professional Paper 500-A. U. S. Department of the Interior, Washington DC. 20 pp.

Leopold, L. B., M. G. Wolman, and J. P. Miller. 1964. Fluvial processes in geomorphology. W. H. Freeman Co., San Francisco. 522 pp.

Mackin, J. H. 1948. Concept of the graded river. Geological Society of America Bulletin 59:463–512.

Magalhaes, L. and T. Chau. In press. Initiation of motion conditions for shale sediments. Canadian Journal of Civil Engineering.

Minshall, G. W., J. T. Brock, and T. W. Lapoint. 1982. Characterization and dynamics of benthic organic matter and invertebrate functional feeding group relationships in the Upper Salmon River, Idaho (USA). Internationale Revue der gesamten Hydrobiologie 67:793–820.

McCullough, G. K. 1978. Approaches to shoreline mapping on Southern Indian Lake, pp. 325–30. In: Application of ecological (biophysical) land classification in Canada. Ecological Land Classification Series No. 7. Environment Canada, Ottawa, Canada. 396 pp.

Oswood, M. E. and W. E. Barber. 1982. Assessment of fish habitat in streams: goals, constraints, and a new technique. Fisheries 7:8–11.

Reice, S. R. 1980. The role of substratum in benthic macroinvertebrate microdistribution and litter decomposition in a woodland stream. Ecology 61:580–90.

Resh, V. H. 1979. Sampling variability and life history features: basic considerations in the design of aquatic insect studies. Journal of the Fisheries Research Board of Canada 36:290–311.

Resh, V. H., D. M. Rosenberg, and A. P. Wiens. In press. Emergence of caddisflies (Trichoptera) from eroding and non-eroding shorelines of Southern Indian Lake, Manitoba, Canada. The Canadian Entomologist.

Schwartz, F. W. 1970. Geohydrology and hydrogeochemistry of groundwater: streamflow systems in the Wilson Creek experimental watershed. M.Sc. thesis, University of Manitoba, Winnipeg, Canada. 120 pp.

Statzner, B. 1981a. The relation between "hydraulic stress" and microdistribution of benthic macroinvertebrates in a lowland running water system, the Schierenseebrooks (North Germany). Archiv für Hydrobiologie 91:192–218.

Statzner, B. 1981b. A method to estimate the population size of benthic macroinvertebrates in streams. Oecologia 51:157–61.

Statzner, B. 1981c. Shannon-Weaver diversity of the microbenthos in the Schierenseebrooks (North Germany) and problems of its use for the interpretation of the community structure. Internationale Vereinigung für Theoretische und Angewandte Limnologie Verhandlungen 21:782–86.

Statzner, B. and T. F. Holm. 1982. Morphological adaptations of benthic invertebrates to streamflow—an old question studied by means of a new technique (Laser Doppler Anemometry). Oecologia 53:290–92.

Tolkamp, H. H. 1980. Organism-substrate relationships in lowland streams. Agricultural Research Report 907. Agricultural University, Wageningen, The Netherlands. 211 pp.

Townsend, C. R. 1980. The ecology of streams and rivers. Edward Arnold (Publishers) Ltd., London. 68 pp.

U. S. Army Corps of Engineers. 1966. Shore protection, planning and design. Coastal

Engineering Research Center Technical Report No. 4. 3rd ed. U. S. Army Corps of Engineers, Washington DC. 580 pp.

Vannote, R. L., G. W. Minshall, K. W. Cummins, J. R. Sedell, and C. E. Cushing. 1980. abundance, structure, and composition of mussel beds. Proceedings of the National Academy of Sciences of the United States of America 79:4103-7.

Vannote, R. L., G. W. Minshall, K. W. Cummins, J. R. Sedell and C. E. Cushing. 1980. The river continuum concept. Canadian Journal of Fisheries and Aquatic Sciences 37:130-37.

Vogel, S. 1981. Life in moving fluids. The physical biology of flow. Willard Grant Press, Boston. 352 pp.

Winterbourn, M. J., J. S. Rounick, and B. Cowie. 1981. Are New Zealand stream ecosystems really different? New Zealand Journal of Marine and Freshwater Research 15:321-28.

Wolman, M. G. 1977. Changing needs and opportunities in the sediment field. Water Resources Research 13:50-54.

chapter 12

AQUATIC INSECT-SUBSTRATUM RELATIONSHIPS

G. Wayne Minshall

INTRODUCTION

The substratum is the stage upon which the drama of aquatic insect ecology is acted out. It is the medium upon which aquatic insects move, rest, find shelter, and seek food. They share the substratum with annelids, molluscs, crustaceans, fish, and other animals. The substratum determines to a large extent the micro-environmental conditions under which the insects live, and thus it profoundly affects their growth and survival. It is also the arena in which the outcome of competition and predation is determined.

The substratum consists of various types of organic and inorganic materials, and can be virtually anything sufficiently stable for insects to crawl on, cling to, or burrow in. Organic substrates commonly consist of leaves, branches, or grass that are derived from the land; filamentous algae, moss, and vascular hydrophytes that originate in the water; and even other aquatic insects. The nature of inorganic substrates is determined largely by the composition of the underlying parent materials, but may be modified by transport of materials from the surrounding watershed or by human activities. Inorganic substrates usually are granitic or sedimentary materials ranging in size from microscopic silts to large boulders, but they also may include bottles, cans, concrete, automobiles, and other man-made objects.

In areas of still, deep water the substratum may be relatively homogeneous; but in areas of current, or in still areas where rooted plants occur because sunlight penetrates to the bottom, the substratrum may be more diverse. On wave-swept shores of lakes and reservoirs, and in streams, the substratum is subject to the sorting action of the current. Hence, the structure of the bottom is inseparably linked to variations in flow, and this results in a constantly varying mosaic (both temporal and spatial) of substrate types, each with different environmental conditions for insects. In ponds and lowland streams, and on the sloping shores of sheltered coves in larger bodies of water, a diverse assemblage of aquatic macrophytes may develop and this may provide a variety of substrate types under otherwise homogeneous conditions.

The purpose of this chapter is to provide a comprehensive review of aquatic insect-substratum relationships. I have tried to illustrate these relationships with examples from published works where appropriate, and to provide sufficient documentation, particularly to the literature since 1965, to allow easy access to additional information. Cummins (1966) briefly reviewed most of the literature on insect-substratum relationships through about 1964; the most thorough and recent review to date specifically on this topic was by Hynes (1970), covering the literature through early 1966.

One goal of this chapter is to identify and examine the important components of the substratum that affect the ecology of aquatic insects. A second goal has been to illustrate how these components interact with and/or are affected by other factors in the insect's environment. Finally, I have tried to indicate where information is lacking or is in conflict, or where alternative explanations for reported phenomena are possible. The critical reader should be able to determine from this presentation in what directions the most promising lines for future research lie.

My coverage of the literature and choices of examples may seem biased toward flowing waters. This could stem from the fact that, like those of previous reviewers of this topic (Cummins 1966; Hynes 1970), my own interests lie largely in this area; however, it is also true that substratum relationships have been much more thoroughly explored in lotic than in lentic habitats. The latter may be due to the relatively greater ease of obtaining samples in environments that can be waded, and to the more heterogenous (and therefore generally more interesting) nature of lotic, as compared with lentic, substrates and fauna. This is not intended to imply that investigation of lentic habitats should be avoided. Benthic habitats in these waters are easily and directly accessible by means of SCUBA equipment, and the taxonomically restricted fauna permits more concentrated study of species-specific relations (especially among the Chironomidae—e.g. see McLachlan 1969; Cantrell and McLachlan 1977) without the confounding effects of current. In fact, the seemingly homogeneous distribution of the bottom fauna may actually be more heterogeneous (e.g. see Patterson and Fernando 1971; Shiozawa and Barnes 1977) than has been assumed (as is the case with zooplankton above it, George 1981). It already is known that the presence of aquatic plants or mixed substrates can lead to increased diversity of the fauna in lentic habitats (e.g. see Rosine 1955; Lyman 1956; Macan 1965, 1975, 1977; Macan and Maudsley 1968, 1969).

APPROACHES TO THE STUDY OF INSECT-SUBSTRATUM RELATIONSHIPS

Although recognition of the importance of substratum to aquatic insects may be traced back to the turn of this century, the first significant research on aquatic insect-substratum relationships was published by Percival and White-head (1929). In streams of northeastern England, they examined the macroin-vertebrate fauna associated with seven types of substratum that differed in degree of stability and in type and amount of plant cover. Although there are several shortcomings to their study (see Macan 1963), it established a standard of quality that was not exceeded for at least 30 years. A less well-known, but comparable, study was done by Hunt (1930) between 1925 and 1928 on several streams in the southeastern United States. Percival and Whitehead (1926) also examined the distribution of the burrowing mayfly *Ephemera danica* in relation to substratum particle size, as well as to light and current velocity. Over the next 25 years (1930–1955) only a few investigations dealt specifically with insect-substratum relationships, most notably those by Moon (1939), Wene (1940), Wene and Wickliff (1940), Linduska (1942), Pennak and Van Gerpen (1947), Sprules (1947), and Berg et al. (1948), in flowing waters; and Krecker and Lancaster (1933), Krecker (1939), Moon (1935a, 1935b, 1940), and Rosine (1955), in standing waters. The works of Moon and of Wene and Wickliff are especially noteworthy because they introduced the technique of using substrate-filled baskets for studying the effects of contrasting types of substrate on the insect fauna; it was to become the forerunner of much experimental work from the mid-1960s to the present. These early studies demonstrated clear differences in species' preference for, and the carrying capacities of, different substrata. However, they were hampered by an inability to isolate the effects of various confounding factors, especially food and current.

An alternative to the descriptive approach, the experimental manipulation of substrate conditions, was used early on by a few investigators (e.g. see Percival and Whitehead 1926; Cianficconi and Riatti 1957), and by the mid-1960s many researchers had begun to control the confounding variables. Sometimes this has involved the use of laboratory microcosms in conjunction with field observations (e.g. see Cummins and Lauff 1969; Brusven and Prather 1974; Luedtke and Brusven 1976; Mackay 1977; Tolkamp 1980). However, in order to maintain more realistic conditions, most investigators accepted the challenge to conduct experiments directly in the natural environment (e.g. see Lillehammer 1966; Allan 1975; Luedtke et al. 1976; Macan and Kitching 1976; Minshall and Minshall 1977; Rabeni and Minshall 1977; Hart 1978; Shelly 1979; Trush 1979; Khalaf and Tachet 1980; Reice 1980; Williams 1980; Gregg 1981; Erman and Erman, in press). As a result of these efforts, substratum relationships have become one of the most intensively studied areas of freshwater benthic invertebrate ecology, and the researchers on this topic have pioneered the application of the experimental approach to a wide variety of questions. This represents a significant change in direction in the years since Cummins' (1966) survey of the literature on insect-substratum relationships; and, as the research reviewed in this chapter will show, the progress made in that time not only has been phenomenal, but the prospects for the foreseeable future are equally exciting.

IMPORTANCE OF SUBSTRATUM

Substratum acts directly on insects as a medium for their existence, and indirectly as a major modifier of their environment. As a direct influence, substratum can either restrict or enhance an insect's ability to adhere, cling, or burrow, along with its ability to escape from predators, be protected from current or disturbance, construct cases, or deposit eggs. Sedentary insects (e.g. *Leucotrichia, Parargyractis, Rheotanytarsus,* and *Simulium*) require attachment sites for feeding, growth, and pupation. Suction discs, which function only on relatively smooth surfaces, enable some mayfly nymphs (e.g. *Epeorus, Rhithrogena,* and *Ephemerella doddsi*) and blepharocerid fly larvae to withstand the force of the current. A number of aquatic insects in the orders Ephemeroptera, Odonata, and Trichoptera are restricted to vascular plants partly because these provide an especially satisfactory substrate for clinging. Several mayflies (e.g. *Ephemera, Hexagenia*) and midges (e.g. *Chironomus plumosus*) burrow into the substrate, and thus are restricted to substrates they can penetrate. Other insects, such as the stoneflies *Capnia* and *Leuctra,* require looser particles to slither between. Some aquatic insects (e.g. the stonefly *Hesperoperla pacifica,* the mayfly *Ephemerella grandis,* and the caddisfly *Rhyacophila vaccua*) are more susceptible to predation by fish (e.g. the sculpin *Cottus*) when on bare sand substrate than when pebbles or cobbles are present (Brusven and Rose 1981).

Case-building insects need substratum materials to build their shelters. Trichoptera are the major group of benthic invertebrates that use mineral and / or organic material to build their cases, although case- and tube-building Chironomidae and some case-building Lepidoptera also occur. The composition of the substratum determines the availability of materials for case construction and attachment, and thereby can control the success of these organisms. For example, *Sericostoma personatum* uses mainly 0.25- to 0.50-mm grains for its case, and the range that can be used is restricted to 0.125 to 1.0 mm. Furthermore, it prefers to live in areas of predominantly coarse substrates, which means that a mixture of appropriately sized materials must be available (Tolkamp 1980). Larvae of another caddisfly, *Pycnopsyche scabripennis,* are so selective that even when they are able to inhabit a broad range of stream types their abundance within these streams may be limited by the extent of 4- and 8-mm gravel available for aestivation and pupation sites (Mackay 1977).

Another direct effect of substratum is its influence on the respiration rate of certain insects (Wautier and Pattée 1955; Eriksen 1963, 1964, 1966). For example, Eriksen (1963, 1964) found that oxygen uptake by the mayflies *Ephemera simulans* and *Hexagenia limbata* showed distinct maxima and minima when tested over a range of particle sizes. However, when nymphs were given an opportunity to select substrates, preference was shown for those that produced the lowest rate of respiration. Reasons for the variation in oxygen consumption probably differ among aquatic insect species, but in these mayflies it appears to be due to differences in activity associated with the ease of burrowing.

The substratum also exerts a number of influences that can indirectly affect the suitability of the aquatic environment for insects. Small inorganic particles may fill the spaces between larger particles, thereby limiting both access by the insects and the movement of water and dissolved gases through the substratum. These finer particles may also blanket the surface of the bed, thereby restricting algal growth or diluting the organic content of the food supply, as well as directly impeding the movement, feeding activities, or respiration of the insects. An excess of decaying organic particles may lead to a severe reduction in oxygen concentrations. The substratum also can indirectly impinge upon insects through alteration of the direction, force, and turbulence of water flow (see also Newbury, Chapter 11). Finally, bottom materials can influence the type, particle size composition, and quantity of plant organic matter, and the abundance and availability of prey items. Any of these alterations in trophic conditions can significantly affect the structure of the benthic insect community.

INSECT RESPONSES TO DIFFERENCES IN COMPOSITION AND SIZE OF SUBSTRATE

One of the earliest discoveries concerning insect- (and other macroinverte-brate-) substration relationships was that total abundance of animals in the community varied with substrate type, and most (but not all) taxa showed individual "preferences" for different types of substrate. Initially most of these observations concerned substrates that varied considerably (e.g. stones compared to plants: Percival and Whitehead 1929; Lyman 1956), but gradually it became apparent that marked differences in the fauna occurred even on different sizes of inorganic particles (e.g. see Linduska 1942; Pennak and Van Gerpen 1947) or on different types of vascular plants (e.g. Krecker 1939; Rosine 1955; Harrod 1964).

At first the influence of different types and sizes of substratum could not be clearly separated from the influence of current and other variables. But as evidence accumulated from a number of independent sources (Cummins 1964; Cummins and Lauff 1969; Mackay and Kalff 1969; McLachlan 1969; Allan 1975; Higler 1975; Ward 1975; Ali and Mulla 1976; de March 1976; Mackay 1977; Minshall and Minshall 1977; Rabeni and Minshall 1977; Williams and Mundie 1978; Khalaf and Tachet 1980; Reice 1980), the importance of substratum (as distinct from that of other factors examined) was strongly implicated. Later field studies have served to extend documentation of insects' substrate preferences to additional materials (e.g., mosses: Thorup and Linde-gaard 1977; and macrophytes: Gregg 1981), and specific environments (e.g. lowland streams; Tolkamp 1980).

Several generalizations on substrate composition and size have developed from the research just cited. These are (1) aquatic plant substrates support higher densities of animals than do mineral substrates, and plant species may differ considerably in this regard; (2) large inorganic substrate particles are more productive than small-sized ones; and (3) preferences for a given substratum differ among insect species. It is useful to examine the information bearing on each of these generalizations, and to note some important qualifications.

Abundance on Aquatic Plant vs. Mineral Substrates

Percival and Whitehead (1929) found the lowest densities of insects and other invertebrates on bare stones, intermediate densities on the alga *Cladophora* and clumps of loose moss, and highest densities on thick moss and river weed. Minckley (1963) also found lowest densities on bare substrates (sand, stones), intermediate densities on the vascular hydrophytes *Nasturtium, Myriophyllum,* and *Myosotis,* and highest densities on moss. Few other investigators have examined insect abundances on such a wide range of materials, but the evidence pieced together from more restricted analyses (e.g. see Whitehead 1935; Harrod 1964; Lillehammer 1966; Barber and Kevern 1973; Lindegaard et al. 1975; Gregg 1981) tends to support their observations. For example, Lillehammer (1966) found more insects $(606/m^2)$ on stones with thick moss than on those without $(471/m^2)$. This is supported by the results of Hynes (1961), Minckley (1963), and Egglishaw (1969). Whitehead (1935) observed denser populations of invertebrates on the vascular plant *Ranunculus* than on the macrophyte *Sium erectum,* which in turn had higher densities than loose stones. Gregg (1981) recorded higher densities (numbers/sample) of invertebrates on *Ranunculus* than on bare substrates, but found no significant difference between numbers on watercress (*Rorippa*) and on substrates lacking plants. However, a more important finding was that when the results were expressed as numbers per unit of available surface, significantly fewer invertebrates were found on either *Rorippa* (1,933; 26% insects) or *Ranunculus* (2,455; 34% insects) than on bare substrates (12,183; 43% insects). These results suggest that the main factor responsible for the widely observed differences between aquatic plant and mineral substrates is the increased surface area afforded by the plants.

Relation of Abundance to Particle Size

It is generally believed that benthic insects and other invertebrates increase in numbers over the sequence of increasing particle sizes from sand through large rocks or boulders (e.g. see Needham 1927; Tarzwell 1938; Wene and Wickliff 1940; Sprules 1947; Pennak and Van Gerpen 1947; Allan 1975). This generalization has a sound ecological basis and is supported by recent experiments under controlled conditions. However, under actual conditions found in streams, seemingly contradictory results have been obtained, suggesting that other factors related to substratum characteristics (especially substrate heterogeneity and food) also may be involved.

The findings of Pennak and Van Gerpen (1947) are often cited as a prime example in support of the pattern whereby invertebrate abundance increases with increasing particle size. Pennak and Van Gerpen found a progressive increase in total numbers and biomass from sand through rubble (Table 12.1), and this was substantiated by work on the same site done 19 years later (Ward 1975). However, both Pennak and Van Gerpen and Ward found that numbers and biomass decreased when substratum size increased to bedrock (i.e. surfaces of large boulders). This latter finding, coupled with the low numbers found in the smaller (less diverse) substrates, suggests that the observed changes may be due to (or at least complicated by) changes in substratum heterogeneity. For example, Uresk (1967) noted that natural stream substrate (comparable to the rubble category in Table 12.1) was actually a heterogeneous mixture of materials ranging from sand ($\leq$ 3.2 mm) through cobble (150 mm). In a series of tests comparing the fauna in this mixed substrate with that in uniform substrates, he found that the uniform-sized material supported fewer insects than the mixed. When different sizes of uniform substrate were examined, he found that the populations decreased as the size of the substrate increased (e.g. densities on 7.5 mm > on 50 mm > on 100 mm).

However, even when differences in heterogeneity are removed through the use of uniform particle sizes, a direct relationship between rock size and numbers of invertebrates is not always found. For example, Minshall and Minshall (1977) consistently found higher total densities on their small substratum (25 mm) than on their large one (60 mm), and Rabeni and Minshall (1977) found that the

TABLE 12.1: **Mean Abundance and Biomass of Macroinvertebrates Collected in North St. Vrain Creek, Colorado, in the Summers of 1945 by Pennak and Van Gerpen (1947), and 1974 by Ward (1975)**

PARTICLE SIZE	YEAR OF STUDY	SAND (1.5–3 mm)	GRAVEL (6–25 mm)	RUBBLE (30–200 mm)	BEDROCK (boulder surfaces)
Number of	1945	202	575	610	551
individuals/m²	1974	141	261	274	136
Biomass	1945	0.6	1.3	2.5	1.7
(g wet wt)/m²	1974	2.2	2.3	3.4	0.7

amount of colonization increased over their 6-, 15-, and 30-mm particle classes, but was markedly reduced on the largest ($\sim$45 mm) size. The higher densities on the 30-mm material were attributed to this size substratum being a more efficient collector of fine particulate organic matter than those substrata that were larger or smaller. Support for the idea that intermediate-sized materials maintain the highest densities also comes from the work of Williams and Mundie (1978). However, unlike Pennak and Van Gerpen (1947) and Ward (1975), they found that biomass was greatest on small (12-mm) particles, and progressively less on the large (41-mm), and medium (24-mm) sizes.

Insect Associations with Specific Substrates

A number of investigators have reported that more individuals of a certain species (or higher taxon, e.g. genus, family) occur on one type or size of substratum than on others, and have concluded that the taxon "prefers" that substratum. However, as noted by Tolkamp (1980), relationships with certain substrate types that are deduced from over-representation on these materials are not necessarily the result of preferences for the substrates because "preference" implies active choice of optimal conditions. The matter is further complicated because not all species show preferences (Shelly 1979; Khalaf and Tachet 1980; Tolkamp 1980), or show them consistently. The association of at least some species with different particle sizes or materials can change during the life cycle (Egglishaw 1969; Mackay and Kalff 1969; Thorup and Lindegaard 1977), and this is especially common among the case-building Trichoptera (Scott 1958; Hanna 1961; Cummins 1964; Higler 1975; Otto 1976; Hildrew and Townsend 1977; Mackay 1977; Tolkamp 1980). Even those species that show a preference usually are found over a wide range of substrates, and they presumably can grow and develop on them (e.g. see Ali and Mulla 1976). Also, since some substrates clearly are unsatisfactory (e.g. too unstable or too small relative to the size of the animal), they automatically skew the distributions toward certain substrate sizes. Finally, in all of the studies done to date, it is not certain that other microscale factors may in fact have produced the observed patterns. This is especially evident when taxa that are known to be regulated primarily by factors other than substratum (e.g. as the black fly *Simulium* is by current velocity and/or food) are purported to show a preference for a particular particle size or type of material.

In the face of these uncertainties, one approach would be to seek confirmation of these inferred preferences from studies done by different workers under a variety of conditions. However, this is virtually impossible at present due to (1) a failure or inability to identify or name certain test organisms to species, (2) the wide variety of species that have been (and have to be) dealt with, and (3) the lack of standardization in the size and range of particle sizes used or distinguished by the researchers. For example, even in the Rocky Mountain area of the United States, where the largest number of studies on insect-substratum relationships has been done, not a single unequivocal comparison is possible for the aforementioned reasons.

Some of the strongest evidence of preference comes from the laboratory experiments of Cummins and Lauff (1969). However, even here the results are not totally satisfying because (1) relatively few (i.e. ten) species were examined, and then only at certain points in their life cycles; (2) current velocities were abnormally low (3 cm/s); (3) some of the results varied depending on placement of the different particle sizes relative to the current; and (4) there was not always agreement between results obtained in the laboratory and those found in nature. Nevertheless, although these qualifiers illustrate the magnitude of the problems facing future investigators, the approach used by Cummins and Lauff is undoubtedly a promising one. For example, in spite of these difficulties, Cummins and Lauff were able to demonstrate primary habitat selection on the basis of substrate size for the stonefly *Perlesta placida*, the riffle beetle *Stenelmis crenata*, and the caddisflies *Pycnopsyche guttifer* and *Pycnopsyche lepida*. Although they examined a much broader range of species, both Khalaf and Tachet (1980) and Tolkamp (1980) also found that only about half of the species showed a definite preference for substratum size.

Mackay (1977) reported substratum preference for three other species of *Pycnopsyche (gentilis, luculenta, scabripennis)* over most of their life cycles, but only the two species that use mineral particles for cases or burrows demonstrated a particle size preference, and these only for short periods (e.g. when the 5th instar leaf-disc cases were transformed to sand-grain cases by *P. gentilis* or when mature larvae of *P. scabripennis* burrowed into gravel for aestivation). Otherwise, all three species responded positively only to organic substrates, and showed no mineral substrate selection that was coincident with larval distribution. Brusven and Prather (1974) examined the preference of five species of aquatic insects under laboratory conditions. The stonefly *Pteronarcys californica* and the caddisfly *Arctopsyche grandis* preferred a substrate of large pebble (12–25 mm) over small pebble (6–12 mm), coarse sand (2.5–6 mm), or fine sand (1–1.5 mm). The mayfly *Ephemerella grandis* and the caddisfly *Brachycentrus* sp. showed moderate preference for large pebble over coarse and fine sand, but little differentiation between large and small pebble substrates. The dipteran *Atherix variegata* showed little preference for one particular size over another.

McLachlan (1969) is one of the few to have examined particle size selection by lentic insects. He experimentally demonstrated the preference of the chironomid *Nilodorum brevibucca* for silt, and concluded that substrate selection depended on the suitability of the material for tube construction.

For the present, it appears that although some aquatic insects actively choose specific substrates, most are substratum generalists. Even where definite preference exists, these may change during the life cycle and thereby go undetected with the usual methods of analysis, especially with species where overlapping cohorts occur (which is the common case). There clearly is room for additional research on substratum preferences, such as when an array of different insect species is examined under carefully documented or controlled conditions, possibly using direct observation, time-lapse photography, or the tagging of individual insects.

SUBSTRATE CHARACTERISTICS AND THEIR MEASUREMENT

We are not certain which features of the substratum are important to an aquatic insect, but they certainly include some aspect of size, stability, and heterogeneity. These three features are strongly implicated as important ecological factors from theoretical considerations (e.g. see MacArthur and Wilson 1967; Smith 1972; Southwood 1977; Huston 1979), and from recent work in marine (Kohn 1967, 1968; Osman 1978; Sousa 1979) and freshwater (Hart 1978; Shelly 1979; Trush 1979; Reice 1980) environments. Other characteristics of the substratum likely to be of importance to the insect fauna are its texture and porosity, and the amount of silt and organic matter. If the substratum is primarily organic, then additional features including the type of organic matter, the degree of conditioning by microbial and physical processes, and its position (e.g. floating, submerged, buried) may be important. At present, it appears that the following features are necessary to adequately characterize the substratum, and therefore should be controlled or measured in any study of insect-substratum relationships: particle size, surface area, stability, heterogeneity, texture, pore space, amount of silt, and quantity and kind of organic matter.

Particle size (i.e. diameter), surface area, and stability are three factors that are closely related, yet each is distinct. Particle size as determined by particle diameter is the most common measurement made in defining substratum conditions. However, it is not the only important aspect of size and, with the exception of the rather precise particle selection generally found for case-building species (e.g. see Hanna 1961; McLachlan 1969) and a few burrowing forms (e.g. see Eriksen 1968), it probably is not a primary factory on which organisms key. However, particle diameter is an index to surface area and stability. The larger the diameter of a particle, the larger its surface area will be, and (in general) the less likely the particle will move. Likewise, some measure of particle size diversity can serve as a measure of substratum heterogeneity.

Particle Diameter

Particle diameter is determined by direct measurement, mechanical sieving, determination of settling velocities, or by a combination of these methods (Guy 1969; Leopold 1970). A standard categorization of particle diameters (Table 12.2) has been available for about 60 years (Wentworth 1922), although it was not widely used by freshwater ecologists before the mid-1960s (Cummins 1962). Even then its use was restricted largely to North America, and only recently has it begun to be applied in Europe (Friberg et al. 1977; Gee 1979; Tolkamp 1980). The Wentworth classification is recommended because it is possible to convert the geometric particle size categories, in which the diameter of each particle size fraction is twice the preceeding one, into an arithmetic one with equal class intervals (i.e. the phi scale). Phi is defined as the negative log to the base 2 of the particle size diameter in millimeters; the divisions of the Wentworth classification are whole integers on the phi scale (Table 12.2).

The most commonly reported value for particle diameter is the median diameter (Md). This is equal to the 50th percentile and may be obtained from a plot of cumulative percent composition against particle size (Krumbein 1939; Inman 1952), or it can be calculated directly (Tolkamp 1980). Other measures of central tendency are the mean (= 16th + 84th percentiles) and mode of the

TABLE 12.2: Wentworth Classification of Substratum Particle Size*

SIZE CATEGORY	PARTICLE DIAMETER (range in mm)	PHI VALUE ($-\log_2$ particle diameter in mm)
Boulder	>256	≤ -8
Cobble		
Large	128–256	-7
Small	64–128	-6
Pebble		
Large	32–64	-5
Small	16–32	-4
Gravel		
Coarse	8–16	-3
Medium	4–8	-2
Fine	2–4	-1
Sand		
Very coarse	1–2	0
Coarse	0.500–1	1
Medium	0.250–0.500	2
Fine	0.125–0.250	3
Very fine	0.063–0.125	4
Silt	<0.063	≥ 5

*After Cummins 1962; Tolkamp 1980

diameter (= the 50% value of the class interval or intervals containing the most material) (Williams 1972). However, comparison of results obtained from laboratory and field determinations (e.g. see Cummins and Lauff 1969; Tolkamp 1980) indicates that the median particle diameter (or other measures of central tendency) alone will not permit accurate description of particle size requirements. This is because many organisms are found in small patches of material that are scattered among other particles, although the amounts of material in these patches are insufficient to have an appreciable effect on the median. Furthermore, since a mixture of particles may be needed to meet the needs of an animal (see Tolkamp 1980), more information than is available from just a measure of the predominant size class is often necessary. Finally, although the Wentworth classification is convenient to use and some standard form is needed to facilitate comparison with results of different investigators, it is uncertain (even unlikely) that the values have precise meaning to the insects.

Particle Area

In addition to particle diameter, the surface area available for occupancy may be an important determinant of the distribution and abundance of aquatic insects. Several methods are available for direct assessment of the surface areas of irregularly shaped substrate particles (Calow 1972; Dahl 1973; Minshall and Minshall 1977; Shelly 1979; Trush 1979; Reice 1980). A common method for rocks is to obtain an impression of the surface using latex or aluminum foil and then, by using a planimeter or by weighing (with appropriate conversion), the area can be determined. On relatively uniform particles, a satisfactory estimate of area can be made using the mean diameter and a standard formula for the appropriate geometric shape (e.g. see Khalaf and Tachet 1980).

In an early study of the influence of surface area on benthic organisms, Sprules (1947) compared numbers of insects emerging from areas with rocks imbedded in gravel with the numbers emerging from rocks piled upon one another; six times as many insects emerged from the latter. Likewise, in a study by Glime and Clemons (1972) the surface-rich moss supported more insects than did strands of plastic or string. Gregg (1981) showed that increased surface area explained the higher densities of invertebrates on aquatic plants than on bare inorganic materials. However, the clearest evidence for the importance of surface area (and surface complexity) to aquatic insects comes from the experiments of Hart (1978) and Trush (1979). Hart mounted individual standardized substrates of different sizes, shapes, and degrees of surface complexity on a board, placed it in a stream, and examined the number of species and individuals colonizing the different substrates. Trush obtained abundance, species richness, and surface area measurements from naturally occurring rocks in a small stream. Using uniformly shaped particles formed from concrete and placed in the same stream, he determined colonization, immigration, and extinction rates. Hart and Trush both found that significantly higher numbers of species and total individuals occur on large individual "rocks" than on small ones (Table 12.3A, B). Trush also found that communities on larger substrate particles have higher overall immigration, invasion, replacement, and colonization rates than do those on smaller substrates. However, when uniform particles are placed in groups and analyzed together, the effect of surface area is often obscured by other factors.

TABLE 12.3A, B: **Mean Richness and Numbers of Invertebrates on Individual Rocks of Different Sizes Expressed as Raw Values and Standardized for Surface Area. A: Data from Hart (1978). B: Data from Trush (1979).**

A	SPHERE		CUBE		IRREGULAR	
	Small	Large	Small	Large	Small	Large
	($105cm^2$)	($710cm^2$)	($89cm^2$)	($547cm^2$)	($93cm^2$)	($550cm^2$)
Richness/rock	2.33	4.17	2.58	4.75	2.67	7.42
Richness/ 1000 cm^2	22.22	5.87	29.03	8.68	28.99	13.48
Number of individuals/rock	3.75	10.08	3.08	11.42	4.58	18.42
Number of individuals/ 1000 cm^2	35.71	14.20	34.61	20.88	49.78	33.49

B	NATURAL (rock)			ARTIFICIAL (concrete)		
	Small	Medium	Large	Small	Medium	Large
	($\sim31cm^2$)	($\sim274cm^2$)	($\sim1013cm^2$)	($31cm^2$)	($274cm^2$)	($1013cm^2$)
Richness/rock	6.72	20.93	29.45	11.25	20.50	26.67
Richness/ 1000 cm^2	216.77	76.39	29.07	362.90	74.82	26.33
Number of individuals/ rock	34.38	171.80	737.46	66.03	295.92	506.50
Number of individuals/ 1000 cm^2	1109.03	627.01	728.00	2130.00	1080.00	500.00

TABLE 12.3C, D: Numbers of Invertebrates per Unit Area of Available Surface on Patches of Different Sized Rocks. *C*: Mink Creek, Idaho (Minshall and Minshall 1977). *D*: New Hope Creek, North Carolina (Reice 1980)

C	TRIAL 1		TRIAL 2		TRIAL 3	
	Small	*Large*	*Small*	*Large*	*Small*	*Large*
Number of individuals/m^2	3287	5459	2509	3756	1757	2848

D	WITH LEAF PACKS			WITHOUT LEAF PACKS		
	Gravel	*Pebble*	*Cobble*	*Gravel*	*Pebble*	*Cobble*
Number of individuals/m^2	37.47	85.77	177.40	25.51	129.73	133.33
Number of species/m^2	2.82	10.12	16.64	2.13	9.15	21.17

For example, using substratum-filled trays or baskets, Minshall and Minshall (1977) found only three taxa that increased in direct proportion to surface area, whereas Reice (1980) found only two, and Khalaf and Tachet (1980) apparently found none.

If there were a strict relationship between invertebrate abundance and surface area, one would expect that densities on the large and small rocks would be similar when expressed on a unit area basis. Higher densities on the larger rocks would indicate enhancement of living conditions over that supplied solely by increased space (e.g. increased complexity), whereas lower densities on the larger substrate would suggest a possible deterioration of conditions. However, when the results of studies with individual rocks and studies with groups of rocks are examined in this way, an interesting dichotomy emerges. On individual rocks, small substrates support higher densities than large ones (Table 12.3*A, B*), but on combined rocks the reverse is true (i.e. densities are highest on the large substrates; Table 12.3*C, D*). These findings indicate that when rocks are combined, some feature that is not associated with individual rocks increases disproportionately with change in rock size. One possibility is that as the substrate particles become larger, the inhabitability of the spaces between the rocks increases.

Substrate Stability

Stability of the substrate, in the context used here, refers to the degree of resistance to movement. Because smaller rocks can be disturbed or overturned more frequently, stability generally will be proportional to the size of the particle. In general, intermediate physical stability is expected to support a larger, more diverse fauna (Connell 1978; Stanford and Ward 1983). In addition to affecting the degree to which a given particle or patch of substratum undergoes disturbance, stability influences the amount of detritus trapped between the particles (Tolkamp, unpublished data).

Stability of the substratum has long been recognized as an important factor in the distributional patterns of insects, particularly in streams (Moon 1939; Chutter 1969; Luedtke and Brusven 1976; Malmqvist et al. 1978). However, its measurement remains largely subjective and usually is equated with size. Reduced species richness and abundance are commonly associated with areas of shifting sand, although certain species (e.g. the mayflies *Baetis rhodani* and *Rhithrogena semicolorata,* and the midge larva *Chironomus*) apparently prefer this substratum (Nuttall 1972; Ali and Mulla 1976). Luedtke and Brusven (1976) found that many common riffle insects are unable to move upstream on sand substrates. Insects with relatively long life cycles (2–3 years), such as some dragonflies (e.g. *Aeshna*) and stoneflies (e.g. *Paragnetina, Pteronarcys*) are commonly associated with boulder substrates, possibly because they are more likely to find shelter there during high flow (de March 1976; G. W. Minshall, personal observations). McAuliffe (1983) has observed that the purse-case caddisfly *Leucotrichia* is restricted to unstable substrates, where moss is unable to overgrow the rocks and thereby prevent the occurrence of these larvae. On stable substrates with a heavy moss growth, the insect assemblage shifts from one predominated by *Leucotrichia* to one predominated by filter-feeding caddisflies of the genus *Hydropsyche.* Although no specific evaluation of the effect of substratum stability on either freshwater insects or other invertebrates has been published (however, see Petran 1977), work on marine invertebrates has indicated its potential importance, as well as some possible research approaches (e.g. see Osman 1978; Sousa 1979; Paine 1980).

Substrate Heterogeneity

The idea that a mixed substratum provides more kinds of living places, and therefore can support a greater variety of insects than a simple one, is evident in the writings of many freshwater ecologists (e.g. see Sprules 1947; Mackay and Kalff 1969; Hynes 1970; Tolkamp 1980), and is consistent with the belief that species differ in their substrate preferences or requirements. There is also support for the importance of substrate heterogeneity from studies in the marine environment (e.g. see Kohn 1967, 1968; Abele 1974).

It is generally true that as the median particle size of sediments increases, their physical complexity also increases (Hynes 1970; Reice 1974; Osman 1978; Fig. 12.1, *dashed line*). On wave-swept shores, this relationship occurs because maximum velocities may be insufficient to remove gravel and pebbles from among the boulders. However, in streams there appears to be a binominal relationship between median particle size and complexity, with a decrease in substrate variety occurring beyond some point, as median particle size continues to increase (Fig. 12.1, *solid line*). This occurs about where large pebbles (−5 phi) or cobbles (−6 phi) begin to predominate, and coincides with regions where current velocity is able to move about 75 percent of all particle sizes (see also Lamberti and Resh 1979). Erman and Erman (in press) generally found no statistical interaction between median particle size and substrate heterogeneity (measured as either number of particle size classes or distribution of size classes about a median), but they used prepared substrates and a limited range of particle sizes (Md 2, 8, 32 mm).

As with the study of surface area, investigations on the role of substratum heterogeneity have proceeded along two lines. One group (Hart 1978; Trush 1979), working with individual rocks, has focused on conditions occurring on each substratum particle, whereas the other group (e.g. see Allan 1975; Wise and Molles 1979; Williams 1980), working with patches of substrate materials, has examined the effect that various mixtures of different-sized particles have on the invertebrate community. Therefore, it is useful to distinguish between these two

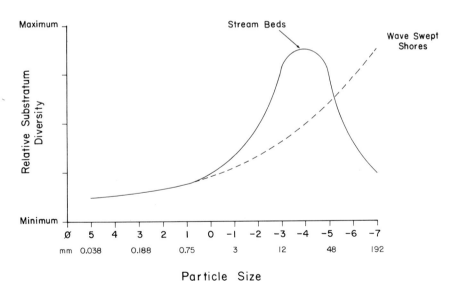

Figure 12.1 Hypothetical relationship between substratum diversity and predominant particle size (median diameter).

types of heterogeneity as *surface complexity* and *particle diversity*, respectively. With either type, the scale of irregularities must be suitably related to the size and habits of the organism (Allen 1959).

Surface complexity has been treated mainly as function of the degree of roughness (number and variety of crevices and smooth areas) on the surface of an individual particle (Hart 1978; Trush 1979; Erman and Erman, in press), but also includes traits such as shape and roundness (Erman and Erman, in press). Trush (1979) observed that algal mats can reduce surface complexity, and tend to make rough and smooth surfaces ecologically similar. As yet, no satisfactory method has been found to quantify surface complexity, although Erman and Erman (in press) provide a method for measuring relative surface roughness on nonporous rocks (see following section on substrate texture).

Studies of surface complexity (Hart 1978; Trush 1979) generally support the notion that species richness is greater on complex (irregular) surfaces than on simple ones (e.g. see Table 12.3A). However, Erman and Erman (in press) found similar species composition on the three rock types of differing roughness (granite, sandstone, quartzite) that they used in their experiments. Hart and Trush both concluded that increased richness was due to the fact that complex habitats provide resources that are unavailable in simpler habitats. In addition, Trush found that the relationship between richness and increasing surface area had a steeper slope for rocks with relatively smooth surfaces than for rocks with more complex surfaces. He reasoned that in a series of different-sized rocks with complex surfaces, little change in the types of structural surface variations would be exhibited. Thus, most increases in species richness with increasing surface area would be the result of a gradual increase in the abundance of microhabitats that are present on all sizes of complex rocks. He also argued that a series of rocks having relatively simple surfaces would show a progressive addition of new habitats with increasing area. Thus, for a series of substrates with simple surfaces, most increases in species richness with increasing area would be the result of an addition of new habitats not found on the smaller rocks in the series. This explanation seems quite plausible. However, when the numerical data are expressed as densities (i.e. per unit area) rather than abundances (i.e. per rock), it becomes apparent that the small (simple-surfaced) substrates support higher values than do large substrates (Table 12.3B; data could not be found to examine the situation for complex surfaces). This suggests that surface complexity on rocks with simple surfaces does not increase proportionately with surface area.

Apparently conflicting results for total number of individuals were obtained by Hart (1978) and Trush (1979). Hart found that densities (and species richness) were higher on complex rocks than on simple ones; however, Trush observed no significant differences between simple and complex rocks.

In contrast to the situation with surface complexity, several methods have been used to measure particle diversity. Harman (1972) applied the simple procedure of recording the number of substrate types observed in each habitat. Allan (1975) determined Shannon-Weiner diversities, whereas de March (1976),

Lamberti and Resh (1979), and Erman and Erman (in press) used the sorting coefficient or quartile deviation described by Krumbein (1939) and Inman (1952). Friberg et al. (1977) and Williams (1980) used a technique attributed to Schwoerbel (1961) in which the grain size (starting with the fine particles) making up 60 percent of the particles by weight is divided by the grain size containing 10 percent of the particles ($Q_{60/10}$ index). No tests of the relative efficacy of these various techniques have been published.

Results from the studies of particle diversity are similar to those just described for surface complexity. Generally, more taxa inhabited mixed, high-diversity substrates than simpler ones (Allan 1975; Wise and Molles 1979; Williams 1980), although Erman and Erman (in press) found that only particle size (and the covariates current and detritus), and not particle diversity, accounted for differences in number of taxa. Also, there was lack of agreement between studies on the relationship between abundance of organisms and particle diversity. For example, Williams, and Erman and Erman, found no significant difference in total numbers (and biomass) for any of the substrates they examined, whereas Wise and Molles found that total numbers on their mixed substrates were intermediate between those on their large and small uniform substrates.

Substrate Texture

Evidence for the importance of substratum texture to aquatic insects is largely circumstantial. Interpretation is complicated by the fact that a substratum may be suitable mainly because it is a source of food rather than because it is hard, soft, rough, or smooth. Also, as used here, "texture" is considered separate from "surface complexity," yet the two are obviously interrelated.

Methods for the measurement of substrate texture have been largely subjective and are in need of refinement. A penetrometer could be devised for measuring the relative hardness of different substrate materials. Erman and Erman (in press) quantified differences in the roughness of three rock types by measuring the rate of spread of a drop of water that was placed on the dry surface.

In spite of its circumstantial and subjective nature, the information presently available suggests that substratum texture may be important for aquatic insects. For example, some aquatic insects are known to burrow into only certain materials (e.g. *Heteroplectron*, *Lara*, and some species of *Brillia* only burrow into wood; Anderson and Cummins 1979), and algal and detritus mats are thought to provide better surfaces for organisms to cling to than are provided by bare substrates (Nilsen and Larimore 1973). Some dragonfly nymphs are thought to be restricted to vascular hydrophytes by their clinging ability (Hynes 1970), and Glime and Clemons (1972) found higher numbers of insect taxa (25) on moss (*Fontinalis*) than on string (23) or on plastic (13), possibly reflecting differences in substrate textures. Of the sessile insects studied by McAuliffe (1983), hydropsychids predominated on moss-covered or rough substrates, wherea *Leuco-trichia, Nanocladius, Parargyractis,* and *Rheotanytarsus* predominated on smooth surfaces. Erman and Erman (in press) collected more individuals (but not more taxa) from containers of rocks (64-mm diameter) having rough texture (granite and sandstone) than from those with a smooth texture (quartzite). These examples indicate that some property of the substratum associated with texture influences the distribution of at least some insects.

Pore Space

Pore space is another feature of the substratum that is likely to be important to aquatic insects, but there is little direct evidence to support this idea. Pore space includes both the dimensions of the opening and the extent to which the substratum is perforated (porosity).

Pore size may be measured by microscopic examination of substrates imbedded in resin (Williams 1972), whereas porosity can be determined by volume displacement (Pollard 1955). Presumably, mixed substrates have a more heterogeneous array of pore spaces than do homogeneous substrates. In addition to controlling access by aquatic insects, pore space will alter the subsurface movement of water and the rate at which organic and mineral particles accumulate (Khalaf and Tachet 1980). Williams (1972) found a general relationship between pore size and the diameters of marine invertebrates. In addition to the tendency for large forms of certain groups to be excluded from deposits with small pore sizes, the smaller forms tended to be absent from the coarser materials. Williams suggested that the avoidance of large spaces was due possibly to thigmotactic behavior of the animals, or the absence of suitable food from well-flushed deposits. Certain substratum particle sizes collect detritus more efficiently than others, presumably due to the attainment of optimal spacing of the substrates (Rabeni and Minshall 1977; Williams and Mundie 1978).

Reice (1978) attributed the concentration of many insect species on slowly, rather than rapidly, decomposing leaves to the relative ease of access by larger animals. This may occur because packets of slowly decomposing leaves generally are more loosely constructed than those of more rapidly decomposing ones. Corkum et al. (1977) concluded that the pore sizes of the larger particles in their experiments did not provide a suitable habitat for *Paraleptophlebia mollis* mayfly nymphs at high current velocities (~50 cm/s). Brusven and Rose (1981) found that vulnerability to predation from sculpin (*Cottus*) was markedly greater for active forms of stream insects (*Hesperoperla* and *Rhyacophila*) when the substrata on which they are normally found (pebbles and cobbles) were half imbedded in sand than when the substrata merely rested on the sand. However, except for scattered pieces of information such as these, there is little evidence of the importance of pore space to freshwater invertebrates. None of the studies in which pore space has been determined in relatively natural settings (Williams and

Mundie 1978; Khalaf and Tachet 1980; Williams 1980) has shown a significant correlation between pore space and the numbers or kinds of insects, indicating that under natural conditions this feature probably is complicated by other factors.

Silt

Apart from its importance as a medium in which some insects live (e.g. see Hynes 1970), silt (and fine sand) may significantly affect the inhabitability of a substratum by altering water movement, food quality, oxygen availability, and interstitial spacing (Cordone and Kelley 1961; Eriksen 1966; Chutter 1969; Williams 1972; de March 1976; Williams, Chapter 14). In addition, the transport of silt and sand by the current may result in the scouring of more stable substrates, and the clogging of the filtering apparatus of insects like the black fly *Simulium*. For example, Wu (1931) observed that the deposition of silt caused *Simulium* larvae to move from the tops to the undersides of leaves and sticks. Certain net-spinning caddisflies could also be affected in this way.

Light silting ($\sim$1 mm thick) on rock surfaces appears to have a variable effect on the insect fauna. For example, Cummins and Lauff (1969) found that it enhanced the selection of the interstices of coarse sediments by the mayfly *Caenis latipennis* and the stonefly *Perlesta placida*, but it had no discernable effect on eight other species they examined. However, Rabeni and Minshall (1977) found that a light layer of silt reduced the abundance of six taxa (*Alloperla, Arcynopteryx parallela*, Chironomidae, *Ephemerella grandis, Optioservus quadrimaculatus, Paraleptophlebia heteronea*) when added to trays of coarse substrata placed in a stream. De March (1976) found distinct groupings of insects based on the presence or absence of silt, but in general the number and types of animals were limited by the presence of silt. She concluded that the presence of silt was more important than mean particle size in regulating the distribution of insects. Reice noted that the breakdown of leaf litter (Reice 1974), and the kinds of abundances of leaf-processing insects (Reice 1977), were both consistently and significantly less on silt than on sand, gravel, or rock.

Heavy silting generally results in lower insect species diversity and productivity than occurs in unsilted areas. Chutter (1969) observed that the presence of large amounts of silt and sand coincided with a marked reduction in the variety and abundance of stream insects. Nuttall and Bielby (1973) recorded a similar response to china-clay wastes. Luedtke et al. (1976) found increased diversity and total abundance, and shifts in numerical dominance of insects, following removal of fine sediments from portions of Emerald Creek, Idaho, by means of experimental in-stream alterations. Nuttall (1972) observed comparable increases along the length of the River Camel, England, as the effects of a large influx of sand near the headwaters dissipated downstream. He also noted a distinct difference in the response of the macroinvertebrate community to the deposition of inert sand particles compared to that resulting from the deposition of fine-grained, organically rich mud.

The results of these studies indicate the necessity for both collecting the silt and sand fractions, and then analyzing them apart from the other particle sizes. The need for measuring the silt fraction is also implicit in procedures for measuring particle diversity (e.g. $Q_{60/10}$) and "cobble embeddedness" (Luedtke et al. 1976). In spite of its apparent importance, the silt fraction is inadequately sampled by most techniques (Eriksen 1968; Mundie 1971).

Organic Matter

The principal importance of organic matter to aquatic insects is its use as a food source, and hence it can serve as an attractant to, or a limiting factor in, different mineral substrates. However, various types of organic matter also can provide case- or tube-building materials for certain caddisflies and chironomids (e.g. Hanna 1961; Cummins 1964; Higler 1975; Mackay 1977; Brennan et al. 1978; Otto and Svensson 1980; Tolkamp 1980) as well as a substratum on which to live (e.g. see Macan 1965, 1977; Nilsen and Larimore 1973; Gregg 1981). Various authors (notably Egglishaw 1964, 1968, 1969; Rabeni and Minshall 1977; Williams and Mundie 1978) have observed a strong association between the abundance of insects and the amount of terrestrial plant litter, and have attributed this to the food requirements of the insects.

However, studies by Reice (1974, 1978, 1980) suggest that at least in some cases the material may serve primarily as a substratum. For example, in one experiment in which abundances of individual invertebrate species on bare substrates were compared with those on substrates to which leaf packs had been added, 14 of the 16 common species showed no significant differences between the two conditions. The stonefly *Allocapnia* was the only common taxon that preferred substrata with leaf packs (Reice 1980).

Other workers also have found no correlation between the amount of leaf detritus present and the abundance of insects (e.g. see Minshall and Minshall 1977; Williams 1980; Peckarsky 1980a). However, in these cases, the amounts of detritus were always in excess of the insects' requirements and thus did not serve as a limiting factor. It also may be that the organic matter consisted primarily of large particles and that, other than for shredders, these serve mainly as a substratum. In studies where the fine particles of detritus (< 1 mm) have been analyzed separately, a strong correlation has been found between the amount of detritus and the abundance of a large portion of the community (e.g. see Rabeni and Minshall 1977; Short et al. 1980).

Given the pronounced differences that exist between aquatic insects in food-gathering mechanisms and food preferences (e.g. see Cummins 1973; Cummins and Klug 1979), it is to be expected that insect species will respond differently to organic matter. For example, larvae of the chironomid genus *Cricotopus* show a strong association with detritus (60–80% of its total abundance on five different substrate types), whereas the highest numbers of larvae in the cofamilial genus *Chironomus* were correlated with fine sand (40–80% of its total) or filamentous algae (10–45%). In addition, differential responses to different particle sizes of detritus (as just noted, see also McLachlan et al. 1978) and to food quality (e.g. see Anderson and Cummins 1979; Cummins and Klug 1979) are likely to occur.

SUBSTRATE INTERACTIONS WITH OTHER FACTORS

As noted in the Introduction, there are a number of ways in which substratum can act to modify the microclimate in which aquatic insects live. In turn, there are environmental factors acting at either the macrolevel or microlevel scale that may alter an insect's response to the substratum. Thus, because of the imposition of other limiting factors, a species might be confined to less preferred substrates, or it may not attain maximum individual growth or maximum population densities, even when substratum characteristics are ideal for that particular species. Among the most important modifiers of an insect's response to substratum are water temperature, flow regime, and chemical composition of the water (all of which usually operate on a broad or "macro" scale), and light, food, current, oxygen content, and interactions with other organisms (which are imposed primarily at the local or microhabitat level). It is the latter group of modifiers, in particular, that most often complicates interpretation of observations and experimental results intended to elucidate insect-substratum relationships. Although all of these factors commonly act simultaneously and therefore are not easily separated, the intention here is to briefly illustrate their individual roles and to provide access to further information. Meadows and Campbell (1972) reviewed the responses of marine and freshwater invertebrates to a variety of environmental factors and developed the thesis that the local distribution of most aquatic invertebrates is determined by their behavioral reactions (i.e. there is active selection of sites). Especially noteworthy is their contention that chemical cues produced by the biota may serve to attract or repel invertebrates to or from a particular substratum.

Water Temperature

Ide (1935) observed that the increase in the number of species of mayflies along a stream was associated with increasing magnitude of temperature fluctuations in downstream reaches. An extension of these findings was used in formulating the idea that, over an entire river system, biotic diversity will be highest in the mid-reaches (Vannote et al. 1980). Temperature is believed to superimpose limits on aquatic insect distribution within which the effects of substratum, current velocity, and other factors are expressed (e.g. see Ide 1935; Ulfstrand 1967; Minshall 1968; Rabeni and Minshall 1977; Sweeney, Chapter 4). Sweeney and Vannote (1981) suggest that temperature affects growth and controls the endocrine system of insects, which in turn determines the ultimate size and reproductive capacity of the adults. These effects of temperature are seen to have geographic as well as within-habitat implications for the distribution of aquatic insects (Vannote and Sweeney 1980). Additional treatment of the influence of temperature can be found in Sweeney (Chapter 4) and in Ward and Stanford (1982).

Flow Regime

The volume of flow, the relationship of velocity to depth, the periodicity in timing of high and low flow, and so forth, can have important effects on the particle size, composition, and relative stability of the substratum, the amount of channel that is under water, food availability, and several other factors that occur on a macroscale level (Leopold et al. 1964). For example, because of differences in the hydraulic regimen, as well as other factors, bed materials nearer the source will not be as well sorted as those near the mouth of the river (Inman 1952). Also, along with the increasing fluctuations (and therefore diversity) of temperature, substratum heterogeneity is expected to be greatest in the middle reaches of a river system (Vannote et al. 1980). In the headwaters, large particles predominate as a result of the proximity of bedrock materials and the ability of relatively rapid, turbulent flow to move the smaller particles. Reduced substratum heterogeneity in larger streams is attributed to reduction in river competence (i.e. the maximum size of particles that the stream will move), and leads to a more uniform distribution of fine substrates (Vannote 1981). Thus, patterns of erosion and deposition which can be seen on a local level (Moon 1939) also may be found on a larger scale, and shifts in the composition of the insect community will follow accordingly (see also Newbury, Chapter 11).

Chemical Composition of the Water

Minshall and Kuehne (1969) documented the virtual absence of mayflies and certain species of stoneflies, caddisflies, and other taxa from an extensive portion of the River Duddon, England. Later, Sutcliffe and Carrick (1973) and Minshall and Minshall (1978) established that this peculiar pattern of distribution was due to a chemical imbalance associated with elevated hyrogen ion levels and reduced concentrations of other chemicals, including potassium. Egglishaw (1968) demonstrated that the chemical content of the water could override the primary influence of food on the abundance of insects in several moorland streams. For example, a regression of weight of bottom fauna per gram of detritus against calcium concentration showed a significant positive relationship for the eight streams examined.

Light

Light (or shade) usually is not given careful consideration in studies of benthic insect distribution, but there is evidence to indicate that it can be important (e.g. see Percival and Whitehead 1926; Scherer 1962; Hughes 1966). For example, Thorup (1966) reported that, over the 400-m length of the springbrook Rold Kilde in Denmark, there is sufficient variation in flow and substratum to explain differences in the distribution of several insects. In spite of this, *Baetis rhodani* showed a consistent pattern of highest abundance in the unshaded areas, and lowest abundance in the most heavily shaded regions. The same pattern was found for the trichopterans *Agapetus fuscipes, Eccliopteryx guttulata,* and *Silo pallipes,* and the coleopteran *Helodes minuta,* whereas *Wormaldia occipitalis* (Trichoptera) showed the opposite reaction to light. More recently, McLachlan (1969) showed that larvae of the chironomid *Nilodorum brevibucca* were negatively phototropic under laboratory conditions; and Higler (1975) demonstrated a distinct preference by the caddisflies *Limnephilus rhombicus* and *Potamophylax rotundipennis* for shaded areas when offered a choice between shaded and unshaded sections of an experimental stream having identical substrata and flow. Lyman (1956) provides evidence for the importance of light in the distribution of several lake-dwelling mayflies.

Food

In order to continue inhabiting a given substratum, an insect must be able to obtain food. The presence or absence of food can alter the attractiveness of a given substratum for an aquatic insect (e.g. see Keller 1975; Short et al. 1980; Tolkamp 1980). However, once food requirements are met, other factors will be the primary determinants of the microdistributional pattern of macroinvertebrates. Thus, it is not surprising to find situations in which there is no strong positive correlation between insect abundance or biomass and the amount of plant litter (e.g. see Minshall and Minshall 1977; Peckarsky 1980a; Williams 1980), or even cases where an excess of organic matter appears to have resulted in a negative relationship (e.g. see Buscemi 1966; Barber and Kevern 1973).

Egglishaw (1964, 1968, 1969) concluded that plant detritus is a causal factor in the distribution of several benthic species. In a study of Shelligan Burn and River Almond in Scotland, Egglishaw (1969) showed that different species have different quantitative relationships with plant detritus, and that the relationships differ from one stream to another. Thus, the structure of the community at a site depends to a large extent on the amount of detritus present, but other factors undoubtedly are involved as well. In an analogous situation, Reice (1978) observed that different members of the stream macrobenthic community specialize on particular kinds of leaves, but that the specialists tend to counterbalance one another. Thus, although the structure of the community varied with type of leaf, the overall pattern of distribution appeared uniform, and the number of individuals per gram of leaf was not significantly different among leaf species.

Frequently there may be an interplay between the substratum and the amount of food available. For example, Rabeni and Minshall (1977) and Williams and Mundie (1978) found that the greatest abundance of insects occurred on intermediate-sized substrate particles, probably because more small-sized detritus particles accumulated there than on larger or smaller substrates.

Current

Several studies have demonstrated the importance of current to the distribution of benthic insects (e.g. see Linduska 1942; Scott 1958; Ambühl 1959; Edington 1968). Current velocity affects an insect's ability to gather food (Wallace and Merritt 1980), meet its respiratory requirements (Philipson 1954; Knight and Gaufin 1963, 1964; Jaag and Ambühl 1964), avoid competition and predation (e.g. see Corkum and Clifford 1980; Peckarsky 1980a, 1980b; Wiley and Kohler 1981), leave unfavorable environmental conditions (e.g. see Minshall and Winger 1968; Hildebrand 1974; Corkum et al. 1977), or colonize favorable ones (e.g. see Townsend and Hildrew 1976; Shaw and Minshall 1980; Minshall et al. 1983). However, the complex features of flow (force, direction, degree of turbulence, etc.) and substratum (particle size, particle composition, stability, silt and organic content, etc.) are so closely interrelated that under natural conditions it frequently is impossible to separate the individual effects of the two. For example, coarse substrates, open interstitial spaces, and low accumulations of organic matter predominate where the current is fast, whereas fine substrates, tightly packed particles, and large amounts of organic matter are typical of slow currents or standing waters. Current also can be responsible for local variations of particle size composition that give rise to mosaic patterns and longitudinal substratum bands (Lamberti and Resh 1979; Tolkamp 1980).

Variations in the competence of moving water, whether along a lake shore or within a stream, result in distinct areas of erosion or deposition, which have been used in broadly subdividing aquatic habitats (e.g. see Moon 1939; Hynes 1970; Cummins 1975). Erosional subhabitats (e.g. riffles) generally support more kinds of animals, but lower population densities, than do depositional subhabitats (e.g. pools). Also, the two subhabitats commonly harbor insects with distinctive morphological and behavioral characteristics. For example, in erosional sub-habitats the fauna is characterized by adaptation for attachment, clinging, or avoiding direct contact with the current, whereas in depositional areas the animals are adapted for sprawling, climbing, or burrowing (e.g. see Hora 1930; Lyman 1956; Bournaud 1963; Hynes 1970).

Current velocity largely determines the median size and heterogeneity of the mineral substratum in streams (Leopold et al. 1964). The substratum, in turn, influences the amount and particle size of allochthonous organic matter available, and this in turn affects the size of benthic insect populations (Rabeni and Minshall 1977). Reice (1974, 1977) also has shown that current velocity interacts with substratum to determine the specific animal associations found in accumulations of leaf litter.

Lehmkuhl and Anderson (1972) found that high discharge can displace insects from one subhabitat to another; for example, the mayfly *Paralepto-phlebia temporalis* occupied riffle areas of rapid current and clear substratum in October, but was displaced into glide (i.e. run) and backwater areas in November and remained there. *Baetis parvus*, which normally inhabits areas of relatively slow current, was displaced more rapidly than another mayfly, *Baetis tri-caudatus*, which occupies areas of faster current. *B. parvus* apparently remained in the slower currents until emergence, whereas *B. tricaudatus* returned to the riffles. A fourth mayfly, *Cinygmula reticulata*, normally is a fast-water form but also was displaced to slower water by freshets. This latter species then returned to riffle areas in April and May.

Oxygen

Variations in the supply of dissolved oxygen commonly occur on a local scale (although limiting concentrations may extend over the entire bed of a pond, or over a considerable length of stream, especially in cases of organic pollution). Thus, the influence of oxygen can be superimposed on that of substratum and other factors in ultimately determining the abundance and distribution of insects (e.g. see Eriksen 1964, 1966). In addition, as suggested by Kovalak (1976) for *Glossosoma nigrior*, changes in distribution on the surface of individual rocks may occur in response to respiratory needs.

Oxygen concentrations are regulated by the degree of chemical and biological oxygen demand, temperature (which affects both oxygen solubility and insect respiration rate), amount of water circulation or downstream movement, and the substratum itself (especially bed roughness and porosity). Water movement determines the degree of interchange between air and water, and between water and the substratum. Water movement also interacts with the substratum and produces turbulence. This thereby affects both the thickness of the boundary layer that envelops an insect, and the steepness of the diffusion gradient between this layer and the overlying water (Ambühl 1959; Jaag and Ambühl 1964; Décamps et al. 1972; Eriksen et al., in press). Porosity regulates the rate of water circulation, and thus the oxygen content in the interstitial spaces. For example, Eriksen (1966) and others have shown that the presence of more than 10 to 20 of fine sand and silt (<0.25 mm) can cause a signif-icant reduction in the oxygen content within the substratum. Leaf packs pro-vide an environment where oxygent content, current velocity, and amount of fine organic matter may vary considerably, depending on their position in the habitat (Tolkamp 1980). For example, if the leaf packs are loosely piled against objects projecting from the stream bed, conditions will be similar to those in coarse mineral substrates. If leak packs are accumulated near the banks or on the bottom of pools or in lakes, conditions will be more like those in fine mineral substrates.

The minimum level of oxygen that an insect can tolerate varies with each species. In general, lotic species require higher concentrations than lentic species. Commonly, insects that live in slowly moving or still water have adaptations (e.g. the ability to use gills for ventilation) for coping with these more strenuous respiratory conditions. But even forms that normally occupy locations of more abundant oxygen supply may possess means of surviving temporary periods of oxygen stress; for example, by ventilatory movement or escape (Knight and Gaufin 1964; Ericksen 1966, 1968; Minshall and Winger 1968; Eriksen et al., in press).

Interactions With Other Organisms

Even though substratum and other conditions may be suitable, the abundance and distribution of insect species may vary from one substratum patch to another, or even over the surface of a single rock due to the effects of competition for food, space, or predation. Territorial "spacing" of individuals, which indicates control of distribution through competition, has been reported by Zahar (1951) for simuliid larvae and by Glass and Bovbjerg (1969) for net-spinning *Cheumatopsyche* larvae.

Benthic invertebrates seem to prefer substrate areas with low faunal populations compared to physically and chemically comparable areas with high populations (Peckarsky 1979, 1981). At high densities, aggressive encounters enforce spacing of individual *Cheumatopsyche* caddisfly larvae (Glass and Bovbjerg 1969) and *Paraleptophlebia mollis* mayfly nymphs (Corkum 1978). However, *Baetis vagans* nymphs emigrate in response to crowding (Corkum 1978). McAuliffe (1983) found that fewer *Baetis* and *Simulium* colonized rocks with high densities of the sessile, grazing caddisfly *Leucotrichia*, than colonized rocks without *Leucotrichia*. He also demonstrated the ability of certain grazers (e.g. *Glossosoma*) to exploit and depress periphyton abundances, and suggested that this could lead to lowered colonization rates and equilibrium densities of other grazers as well (e.g. *Baetis, Oligoplectrum*).

McLachlan (1969) examined the distribution of the chironomid *Nilodorum brevibucca* at different densities. Introduction of a small number of first-instar larvae onto a uniform substratum resulted in a random distribution, whereas a highly clumped distribution of the larvae occurred if an egg mass was introduced instead. As more and more larvae were added, the distribution became uniform, and the spacing between the larvae gradually decreased. The critical point was reached when the "feeding zones" of adjacent larvae touched. Larvae added to the substratum then were unable to settle and eventually died. As the larvae grew, some individuals were forced out of their tubes.

The predaceous caddisfly *Plectrocnemia conspersa* responds directly to prey distribution, and may aggregate in patches when prey is abundant (Townsend and Hildrew 1980). Predaceous insects such as *Acroneuria, Megarcys,* and *Rhyacophila* have been observed to reduce the densities of insect-prey populations through direct consumption of the prey, and through predator avoidance by the prey (Peckarsky 1979, 1980b; Wiley and Kohler 1981). In terms of fish predation, Allan (1975) reported that invertebrate densities were two to six times greater in sections of a stream from which trout were absent than in adjacent ones where trout were present (see also Peckarsky, Chapter 8; Healey, Chapter 9). However, Allan (1982; Chapter 16) was later unable to demonstrate the statistical significance of this difference in densities.

SPATIAL ASPECTS OF INSECT-SUBSTRATUM RELATIONSHIPS

Vertical Distribution

In the presentation thus far, emphasis has been placed on aquatic insect-substratum relationships in surface sediments (upper 5–10 cm). In most standing waters, insects usually do not penetrate much below this level (Rawson 1930; Cole 1953; Shiozawa and Barnes 1977), although in some cases a few species of chironomids may extend down to 20 to 40 cm (e.g. see Berg 1938). The depth of penetration is related to the degree of compaction of sediments. In closely packed materials, such as the mud bottoms of streams, penetration does not exceed about 12 cm (Ford 1962). But where interstitial flow occurs (e.g. clear streams flowing over deep-bedded glacial or alluvial deposits), distributions may extend to a meter or more under suitable conditions. Most of the research on vertical distribution has been directed toward the documentation and description of the vertical profile of insect occurrence and abundance, but little attention has been given to causal agents or the extent of exchange between layers (see Williams, Chapter 14). Differences in species and abundances have been shown to occur with depth, but at present it is not known to what extent vertical stratification of the substratum is responsible for these patterns.

Horizontal Distribution

A patchy (contagious) distribution of insects is a common feature of aquatic habitats (e.g. see Needham and Usinger 1956; Elliott 1977; Minshall and Minshall 1977) and probably is frequently a result of the patchiness of the substrate. Little effort has been made to document this for either plant or mineral substrates, but the works by Jónasson (1948), Ulfstrand (1967), Egglishaw (1969), Reice (1974), Resh (1977, 1979), and Lamberti and Resh (1979) are notable exceptions and generally support this causal hypothesis. If the substratum acts as a templet on which the benthic community is formed, then the mosaic pattern that results in the community is simply a reflection of this underlying influence. It is this large-scale environmental heterogeneity that may be responsible for maintaining relatively high levels of species richness in natural aquatic habitats (Tolkamp 1980). Other factors, especially food and current velocity, at times may override or further modify the insect distribution patterns set by the substratum (Ulfstrand 1967; Egglishaw 1969; Reice 1980). But by taking environmental patchiness into account, it generally should be possible to devise sampling schemes that attain acceptable statistical precision without requiring inordinate numbers of samples (Resh 1979). Furthermore, a clearer view of the causes of the horizontal patterns of benthic insect distributions should emerge from such efforts.

Longitudinal Distribution

In addition to the vertical and horizontal components that produce a mosaic pattern of insect distribution that is seen locally in lakes and streams, and over the entire area of ponds, there is also a longitudinal component in which insect communities along the length of a lake or river change predictably (e.g. see Ide 1935; Maitland 1964; Macan and Maudsley 1968, 1969; Allan 1975). In these cases, factors other than substratum usually are of primary importance. However, it is important to remember that insect responses to the substrate may differ longitudinally along a lake or stream as the result of other environmental factors, even when the substratum remains fairly constant. Thus, insect richness and abundance on a given substratum size may differ among sites even though substratum may be a primary factor controlling community structure within each site. As a result, information on substratum preferences needs to be considered in accordance with the location of the habitat within the aquatic system being examined.

TEMPORAL CHANGES IN INSECT-SUBSTRATUM RELATIONSHIPS

Even on the same substratum, the occurrence and abundance of aquatic insects vary with time as a consequence of changes due to life-cycle stages and the physical conditions of the environment. These temporal variations occur on scales ranging from minutes to years, and are the result of factors such as changing metabolic needs, and behavioral responses that vary as the insects pass from eggs to adults. They also are due to both continual and periodic (seasonal, annual) shifts in conditions in and around the substratum. These conditions include water level, current, substratum movement, erosion or deposition of sediments, oxygen concentrations, growth and removal of algae and vascular plants, and supplies of allochthonous detritus.

For example, using time-lapse cinematography, Wiley and Kohler (1981) observed up to two-fold variations in the total number of insects on the surface of a rock within a 24-hour period. In one case, the number of *Baetis vagans* nymphs ranged from 1 to 9 in a 100-cm^2 area in less than seven hours. Some of the variations in density, such as with *B. vagans*, appear to reflect simple volitional movements on the part of the insects, but others are due to the displacement of one insect by another. For example, grazing *Glossosoma nigrior* larvae were observed to displace and cause the eventual emigration of another caddisfly (*Brachycentrus americanus*), and predatory *Rhyacophila* were seen to consume, locally displace, and cause the emigration of simuliid larvae. In streams, most emigrating insects enter the drift (Townsend and Hildrew 1976). The composition of the drift is known to vary markedly with time, and probably reflects comparable changes taking place on the substratum. Most insects show a diel perodicity in drift, being more active either during the day or at night, and frequently show distinct peaks near sunset and sunrise (Waters 1972). Moon (1940) found that, in both lakes and streams, insects generally move much more at night than in the day. Differences in drift activity (Anderson 1967; Elliott 1967a, 1967b; Waters 1972) or other movements (Moon 1935b) with life-cycle stage have also been observed, and these differences may be due to adjustments in density as food and space requirements increase and to movements in preparation for emergence (Wiley and Kohler, Chapter 5).

A variety of aquatic insects, including species of Diptera, Ephemeroptera, and Trichoptera, change their spatial distribution patterns over their life cycles (Thut 1969; Lehmkuhl and Anderson 1972; Elliott 1977; Macan 1977; McLachlan 1977; Resh 1977, 1979; Shiozawa and Barnes 1977). In cases where closely related taxa occupy the same habitat, competition for food and space may be reduced through temporally staggered life cycles (e.g. see Minshall 1968; Grant and Mackay 1969; Allan 1975; Tolkamp 1980). Mackay (1969) recorded seasonal differences in the diversity of benthic insects, and attributed these to changes in food supply and life-history patterns. During their larval life, some caddisflies change the size and/or type of material used in the construction of their cases. Frequently, these changes reflect shifts in microhabitat locations on the substratum (e.g. see Cummins 1964; Otto 1976; Mackay 1977; Resh 1979; Tolkamp 1980), at least some of which may be in response to selection pressure from fish predation (Otto and Svensson 1980).

A number of physical conditions affecting the substratum have a distinct seasonal periodicity (e.g. ice scouring, rocks overturning during spring runoff, spring and fall turnover in lakes). De March (1976) concluded that the seasonal differences in species richness that she observed were related to sediment changes and accompanying variations in habitat heterogeneity. In spring, when high current velocities produced well-sorted, distinct sediment types, species richness was highly correlated with mean particle size. As the interstices of substrates were filled in with progressively finer particles during the summer, habitat heterogeneity was gradually reduced and mean particle size was less efficient in predicting the number of species. Even greater shifts in substratum composition were seen by Ulfstrand (1967, Figures 1–6) and Tolkamp (1980, Figure 7). Reice (1980) detected a spatial shift in which insects moved from leaf packs to the underlying substrata as the leaves decomposed. He interpreted this as a response to a changing supply of suitable food. Changes such as these can result in dramatic differences in population responses at a single site during a year, and point up the difficulty of interpreting field data on insect densities when critical substratum characteristics are not measured concurrently.

Whenever disturbances occur in benthic habitats, they are followed by periods of recolonization that vary in length from a few days to several years, depending on the nature and extent of the disturbance. For example, recolonization and a return to equilibrium may occur in only a few days when a small number of rocks are introduced into an otherwise undisturbed lake or stream; commonly, however, several weeks or more are required for the community to stabilize (e.g. see Moon 1935a, 1935b; Nilsen and Larimore 1973; Meier et al. 1979; Morris and Brooker 1979; Shaw and Minshall 1980; Rosenberg and Resh 1982). Large sections of a stream of lake may be severely impacted by physical disturbances (floods, fires, etc.), and recovery may take months or years, during which time continual changes in species composition and density may occur (e.g. see Williams and Hynes 1977; Gore 1979; Minshall et al. 1983). The final structure of the community may be determined to a large extent by which species become established first (e.g. see Trush 1979; McAuliffe 1983). Thus, the number and kinds of insects inhabiting a substratum will vary as a function of the time since the disturbance took place (see also Sheldon, Chapter 13). Consequently, conclusions about insect-substratum relationships may differ depending on when the sample is collected. This is especially problematic in field studies because prior history of the site is rarely known in sufficient detail (see Cummins et al. 1983).

CONCLUSIONS

It is clear that substratum is a primary factor influencing the abundance and distribution of aquatic insects. It acts directly on insects as a medium for their existence, and indirectly as a major modifier of their environment. However, it also is true that the substratum rarely, if ever, exerts its influence in isolation. The approach used in this review (i.e. considering each factor individually), therefore, is grossly oversimplified. Future researchers will need to examine the interaction of the substratum with various factors (especiaily the biotic ones) in more detail and under less artificial circumstances than has been possible to date.

It should also be evident from this review that the study of aquatic insect-substratum relationships has progressed to a relatively high level of refinement compared with other areas of aquatic insect ecology, and that knowledge of in-sect-substratum interactions has matured to the point that useful generalizations are beginning to emerge. However, exceptions to many currently perceived ideas are commonly found and, in some cases, outright conflicts in the findings of different investigators seem to exist. Therefore, researchers in this area of aquatic ecology should proceed with an open mind. It is certain that contemporary, accepted ideas will evolve as techniques improve, and as additional information and insight become available.

Further research is needed to resolve existing inconsistencies, especially in the areas of substrate particle size, area, and diversity. Likewise, the topics of substrate stability, surface complexity, and pore space, although they are just beginning to be explored quantitatively and experimentally, already show promise of yielding valuable information regarding insect-substratum relationships. More detailed information is needed regarding the apparent importance of the plant matter within the substratum in determining the carrying capacity and community structure of the aquatic macroinvertebrates that occur on different substrate particle sizes. Studies on this topic should include information on the role of food size, composition, quality, and amount.

Furthermore, there seems to be have been a hesitancy on the part of researchers of insect-substratum relationships to explore the topic in ways that could lead to meaningful generalizations, and thus provide insights that can be applied beyond the specific circumstances under which these relationships were studied. A good deal of effort continues to be expended on routine or poorly focused questions which, in spite of the work involved, contribute little to our understanding of insect-substratum relationships. The goal of developing a reliable set of general principles in this field will continue to elude us unless researchers identify the species of insects they study, and arrive at a set of standard measurements (and their units of expression) of substratum characteristics. Universal acceptance and use of the Wentworth classification for substrate particle size is only a first step toward the latter. Efforts at standardization should also address the other key substrate characteristics covered in this chapter, such as stability, surface complexity, particle diversity, texture, pore space, silt, and organic matter content.

A real challenge awaits aquatic ecologists in applying knowledge from studies of insect-substratum relationships to the management of aquatic habitats. Although some effort is beginning to be expended in this direction (e.g. Luedtke et al. 1976; Williams and Mundie 1978; Ward, Chapter 18), the mixed results of these efforts indicate that much remains to be learned. In the future, researchers studying insect-substratum relationships will also need to account for a greater array of factors than has been done in the past, and to conduct experiments in which two or more factors are varied systematically in order that the strength of various interactions can be determined. In addition, researchers will have to rely much more heavily on direct observation, or photographic and video techniques, to observe the responses of insects to different substrate conditions and interactions. Thus, although the insect-substratum relationship has become one of the most intensively studied areas in aquatic insect ecology, much remains to be done, and the topic promises to continue to attract researchers—and to reward their efforts—for some time to come.

Acknowledgments

This chapter was written while I was on sabbatical leave from Idaho State University at the Stroud Water Research Center for the Academy of Natural Sciences of Philadelphia. I thank both institutions for this opportunity, especially Dr. R. L. Vannote. Judy N. Minshall aided in innumerable ways, including helping to make this review more intelligible. I am indebted to Drs. S. R. Reice and H. H. Tolkamp for their thorough review and suggestions for improvement of an earlier version of this chapter.

References

Abele, L. G. 1974. Species diversity of decapod crustaceans in marine habitats. Ecology 55:156–61.

Ali, A. and M. S. Mulla. 1976. Substrate type as a factor influencing spatial distribution of chironomid midges in an urban flood control channel system. Environmental Entomology 5:631–36.

Allan, J. D. 1975. The distributional ecology and diversity of benthic insects in Cement Creek, Colorado. Ecology 56:1040–53.

Allan, J. D. 1982. The effects of reduction in trout density on the invertebrate community of a mountain stream. Ecology 63:1444–55.

Allen, K. R. 1959. The distribution of stream bottom fauna. Proceedings of the New Zealand Ecological Society 6:5–8.

Ambühl, H. 1959. Die Bedeutung der Strömung als ökologischer Faktor. Schweizerische Zeitschrift für Hydrologie 21:133–264.

Anderson, N. H. 1967. Biology and downstream drift of some Oregon Trichoptera. The Canadian Entomologist 99:507–21.

Anderson, N. H. and K. W. Cummins. 1979. Influences of diet on the life histories of aquatic insects. Journal of the Fisheries Research Board of Canada 36:335–42.

Barber, W. E. and N. R. Kevern. 1973. Ecological factors influencing macroinvertebrate standing crop distribution. Hydrobiologia 43:53–75.

Berg, K. 1938. Studies on the bottom animals of Esrom Lake. Kongelige Danske Videnskabernes Selskabs Skrifter, Copenhagen 9:1–255.

Berg, K., S. A. Boisen-Bennike, P. M. Jónasson, J. Keiding, and A. Nielsen. 1948. Biological studies on the River Susaa. Folia Limnologica Scandinavica 4:1–318.

Bournaud, M. 1963. Le courant, facteur écologique et éthologique de la vie aquatique. Hydrobiologia 21:125–65.

Brennan, A., A. J. McLachlan, and R. S. Wotton. 1978. Particulate material and midge larvae (Chironomidae: Diptera) in an upland river. Hydrobiologia 59:67–73.

Brusven, M. A. and K. V. Prather. 1974. Influence of stream sediments on distribution of macrobenthos. Journal of the Entomological Society of British Columbia 71:25–32.

Brusven, M. A. and S. T. Rose. 1981. Influence of substrate composition and suspended sediment on insect predation by the torrent sculpin, *Cottus rhotheus*. Canadian Journal of Fisheries and Aquatic Sciences 38:1444–48.

Buscemi, P. A. 1966. The importance of sedimentary organics in the distribution of benthic organisms. Pymatuning Laboratory of Ecology Special Publication 4:79–86.

Calow, P. 1972. A method for determining the surface area of stones to enable quantitative density estimates of littoral stonedwelling organisms to be made. Hydrobiologia 40: 37–50.

Cantrell, M. A. and A. J. McLachlan. 1977. Competition and chironomid distribution patterns in a newly flooded lake. Oikos 29:429–32.

Chutter, F. M. 1969. The effects of silt and sand on the invertebrate fauna of streams and rivers. Hydrobiologia 34:57–76.

Cianficconi, F. and M. Riatti. 1957. Impiego di pietre artificiali per l'analisi quantitativa delle colonizzazioni faunistiche dei fondi potamici. Bollettino di Pesca, Piscicoltura e Idrobiologia 12:299–334.

Cole, G. A. 1953. Notes on the vertical distribution of organisms in the profundal sediments of Douglas Lake, Michigan. American Midland Naturalist 49:252–56.

Connell, J. H. 1978. Diversity in tropical rainforests and coral reefs. Science 199:1302–10.

Cordone, A. J. and D. W. Kelley. 1961. The influence of inorganic sediment on the aquatic life of streams. California Fish and Game 47:189–228.

Corkum, L. D. 1978. The influence of density and behavioural type on the active entry of two mayfly species (Ephemeroptera) into the water column. Canadian Journal of Zoology 56:1201–6.

Corkum, L. D. and H. F. Clifford. 1980. The importance of species associations and substrate types to behavioural drift, pp. 331–41. *In*: J. F. Flannagan and K. E. Marshall (eds.). Advances in Ephemeroptera biology. Plenum Press, New York, NY. 552 pp.

Corkum, L. D., P. J. Pointing, and J. J. H. Ciborowski. 1977. The influence of current velocity and substrate on the distribution and drift of two species of mayflies (Ephemeroptera). Canadian Journal of Zoology 55:1970–77.

Cummins, K. W. 1962. An evaluation of some techniques for the collection and analysis of benthic samples with special emphasis on lotic waters. American Midland Naturalist 67:477–504.

Cummins, K. W. 1964. Factors limiting the microdistribution of larvae of the caddisflies *Pycnopsyche lepida* (Hagen) and *Pycnopsyche guttifer* (Walker) in a Michigan stream. Ecological Monographs 34:271–95.

Cummins, K. W. 1966. A review of stream ecology with special emphasis on organism-substrate relationships. Pymatuning Laboratory of Ecology Special Publication 4:2–51.

Cummins, K. W. 1973. Trophic relations of aquatic insects. Annual Review of Entomology 18:183–206.

Cummins, K. W. 1975. Macroinvertebrates, pp. 170–98. *In*: B. A. Whitton (ed.). River ecology. Blackwell Scientific Publications, London. 725 pp.

Cummins, K. W. and M. J. Klug. 1979. Feeding ecology of stream invertebrates. Annual Review of Ecology and Systematics 10:147-72.

Cummins, K. W. and G. H. Lauff. 1969. The influence of substrate particle size on the microdistribution of stream macrobenthos. Hydrobiologia 34:145-81.

Cummins, K. W., J. R. Sedell, F. J. Swanson, G. W. Minshall, S. G. Fisher, C. E. Cushing, R. C. Petersen, and R. L. Vannote. 1983. Organic matter budgets for stream ecosystems: problems in their evaluation, pp. 299-353. In: J. R. Barnes and G. W. Minshall (eds.). Stream ecology: application and testing of general ecological theory. Plenum Press, New York, NY. 399 pp.

Dahl, A. L. 1973. Surface area in ecological analysis: quantification of benthic coral-reef algae. Marine Biology 23:239-49.

Décamps, H., J. Capblancq, and J. P. Hirigoyen. 1972. Etude des conditions d'écoulement près du substrat en canal expérimental. Internationale Vereinigung für Theoretische und Angewandte Limnologie Verhandlungen 18:718-25.

de March, B. G. E. 1976. Spatial and temporal patterns in macrobenthic stream diversity. Journal of the Fisheries Research Board of Canada 33:1261-70.

Edington, J. M. 1968. Habitat preferences in net-spinning caddis larvae with special reference to the influence of water velocity. Journal of Animal Ecology 37:675-92.

Egglishaw, H. J. 1964. The distributional relationship between the bottom fauna and plant detritus in streams. Journal of Animal Ecology 33:463-76.

Egglishaw, H. J. 1968. The quantitative relationship between bottom fauna and plant detritus in streams of different calcium concentrations. Journal of Applied Ecology 5: 731-40.

Egglishaw, H. J. 1969. The distribution of benthic invertebrates on substrata in fast-flowing streams. Journal of Animal Ecology 38:19-33.

Elliott, J. M. 1967a. The life histories and drifting of the Plecoptera and Ephemeroptera in a Dartmoor stream. Journal of Animal Ecology 36:343-62.

Elliott, J. M. 1967b. Invertebrate drift in a Dartmoor stream. Archiv für Hydrobiologie 63:202-37.

Elliott, J. M. 1977. Some methods for the statistical analysis of samples of benthic invertebrates. 2nd ed. Freshwater Biological Association Scientific Publication No. 25. 160 pp.

Eriksen, C. H. 1963. The relation of oxygen consumption to substrate particle size in two burrowing mayflies. Journal of Experimental Biology 40:447-53.

Eriksen, C. H. 1964. The influence of respiration and substrate upon the distribution of burrowing mayfly naiads. Internationale Vereinigung für Theoretische und Angewandte Limnologie Verhandlungen 15:903-11.

Eriksen, C. H. 1966. Benthic invertebrates and some substrate-current-oxygen interrelationships. Pymatuning Laboratory of Ecology Special Publication 4:98-115.

Eriksen, C. H. 1968. Ecological significance of respiration and substrate for burrowing Ephemeroptera. Canadian Journal of Zoology 46:93-103.

Eriksen, C. H., V. H. Resh, S. S. Balling, and G. A. Lamberti. In press. Aquatic insect respiration. In: R. W. Merritt and K. W. Cummins (eds.). An introduction to the aquatic insects of North America, 2nd ed. Kendall/Hunt Publishing Co., Dubuque, IA.

Erman, D. C. and N. A. Erman. In press. The response of stream macroinvertebrates to substrate size and heterogeneity. Hydrobiologia.

Ford, J. B. 1962. The vertical distribution of larval Chironomidae (Dipt.) in the mud of a stream. Hydrobiologia 19:262–72.

Friberg, F., L. M. Nilsson, C. Otto, P. Sjöström, B. W. Svensson, Bj. Svensson, and S. Ulfstrand. 1977. Diversity and environments of benthic invertebrate communities in south Swedish streams. Archiv für Hydrobiologie 81:129–54.

Gee, J. H. R. 1979. A comparison of gravimetric and photographic methods of stream substrate analysis in a study of benthos microdistribution. Oikos 33:74–79.

George, D. G. 1981. Zooplankton patchiness. Annual Report of the Freshwater Biological Association 49:32–44.

Glass, L. W. and R. V. Bovbjerg. 1969. Density and dispersion in laboratory populations of caddisfly larvae (*Cheumatopsyche*, Hydropsychidae). Ecology 50:1082–84.

Glime, J. M. and R. M. Clemons. 1972. Species diversity of stream insects on *Fontinalis* spp. compared to diversity on artificial substrates. Ecology 53:458–64.

Gore, J. A. 1979. Patterns of initial benthic recolonization of a reclaimed coal strip-mined river channel. Canadian Journal of Zoology 57:2429–39.

Grant, P. R. and R. J. Mackay. 1969. Ecological segregation of systematically related stream insects. Canadian Journal of Zoology 47:691–94.

Gregg, W. W. 1981. Aquatic macrophytes as a factor affecting the microdistribution of benthic stream invertebrates. Unpublished M. S. thesis, Idaho State University, Pocatello, ID. 163 pp.

Guy, H. P. 1969. Laboratory theory and methods for sediment analysis. Techniques of water-resources investigations of the United States Geological Survey. Book 5, Chapter CI. United States Department of the Interior, Washington DC. 58 pp.

Hanna, H. M. 1961. Selection of materials for case-building by larvae of caddisflies (Trichoptera). Proceedings of the Royal Entomological Society of London 36:37–47.

Harman, W. N. 1972. Benthic substrates: their effect on fresh-water Mollusca. Ecology 53:271–77.

Harrod, J. J. 1964. The distribution of invertebrates on submerged aquatic plants in a chalk stream. Journal of Animal Ecology 33:335–48.

Hart, D. D. 1978. Diversity in stream insects: regulation by rock size and microspatial complexity. Internationale Vereinigung für Theoretische und Angewandte Limnologie Verhandlungen 20:1376–81.

Higler, L. W. G. 1975. Reactions of some caddis larvae (Trichoptera) to different types of substrate in an experimental stream. Freshwater Biology 5:151–58.

Hildebrand, S. G. 1974. The relation of drift to benthos density and food level in an artificial stream. Limnology and Oceanography 19:951–57.

Hildrew, A. G. and C. R. Townsend. 1977. The influence of substrate on the functional response of *Plectrocnemia conspersa* (Curtis) larvae (Trichoptera: Polycentropodidae). Oecologia 31:21–26.

Hora, S. L. 1930. Ecology, bionomics, and evolution of the torrential fauna, with special reference to the organs of attachment. Philosophical Transactions of the Royal Society (B) 218:171–282.

Hughes, D. A. 1966. The role of responses to light in the selection and maintenance of microhabitat by the nymphs of two species of mayfly. Animal Behaviour. 14:17–33.

Hunt, J. S. 1930. Bottom as a factor in animal distribution in small streams. Journal of the Tennessee Academy of Sciences 5:11–18.

Huston, M. 1979. A general hypothesis of species diversity. The American Naturalist 113:81–101.

Hynes, H. B. N. 1961. The invertebrate fauna of a Welsh mountain stream. Archiv für Hydrobiologie 57:344–88.

Hynes, H. B. N. 1970. The ecology of running waters. University of Toronto Press, Toronto. 555 pp.

Ide, F. P. 1935. The effect of temperature on the distribution of the mayfly fauna of a stream. University of Toronto Studies, Biological Series 39:1–76.

Inman, D. L. 1952. Measures for describing the size distribution of sediments. Journal of Sedimentary Petrology 22:125–45.

Jaag, O. and H. Ambühl. 1964. The effect of the current on the composition of bio-coenoses in flowing water streams, pp. 31–44. In: First International Conference on Water Pollution Research, London. Pergamon Press, Oxford. 341 pp.

Jónasson, P. M. 1948. Quantitative studies of the bottom fauna, pp. 204–85. In: K. Berg (ed.). Biological studies of the River Susaa. Folia Limnologica Scandinavica No. 4. 318 pp.

Keller, A. 1975. Die Drift und ihre ökologische Bedeutung. Experimentelle Untersuchung an Ecdyonurus venosus (Fabr.) in einem Fliesswassermodell. Schweizerische Zeitschrift für Hydrologie 37:294–331.

Khalaf, G. and H. Tachet. 1980. Colonization of artificial substrata by macroinvertebrates in a stream and variations according to stone size. Freshwater Biology 10:475–82.

Knight, A. W. and A. R. Gaufin. 1963. The effect of water flow, temperature, and oxygen concentration on the Plecoptera nymph, Acroneuria pacifica Banks. Proceedings of the Utah Academy of Sciences, Arts, and Letters 40:175–84.

Knight, A. W. and A. R. Gaufin. 1964. Relative importance of varying oxygen concentration, temperature, and water flow on the mechanical activity and survival of the Plecoptera nymph, Pteronarcys californica Newport. Proceedings of the Utah Academy of Sciences, Arts, and Letters 41:14–28.

Kohn, A. J. 1967. Environmental complexity and species diversity in the gastropod genus Conus on Indo-West Pacific reef platforms. The American Naturalist 101: 251–60.

Kohn, A. J. 1968. Microhabitats, abundance, and food of Conus on atoll reefs in the Maldive and Chagos Islands. Ecology 49:1046–62.

Kovalak, W. P. 1976. Seasonal and diel changes in the positioning of Glossosoma nigrior (Trichoptera: Glossosomatidae) on artificial substrates. Canadian Journal of Zoology 54:1585–94.

Krecker, F. H. 1939. A comparative study of the animal population of certain submerged aquatic plants. Ecology 20:553–62.

Krecker, F. H. and L. Y. Lancaster. 1933. Bottom shore fauna of western Lake Erie: a population study to a depth of six feet. Ecology 14:79–93.

Krumbein, W. C. 1939. Graphic representation and statistical analysis of sedimentary data, pp. 588–91. In: P. D. Trask (ed.). Recent marine sediments. Thomas Murby and Co., London. 736 pp.

Lamberti, G. A. and V. H. Resh. 1979. Substrate relationships, spatial distribution patterns, and sampling variability in a stream caddisfly population. Environmental Entomology 8:561–67.

Lehmkuhl, D. M. and N. H. Anderson. 1972. Microdistribution and density as factors affecting the downstream drift of mayflies. Ecology 53:661–67.

Leopold, L. B. 1970. An improved method for size distribution of stream bed gravel. Water Resources Research 6:1357–66.

Leopold, L. B., M. G. Wolman, and J. P. Miller. 1964. Fluvial processes in geomorphology. W. B. Freeman and Co., San Francisco. 522 pp.

Lillehammer, A. 1966. Bottom fauna investigations in a Norwegian river: the influence of ecological factors. Nytt Magasin for Zoologi 13:10–29.

Lindegaard, C., J. Thorup, and M. Bahn. 1975. The invertebrate fauna of the moss carpet in the Danish spring Ravnkilde and its seasonal, vertical, and horizontal distribution. Archiv für Hydrobiologie 75:109–39.

Linduska, J. P. 1942. Bottom type as a factor influencing the local distribution of mayfly nymphs. The Canadian Entomologist 74:26–30.

Luedtke, R. J. and M. A. Brusven. 1976. Effects of sand sedimentation on colonization of stream insects. Journal of the Fisheries Research Board of Canada 33:1881–86.

Luedtke, R. J., M. A. Brusven, and F. J. Watts. 1976. Benthic insect community changes in relation to in-stream alterations of a sediment-polluted stream. Melanderia 23: 21–39.

Lyman, F. E. 1956. Environmental factors affecting distribution of mayfly nymphs in Douglas Lake, Michigan. Ecology 37:568–76.

Macan, T. T. 1963. Freshwater ecology. John Wiley and Sons. Inc., New York, NY. 338 pp.

Macan, T. T. 1965. The fauna in the vegetation of a moorland fishpond. Archiv für Hydrobiologie 61:273–310.

Macan, T. T. 1975. Structure of the community in the vegetation of a moorland fishpond. Internationale Vereinigung für Theoretische und Angewandte Limnologie Verhandlungen 19:2298–304.

Macan, T. T. 1977. The fauna in the vegetation of a moorland fishpond as revealed by different methods of collecting. Hydrobiologia 55:3–15.

Macan, T. T. and A. Kitching. 1976. The colonization of squares of plastic suspended in midwater. Freshwater Biology 6:33–40.

Macan, T. T. and R. Maudsley. 1968. The insects of the stony substratum of Windermere. Transactions of the Society of British Entomologists 18:1–18.

Macan, T. T. and R. Maudsley. 1969. Fauna of the stony substratum in lakes in the English Lake District. Internationale Vereinigung für Theoretische und Angewandte Limnologie Verhandlungen 17:173–80.

MacArthur, R. H. and E. O. Wilson. 1967. The theory of island biogeography. Monographs in Population Biology 1. Princeton University Press, Princeton, NJ. 203 pp.

Mackay, R. J. 1969. Aquatic insect communities of a small stream on Mont St. Hilaire, Quebec. Journal of the Fisheries Research Board of Canada 26:1157–83.

Mackay, R. J. 1977. Behavior of *Pycnopsyche* (Trichoptera: Limnephilidae) on mineral substrates in laboratory streams. Ecology 58:191–95.

Mackay, R. J. and J. Kalff. 1969. Seasonal variation in standing crop and species diversity of insect communities in a small Quebec stream. Ecology 50:101–9.

Maitland, P. S. 1964. Quantitative studies on the invertebrate fauna of sandy and stony substrates in the River Endrick, Scotland. Proceedings of the Royal Society of Edinburgh 68:277–301.

Malmqvist, B., L. M. Nilsson, and B. S. Svensson. 1978. Dynamics of detritus in a small stream in southern Sweden and its influence on the distribution of the bottom animal communities. Oikos 31:3–16.

McAuliffe, J. R. 1983. Competition, colonization patterns, and disturbance in stream benthic communities, pp. 137–56. *In*: J. R. Barnes and G. W. Minshall (eds.). Stream ecology: application and testing of general ecological theory. Plenum Press, New York, NY. 399 pp.

McLachlan, A. J. 1969. Substrate preferences and invasion behaviour exhibited by larvae of *Nilodorum brevibucca* Freeman (Chironomidae) under experimental conditions. Hydrobiologia 33:237–49.

McLachlan, A. J. 1977. Density and distribution in laboratory populations of midge larvae (Chironomidae: Diptera). Hydrobiologia 55:195–99.

McLachlan, A. J., A. Brennan, and R. S. Wotton. 1978. Particle size and chironomid (Diptera) food in a upland river. Oikos 31:247–52.

Meadows, P. S. and J. I. Campbell. 1972. Habitat selection by aquatic invertebrates. Advances in Marine Biology 10:271–382.

Meier, P. G., D. L. Penrose, and L. Polak. 1979. The rate of colonization by macroinvertebrates on artificial substrate samplers. Freshwater Biology 9:381–92.

Minckley, W. L. 1963. The ecology of a spring stream: Doe Run, Meade County, Kentucky. Wildlife Monographs 11:1–124.

Minshall, G. W. 1968. Community dynamics of the benthic fauna in a woodland spring-brook. Hydrobiologia 32:305–39.

Minshall, G. W., D. A. Andrews, and C. Y. Manuel-Faler. 1983. Application of island biogeographic theory to streams: macroinvertebrate recolonization of the Teton River, Idaho, pp. 279–297. *In*: J. R. Barnes and G. W. Minshall (eds.). Stream ecology: application and testing of general ecological theory. Plenum Press, New York, NY. 399 pp.

Minshall, G. W. and R. A. Kuehne. 1969. An ecological study of invertebrates of the Duddon, an English mountain stream. Archiv für Hydrobiologie 66:169–91.

Minshall, G. W. and J. N. Minshall. 1977. Microdistribution of benthic invertebrates in a Rocky Mountain (U.S.A.) stream. Hydrobiologia 55:231–49.

Minshall, G. W. and J. N. Minshall. 1978. Further evidence on the role of chemical factors in determining the distribution of benthic invertebrates in the River Duddon. Archiv für Hydrobiologie 83:324–55.

Minshall, G. W. and P. V. Winger. 1968. The effect of reduction in stream flow on invertebrate drift. Ecology 49:580–82.

Moon, H. P. 1935a. Methods and apparatus suitable for an investigation of the littoral region of oligotrophic lakes. Internationale Revue der gesamten Hydrobiologie und Hydrographie 32:319–33.

Moon, H. P. 1935b. Flood movements of the littoral fauna of Windermere. Journal of Animal Ecology 4:216–28.

Moon, H. P. 1939. Aspects of the ecology of aquatic insects. Transactions of the British Entomological Society 6:39–49.

Moon, H. P. 1940. An investigation of the movements of freshwater invertebrate faunas. Journal of Animal Ecology 9:76–83.

Morris, D. L. and M. P. Brooker. 1979. The vertical distribution of macro-invertebrates in the substratum of the upper reaches of the River Wye, Wales. Freshwater Biology 9:573–83.

Mundie, J. H. 1971. Sampling benthos and substrate materials, down to 50 microns in size, in shallow streams. Journal of the Fisheries Research Board of Canada 28:849–60.

Needham, P. R. 1927. A quantitative study of the fish food supply in selected areas. A biological survey of the Oswego River System. New York State Conservation Department Annual Report 17:192–206.

Needham, P. R. and R. L. Usinger, 1956. Variability in the macrofauna of a single riffle in Prosser Creek, California, as indicated by the Surber sampler. Hilgardia 24:383–409.

Nilsen, H. C. and R. W. Larimore. 1973. Establishment of invertebrate communities on log substrates in the Kaskaskia River, Illinois. Ecology 54:366–74.

Nuttall, P. M. 1972. The effects of sand deposition upon the macroinvertebrate fauna of the River Camel, Cornwall. Freshwater Biology 2:181–86.

Nuttall, P. M. and G. H. Bielby. 1973. The effect of china-clay wastes on stream invertebrates. Environmental Pollution 5:77–86.

Osman, R. W. 1978. The influence of seasonality and stability on the species equilibrium. Ecology 59:383–99.

Otto, C. 1976. Habitat relationships in the larvae of three Trichoptera species. Archiv für Hydrobiologie 77:505–17.

Otto, C. and B. S. Svensson. 1980. The significance of case material selection for the survival of caddis larvae. Journal of Animal Ecology 49:855–65.

Paine, R. T. 1980. Food webs: linkage, interaction strength and community infrastructure. Journal of Animal Ecology 49:667–85.

Paterson, C. G. and C. H. Fernando. 1971. Studies on the spatial heterogeneity of shallow water benthos with particular reference to the Chironomidae. Canadian Journal of Zoology 49:1013–19.

Peckarsky, B. L. 1979. Biological interactions as determinants of distributions of benthic invertebrates within the substrate of stony streams. Limnology and Oceanography 24:59–68.

Peckarsky, B. L. 1980a. Influence of detritus upon colonization of stream invertebrates. Canadian Journal of Fisheries and Aquatic Sciences 37:957–63.

Peckarsky, B. L. 1980b. Predator-prey interactions between stoneflies and mayflies: behavioral observations. Ecology 61:932–43.

Peckarsky, B. L. 1981. Reply to comment by Sell. Limnology and Oceanography 26: 982–87.

Pennak, R. W. and E. D. Van Gerpen. 1947. Bottom fauna production and physical nature of the substrate in a northern Colorado trout stream. Ecology 28:42–48.

Percival, E. and H. Whitehead. 1926. Observations on the biology of *Ephemera danica* Mull. Proceedings of the Leeds Philosophical and Literary Society. 1:136–48.

Percival, E. and H. Whitehead. 1929. A quantitative study of the fauna of some types of stream-bed. Journal of Ecology 17:282–314.

Petran, M. 1977. Ökologische Untersuchungen an Fliessgewässern über die Beziehungen zwischen Makrobenthos, substrat und Geschiebetrieb. Unpublished Ph.D. Dissertation, Rheinische Friedrich-Wilhelms University, Bonn, Germany.

Philipson, G. N. 1954. The effect of waterflow and oxygen concentration on six species of caddisfly (Trichoptera). Proceedings of the Zoological Society of London 124: 547–64.

Pollard, R. A. 1955. Measuring seepage through salmon spawning gravel. Journal of the Fisheries Research Board of Canada 12:706–41.

Rabeni, C. F. and G. W. Minshall. 1977. Factors affecting microdistribution of stream benthic insects. Oikos 29:33–43.

Rawson, D. S. 1930. The bottom fauna of Lake Simcoe and its role in the ecology of the lake. University of Toronto Studies, Biological Series 34:1–183.

Reice, S. R. 1974. Environmental patchiness and the breakdown of leaf litter in a woodland stream. Ecology 55:1271–82.

Reice, S. R. 1977. The role of animal association and current velocity in sediment-specific leaf litter decomposition. Oikos 29:357–65.

Reice, S. R. 1978. Role of detritivore selectivity in species-specific litter decomposition in a woodland stream. Internationale Vereinigung für Theoretische und Angewandte Limnologie Verhandlungen 20:1396–400.

Reice, S. R. 1980. The role of substratum in benthic macroinvertebrate microdistribution and litter decomposition in a woodland stream. Ecology 61:580–90.

Resh, V. H. 1977. Habitat and substrate influences on population and production dynamics of a stream caddisfly, *Ceraclea ancylus* (Leptoceridae). Freshwater Biology 7:261–77.

Resh, V. H. 1979. Sampling variability and life history features: basic considerations in the design of aquatic insect studies. Journal of the Fisheries Research Board of Canada 36:290–311.

Rosenberg, D. M. and V. H. Resh. 1982. The use of artificial substrates in the study of freshwater benthic macroinvertebrates, pp. 175–235. *In*: J. Cairns, Jr. (ed.). Artificial substrates. Ann Arbor Science Publishers, Ann Arbor, MI. 279 pp.

Rosine, W. N. 1955. The distribution of invertebrates on submerged aquatic plant surfaces in Muskee Lake, Colorado. Ecology 36:308–14.

Scherer, E. 1962. Phototaktisches Verhalten von Fliesswasser-Insectenlarven. Naturwissenschaften 49:477–78.

Schwoerbel, J. 1961. Über die Lebensbedingungen und die Besiedlung des hyporheischen Lebensraumes. Archiv für Hydrobiologie Supplement 25:182–214.

Scott, D. 1958. Ecological studies on the Trichoptera of the River Dean, Cheshire. Archiv für Hydrobiologie 54:340–92.

Shaw, D. W. and G. W. Minshall. 1980. Colonization of an introduced substrate by stream macroinvertebrates. Oikos 34:259–71.

Shelly, T. E. 1979. The effect of rock size upon the distribution of species of Orthocladiinae (Chironomidae: Diptera) and *Baetis intercalaris* McDunnough (Baetidae: Ephemeroptera). Ecological Entomology 4:95–100.

Shiozawa, D. K. and J. R. Barnes. 1977. The microdistribution and population trends of larval *Tanypus stellatus* Coquillett and *Chironomus frommeri* Atchley and Martin (Diptera: Chironomidae) in Utah Lake, Utah. Ecology 58:610–18.

Short, R. A., S. P. Canton, and J. V. Ward. 1980. Detrital processing and associated macroinvertebrates in a Colorado mountain stream. Ecology 61:727–32.

Smith, F. E. 1972. Spatial heterogeneity, stability, and diversity in ecosystems. Transactions of the Connecticut Academy of Arts and Sciences 44:307–35.

Sousa, W. P. 1979. Disturbance in marine intertidal boulder fields: the nonequilibrium maintenance of species diversity. Ecology 60:1225–39.

Southwood, T. R. E. 1977. Habitat, the templet for ecological strategies? Journal of Animal Ecology 46:337–65.

Sprules, W. M. 1947. An ecological investigation of stream insects in Algonquin Park, Ontario. University of Toronto Studies, Biological Series 56:1–81.

Stanford, J. A. and J. V. Ward. 1983. Insect species diversity as a function of environmental variability and disturbance in stream systems, pp. 265–78. *In*: J. R. Barnes and G. W. Minshall (eds.). Stream ecology: application and testing of general ecological theory. Plenum Press, New York, NY. 399 pp.

Sutcliffe, D. W. and T. R. Carrick. 1973. Studies on mountain streams in the English Lake District. I. pH, calcium, and the distribution of invertebrates in the River Duddon. Freshwater Biology 3:437–62.

Sweeney, B. W. and R. L. Vannote. 1981. *Ephemerella* mayflies of White Clay Creek: bioenergetic and ecological relationships among six coexisting species. Ecology 62:1353–69.

Tarzwell, C. M. 1938. Factors influencing fish food and fish production in southwestern streams. Transactions of the American Fisheries Society 67:246–55.

Thorup, J. 1966. Substrate type and its value as a basis for the delimitation of bottom fauna communities in running waters. Pymatuning Laboratory of Ecology Special Publication 4:59–74.

Thorup, J. and C. Lindegaard. 1977. Studies on Danish springs. Folia Limnologica Scandinavica 17:7–15.

Thut, R. N. 1969. A study of the profundal bottom fauna of Lake Washington. Ecological Monographs 39:79–100.

Tolkamp, H. H. 1980. Organism-substrate relationships in lowland streams. Agricultural Research Report 907, Agricultural University, Wageningen, The Netherlands. 211 pp.

Townsend, C. R. and A. G. Hildrew. 1976. Field experiments on the drifting, colonization, and continuous redistribution of stream benthos. Journal of Animal Ecology 45:759–72.

Townsend, C. R. and A. G. Hildrew. 1980. Foraging in a patchy environment by a predatory net-spinning caddis larva: a test of optimal foraging theory. Oecologia (Berlin) 47:219–21.

Trush, W. J., Jr. 1979. The effects of area and surface complexity on the structure and formation of stream benthic communities. Unpublished M. S. thesis, Virginia Polytechnic Institute and State University, Blacksburg, VA. 149 pp.

Ulfstrand, S. 1967. Microdistribution of benthic species (Ephemeroptera, Plecoptera, Trichoptera, Diptera: Simuliidae) in Lapland streams. Oikos 18:293–310.

Uresk, D. W. 1967. The substrate preference of aquatic insects in Trout Creek, Wasatch County, Utah. M. S. thesis, University of Utah, Salt Lake City, UT. 42 pp.

Vannote, R. L. 1981. The river continuum: a theoretical construct for analysis of river ecosystems, pp. 289–304. *In*: R. Cross (ed.). Proceedings of the National Symposium on Freshwater Inflow to Estuaries. Vol. 2. U. S. Fish and Wildlife Service, Washington DC. 528 pp.

Vannote, R. L., G. W. Minshall, K. W. Cummins, J. R. Sedell, and C. E. Cushing. 1980. The river continuum concept. Canadian Journal of Fisheries and Aquatic Sciences 37:130–37.

Vannote, R. L. and B. W. Sweeney. 1980. Geographic analysis of thermal equilibria: a conceptual model for evaluating the effect of natural and modified thermal regimes on aquatic insect communities. The American Naturalist 115:667–95.

Wallace, J. B. and R. W. Merritt. 1980. Filter-feeding ecology of aquatic insects. Annual Review of Entomology 25:103–32.

Ward, J. V. 1975. Bottom fauna-substrate relationships in a Northern Colorado trout stream: 1945 and 1974. Ecology 56:1429-34.

Ward, J. V. and J. A. Stanford. 1982. Thermal responses in the evolutionary ecology of aquatic insects. Annual Review of Entomology 27:97-117.

Waters, T. F. 1972. The drift of stream insects. Annual Review of Entomology 17:253-72.

Wautier, J. and E. Pattée. 1955. Expérience physiologique et expérience écologique. L'influence du substrat sur la consommation d'oxygène chez les larves d'Ephéméroptères. Bulletin mensuel de la Société linnéenne de Lyon 24:178-83.

Wene, G. 1940. The soil as an ecological factor in the abundance of chironomid larvae. Ohio Journal of Science 40:193-99.

Wene, G. and E. L. Wickliff. 1940. Modification of a stream bottom and its effect on the insect fauna. The Canadian Entomologist 72:131-35.

Wentworth, C. K. 1922. A scale of grade and class terms for clastic sediments. Journal of Geology 30:377-92.

Whitehead, H. 1935. An ecological study of the invertebrate fauna of a chalk stream near Great Driffield, Yorkshire. Journal of Animal Ecology 4:58-78.

Wiley, M. J. and S. L. Kohler. 1981. An assessment of biological interactions in an epilithic stream community using time-lapse cinematography. Hydrobiologia 78: 183-88.

Williams, D. D. 1980. Some relationships between stream benthos and substrate heterogeneity. Limnology and Oceanography 25:166-72.

Williams, D. D. and H. B. N. Hynes. 1977. Benthic community development in a new stream. Canadian Journal of Zoology 55:1071-76.

Williams, D. D. and J. H. Mundie. 1978. Substrate size selection by stream invertebrates and the influence of sand. Limnology and Oceanography 23:1030-33.

Williams, R. 1972. The abundance and biomass of the interstitial fauna of a graded series of shell-gravels in relation to the available space. Journal of Animal Ecology 41: 623-46.

Wise, D. H. and M. C. Molles, Jr. 1979. Colonization of artificial substrates by stream insects: influence of substrate size and diversity. Hydrobiologia 65:69-74.

Wu, Y. F. 1931. A contribution to the biology of *Simulium* (Diptera). Papers of the Michigan Academy of Sciences, Arts and Letters 13:543-99.

Zahar, A. R. 1951. The ecology and distribution of black-flies (Simuliidae) in south-east Scotland. Journal of Animal Ecology 20:33-62.

chapter 13

COLONIZATION DYNAMICS OF AQUATIC INSECTS

Andrew L. Sheldon

For aquatic insects, the world is an everchanging mosaic of environments in which elements arise, vanish, and reappear with different periodicities and predictabilities. Islands rise from the sea and receding glaciers lay bare landscapes rich in aquatic habitats. Reservoirs are constructed. Ephemeral pools fill after rare desert rains or, more predictably, with the melting of winter snow. Dry streambeds bake in the sun, then fill and flow for days or years, and then dry again. Spates may destroy most of a stream's fauna or merely overturn a few stones to expose new surfaces for occupancy. Episodic pollution or insecticide applications create vacant environments. Emergence of a cohort of insects leaves space that larvae of other cohorts or species may occupy. If this kaleidoscopic view of the world is realistic, then colonization, broadly defined, must occur so frequently that it is an integral part of the life histories of all but a few extraordinarily specialized aquatic insects. Since colonization includes events on scales from intercontinental distances to centimeters, and from millennia to seconds, it seems reasonable to seek general principles over the entire spectrum of examples. Inevitably, certain taxa and situations will be over- or under-represented and any generalizations will be imperfect. Fortunately, there exists a body of literature on the ecology and genetics of colonization by other animal groups and by plants, which supplements the existing knowledge of aquatic insects and suggests profitable lines of future research.

Colonization can be viewed as the sequence of events that leads to the establishment of individuals, populations, species, or higher taxa in places from which they were, however temporarily, absent. This definition encompasses the entire spectrum of concern and excludes the alternative to colonization, which is persistence in dormant, resistant life-cycle stages. Insects of some habitats such as ephemeral ponds use both tactics (Wiggins et al. 1980).

SCALE OF COLONIZATION

The spatial and temporal framework of Table 13.1 simplifies discussion of colonization. The concentration of examples in the middle range of time and distance reflects the life span, mobility, and technical capabilities of ecologists as much as ecological factors.

Events involving the greatest times and distances are in the domain of the

TABLE 13.1: Temporal and Spatial Scales of Colonization Events

| | SPATIAL | | |
	Macro-level (>10^2 - 10^3 km)	*Meso-level* (10^{-1} - 10^2 km)	*Micro-level* (10^{-2} - 10 m)
Long (10^2 - 10^6 years)	Continental and insular invasions; introductions	Some introductions?	—
Moderate (one generation—10 yrs)	Species migration	Artificial lakes; rock pools; age-specific movements (e.g. "colonization cycle"); new stream channels; container plants	Littoral zones; animal hosts
Short (minutes–weeks)	—	Rock pools; littoral zones	Adjacent stones; artificial substrates; animal hosts

TEMPORAL

biogeographer, but they set the stage for ecological investigations of existing aquatic communities. At the far end of the spectrum are intercontinental exchanges such as those proposed by Illies (1965) for higher taxa of Plecoptera. However, post-Pleistocene colonizations of deglaciated areas (Brinck 1949; Ricker 1964; Ross et al. 1967; Lehmkuhl 1980) are better dated and provide conservative estimates of the rate at which populations of stoneflies and other aquatic insects can spread.

Oceanic islands have derived their faunas by overwater dispersal. Some aquatic orders such as Ephemeroptera, Plecoptera, and Trichoptera, seldom occur in these islands (Laird 1956; Illies 1965; Edmunds 1972; Winterbourn 1980), whereas others such as Odonata, Hemiptera, Coleoptera, and Diptera, have dispersed to them successfully, although the diversity of these latter groups is progressively attenuated with increasing distance from source areas. The low diversity of aquatic insects on oceanic islands has allowed invasion by marine forms, including the marine midge *Telmatogeton* in which the sequence of colonization and subsequent speciation has been investigated with cytological techniques (Newman 1977). The unbalanced faunas of oceanic islands have received recent introductions of mosquitoes (Laird 1956; Elton 1958) and odonates (Paulson 1978a). An introduced hydropsychid caddisfly (*Cheumatopsyche pettiti = analis*) now occupies mountain streams in several of the Hawaiian Islands (Maciolek 1975; J. Maciolek, personal communication). The effect of the addition of an efficient filter-feeder to these simple communities is unknown. Earlier, Needham and Welsh (1953) advocated the introduction of mayflies to Hawaii as food for introduced trout. Armstrong (1978) documented the establishment of an Australian dragonfly in New Zealand and its apparent replacement of a native species.

In contrast to the situation in terrestrial insects, introductions and recent invasions of continents by aquatic insects seem to be uncommon. Exceptions include an Asiatic dragonfly that may be breeding in Florida (Paulson 1978a, 1978b), and the highly dangerous malaria vector *Anopheles gambiae* that was eradicated from Brazil after it had spread 300 km from its point of introduction (Elton 1958). Recent attempts to control aquatic plants such as the mat-forming fern *Salvinia* have resulted in the establishment of a neotropical aquatic grasshopper in Africa and Asia (Bennett 1977), and further introductions of aquatic and semi-aquatic insects for weed control are likely.

Within continents, distributional data are seldom good enough to verify range extensions by native species or to distinguish true introductions from supplements to existing populations; examples are therefore limited. Anglers, continuing the British tradition of intensive management of trout waters, introduced ephemerid mayflies into several Pennsylvania streams (Marinaro 1970). Patterson and Vannote's (1979) introduction of the eggs of a caddisfly into a stream void of that species simplified cohort analysis and verified a provisional interpretation of its life history.

Long-distance migrations are uncommon among aquatic insects. Corbet (1980) summarized the literature on a few southern U.S. dragonflies whose adults move as far north as southern Canada. Their progeny emerge in late summer and migrate southward. This situation, comparable to the migration of the well-known Monarch butterfly (Johnson 1969) and the milkweed bug *Oncopeltus fasciatus* (Dingle 1978), may occur in other vagile aquatic orders such as Hemiptera and Coleoptera.

Colonization events occurring in the mid-range of the temporal and spatial scales are better documented. The following examples illustrate the variety of habitats in which colonization is important, and related papers that clarify the mechanisms and interactions involved are discussed in later sections. Catastrophic events such as flash floods in desert regions (Bruns and Minckley 1980) or the wet tropics (Stout 1981), dam failures (Minshall et al. 1983), and pesticide applications (Hynes and Williams 1962; Ide 1967; Hilsenhoff 1971) devastate aquatic communities and initiate colonization sequences. Likewise, human-made lakes are colonized rapidly (Paterson and Fernando 1970; Voshell and Simmons 1978; Ertlová 1980), as are newly created streams (Williams and Hynes 1977; Gore 1979), and rehabilitated segments of rivers (Boles 1981). The water-filled parts of plants (Istock et al. 1975; Seifert 1975) and even individual animals, such as the mayfly host of a commensal chironomid (Svensson 1980), are initially vacant colonizable units.

Habitats that dry and fill repeatedly may be reoccupied by colonists or by reactivation of resistant stages, and often the mechanism of repopulation is difficult to determine. Ephemeral pools derive part of their fauna by aerial colonization (McLachlan and Cantrell 1980), and streams in drought-prone regions are repopulated rapidly (Larimore et al. 1959; Harrison 1966). Irrigation ditches (Fredeen and Shemanchuk 1960) and rice fields (Clement et al. 1977) are artificial analogs of intermittent streams and seasonal marshes.

A whole series of seasonal and ontogenetic changes may be considered colonization events. Müller (1954, 1982) proposed the term "colonization cycle" for downstream drift and compensatory upstream flight by adult females, and later generalized the concept to include movements into seasonably favorable habitats (e.g. see Mendl and Müller 1978; Clifford et al. 1979; Lingdell and Müller 1979; Müller and Mendl 1979; Malicky 1980; Müller 1982). Most studies have demonstrated directional movement but have failed to quantify the distances moved or the number of individuals involved. However, Neves (1979) recovered marked caddisfly larvae 0.4 to 1.5 km downstream from release points and trapped adult females (unmarked) flying upstream. Schwarz (1970) showed that net downstream displacement occurred in a population of one perlodid stonefly, but was balanced by upstream flight; in contrast, another species exhibited lower drift rates and no net displacement. Monitoring of marked individuals and data on spatial distribution and density of entire populations are necessary if the generality of the colonization cycle is to be confirmed. Other age and season-specific movements such as local movements of beetles in the littoral zone of a lake (Landin 1976) and similar directional movements (Cummins 1964; Gibbs 1979; Tozer et al. 1981) are small-scale examples of a colonization cycle.

Transitory occupation of adjacent habitat units by diffusion-like movements is dismissed by some as trivial (e.g. see Smith 1972), but it is an integral part of the life-history tactics of corixids inhabiting rock pools (Pajunen 1977), and it is of special interest to users of artificial substrates (Rosenberg and Resh 1982) as either sampling or experimental devices. Townsend and Hildrew (1976) suggested that stream benthos is continually redistributed through drift and recolonization, and Sheldon (1977) proposed that the fauna of each section of a stream is in a transitory equilibrium maintained by high rates of immigration and emigration. The consequences of this dynamic perspective are many. For example, Stout and Vandermeer (1975) suggest that interspecies interactions may be governed by the rate at which new individuals are delivered to an area of stream bottom. Although there are many unanswered questions and questionable assumptions, events at this scale are tractable subjects for experimental analysis since small areas (e.g. single stones, substrate baskets) and short exposure times permit replication to an extent that is quite impractical in larger systems.

COLONIZATION BY ADULTS

Adult insects capable of flight may oviposit in their natal environments or disperse to other locations. Southwood (1977) characterized these options as "here versus elsewhere" and "now versus later." The selective advantage of the alternatives depends on availability and quality of habitat units, and the risks and costs of delayed reproduction. Dispersal and reproductive patterns should be subject to strong selective pressures and are of general ecological interest.

Most aquatic insect orders have species that are known to disperse as adults. For example, odonates disperse widely (Corbet 1980), mayflies colonize new reservoirs (Burks 1953), both mayflies and stoneflies occur in recolonization traps (Williams and Hynes 1976a), and the large stonefly *Pteronarcys californica* swarms over highways several kilometers from water (A. L. Sheldon, unpublished data). In fact, *Pteronarcys* swarms have been encountered 1000 m above the ground (R. Marcoux, personal communication). Diptera, especially Chironomidae, disperse widely (Johnson 1969; Cheng and Birch 1977), and also occur in large numbers in recolonization traps (Williams and Hynes 1976a). Many Trichoptera are strong fliers and light-trap catches include a major fraction of the regional species pool (Ulfstrand 1970; Svensson 1974; Crichton et al. 1978; Roy and Harper 1981), although both local and large-scale heterogeneity may be pronounced (Svensson 1974; Crichton et al. 1978).

Resh et al. (1981) showed that adult oviposition was responsible for recolonization of a caddisfly population in a California spring following complete elimination and subsequent recovery of the habitat. In a desert stream, most of the colonizing insects arrived as adults, and the life histories of the majority of species in this habitat featured long-lived adults or continuous emergence (Gray and Fisher 1981; Gray 1981). Repopulation of an artificial stream was, except for a few accidental introductions when the system was established, entirely through oviposition (Ladle et al. 1980). While taxa differ in their dispersal abilities, it is obvious that adult insects do move about (Bird and Hynes 1981), and that the pool of potential colonists is large. However, restrictions imposed by adult behavior (e.g. in the selection of oviposition sites) and larval requirements drastically reduce the numbers of successful colonists.

In a review of habitat selection and oviposition behavior, Macan (1974) noted that there was evidence for discrimination by aquatic insects, and that there is a generalized response to reflecting surfaces. For example, Belostomatidae attempt to colonize parking lots wet by rain (A. L. Sheldon, unpublished data), and mirror traps (Fernando 1958) have provided data on flight periods and colonization. Recent experimental studies on habitat selection by adults include those of Golini and Davies (1975), Williams and Hynes (1976b), Angerilli and Beirne (1980), and Spence (1981). Spence's (1981) work is notable for its use of marked animals and the measurement of immigration and emigration in response to habitat features and population density.

The most coherent, although far from complete, picture of colonization by adults comes from studies on Coleoptera and Hemiptera. Long-lived aquatic adults in these orders disperse widely and persist, with or without reproducing, for some time in the invaded habitats. Dispersal flights (Fernando 1959; Popham 1964; Pajunen and Jansson 1969; Fernando and Galbraith 1973; Landin and Vepsäläinen 1977; Zalom et al. 1979, 1980; Landin 1980) are often synchronous within species and proximately controlled by temperature, light intensity, and wind velocity.

Beetles and bugs are strikingly variable in wing morphology, musculature, and flying ability (Harrison 1980). Vepsäläinen (1978) reviewed the complex alary polymorphism in Gerridae and applied ecological and genetic models in his analysis. Water striders from permanent habitats are short-winged or polymorphic for wing length, whereas long-winged species or individuals of polymorphic species dominate in less persistent environments. Isolated but persistent habitat units often support monomorphic short-winged populations. Wing length is subject to environmental influences as well since, in some multivoltine species, long days and high temperatures are the proximate cues for development of long wings in genotypes that are short-winged under early season conditions. Presumably, the summer generation is at greater risk from drought and a developmental response that facilitates dispersal is selectively advantageous. Wing length appears to be determined by a pair of alleles (with modifiers) at a single locus. Vepsäläinen (1978) suggests that the allele for long wings should be recessive in those species that colonize temporary habitats, since long-winged colonists will be homozygous and all their descendants will be long-winged. Conversely, dominance of the allele for long wings should evolve in persistent populations. Brown (1951), Young (1965), and Calabrese (1979) also have documented greater abundance of flightless morphs in permanent habitats. However, Scudder (1975) has shown considerable variability in the proportion of flightless individuals of Corixidae in permanent lakes. In a less studied order, brachyptery of Plecoptera (Brinck 1949) is associated with large permanent lakes, although Müller and Mendl (1978) suggest that short wings are a developmental consequence of lower temperatures in lakes. Thus, in spite of varied and sometimes ambiguous control mechanisms, the generalization that colonization potential is inversely related to habitat stability appears to be sound (Southwood 1962).

The system studied for many years by Pajunen (1971, 1982) illustrates the importance of colonization in life histories. Two corixid species overwinter in deeper pools, but disperse to smaller and often temporary pools for reproduction. In addition to these regular seasonal movements, continuous diffusion-like dispersal between breeding pools probably minimizes the risk that a female will place all her eggs in a single pool that will dry before her offspring mature. One corixid species is more mobile than the other and effectively exploits large populations of larval chironomids, which themselves are rapidly colonizing species, in pools which refill after drought. Pajunen's work and that of McLachlan and Cantrell (1980) illustrate the potential of rock-pool systems for comparative and experimental studies of colonization dynamics.

The related subjects of quantitative demography and evolutionary genetics are, with the notable exception of wing polymorphism, poorly known for colonizing aquatic insects. Many of the life-history studies cited here provide some data, but rigorous life-table analyses do not seem to exist for colonizers. Vepsäläinen's (1978) fitness set analysis (*sensu* Levins 1968) of wing polymorphism and his models of allelic dominance under premigration and postmigration mating strategies are unique. Landin (1980) suggested that premating dispersal by hydrophilids occurs because outbreeding is selectively advantageous even though mated females are potentially better colonists. Life-history tactics, breeding systems, and genetic variability are important attributes of all organisms, and reviews (Baker and Stebbins 1965; Parsons 1982) and general models (e.g. see Roff 1975) suggest questions that aquatic entomologists might ask.

COLONIZATION BY LARVAE—ADAPTATION OR ACCIDENT?

Both the reality and the magnitude of colonization by larval movements are well documented (Williams and Hynes 1976a; Rosenberg and Resh 1982), but the mechanisms and adaptive significance of colonization are problematical. The simplest view is that larval movements are comparable to molecular diffusion.

Areas with many larvae will contribute to empty areas until once-empty sites contain enough larvae to provide a balancing counterflow whereby densities equilibrate at all sites. This *neutral hypothesis* is developed further in the section on models (see following), but it serves here to introduce a series of questions. Why do larvae leave a site? Are some individuals or species more likely to leave than others? Is dispersal between sites passive or active? Do larvae select sites for settlement? How frequently do larvae change location and how far do they travel in each relocation? Are there evolutionary advantages in colonization, or is continual redistribution (Townsend and Hildrew 1976) an accident or a stochastic process akin to Brownian movement? Regardless of whether colonization is an adaptation or either an accidental or statistical (stochastic) process, what are its consequences for the distribution and abundance of aquatic insects and the resources they use?

Larvae may leave sites for many reasons. Density-independent (constant probability) removal is exemplified by an insect that loses its anchoring to the substrate and is swept away. Physical contact or resource reduction by conspecific organisms may lead to density-dependent emigration. The evidence for density dependence is equivocal since unsuitable statistical tests have been used in some cases (e.g. see Sell's 1981 comments on Peckarsky 1979) and, even when appropriate analytical procedures were used (Peckarsky 1981), colonization rates of many species were density independent. Some laboratory studies (Walton et al. 1977; Walton 1980a; Wiley 1981) have demonstrated density-dependent emigration, whereas others (Hildebrand 1974; Corkum 1978) failed to detect density dependence. Likewise, interspecific effects on emigration and colonization range from being undetectable between two potentially competitive mayflies (Corkum and Clifford 1980) to spectacular, as when mayfly prey evade stonefly predators (Corkum and Clifford 1980; Peckarsky 1980a; Peckarsky and Dodson 1980). Walton (1980a) also described mixed results in which a threshold density of the stonefly *Acroneuria abnormis* was required before drift of another stonefly, *Taeniopteryx*, was enhanced. In contrast, *A. abnormis* did not increase the drift rate of *Stenacron* mayfly nymphs.

In general, effects of food supply on emigration or colonization are more easily demonstrated than are direct density-dependent responses. Hildebrand (1974), Bohle (1978), and McAuliffe (1983) obtained strong responses to manipulated densities of periphyton. Bohle (1978) noted that maximum drift of *Baetis* occurred 12 to 36 hours after food depletion and Walton (1978) noted both delayed and immediate responses to sterile substrates. Periphyton growth had a converse effect on colonizing black fly larvae, which accumulated on clean artificial substrates and vacated those that were overgrown by algae (Gersabeck and Merritt 1979). Aggregation of black fly larvae in regions of high seston density (Erman and Chouteau 1979) may be caused by reduced emigration of well-fed larvae. It would seem advantageous for detritivores to aggregate on their food supply, which is often patchily distributed, but Peckarsky (1980b) obtained extremely variable responses to added detritus in colonization cages. A cautionary note was provided by the work of Winterbourn (1978), who collected almost identical numbers and kinds of insects on empty mesh bags as on those that contained dead leaves. Similar tactile responses may explain colonization of vegetation (Corkum and Clifford 1980) and other substrates (Minshall, Chapter 12). Along with food and substrate, chemical and tactile cues from conspecific organisms may enhance settling and aggregation. Such responses by larval marine invertebrates are well known (Meadows and Campbell 1972), but the possibility that aquatic insects settle preferentially near others of their kind remains uninvestigated.

Modes of dispersal and causes of emigration are varied. For example, Crisp and Gledhill (1970) and Gore (1979) reported passive dispersal by rafting of larvae on dislodged algal mats. Drifting individuals may be the major source of colonizing larvae (Williams and Hynes 1976a), and many authors have assumed that increased activity induced by competition or food shortage (e.g. see Bohle 1978) increases the liklihood of removal by currents or of active movement away from a site (Wiley and Kohler, Chapter 5). Walton (1980b) assumed active entry into the water column, and the escape responses described by Peckarsky (1980a) are clear examples of active emigration. However, not all colonization is by drift, and simple experiments using artificial substrates placed on the bottom and in the water column demonstrate colonization by crawling animals (Townsend and Hildrew 1976; Bournaud et al. 1978). Gore (1977) postulated upstream movement by larvae in a thermally modified stream, and approximately 20 percent of the colonists censused by Williams and Hynes (1976a) arrived from downstream.

Lentic insects are found in significant numbers in the water column (Mundie 1959; Davies 1976), and they readily colonize artificial substrates suspended in midwater (Macan and Kitching 1976). Bray (1971), extending work by Weerekoon (1956), studied movements of phryganeid caddisfly larvae in an English tarn. Eggs and first instar larvae of one species were found to be restricted to a small portion of the shore, whereas older larvae were found in vegetation throughout the deeper part of the lake. Bray postulated that passive removal from the shoreline and subsequent dispersal by wind-induced currents were responsible because he considered that the swimming ability of first instar larvae was too weak to function in active dispersal. McLachlan (1970) also implicated drift of first instar larvae in the colonization of submerged trees in reservoirs by Chironomidae. Reduced densities on some trees were correlated with distance from oviposition sites.

Faunas established by dispersal and colonization appear to be non-random subsets of the source faunas. Differential dispersal of black fly larvae (Colbo 1979) resulted in seasonal occupancy of downstream areas by two species, whereas two others remained near the oviposition sites at lake outlets. Disney (1972) and Gersabeck and Merritt (1979) have noted the overrepresentation of small instars of Simuliidae on newly exposed substrates or in the drift, and have suggested reasons such as size-related dominance in aggressive encounters for the bias.

Experiments by Nilsson and Sjöström (1977) on the crustacean *Gammarus* included measurements of size composition of both the drifting and colonizing components of the population. Although small animals colonized more rapidly than large ones, there was no significant difference between the size frequency distributions of drifters and colonizers. Colonization by insects should be examined in the same way, including size measurements of the benthic population on natural substrates that are the source for drifting animals.

Observed rates of colonization and drift imply substantial turnover of individuals in the benthos. Townsend and Hildrew (1976) estimated that an average of 3.6 percent of all animals left experimental substrates each day, but some taxa exhibited much higher turnovers (e.g. 20% d^{-1} for *Plectrocnemia*, 43% d^{-1} for *Nemurella*). Sheldon (1977), using different mathematics, estimated an emigration rate of 20 percent d^{-1} for *Baetis*. Although turnover is high, most colonists seem to move only a few meters before establishing themselves again (McLay 1970; Elliott 1970; Townsend and Hildrew 1976). The possibility that individual animals may be exceptionally mobile, whereas other conspecific animals are sedentary, can be investigated only by marking experiments directed to this purpose.

The adaptive significance of larval dispersal and colonization is uncertain. As suggested by Moon (1940), there is a large random component to movement and under many circumstances the neutral diffusion model introduced above, and developed formally in the next section, may be realistic. However, responses to predators, competitors, and resource levels suggest that the Fretwell-Lucas model (Fretwell 1972) of habitat use may apply to larvae of aquatic insects. This model suggests that larvae preferentially colonize the best sites until densities become high enough to reduce the marginal suitability for subsequent colonists to zero; inferior sites will then offer higher returns and will be colonized. The end result will be an "ideal free distribution" (Fretwell 1972) in which densities vary with suitability of site and the poorest sites will be occupied only when populations are very large. Wiley (1981) tested this hypothesis in a two-compartment experimental chamber that allowed midge larvae to emigrate from densely inhabited areas or less-preferred substrates, and he found that suitability was an interactive function of substrate type and population density.

Townsend and Hildrew (1976) noted that *Leuctra*, which fed on uniformly distributed iron bacteria, was less common in the drift than another stonefly, *Nemurella*, which fed on patchily distributed leaves. Dispersal and settling behavior may be adaptive tactics that allow *Nemurella* to locate and deplete resource rich sites. This is an attractive idea but supporting comparative or experimental data are unavailable.

In summary, the variety of dispersal and colonization behaviors that have been observed probably have adaptive bases, but simple physical explanations play a large part in larval colonization dynamics and must be accounted for in analyses of colonization. This contention is developed more formally in the next section of this chapter. Adaptive or not, larval movement ensures occupancy of all sites by a numerous and varied fauna. Colonization will maintain individuals on suboptimal substrates, elevate diversity of local units, and cause continued and unpredictable contacts and interactions between individuals of different species.

MODELS OF LARVAL COLONIZATION

Colonization is a dynamic process that can be studied with marked animals, or in non-equilibrium situations such as post-disturbance recovery, or with artificial substrates, using numbers (N) and time (t) as the measured variables. The simplest descriptive model is a least squares regression of N on t, t^2, t^3 . . . with significance tests of the linear and higher order terms in t. Abundance data usually require a normalizing and variance-stabilizing transformation such as the logarithmic one (Allan, Chapter 16) so that the regression of log N on t will be more appropriate. Benthic ecologists have been reluctant to compute polynomial regressions, yet some have interpreted inflections in colonization curves as biologically significant without first testing for statistical nonlinearity.

An empirical model (Sheldon 1977) that fits some colonization data (Gore 1979) is the power function

$$N_t = at^b \tag{1}$$

with its linear form as:

$$\log N_t = \log a + b \log t \tag{2}$$

where

N_t = the number of animals present at time t, and a and b are constants

In biological terms, a is the geometric mean abundance of colonists at time 1, and b is a measure of curvature. If $b < 1$ the colonization curve flattens over time (the usual case), if $b > 1$ colonization rate increases with time, and if $b = 1$ the power function is a linear function on t. Equation (2) is readily extended to tests of higher order terms in t or the effects of substrate, water velocity, or other variables. The use of logarithmic time intervals (1, 2, 4, 8 . . .) in colonization experiments is now quite general and is consistent with the power relationship.

Models incorporating mechanisms and component processes provide greater insight into colonization dynamics, although they may not fit data any better than descriptive regressions. A basic model (Sheldon 1977) formalizes the statement that net colonization rate is the difference between immigration (I) and emigration (E):

$$\frac{dN}{dt} = I - mN \tag{3}$$

where

I = immigration rate (number of animals/unit time)
N = number of animals present
m = proportion emigrating/unit time
so: $E = mN$ = emigration rate (number of animals/unit time).
Integration then yields

$$N_t = \frac{I}{m}(1 - e^{-mt}) \tag{4}$$

where

e = base of Napierian logarithms

Equations (3) and (4) are density independent (m is constant) and invoke no biological interactions; yet, they yield an asymptotic equilibrium density, I/m. These equations provide a parsimonious model of colonization dynamics and show that the behavior of independent individuals can generate realistic colonization curves. Representative curves generated by combinations of I and m (Fig. 13.1) show the sensitivity of equilibrium density (a, c, d) to changes in either parameter, and similar asymptotic densities (a, d) may reflect very different component rates and temporal dynamics. Equation (4) is a general nonlinear function and most computer centers have programs for fitting it to data. In theory, one could estimate I and m from the regression, although independent experimental estimates would be helpful.

The following discussion explores the consequences of incorporating biologically reasonable features into the basic model. Since there is a real danger that models will outstrip their factual basis (Pielou 1981), models are used here only to show that simple mechanisms will generate most of the commonly observed patterns.

Difference equations, which lend themselves to numerical simulations and better represent experimental data collected over discrete time periods, are used in place of differential equations. The difference equivalent of equation (3) is

$$N_{t+1} = I + (1 - m) N_t \tag{5}$$

where

N_t and N_{t+1} = numbers present in two consecutive periods
    $1 - m$ = the proportion of the initial population persisting over the interval.

Discrete time models do have a disadvantageous feature since they incorporate time lags, and equation (5) implies that there will never be fewer than I animals present. A more general form,

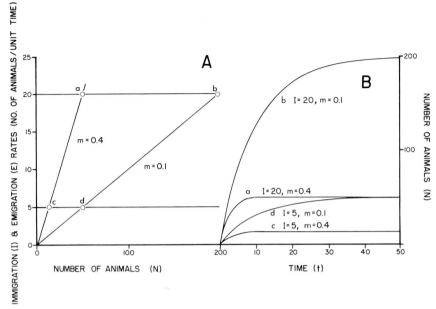

Figure 13.1 Graphical forms of equations 3 (A) and 4 (B) illustrating the effects of different combinations of immigration (I) and proportion emigrating/unit time (m). The number emigrating/unit time is $E = mN$. Equilibrium densities are defined by the intersection of the I and E curves. Time courses of colonization are shown in part B. (Curve b does not reach equilibrium within the time shown although it will eventually do so.)

$$N_{t+1} = f(I) + g(N_t) \qquad (6)$$

in which f and g could be functions of population density, substrate, location, or food supply, would permit modeling and experimental tests of modifications of both immigration and emigration. For simplicity, extensions of equation (5) are used here, and competition and other factors are assumed to influence only the emigration rate.

Density dependence is easily introduced to the model, i.e.,

$$N_{t+1} = I + (1 - m - \gamma N_t) N_t \qquad (7)$$

where

γ = a self-inhibition coefficient

Given the statistical variability of most colonization data, representative curves (Fig. 13.2) would be difficult to distinguish from the density independent case (Peckarsky 1981; Sell 1981). Note that immigration rates have a major impact on equilibrium densities.

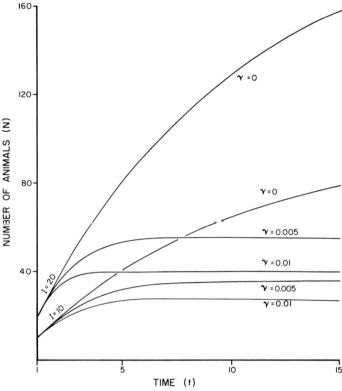

Figure 13.2 Simulations of density dependent colonization as in equation 7 with parameter γ equal to values shown, and two levels of immigration ($I = 10$, 20) and $m = 0.10$ for all curves.

Competition between species 1 and 2 is modeled by

$$N_{1,t+1} = I_1 + (1 - m_1 - \alpha N_{2,t}) N_{1,t} \tag{8}$$

$$N_{2,t+1} = I_2 + (1 - m_2 - \beta N_{1,t}) N_{2,t} \tag{9}$$

where

α and β are competition coefficients

Competition can produce, within appropriate times, a decline of one species and subsequent dominance by another (Fig. 13.3). Waters' (1964) data on rapid colonization by *Baetis* and subsequent decline as populations of slowly colonizing *Gammarus* increased are compatible with the model. The model implies that colonization rates can be as important as competitive ability. Maguire and Porter (1977) simulated interactions in coral communities and also identified colonization rates as key determinants of competitive outcomes.

New habitats such as artificial substrates or stones overturned by spates may become more favorable as periphyton develops. The dynamics of a colonizer that responds to resource (periphyton) density are given by

$$N_{t+1} = I + (1 - m + RP_t) N_t \qquad (10)$$

where

P_t = resource density
R = the reduction in the proportion emigrating attributable to the presence of a unit of resource P

The behavior of the resource is described by

$$P_{t+1} = (\lambda - GN_t) P_t \qquad (11)$$

where

λ = the finite rate of increase of the resource
G = grazing rate per individual of the insect colonizer

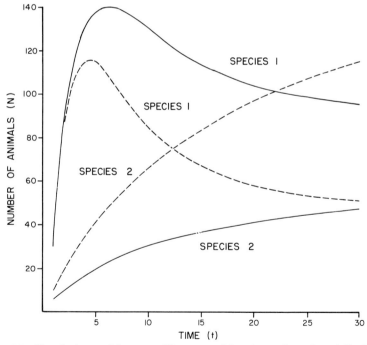

Figure 13.3 Simulations of interspecific competition (equations 8 and 9). In the first example (solid line), species 1 has parameters $I = 50$, $m = 0.20$ and $\alpha = 0.007$; for species 2, $I = 5$, $m = 0.05$, and $\beta = 0.001$. The second example (dashed line) is identical except that species 2 colonizes at a higher rate ($I = 10$).

This overly simple resource tracking model requires initial and boundary conditions for realistic simulations. The habitat unit is assumed to contain P_o units of resource at the beginning of each run. The resource P grows exponentially at rate λ up to its carrying capacity P_∞ (i.e. $\lambda = O$ for $P \geq P_\infty$). The colonizer model is constrained to cases where $m \geq RP$. The simulations of this model (Fig. 13.4) imply that slow colonizers will overshoot the food supply and decline abruptly, whereas fast colonizers control their resource base and approach equilibrium. Note the high resource densities attainable when colonization is slow. For nonrenewable resources Cancela da Fonseca (1979) has developed a more elaborate set of models, which should be very useful in studies of colonization of leaf packs and other detritus.

These basic models could be combined or refined. Simulations using wider ranges of parameter values undoubtedly would yield surprising outcomes, but these simple exercises accomplish two things. First, they clearly identify components and mechanisms that must be evaluated before colonization is understood. Second, the models demonstrate the importance of I and m in determining the outcome of biological interactions. The parameter m is of special interest as a measurable property or tactic of individuals, which is determined by

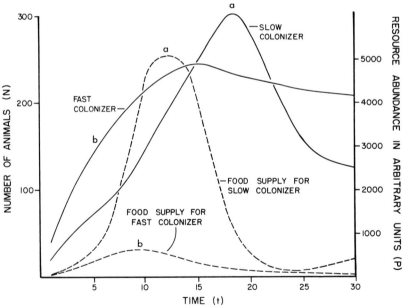

Figure 13.4 Simulations of the resource-tracking model (equations 10 and 11). Colonizer parameters (equation 10) are $m = 0.2$, $R = 0.001$, and $I = 40$ for the fast colonizer example (a) and differ only in the immigration rate, $I = 20$, for the slow colonizer example (b). Resource parameters (equation 11) are $\lambda = 2$, $G = 0.005$, $P_o = 50$, and $P_\infty = 5000$ for both examples.

habitat use and exploitation strategies of species. Immigration depends on population density and spatial arrangement of habitat, and Sheldon (1977) suggested that continual colonization held patches of stream bottom in transitory equilibrium with other patches. Pegel (1980) incorporated experimental data on drift and colonization by larval Simuliidae into a simple model, but more elaborate mathematics such as those of Okubo (1980) probably will be required for an adequate model of colonization in spatially complex habitats. Recent ideas on patch dynamics (e.g. see Gilbert and Singer 1973; Grassle 1977; DeAngelis et al. 1979; Taylor and Taylor 1979; Vandermeer et al. 1980; Paine and Levin 1981) and metapopulation structure (Addicott 1978; Gill 1978) may hold the key to understanding colonization rates and mechanisms in heterogeneous stream habitats, pond complexes, and other large-scale systems.

COLONIZATION, INTERACTION, AND SUCCESSION

Colonization dynamics are an integral part of succession, and the two are often equated. However, the relative roles of exogenously driven environmental change, differential colonization, and biological interactions are seldom separable. Newly constructed reservoirs may be colonized so rapidly that the fauna is in equilibrium with prevailing environmental conditions within a few months (Paterson and Fernando 1969a, 1969b). Voshell and Simmons (1978) attributed dominance of the damselfly *Enallagma civile*, in the first postimpoundment year of a reservoir, to an abundance of flooded terrestrial vegetation suitable for oviposition. *E. civile* disappeared as this vegetation decomposed, and other odonates appeared. The replacement of *E. civile* by a congener, *E. basidens*, could be construed as evidence for competition, but the latter's preference for aquatic macrophytes, which developed in later years in the reservoir, seems to be a better explanation. Street and Titmus (1979) also explained increasing diversity of Chironomidae in gravel pits by development of vegetation. In contrast, Cantrell and McLachlan (1977) demonstrated competitive interactions between Chironomidae in a new lake, but a less thorough or nonexperimental study might have correlated the observed pattern with exogenously determined sediment characteristics, and overlooked significant interspecific interactions.

The temporal pattern of recovery following extensive devastation in streams indicates differential colonization and reproduction by various taxa. After pesticide application, Ide (1967) noted early development of chironomid populations, whereas larger insects, especially Trichoptera, did not recover until four or more years had passed. Recovery times were shorter for sites nearer unsprayed sources of colonists. Ide observed very large standing crops of periphyton in the year following spraying; but with increased density and diversity of insects, periphyton declined to prespray levels. These observations are consistent with the resource tracking model proposed in this chapter.

418 / ECOLOGY OF AQUATIC INSECTS

Fisher et al. (1982) monitored ecosystem properties and the development of insect assemblages in a desert stream after flash flooding. Aerial colonization, high reproductive rates, and short generation times restored insect biomass to 50 percent of its ultimate value within 13 days. Colonization by ovipositing adults sometimes produces tightly synchronized cohorts of larvae (Ladle et al. 1980), and synchrony may be retained for several generations (Resh 1982). This situation may be analogous to the establishment of dominant even-aged stands in forest succession (West et al. 1981), and may have similar consequences, although this suggestion is speculative.

Williams and Hynes (1977) examined sources of colonists in relation to the MacArthur and Wilson (1967) immigration-extinction model in a newly dredged stream. The model held, but poorly so, since the pool of colonists varied seasonally (e.g. see Williams 1980), and this prevented the colonized sites from ever attaining equilibrium. Huckin (1982) described similar assemblages in which marine benthic copepods tracked, but never reached, an ever-shifting equilibrium composition and diversity. In a reclaimed section of the Tongue River, Wyoming, time to faunal recovery (achievement of 85% similarity of species composition relative to a control site) was approximately doubled with each 200 m increase in distance from the primary source of colonists, and trophic groups colonized at different rates with collector-detritivores appearing first, scraper-grazers next, and predators colonizing very late in the sequence (Gore 1982). Only part of the species data was published, but differential drift rates seem to have produced waves of succession along the channel in the same way that substances of differing mobility are separated in a chromatographic column (Gore 1979). Small semi-natural substrates are colonized rapidly and illustrate successional features such as increased diversity and species replacements (Nilsen and Larimore 1973; Ulfstrand et al. 1974; Tevesz 1978; Shaw and Minshall 1980). Similarly, McAuliffe (1983) described a successional sequence in which newly overturned stones are colonized by hydroptilid caddisfly larvae and, more slowly, by moss. This plant covers the smooth surface needed by the hydroptilids, but also provides attachment for hydropsychid larvae and their nets.

Truly interactive succession has been difficult to demonstrate. McAuliffe's (1983) work is one example and Pajunen (1982) described the changing community background against which competition among corixids occurred. A related paper, although it deals with alternative system states rather than succession, is McLachlan (1981). A colonizing midge in rock pools containing tadpoles produced fewer, but larger, adults in a much shorter developmental time than when in pools without tadpoles. McLachlan suggested that tadpole-induced rapid development allows the midge to escape and colonize new habitats before the pool dries. Unfortunately, the demographic advantages of shorter generation times and larger, more fecund adults have not been analyzed. Case and Washino (1979) showed that flatworms could control mosquitoes in rice fields, but that newly established fields lacked flatworms and supported high densities of mosquitoes. Eventual colonization by flatworms initiated a new successional stage in which mosquitoes were uncommon. Seifert's (1982) review of the ecology of insect communities in *Heliconia* bracts is an outstanding synthesis of observational and experimental ecology and biogeography as related to colonization and succession. Especially important results included the dominant effect of habitat age, the weakness of competitive interactions in the system, and the important roles that positive interactions and predation played in succession.

PROSPECT

Colonization is such a multifaceted phenomenon that dogmatism about principles and profitable approaches is unwarranted. Investigations at all hierarchical levels from reproductive physiology to biogeography contribute to an understanding of colonization ecology. However, synthesis will be difficult unless observations are embedded in a larger structure of life history and spatial patterns. Colonization behavior is only one of a package of tactics that constitute a successful life history (Stearns 1976, 1980), and it must be placed in a demographic context if it is to make adaptive sense. Similarly, the colonized localities should be viewed as part of a metapopulation (Gill 1978) made up of sources and sinks of colonists.

Although colonization seems to be adaptive and explainable by natural selection, there is a large random component. Neutral or random concepts provide a standard against which adaptive or deterministic explanations can be tested. Simple neutral models also serve to isolate components such as probability of emigration, which may be under selective control. Models also emphasize the need to know both rates and turnovers, and their variances with time, location, individuals, and species. Likewise, theory and techniques should be borrowed freely from marine biology (e.g. see Schoener 1974; Sousa 1979; Sale 1980; Sutherland 1981) and plant ecology (Harper 1977a, 1977b).

Methodologies for colonization studies will remain varied, with physiological and behavioral research complementing population ecology and genetics, and community and ecosystem investigations. Colonization ecology has been an experimental science for more than 40 years (e.g. see Moon 1940), and manipulations of habitats and populations seem to be the most profitable avenue of further research. Experimental studies using artificial substrates or ponds would be much improved if sources, dispersing animals, and colonists could be monitored simultaneously. Marked insects would provide information on individual behavior, and allow one to prepare a balanced budget for the metapopulation. Successional variables such as periphyton, macrophytes, sediment, and associated fauna also need more attention. In summary, the naturalist's synthesis of experience and intuition coupled with selectionist or stochastic theory, proper statistical designs, and ingenious techniques will all be required before colonization ecology becomes an integral part of the population and community ecology of aquatic insects.

References

Addicott, J. F. 1978. The population dynamics of aphids on fireweed: a comparison of local populations and metapopulations. Canadian Journal of Zoology 56:2554–64.

Angerilli, N. P. D. and B. P. Beirne. 1980. Influences of aquatic plants on colonization of artificial ponds by mosquitoes and their insect predators. The Canadian Entomologist 112:793–96.

Armstrong, J. S. 1978. Colonisation of New Zealand by *Hemicordulia australiae*, with notes on its displacement of the indigenous *Procordulia grayi* (Odonata:Corduliidae). New Zealand Entomologist 6:381–84.

Baker, H. G. and G. L. Stebbins (eds.). 1965. The genetics of colonizing species. Academic Press, New York, NY. 588 pp.

Bennett, F. D. 1977. Insects as agents for biological control of aquatic weeds. Aquatic Botany 3:165–74.

Bird, G. A. and H. B. N. Hynes. 1981. Movements of adult aquatic insects near streams in southern Ontario. Hydrobiologia 77:65–69.

Bohle, H. W. 1978. Beziehungen zwischen dem Nahrungsangebot, der Drift und der räumlichen Verteilung bei Larven von *Baëtis rhodani* (Pictet) (Ephemeroptera: Baëtidae). Untersuchungen in künstlichen Fliesswasseranlagen. Archiv für Hydrobiologie 84:500–25.

Boles, G. L. 1981. Macroinvertebrate colonization of replacement substrate below a hypolimnial release reservoir. Hydrobiologia 78:133–46.

Bournaud, M., G. Chavanon, and H. Tachet. 1978. Structure et fonctionnement des écosystèmes du Haut-Rhône français. 5. Colonisation par les macroinvertébrés de substrats artificiels suspendus en pleine eau ou posés sur le fond. Internationale Vereinigung für Theoretische und Angewandte Limnologie Verhandlungen 20: 1485–93.

Bray, R. P. 1971. Factors affecting the distribution of some Phryganeidae (Trichoptera) in Malham Tarn, Yorkshire. Freshwater Biology 1:149–58.

Brinck, P. 1949. Studies on Swedish stoneflies. Opuscula Entomologica Supplementum 11:1–250.

Brown, E. S. 1951. The relation between migration-rate and type of habitat in aquatic insects, with special reference to certain species of Corixidae. Proceedings of the Zoological Society of London 121:539–45.

Bruns, D. A. and W. L. Minckley. 1980. Distribution and abundance of benthic invertebrates in a Sonoran Desert stream. Journal of Arid Environments 3:117–31.

Burks, B. D. 1953. The mayflies, or Ephemeroptera, of Illinois. Illinois Natural History Survey Bulletin 26:1–216.

Calabrese, D. M. 1979. Pterygomorphism in 10 nearctic species of *Gerris*. American Midland Naturalist 101:61–68.

Cancela da Fonseca, J. P. 1979. Species colonization models of temporary ecosystems habitats, pp. 125–95. *In:* G. S. Innis and R. V. O'Neill (eds.). Systems analysis of ecosystems. International Co-operative Publishing House, Fairland, MD. 402 pp.

Cantrell, M. A. and A. J. McLachlan. 1977. Competition and chironomid distribution patterns in a newly flooded lake. Oikos 29:429–33.

Case, T. J. and R. K. Washino. 1979. Flatworm control of mosquito larvae in rice fields. Science 206:1412–14.

Cheng, L. and M. C. Birch. 1977. Terrestrial insects at sea. Journal of the Marine Biological Association of the United Kingdom 57:995–97.

Clement, S. L., A. A. Grigarick, and M. O. Way. 1977. The colonization of California rice paddies by chironomid midges. Journal of Applied Ecology 14:379–89.

Clifford, H. F., H. Hamilton, and B. A. Killins. 1979. Biology of the mayfly *Leptophlebia cupida* (Say) (Ephemeroptera: Leptophlebiidae). Canadian Journal of Zoology 57:1026–45.

Colbo, M. H. 1979. Distribution of winter-developing Simuliidae (Diptera), in eastern Newfoundland. Canadian Journal of Zoology 57: 2143–52.

Corbet, P. S. 1980. Biology of Odonata. Annual Review of Entomology 25:189–217.

Corkum, L. D. 1978. The influence of density and behavioural type on the active entry of two mayfly species (Ephemeroptera) into the water column. Canadian Journal of Zoology 56:1201–06.

Corkum, L. D. and H. F. Clifford. 1980. The importance of species associations and substrate types to behavioural drift, pp. 331–41. *In:* J. F. Flannagan and K. E. Marshall (eds.). Advances in Ephemeroptera biology. Plenum Press, New York, NY. 552 pp.

Crichton, M. I., D. Fisher, and I. P. Woiwod. 1978. Life histories and distribution of British Trichoptera, excluding Limnephilidae and Hydroptilidae, based on the Rothamsted Insect Survey. Holarctic Ecology 1:31–45.

Crisp, D. T. and T. Gledhill. 1970. A quantitative description of the recovery of the bottom fauna in a muddy reach of a mill stream in Southern England after draining and dredging. Archiv für Hydrobiologie 67:502–41.

Cummins, K. W. 1964. Factors limiting the microdistribution of larvae of the caddisflies *Pycnopsyche lepida* (Hagen) and *Pycnopsyche guttifer* (Walker) in a Michigan stream (Trichoptera: Limnephilidae). Ecological Monographs 34:271–95.

Davies, B. R. 1976. A trap for capturing planktonic chironomid larvae. Freshwater Biology 6:373–80.

DeAngelis, D. L., C. C. Travis, and W. M. Post. 1979. Persistence and stability of seed-dispersed species in a patchy environment. Theoretical Population Biology 16:107–25.

Dingle, H. 1978. Migration and diapause in tropical, temperate and island milkweed bugs, pp. 254–76. *In:* H. Dingle (ed.). Evolution of insect migration and diapause. Springer-Verlag, New York, NY. 350 pp.

Disney, R. H. L. 1972. Observations on sampling pre-imaginal populations of blackflies (Dipt., Simuliidae) in West Cameroon. Bulletin of Entomological Research 61: 485–503.

Edmunds, G. F., Jr. 1972. Biogeography and evolution of Ephemeroptera. Annual Review of Entomology 17:21–42.

Elliott, J. M. 1970. The distances travelled by drifting invertebrates in a Lake District stream. Oecologia (Berlin) 6:350–79.

Elton, C. S. 1958. The ecology of invasions by animals and plants. John Wiley and Sons Inc., New York, NY. 181 pp.

Erman, D. C. and W. C. Chouteau. 1979. Fine particulate organic carbon output from fens and its effect on benthic macroinvertebrates. Oikos 32:409–15.

Ertlová, E. 1980. Colonization of the littoral of the Liptovská Mara Reservoir by Chironomidae (Diptera) in the first two years after impoundment. Biológia (Bratislava) 35:311–19.

Fernando, C. H. 1958. The colonization of small freshwater habitats by aquatic insects. I. General discussion, methods and colonization in the aquatic Coleoptera. Ceylon Journal of Science 1:117–54.

Fernando, C. H. 1959. The colonization of small freshwater habitats by aquatic insects. 2. Hemiptera (The water-bugs). Ceylon Journal of Science 2:5–32.

Fernando, C. H. and D. Galbraith. 1973. Seasonality and dynamics of aquatic insects colonizing small habitats. Internationale Vereinigung für Theoretische und Angewandte Limnologie Verhandlungen 18:1564–75.

Fisher, S. G., L. J. Gray, N. B. Grimm, and D. E. Busch. 1982. Temporal succession in a desert stream ecosystem following flash flooding. Ecological Monographs 52:93–110.

Fredeen, F. J. H. and J. A. Shemanchuk. 1960. Black flies (Diptera: Simuliidae) of irrigation systems in Saskatchewan and Alberta. Canadian Journal of Zoology 38:723–35.

Fretwell, S. D. 1972. Populations in a seasonal environment. Monographs in Population Biology 5. Princeton University Press, Princeton, NJ. 217 pp.

Gersabeck, E. F., Jr. and R. W. Merritt. 1979. The effect of physical factors on the colonization and relocation behavior of immature black flies (Diptera: Simuliidae). Environmental Entomology 8:34–39.

Gibbs, K. E. 1979. Ovoviviparity and nymphal seasonal movements of *Callibaetis* spp. (Ephemeroptera: Baetidae) in a pond in southwestern Quebec. The Canadian Entomologist 111:927–31.

Gilbert, L. E. and M. C. Singer. 1973. Dispersal and gene flow in a butterfly species. The American Naturalist 107:58–72.

Gill, D. E. 1978. The metapopulation ecology of the red-spotted newt, *Notophthalmus viridescens* (Rafinesque). Ecological Monographs 48:145–66.

Golini, V. I. and D. M. Davies. 1975. Relative response to colored substrates by ovipositing blackflies (Diptera: Simuliidae). 1. Oviposition by *Simulium (Simulium) verecundum* Stone and Jamnback. Canadian Journal of Zoology 53:521–35.

Gore, J. A. 1977. Reservoir manipulations and benthic macroinvertebrates in a prairie river. Hydrobiologia 55:113–23.

Gore, J. A. 1979. Patterns of initial benthic recolonization of a reclaimed coal strip-mined river channel. Canadian Journal of Zoology 57:2429–39.

Gore, J. A. 1982. Benthic invertebrate colonization: source distance effects on community composition. Hydrobiologia 94:183–93.

Grassle, J. F. 1977. Slow recolonization of deep-sea sediment. Nature (London) 265:618–19.

Gray, L. J. 1981. Species composition and life histories of aquatic insects in a lowland Sonoran Desert stream. The American Midland Naturalist 106:229–42.

Gray, L. J. and S. G. Fisher. 1981. Postflood recolonization pathways of macroinvertebrates in a lowland Sonoran Desert stream. The American Midland Naturalist 106:249–57.

Harper, J. L. 1977a. Population biology of plants. Academic Press, London. 892 pp.

Harper, J. L. 1977b. The contributions of terrestrial plant studies to the development of the theory of ecology, pp. 139–57. *In:* C. E. Goulden (ed.). Changing scenes in natural sciences. Special Publication 12, Academy of Natural Sciences of Philadelphia, Philadelphia, PA. 362 pp.

Harrison, A. D. 1966. Recolonisation of a Rhodesian stream after drought. Archiv für Hydrobiologie 62:405–21.

Harrison, R. G. 1980. Dispersal polymorphisms in insects. Annual Review of Ecology and Systematics 11:95–118.

Hildebrand, S. G. 1974. The relation of drift to benthos density and food level in an artificial stream. Limnology and Oceanography 19:951–57.

Hilsenhoff, W. L. 1971. Repopulation of Hartlaub Lake by *Chaoborus* (Diptera; Chaoboridae) four years after elimination with toxaphene. Annals of the Entomological Society of America 64:308–9.

Hockin, D. C. 1982. Equilibrium insular zoogeography: some tests of the equilibrium theory using meiobenthic harpacticoid copepods. Journal of Biogeography 9:487–97.

Hynes, H. B. N. and T. R. Williams. 1962. The effect of DDT on the fauna of a Central African stream. Annals of Tropical Medicine and Parasitology 56:78–91.

Ide, F. P. 1967. Effects of forest spraying with DDT on aquatic insects of salmon streams in New Brunswick. Journal of the Fisheries Research Board of Canada 24:769–805.

Illies, J. 1965. Phylogeny and zoogeography of the Plecoptera. Annual Review of Entomology 10:117–40.

Istock, C. A., S. A. Wasserman, and H. Zimmer. 1975. Ecology and evolution of the pitcher-plant mosquito: 1. Population dynamics and laboratory responses to food and population density. Evolution 29:296–312.

Johnson, C. G. 1969. Migration and dispersal of insects by flight. Methuen and Co., Ltd., London. 763 pp.

Ladle, M., J. S. Welton, and J. A. B. Bass. 1980. Invertebrate colonization of the gravel substratum of an experimental recirculating channel. Holarctic Ecology 3: 116–23.

Laird, M. 1956. Studies of mosquitos and freshwater ecology in the South Pacific. Bulletin of the Royal Society of New Zealand 6:1–213.

Landin, J. 1976. Seasonal patterns in abundance of water-beetles belonging to the Hydrophiloidea (Coleoptera). Freshwater Biology 6:89–108.

Landin, J. 1980. Habitats, life histories, migration and dispersal by flight of two water-beetles *Helophorus brevipalpis* and *H. strigifrons* (Hydrophilidae). Holarctic Ecology 3:190–201.

Landin, J. and K. Vepsäläinen. 1977. Spring dispersal flights of pondskaters *Gerris* spp. (Heteroptera). Oikos 29:156–60.

Larimore, R. W., W. F. Childers, and C. Heckrotte. 1959. Destruction and re-establishment of stream fish and invertebrates affected by drought. Transactions of the American Fisheries Society 88:261–85.

Lehmkuhl, D. M. 1980. Temporal and spatial changes in the Canacian insect fauna: patterns and explanation. The prairies. The Canadian Entomologist 112: 1145–59.

Levins, R. 1968. Evolution in changing environments. Monographs in Population Biology 2. Princeton University Press, Princeton, NJ. 120 pp.

Lingdell, P.-E. and K. Müller. 1979. Migrations of *Leptophlebia vespertina* L. and *L. marginata* L. (Ins.: Ephemeroptera) in the estuary of a coastal stream. Aquatic Insects 1:137–42.

Macan, T. T. 1974. Freshwater ecology. Longman Group Ltd., London. 343 pp.

Macan, T. T. and A. Kitching. 1976. The colonization of squares of plastic suspended in midwater. Freshwater Biology 6:33–40.

MacArthur, R. H. and E. O. Wilson. 1967. The theory of island biogeography. Monographs in Population Biology 1. Princeton University Press, Princeton, NJ. 203 pp.

Maciolek, J. A. 1975. Limnological ecosystems and Hawaii's preservational planning. Internationale Vereinigung für Theoretische und Angewandte Limnologie Verhandlungen 19:1461–67.

Maguire, L. A. and J. W. Porter. 1977. A spatial model of growth and competition strategies in coral communities. Ecological Modelling 3:249–71.

Malicky, H. 1980. Evidence of seasonal migrations of larvae of two species of philopotamid caddisflies (Trichoptera) in a mountain stream in Lower Austria. Aquatic Insects 2:153–60.

Marinaro, V. C. 1970. A modern dry-fly code. Crown Publishers, New York, NY. 270 pp.

McAuliffe, J. R. 1983. Competition, colonization patterns, and disturbance in stream benthic communities, pp. 137–56. In: J. R. Barnes and G. W. Minshall (eds.). Stream ecology: application and testing of general ecological theory. Plenum Press, New York, NY. 399 pp.

McLachlan, A. 1981. Interaction between insect larvae and tadpoles in tropical rain pools. Ecological Entomology 6:175–82.

McLachlan, A. J. 1970. Submerged trees as a substrate for benthic fauna in the recently created Lake Kariba (central Africa). Journal of Applied Ecology 7:253–66.

McLachlan, A. J. and M. A. Cantrell. 1980. Survival strategies in tropical rain pools. Oecologia 47:344–51.

McLay, C. 1970. A theory concerning the distance travelled by animals entering the drift of a stream. Journal of the Fisheries Research Board of Canada 27:359–70.

Meadows, P. S. and J. I. Campbell. 1972. Habitat selection by aquatic invertebrates. Advances in Marine Biology 10:271–382.

Mendl, H. and K. Müller. 1978. The colonization cycle of *Amphinemura standfussi* Ris (Ins.: Plecoptera) in the Abisko area. Hydrobiologia 60:109–11.

Minshall, G. W., D. A. Andrews, and C. Y. Manuel-Faler. 1983. Application of island biogeographic theory to streams: macroinvertebrate recolonization of the Teton River, Idaho, pp. 279–97. In: J. R. Barnes and G. W. Minshall (eds.). Stream ecology: application and testing of general ecological theory. Plenum Press, New York, NY. 399 pp.

Moon, H. P. 1940. An investigation of the movements of freshwater invertebrate faunas. Journal of Animal Ecology 9:76–83.

Müller, K. 1954. Investigations on the organic drift in North Swedish streams. Drottningholm Institute of Freshwater Research Report 35:133–48.

Müller, K. 1982. The colonization cycle of freshwater insects. Oecologia (Berlin) 52:202–7.

Müller, K. and H. Mendl. 1978. Vingutveckling hos *Capnia atra* Morton i Abiskoområdet (Plecoptera). Entomologisk Tidskrift 99:111–13. (English summary).

Müller, K. and H. Mendl. 1979. The importance of a brackish water area for the stonefly colonization cycle in a coastal river. Oikos 33:272–77.

Mundie, J. H. 1959. The diurnal activity of the larger invertebrates at the surface of Lac la Ronge, Saskatchewan. Canadian Journal of Zoology 37:945–56.

Needham, P. R. and J. P. Welsh. 1953. Rainbow trout (*Salmo gairdneri* Richardson) in the Hawaiian Islands. Journal of Wildlife Management 17:233–55.

Neves, R. J. 1979. Movements of larval and adult *Pycnopsyche guttifer* (Walker) (Trichoptera: Limnephilidae) along Factory Brook, Massachusetts. American Midland Naturalist 102:51–58.

Newman, L. J. 1977. Chromosomal evolution of the Hawaiian *Telmatogeton* (Chironomidae, Diptera). Chromosoma 64:349–69.

Nilsen, H. C. and R. W. Larimore. 1973. Establishment of invertebrate communities on log substrates in the Kaskaskia River, Illinois. Ecology 54:366–74.

Nilsson, L. M. and P. Sjöström. 1977. Colonization of implanted substrates by differently sized *Gammarus pulex* (Amphipoda). Oikos 28:43–48.

Okubo, A. 1980. Diffusion and ecological problems: mathematical models. Springer-Verlag, Berlin, Germany. 254 pp.

Paine, R. T. and S. A. Levin. 1981. Intertidal landscapes: disturbance and the dynamics of pattern. Ecological Monographs 51:145–78.

Pajunen, V. I. 1971. Adaptations of *Arctocorisa carinata* (Sahlb.) and *Callocorixa producta* (Reut.) populations to a rock pool environment, pp. 148–58. *In*: P. J. den Boer and G. R. Gradwell (eds.). Dynamics of populations. Proceedings of the Advanced Study Institute on Dynamics of Numbers in Populations, Oosterbeek, The Netherlands, 7–18 September 1970. Centre for Agricultural Publishing and Documentation, Wageningen, The Netherlands. 611 pp.

Pajunen, V. I. 1977. Population structure in rock-pool corixids (Hemiptera, Corixidae) during the reproductive season. Annales Zoologici Fennici 14:26–47.

Pajunen, V. I. 1982. Replacement analysis of non-equilibrium competition between rock pool corixids (Hemiptera, Corixidae). Oecologia 52:153–55.

Pajunen, V. I. and A. Jansson. 1969. Dispersal of the rock pool corixids *Arctocorisa carinata* (Sahlb.) and *Callocorixa producta* (Reut.) (Heteroptera, Corixidae). Annales Zoologici Fennici 6:391–427.

Parsons, P. A. 1982. Adaptive strategies of colonizing animal species. Biological Reviews of the Cambridge Philosophical Society 57:117–48.

Paterson, C. G. and C. H. Fernando. 1969a. The macro-invertebrate colonization of a small reservoir in Eastern Canada. Internationale Vereinigung für Theoretische und Angewandte Limnologie Verhandlungen 17:126–36.

Paterson, C. G. and C. H. Fernando. 1969b. Macroinvertebrate colonization of the marginal zone of a small impoundment in Eastern Canada. Canadian Journal of Zoology 47:1229–38.

Paterson, C. G. and C. H. Fernando. 1970. Benthic fauna colonization of a new reservoir with particular reference to the Chironomidae. Journal of the Fisheries Research Board of Canada 27:213–32.

Patterson, J. W. and R. L. Vannote. 1979. Life history and population dynamics of *Heteroplectron americanum*. Environmental Entomology 8:665–69.

Paulson, D. R. 1978a. An Asiatic dragonfly, *Crocothemis servilia* (Drury), established in Florida (Anisoptera: Libellulidae). Notulae Odonatologica 1:9–10.

Paulson, D. R. 1978b. Additional record of *Crocothemis servilia* (Drury) from Florida (Anisoptera: Libellulidae). Notulae Odonatologica 1:29–30.

Peckarsky, B. L. 1979. Biological interactions as determinants of distributions of benthic invertebrates within the substrate of stony streams. Limnology and Oceanography 24:59-68.

Peckarsky, B. L. 1980a. Predator-prey interactions between stoneflies and mayflies: behavioral observations. Ecology 61:932-43.

Peckarsky, B. L. 1980b. Influence of detritus upon colonization of stream invertebrates. Canadian Journal of Fisheries and Aquatic Sciences 37:957-63.

Peckarsky, B. L. 1981. Reply to Sell. Limnology and Oceanography 26:982-87.

Peckarsky, B. L. and S. I. Dodson. 1980. Do stonefly predators influence benthic distributions in streams? Ecology 61:1275-82.

Pegel, M. 1980. Zur Methodik der Driftmessung in der Fliessgewässerökologie unter besonderer Berücksichtigung der Simuliidae (Diptera). Zeitschrift für angewandte Entomologie 89:198-214.

Pielou, E. C. 1981. The usefulness of ecological models: a stock-taking. The Quarterly Review of Biology 56:17-31.

Popham, E. J. 1964. The migration of aquatic bugs with special reference to the Corixidae (Hemiptera Heteroptera). Archiv für Hydrobiologie 60:450-96.

Resh, V. H. 1982. Age structure alteration in a caddisfly population after habitat loss and recovery. Oikos 38:280-84.

Resh, V. H., T. S. Flynn, G. A. Lamberti, E. P. McElravy, K. L. Sorg, and J. R. Wood. 1981. Responses of the sericostomatid caddisfly Gumaga nigricula (McL.) to environmental disruption, pp. 311-18. In: G. P. Moretti (ed.) Proceedings of the Third International Symposium on Trichoptera. Series Entomologica, Volume 20. Dr. W. Junk Publishers, The Hague, The Netherlands. 472 pp.

Ricker, W. E. 1964. Distribution of Canadian stoneflies. Gewässer und Abwässer 34/35: 50-71.

Roff, D. A. 1975. Population stability and the evolution of dispersal in a heterogeneous environment. Oecologia 19:217-37.

Rosenberg, D. M. and V. H. Resh. 1982. The use of artificial substrates in the study of freshwater benthic macroinvertebrates, pp. 175-235. In: J. Cairns, Jr. (ed.). Artificial substrates. Ann Arbor Science Publishers Inc., Ann Arbor, MI. 279 pp.

Ross, H. H., G. L. Rotramel, J. E. H. Martin, and J. F. McAlpine. 1967. Postglacial colonization of Canada by its subboreal winter stoneflies of the genus Allocapnia. The Canadian Entomologist 99:703-12.

Roy, D. and P. P. Harper. 1981. An analysis of an adult Trichoptera community in the Laurentian highlands of Quebec. Holarctic Ecology 4:102-15.

Sale, P. F. 1980. Assemblages of fish on patch reefs—predictable or unpredictable? Environmental Biology of Fishes 5:243-49.

Schoener, A. 1974. Experimental zoogeography: colonization of marine mini-islands. The American Naturalist 108:715-38.

Schwarz, P. 1970. Autökologische Untersuchungen zum Lebenszyklus von Setipalpia-Arten (Plecoptera). Archiv für Hydrobiologie 67:103-72.

Scudder, G. G. E. 1975. Field studies on the flight muscle polymorphism in Cenocorixa (Hemiptera: Corixidae). Internationale Vereiningung für Theoretische und Angewandte Limnologie Verhandlungen 19:3064-72.

Seifert, R. P. 1975. Clumps of Heliconia inflorescences as ecological islands. Ecology 56:1416-22.

Seifert, R. P. 1982. Neotropical *Heliconia* insect communities. The Quarterly Review of Biology 57:1–28.

Sell, D. W. 1981. Comment on "Biological interactions as determinants of distributions of benthic invertebrates within the substrate of stony streams" (Peckarsky). Limnology and Oceanography 26:981–82.

Shaw, D. W. and G. W. Minshall. 1980. Colonization of an introduced substrate by stream macroinvertebrates. Oikos 34:259–71.

Sheldon, A. L. 1977. Colonization curves: application to stream insects on semi-natural substrates. Oikos 28:256–61.

Smith, F. E. 1972. Spatial heterogeneity, stability and diversity in ecosystems. Transactions of the Connecticut Academy of Arts and Sciences 44:309–30.

Sousa, W. P. 1979. Experimental investigations of disturbance and ecological succession in a rocky intertidal algal community. Ecological Monographs 49:227–54.

Southwood, T. R. E. 1962. Migration of terrestrial arthropods in relation to habitat. Biological Reviews of the Cambridge Philosophical Society 37:171–214.

Southwood, T. R. E. 1977. Habitat, the templet for ecological strategies? Journal of Animal Ecology 46:337–65.

Spence, J. R. 1981. Experimental analysis of microhabitat selection in water-striders (Heteroptera: Gerridae). Ecology 62:1505–14.

Stearns, S. C. 1976. Life-history tactics: a review of the ideas. The Quarterly Review of Biology 52:3–47.

Stearns, S. C. 1980. A new view of life-history evolution. Oikos 35:266–81.

Stout, J. and J. Vandermeer. 1975. Comparison of species richness for stream-inhabiting insects in tropical and mid-latitude streams. The American Naturalist 109:263–80.

Stout, R. J. 1981. How abiotic factors affect the distribution of two species of tropical predaceous aquatic bugs (Family: Naucoridae). Ecology 62:1170–78.

Street, M. and G. Titmus. 1979. The colonisation of experimental ponds by Chironomidae (Diptera). Aquatic Insects 1:233–44.

Sutherland, J. P. 1981. The fouling community at Beaufort, North Carolina: a study in stability. The American Naturalist 118:499–519.

Svensson, B. S. 1980. The effect of host density on the success of commensalistic *Epoicocladius flavens* (Chironomidae) in utilizing streamliving *Ephemera danica* (Ephemeroptera). Oikos 34:326–36.

Svensson, B. W. 1974. Population movements of adult Trichoptera at a South Swedish stream. Oikos 25:157–75.

Taylor, R. A. J. and L. R. Taylor. 1979. A behavioral model for the evolution of spatial dynamics, pp. 1–27. *In:* R. M. Anderson, B. D. Turner, and L. R. Taylor (eds.). Population dynamics. Blackwell Scientific Publications, Oxford, England. 434 pp.

Tevesz, M. J. S. 1978. Benthic recolonization patterns in the Vermillion River, Ohio. Kirtlandia 27:1–17.

Townsend, C. R. and A. G. Hildrew. 1976. Field experiments on the drifting, colonization and continuous redistribution of stream benthos. Journal of Animal Ecology 45:759–72.

Tozer, W. E., V. H. Resh, and J. O. Solem. 1981. Bionomics and adult behavior of a lentic caddisfly *Nectopsyche albida* (Walker). American Midland Naturalist 106:133–44.

Ulfstrand, S. 1970. Trichoptera from River Vindelälven in Swedish Lapland. A four-year study based mainly on the use of light-traps. Entomologisk Tidskrift 91:46–63.

Ulfstrand, S., L. M. Nilsson, and A. Stergar. 1974. Composition and diversity of benthic species collectives colonizing implanted substrates in a South Swedish stream. Entomologica Scandinavica 5:115–22.

Vandermeer, J., J. Lazarus, C. Ludwig, J. Lyon, B. Schultz, and K. Yih. 1980. Migration as a factor in the community structure of a macroarthropod litter fauna. The American Naturalist 115:606–12.

Vepsäläinen, K. 1978. Wing dimorphism and diapause in *Gerris*: determination and adaptive significance, pp. 218–53. *In*: H. Dingle (ed.). Evolution of insect migration and diapause. Springer-Verlag, New York, NY. 350 pp.

Voshell, J. R., Jr. and G. M. Simmons, Jr. 1978. The Odonata of a new reservoir in the southeastern United States. Odonatologica 7:67–76.

Walton, O. E., Jr. 1978. Substrate attachment by drifting aquatic insect larvae. Ecology 59:1023–30.

Walton, O. E., Jr. 1980a. Invertebrate drift from predator-prey associations. Ecology 61:1486–97.

Walton, O. E., Jr. 1980b. Active entry of stream benthic macroinvertebrates into the water column. Hydrobiologia 74:129–39.

Walton, O. E., Jr., S. R. Reice, and R. W. Andrews. 1977. The effects of density, sediment particle size and velocity on drift of *Acroneuria abnormis* (Plecoptera). Oikos 28:291–98.

Waters, T. F. 1964. Recolonization of denuded stream bottom areas by drift. Transactions of the American Fisheries Society 93:311–15.

Weerekoon, A. C. J. 1956. Studies on the biology of Loch Lomond. 2. The repopulation of McDougal Bank. Ceylon Journal of Science C7:95–133.

West, D. C., H. H. Shugart, and D. B. Botkin (eds.). 1981. Forest succession: concepts and application. Springer-Verlag, New York, NY. 517 pp.

Wiggins, G. B., R. J. Mackay, and I. M. Smith. 1980. Evolutionary and ecological strategies of animals in annual temporary pools. Archiv für Hydrobiologie Supplement 58:97–206.

Wiley, M. J. 1981. Interacting influences of density and preference on the emigration rates of some lotic chironomid larvae (Diptera: Chironomidae). Ecology 62:426–38.

Williams, D. D. 1980. Temporal patterns in recolonization of stream benthos. Archiv für Hydrobiologie 90:56–74.

Williams, D. D. and H. B. N. Hynes. 1976a. The recolonization mechanisms of stream benthos. Oikos 27:265–72.

Williams, D. D. and H. B. N. Hynes. 1976b. Stream habitat selection by aerially colonizing invertebrates. Canadian Journal of Zoology 54:685–93.

Williams, D. D. and H. B. N. Hynes. 1977. Benthic community development in a new stream. Canadian Journal of Zoology 55:1071–76.

Winterbourn, M. J. 1978. An evaluation of the mesh bag method for studying leaf colonization by stream invertebrates. Internationale Vereinigung für Theoretische und Angewandte Limnologie Verhandlungen 20:1557–61.

Winterbourn, M. J. 1980. The freshwater insects of Australasia and their affinities. Palaeogeography, Palaeoclimatology, Palaeoecology 31:235–49.

Young, E. C. 1965. Flight muscle polymorphism in British Corixidae: ecological observations. Journal of Animal Ecology 34:353–90.

Zalom, F. G., A. A. Grigarick, and M. O. Way. 1979. Seasonal and diel flight periodicities of rice field Hydrophilidae. Environmental Entomology 8:938–43.

Zalom, F. G., A. A. Grigarick, and M. O. Way. 1980. Diel flight periodicities of some Dytiscidae (Coleoptera) associated with California rice paddies. Ecological Entomology 5:183–87.

chapter 14

THE HYPORHEIC ZONE AS A HABITAT FOR AQUATIC INSECTS AND ASSOCIATED ARTHROPODS

D. Dudley Williams

Migrations of benthic insects within the substrate of streambeds, where interstices occur, are a relatively recent discovery, although the occurrence of noninsects in the interstices of lake shores (the psammon habitat) has been documented for some time. Lack of sampling technology, without doubt, has contributed to this discrepancy. The aim of this chapter is to describe the biotic and abiotic features of the hyporheic habitat with an attempt to define its boundaries. This will be followed by an account of the biology of interstitial insects and discussion of evolutionary aspects of their colonization of this zone.

Interstitial spaces have long been regarded as refuges for small animals (Racovitza 1907), some of which have become highly specialized to this type of environment. Such habitats occur in both terrestrial and aquatic realms and, in some cases, a transition between them exists, such as in the shore region of lakes (Pennak 1940).

In fresh water, there are two basic types of interstitial environments. These are lentic habitats, where the interstitial flow of water is negligible, and lotic habitats, where the interstices are permeated by flowing water. In lakes, this interstitial habitat has been termed the *psammon* (Wiszniewski 1934) and represents the water-filled spaces between sand grains of lake beaches and, presumably, also between particles on the shoreline of the potamon region of rivers. In running water, this habitat has been termed the *hyporheic zone* (Orghidan 1959) and represents the interstices that are formed in the mixture of coarse sand, gravel, and rocks typically found in the rithron region of streams.

Considerable debate has occurred over the terminology and definition of the hyporheic zone. Orghidan (1959) and Schwoerbel (1961) believed the hyporheic to be a middle zone, bordered by the epigean water of the stream above and by the true groundwater below. Both stated that its physical and chemical characters fluctuate because of flushing by surface water. It cannot, therefore, as others have suggested, be included as part of the groundwater zone, as here these parameters remain fairly constant. The term *hyporheos* has been coined to refer to its fauna (Williams and Hynes 1974). The hyporheos contains some elements found in other underground aquatic habitats such as streams and ponds in caves, and groundwater, as well as elements related to the surface fauna. The study of such hypogean animals constitutes the science of biospeleology, which came into being in the middle 1800s (Viré 1904). Some of the voluminous data on this topic may have application or parallels in the hyporheic zone.

Sassuchin et al. (1927) were perhaps the first to examine the faunas within fluvial deposits, but it was not until 1942, when Chappuis developed a reasonably successful sampling technique, that much impetus was given to this area of study. Among the early investigators of this subject, Kühtreiber (1934) and Aubert (1959) suggested that stream animals could occur elsewhere than on the streambed surface. Aubert, for example, noted that the larvae of Leuctridae (Plecoptera) were found far less often than the adults, and Berthélemy (1966) later showed that these larvae could be found deep within the streambed.

SAMPLING METHODS

The paucity of knowledge on the hyporheic habitat has undoubtedly been due to difficulties in quantitatively sampling this zone. Recently, however, a number of successful devices have been used. No one sampler is successful in all field situations encountered, but, collectively, they are furthering the description of this habitat and its fauna.

The simplest technique, and the one used by the earliest workers, is that of digging a ditch into the gravel substrate at the stream margins or on exposed gravel bars and straining the interstitial water so exposed with a fine net (the Karaman-Chappuis technique, see Schwoerbel 1970). However, this method is not quantitative and it cannot be used under water.

Frozen cores use chemicals such as liquid nitrogen, liquid oxygen, liquid carbon dioxide, or a mixture of crushed, solid carbon dioxide ("dry ice") and alcohol or acetone to freeze the substrate around a standpipe driven into the streambed. The core so formed can be removed and checked for animals. Stocker and Williams (1972) found fewer animals per unit volume of gravel using this approach than with other techniques, and concluded that, although frozen cores lend themselves to an accurate description of the vertical distribution of sediments under stony streams, many of the animals appear to vacate the area immediately around the standpipe before the temperature becomes intolerably low and they are trapped. Recently, however, Danielopol et al. (1980) reported considerable success using this technique in Europe.

Mechanical corers usually are designed to work in soft homogeneous substrates and many are not practical in streams; nevertheless, there are exceptions (Fig. 14.1). Husmann (1971) designed a "Rammpumpe" consisting of a handpump device mounted on top of a standpipe, with hyporheic water and animals sucked up and collected on a sieve. A similar sampler was devised by Bou and Rouch (1967). These methods may be selective in their collection of animals due to a filtering effect in the interstices. Williams and Hynes (1974) devised a standpipe corer capable of sampling small amounts of gravel to depths of

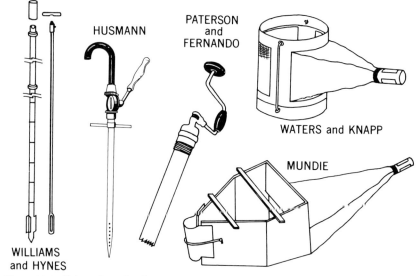

Figure 14.1 Selected mechanical corers suitable for sampling the hyporheos (Waters and Knapp 1961; Paterson and Fernando 1969; Husmann 1971; Mundie 1971; Williams and Hynes 1974).

approximately 1 m in streams of mixed substrate sizes. The method allows instantaneous samples to be taken with minimal disturbance of the streambed. Although it slightly overestimates fanal density, most of the common taxa in the substrate are captured (Williams 1981). Within the category of mechanical corers may be included box samplers such as those of Waters and Knapp (1961) and Mundie (1971), which can be used on hard heterogeneous substrates but which have limited sampling depths.

Artificial substrates are often used for sampling beneath the streambed (Fig. 14.2). This involves placing a cleaned portion of the natural streambed into perforated containers that are sunk into the substrate and then removed after a period of colonization. The problems arising from this method are numerous and include the reestablishment of normal levels of interstitial detritus as a food source, the restructuring of natural pore space, and the time needed for complete recolonization (for further discussions of these problems see Williams and Hynes 1974 and Rosenberg and Resh 1982).

ENVIRONMENTAL CHARACTERISTICS OF THE HYPORHEIC ZONE

Substrate and Pore Space. One of the most important features controlling the nature of the hyporheic zone is the type of substrate forming the streambed. Clearly, in the cases where the bottom is solid bedrock there is no hyporheic zone;

this is true also for streams where the bed is clay or any other type of material that does not have pore spaces capable of containing biologically significant amounts of water. Pore space size is a controlling factor, particularly with respect to the types of animals found. To accomodate small insect larvae and other micro-arthropods, pore diameter should be at least 50 μm, although small nonarthropod metazoans and protozoans may exist in smaller spaces. Pore size depends largely on size, shape, and packing of the substrate particles. For example, a substrate consisting of uniformly sized spheres packed as tightly as possible produces a combined pore volume of approximately 26 percent of the total volume occupied by the mixture (Bruce 1928). In reality though, this never occurs since stream and lake substrate particles are irregular in outline and of different sizes. We should thus expect to see larger volumes of total pore space in nature. Pennak (1940), for example, recorded interstitial volumes that were 37 to 41 percent of the aggregate volume for sand substrates on Wisconsin lakeshores. However, Stocker and Williams (1972) obtained values of 21 to 35 percent for the mixed gravel of the Speed River, a southern Ontario stream. Lower values for the stream study were due to greater substrate particle size diversity (heterogeneity), resulting in smaller particles partially filling the interstices between the larger particles, thus giving a slightly reduced overall pore volume. Although the porosity of homogeneously sized material cannot be changed by compaction (Pollard 1955), this is not true of the more diversely sized substrate materials that

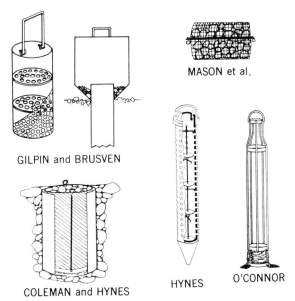

MASON et al.

GILPIN and BRUSVEN

COLEMAN and HYNES

HYNES

O'CONNOR

Figure 14.2 Various designs of artificial substrate samplers suitable for hyporheic work (Mason et al. 1967; Coleman and Hynes 1970; Hynes 1974; O'Connor 1974; Gilpin and Brusven 1976).

are typically found in a stream. Interstitial volume may change on a seasonal basis depending on the degree of reworking of the bed materials by the current during periods of normal and abnormal flow (e.g. see Newbury, Chapter 11; Minshall, Chapter 12).

Little seems known about the relationship between pore space and depth of stream sediments. Several workers (e.g. see Schwoerbel 1967) have shown decreasing porosity with increasing grain size for fluvial sand. However, the picture becomes more complicated in heterogeneous mixtures of sand and gravel due to the partial filling of the interstices discussed previously. Nevertheless, Stocker and Williams (1972) showed that in the Speed River, the substrate below 30 cm showed a substantial decrease in both mean grain size and heterogeneity of grain size, but an increase in porosity as compared with the substrate from 30 cm to the streambed surface. However, the increase in porosity did not necessarily mean that the habitable interstitial space for animals increased with depth, since each organism requires a certain minimum amount of space and the porosity measurement does not take this into account. Although concrete data are lacking, the size of habitable pore spaces (as opposed to the academic measurement, porosity) may well be greater closer to the streambed surface where substrate particles are generally larger.

Water Movement. Flow characteristics through the interstices of the hyporheic zone of streams differ somewhat from that through the psammon of lakes. In the latter, the vertical component dominates, while in streams there is an important horizontal component as well. Williams and Hynes (1974) showed a substantial reduction in current velocity between the surface water of a stream (36 cm/sec) and the water traveling through the interstices at 10 cm beneath the bed (0.1 cm/sec). From 10 cm to 40 cm there was a steady decrease (to approximately 0.04 cm/sec) and beyond that the current was too small to be measured with the apparatus available. Thus, even though the range of surface water velocities may be large in an average stream, the corresponding changes in interstitial velocities of undisturbed sediments are small.

Pollard (1955) calculated permeability values by running water through troughs of various stream gravels. Although natural stream gravels have permeabilities between 2.9 and 1.3 cm/sec he found that these could be reduced to 0.44 cm/sec after compaction.

In the psammon, the fauna is subject to two opposing vertical currents (Pennak 1940). Despite lack of supporting evidence for streams, it seems likely that these two currents are present in the psammonlike habitat found in the lower potamon of rivers; they thus deserve mention here. One is a virtually continuous, slow, upward flow, which is the result of capillary water being drawn up through the substrate; it is driven by evaporation at the substrate surface-air interface. Such a component may be present in streams but probably only at the edges of the hyporheic zone where it extends under the bank. Slope of the bank may affect the extent of this component, with pronounced slopes limiting the aquatic and shoreward extension of the zone (Neel 1948). On a flat beach, the sand may be wet enough to allow the psammolittoral fauna to live as much as 3 m back from the water's edge (Pennak 1940).

The second vertical current described from the psammon by Pennak is a downward one, strong and intermittent in nature and driven by precipitation and wave action. Lee and Hynes (1977) did, in fact, measure such an occurrence in a stream (Hillman Creek, Ontario) where, during an exceptionally heavy storm, the hydraulic gradient was reversed and, for a period of 15 to 30 minutes, downward instead of upward seepage occurred. This phenomenon is part of a normal flood pattern where bank storage of water increases at first and then decreases. As the time since the last rainfall increases, the contribution of streamflow arising from groundwater discharge (baseflow) increases until it becomes maximal during periods of drought. The upward component of baseflow through the bed, under the main channel (not to be confused with the vertical flow under the bank, driven by evaporation), is therefore an important factor in streams and influences conditions in the hyporheic zone. In practice, although we tend to talk in terms of vertical and horizontal current vectors, in streams, the mixed nature of streambed particles causes deflections such that net water movements may occur somewhat obliquely throughout the zone (e.g. see Vaux 1966).

Temperature. The temperature of groundwater is close to that of the mean annual temperature of the surrounding region (Freeze and Cherry 1979) and, in general, is lower than stream temperatures in the summer and higher in the winter (Sheridan 1962). In the hyporheic zone, the latter seems to be true also. Williams and Hynes (1974) found that in the Speed River, Ontario, the upper 20 cm of substrate tended to follow the surface water temperature more closely than did the lower layers. Daily fluctuations of the surface water temperature were greater than those of the substrate temperature, especially during the spring. This was probably due to the buffering effect of the groundwater and the high specific heat of the substrate. The substrate temperature was on the average 3.0 to 4.0°C lower than the surface water temperature during the summer and 0.5 to 1.0°C warmer in mid-winter. Schwoerbel (1961) found that in places continually exposed to the sun, hyporheic temperature was higher than that of the stream (epigean) water, and that daily fluctuations in hyporheic areas could be high. Hyporheic temperature regimes are similar to those of stream water but are different from those of the groundwater. This difference might explain why the faunas of the hyporheic and groundwater zones are different. As an example, many species of the harpacticoid copepod genera *Parastenocaris* and *Elaphoidella* are not found in the hyporheos but are abundant in groundwater (Schwoerbel 1961).

Light. Although penetration of light into the interstices may be controlled by substrate particle size (Schwoerbel 1964), the major portions of the hyporheic and psammon zones are in total darkness as light commonly fails to penetrate the interstices below a depth of 1 to 2 cm (Sassuchin et al. 1927; Neel 1948). As this would preclude photosynthesis, most of the inhabitants of these zones must therefore be detritivores or predators.

Chemical Features

pH, Alkalinity and Dissolved Carbon Dioxide.   In general, the pH of interstitial water is lower than that of surface water (e.g. see Burbanck and Burbanck 1967). Schwoerbel (1961) found that, in streams, pH decreases abruptly within a very short distance from the substrate surface and then remains constant with increasing depth. Often the decrease is in the order of 1 pH unit. This has more recently been confirmed by Husmann (1971) and Williams and Hynes (1974) who obtained surface water values of 7.5 and 8.0 for streams in Germany and Canada, respectively, and corresponding interstitial pH values (below 20 cm) of 6.9 and 7.5, respectively. Pennak (1940) believed that even lower values prevail at around 2 m horizontally from the water's edge, and this has been borne out in part by my own recent observations. However, the pH of interstitial water is probably subject to greater fluctuations than surface water because of greater retention time and local biotic and abiotic factors.

Related to pH is alkalinity [i.e. the quantity and kinds of compounds present which collectively shift the pH to the alkaline side of neutrality (Wetzel 1975)]. Neel (1948) found that psammon water always contained more bicarbonate (and therefore was more alkaline) than did lake water, but that there was considerable variation between sampling stations. This is confirmed by my recent work on streams, with a further increase in alkalinity occurring in that part of the hyporheic zone extending under the stream banks.

Also related to pH and alkalinity is dissolved carbon dioxide content. There are several sources of this gas, including the atmosphere, the dissociation of bicarbonate and carbonate ions, and biological processes. Of the latter, much is derived from aerobic animal respiration and, presumably, CO_2 builds up in the interstices, particularly as there is no photosynthesis occurring to consume it. Decomposition of organic matter trapped in the interstices may also be a contributing source. In general, CO_2 levels in the hyporheic zone are higher than in surface waters (Husmann 1971; Williams and Hynes 1974), although many authors report high variability, particularly if CO_2 is measured in its free state (e.g. see Pennak 1940). Hyporheic levels of CO_2 for Duffin Creek, Ontario, are shown in Fig. 14.3

Dissolved Oxygen.   In the Speed River, Ontario, dissolved oxygen was not detected in subsurface water below a depth of 30 cm (Williams and Hynes 1974). Typically, there was a drop of 20 to 60 percent in saturation values from the surface to the interstitial water at 10 cm. From 10 to 20 cm, there was a further

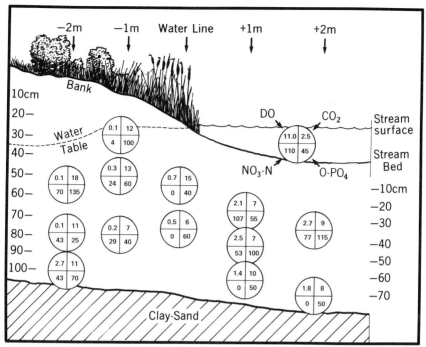

Figure 14.3 Concentrations of selected dissolved chemical parameters at various depths in, and adjacent to, the hyporheic zone of Duffin Creek, Ontario [upper left quadrant, oxygen (ppm); upper right quadrant, carbon dioxide (ppm); lower left quadrant, nitrate-nitrogen ($\mu g/l$); lower right quadrant, orthophosphate ($\mu g/l$)].

drop of 20 to 30 percent saturation, until at 30 cm the oxygen saturation was only about 5 percent of that found in the surface water. These values differ quite markedly from those obtained by Schwoerbel (1961), who found only a slight O_2 decrease with depth, and values 45 to 75 percent of saturation at 30 cm in some German rivers, and Husmann (1971), who obtained an oxygen decrease of more than 50 percent in the first 3 cm of substrate in the River Lautenthal, Germany. In this latter study, a sharp increase back to approximately 80 percent saturation occurred at 6 cm with the O_2 level remaining high down to 30 cm. In Duffin Creek, oxygen occurred in small amounts throughout the hyporheic zone (Fig. 14.3), but it tended to be minimal under the bank where, presumably, groundwater further lowered concentrations. Burbanck and Burbanck (1967) correlated dissolved oxygen levels and type of substrate forming the interstices of a variety of coastal water bodies in the southeastern United States. The amount of oxygen decreased as particle size decreased and the proportion of mud and silt to sand increased.

Oxygen probably originates from surface water and as this water percolates through the interstices, it is subject to a biochemical oxygen demand that varies according to the temperature, number of animals present, and the quantity of organic matter in the surroundings (Sheridan 1962). However, groundwater sometimes contains enough oxygen to sustain the developing eggs of salmon (Krogius and Krokhin 1948, cited in Sheridan 1962).

Other Chemical Parameters. So far as other chemical parameters are concerned, Schwoerbel (1961) found that phosphate was generally higher in the hyporheic zone than on the surface, especially in hardwater streams (see also Fig. 14.3). The same held for levels of nitrate, ammonia, and nitrite nitrogen in the Speed River, although Husmann (1971) found the latter two forms in lower concentration with increasing depth. Schwoerbel (1961) measured little difference between surface and subsurface nitrate concentrations, but in Duffin Creek (see Fig. 14.3) hyporheic levels are substantially lower. Clearly there is variation in the occurrence of these parameters, perhaps controlled by the local extent of faunal metabolism and denitrification processes.

Organic Matter. Particulate organic matter, a principal food material for the hyporheos, is maximal in the top few centimeters of substrate, particularly in highly heterogeneous substrates where trapping of the detritus that moves along the streambed is extensive. Organic matter generally decreases in amount with increasing depth although it seems plentiful even deep down. As it is primarily derived from the same allochthonous material as enters the stream (e.g. see Hynes 1975) its exact nature need not concern us here. Its importance, nevertheless, can be underlined by the findings of Schwoerbel (1961) that a direct relationship exists between the amount of detritus found in the hyporheic zone and the number of animals per unit volume.

Summary. The exact limits of the hyporheic zone are sometimes difficult to define because of the relative instability of its upper and lower boundaries. For example, if there is a surge of groundwater its lower boundary may be pushed nearer the substrate surface. Conversely, during spates or droughts, the zone, along with its associated fauna, might extend downward as the groundwater table retreats. Thus, the limits of the hyporheic zone may vary both spatially and temporally. In many ways, the hyporheic zone is analogous to the marine intertidal zone with its spring and neap tide levels.

THE HYPORHEOS AND ITS ADAPTATIONS

Although the discovery of animals living beneath streambeds dates back several decades, many recent workers in this field have been criticized as to the universality of hyporheic faunas. The existence of hyporheic populations has now been substantiated in many countries such as the United States (Stanford and Gaufin 1974; Poole and Stewart 1976), Malaysia (Bishop 1973), England (Ford 1962; Gledhill 1971), Wales (Hynes et al. 1976), Austria (Danielopol 1976), Germany (Husmann 1971), Yugoslavia (Mestrov and Tavcar 1972), and Canada (Coleman and Hynes 1970; Hynes 1974). Faced with such evidence, we must conclude that many streams with deep gravel beds support such a fauna.

The hyporheos consists of a wide variety of taxa, many of them insects or related arthropods such as mites or crustaceans (including copepods, ostracods, cladocerans, and amphipods). Much work on interstitial faunas in general has concentrated on mites and crustaceans as some species seem to be particularly well adapted to this kind of environment. Thus, for reasons of completeness, I have included information on these noninsect groups in this review, as data on them may apply equally well to the less-studied, though in many cases equally abundant, insect fraction of the hyporheos.

The extent of the hyporheos is largely controlled by physical and chemical conditions in the interstices. In terms of vertical depth, Williams and Hynes (1974) found insect larvae regularly to a depth of 70 cm beneath the surface of the bed of the Speed River. Numbers were greatest around 10 cm and declined gradually with increasing depth until at 80 cm few individuals were present, although interstitial water clearly was present beyond this. In the Matamek River, Quebec, however, these same authors (Williams and Hynes 1974) found animals down to the maximum depth capable of being sampled (100 cm), although at reduced densities. Below about 40 cm, the numbers of insect larvae decreased, but two forms, the oribatid mites and the elongate harpacticoid copepod *Parastenocaris starretti,* became dominant. This copepod genus is typical of true subterranean habitats (Schwoerbel 1961) and both these and the oribatid mites may possibly be indicative of the boundary between hyporheic water and groundwater. Schwoerbel (1961) and Husmann (1966) suggested that the stream interstitial habitat could be subdivided into two zones on the basis of the distribution of mites: an upper layer characterized by the presence of mites in families belonging to the Hydrachnellae, and a lower zone characterized by mites of the subfamily Limnohalacarinae. The point where the limnohalacarids disappear marks the upper boundary of the true groundwater zone (Husmann 1966).

The vertical extent of the hyporheos thus varies from stream to stream. As an example, the hyporheic zone of the St. Anne's River, Quebec, was found to extend a mere 20 cm because it was underlain at this depth by bedrock and highly compacted materials (Williams and Hynes 1974). On the other hand, its extent may be measurable in meters as in the case of rivers with deep graded gravels (e.g. the Tobacco River, Montana; see Stanford and Gaufin 1974).

Dissolved oxygen concentration may be one factor controlling the downward limit of aquatic insects, although they have been recorded below the level of detectable oxygen (Williams and Hynes 1974). Another controlling factor is undoubtedly pore space in relation to the size and shape of the animals (discussed below); groundwater movement is related to pore space, and increases as pore space increases. This can consequently affect the hyporheos. Godbout and Hynes (1982), for example, found that higher hyporheic densities in Salem Creek, Ontario, were correlated with areas of coarser sediments, and upward groundwater flow, which provided increased flushing of the interstices.

The horizontal extent of the hyporheos has not been well studied. The distribution of hyporheos at a depth of 10 cm on a transect from midstream to 2 m into the bank of the Speed River was examined by Williams (1981) who used a combination of core samples and small colonization pots (Fig. 14.4). Results indicate that some of what might be termed typical stream insect taxa (e.g. Ephemeroptera, Trichoptera, Plecoptera) were entirely restricted to the stream interstitial environment (cf. Schwoerbel 1967, who recorded movements of nymphs of the mayfly *Habroleptoides modesta* into the bank of a stream), whereas some of the Chironominae and Elmidae were taken right up to the stream margin, and some of the Orthocladiinae and Tanypodinae were found up to 2 m into the interstitial water under the bank. Among the noninsect arthropods, the Ostracoda, Cyclopoida, Harpacticoida, and Acari all showed fairly continuous distributions from midstream to at least 2 m into the bank. This

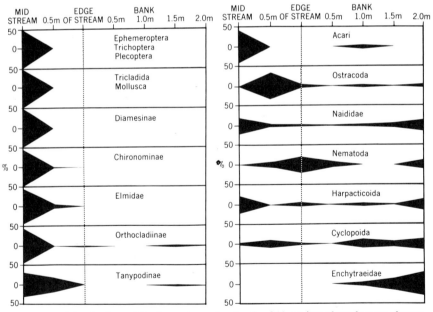

Figure 14.4   Distribution of the hyporheos at a depth of 10 cm into the substrate along a transect from midstream to 2 m into the bank of the Speed River, Ontario (redrawn from Williams 1981, reprinted with permission of Plenum Publishing Corp.).

supports Schwoerbel's (1961) idea that the hyporheic zone extends several meters beyond the margin of the stream. Here again, the hyporheic zone presumably merges with the true groundwater and its associated fauna. The lateral extent also may vary according to local conditions. For example, hyporheic insects have been found as much as 50 m laterally from a river channel (Stanford and Gaufin 1974).

I have suggested previously (Williams and Hynes 1974) that the animals living in the interstices of streambeds can be divided into two types, the occasional and the permanent hyporheos, based on the way in which each uses the habitat. The occasional hyporheos consists of larvae (particularly of aquatic insects) of most of the surface benthos that may seek out this zone as a refuge during their early development. The permanent hyporheos consists of many specialized forms (both larval and adult) of mites, copepods, ostracods, cladocerans, tardigrades, syncarids, and peracarids that complete their life cycle there.

The Occasional Hyporheos

It is in the occasional hyporheos category that many aquatic insect species belong, since most spend part of their life cycle out of water and therefore must leave the hyporheic zone before this time. Williams and Hynes (1974) and, more recently, Godbout and Hynes (1982) have shown seasonal variation in the distribution of insect larvae in the hyporheos. Figures 14.5 and 14.6 give some indication of this variation with depth for different insects from the Speed River.

Figure 14.5*A* shows the distribution of the total benthos (both insect and noninsect fractions) at different depths. Clearly, animal density decreased with increasing depth, although maximum numbers were almost always found around 10 to 20 cm. Peaks occurred in winter, spring, and autumn, the times when most stream species are actively growing. Minimal numbers occurred in summer, which is the emergence period of many insect species. Although larval densities were lower in the interstices at this time, eggs were commonly present. Examination of individual taxa in this three-dimensional manner revealed interesting patterns. Elmid beetle larvae, for example, occurred in large numbers in the hyporheic zone relative to their numbers at the substrate surface. Maximum numbers occurred at 10 to 20 cm in the autumn, but adults were found down to 30 cm, and larvae to 70 cm.

The average penetration depth of the net-spinning caddisfly larvae of *Cheumatopsyche* was 20 cm (Fig. 14.5*B*), although the smallest instars were occasionally found down to 40 cm. Larvae of the caddisfly *Helicopsyche borealis* occurred down to 30 cm with greatest numbers in May and the period of August to October. Their snail-like cases may make them the best adapted of all the cased Trichoptera to an interstitial existence as they are roughly spherical, a shape easily moved between the interstices and very resistant to crushing.

Among the mayfly species of the Speed River, *Caenis* sp. was a common

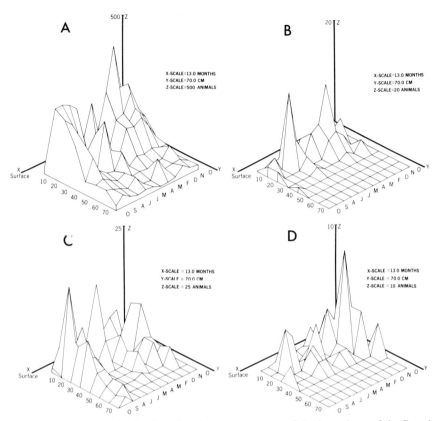

Figure 14.5 Vertical distribution of various components of the hyporheos of the Speed River over time. Values represent the numbers of animals retrieved from a sample of 125 cc of substrate at 10-cm intervals from the surface down to 70 cm; A. total benthos; B. *Cheumatopsyche* sp; C. *Caenis* sp; D. *Ephemerella* spp.

member of the hyporheos. Figure 14.5*C* shows that it regularly occurred down to 40 cm. Very small instars were taken occasionally at 70 cm. The fact that this genus has a pair of large operculate gills on the second abdominal segment which lie on top of, and protect, the other gills might well suit it for an interstitial existence. The maximum penetration depth for *Ephemerella deficiens* and *E. excruciens* was 30 cm (Fig. 14.5*D*) with larger nymphs occurring closer to the surface. Tsuda (1966) has found early instars of this genus deep in the beds of sandy streams. As with *Caenis,* nymphs of this genus are generally robust in appearance. In contrast, the delicate nymphs of *Paraleptophlebia* that occur in the Speed River were most common on the substrate surface, but a few early instars were collected as deep as 30 cm during summer and early fall.

Most of the limited Plecoptera species in the Speed River appeared able to live in the hyporheic zone, especially the early instars. *Allocapnia pygmaea* nymphs were commonly found down to 30 cm and in summer they were in a characteristic diapause position (Harper and Hynes 1970). Stanford and Gaufin (1974) found the hyporheos of the Flathead and Tobacco Rivers, Montana, to be dominated by large stonefly nymphs.

By far the most common constituents of the insect fraction of the hyporheos are the chironomids. In the Speed River, large numbers of larvae occurred at 10 to 30 cm for most of the year but densities were lower from April to August (Williams and Hynes 1974). No pupae were found deeper than 10 cm. Figure 14.6*A* shows the depth distribution for *Cladotanytarsus* sp. The species was

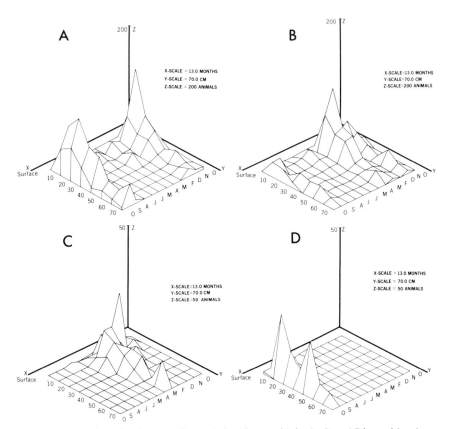

Figure 14.6 Vertical distribution of hyporheic chironomids in the Speed River with values expressed as in Fig. 14.5 (redrawn from Williams and Hynes 1974, with permission of Freshwater Biology); A. *Cladotanytarsus* sp.; B. *Microtendipes* sp.; C. *Orthocladius* sp.; D. *Cricotopus* sp.

univoltine with adults appearing from April to early August. Eggs of a new generation began to hatch in mid-August and early instar larvae were found interstitially at this time. Either this was the result of a downward migration of small animals, coinciding with a drop in the surface water temperature, or it was due to the settling of eggs into the interstices shortly after they had been laid. In the latter case, gentle agitation by the current may have caused the eggs to passively sink in the same way that the smallest particles in a mixture will be sorted towards the bottom. *Cladotanytarsus* larvae overwintered at various depths in the substrate before migrating upwards in March when the surface water began to warm. It seems that the majority of larval growth took place in the substrate.

Figures 14.6*B* to *D* show hyporheic distributions for three other Speed River chironomid species. *Microtendipes* sp. was present in the substrate in high numbers from September to March. This species appeared to be univoltine with adults emerging from April to August, coincident with the presence of many large larvae near the surface. Eggs hatched from August onwards, and the resulting small larvae overwintered in the substrate until the following March or April when their growth accelerated and they began to pupate and emerge. *Orthocladius* was univoltine and had a diapausing egg stage lasting from July to November when a few early instar larvae appeared at about 20 cm. Numbers of larvae increased from December to March and they spread rapidly through the interstices. Growth was fast as pupation and emergence occurred in May. *Cricotopus* sp. also appeared to be univoltine. Last instar larvae were collected on the substrate surface in April and May, presumably in readiness for pupation, emergence, mating, and oviposition by July when early instars appeared. These were common down to 60 cm in later summer, but their whereabouts during winter and spring was not known.

These examples of hyporheic distributions of chironomids show interesting temporal separations. *Cricotopus* and *Cladotanytarsus* dominated the interstices in late summer, with the latter continuing through the fall and winter. *Microtendipes* and *Orthocladius* dominated in the winter. Whether this has any significance in terms of resource partitioning, food or otherwise, remains to be shown.

Most of the chironomid species collected in the Speed River were found at some time in the hyporheos. Similar findings have been made by Tavcar and Mestrov (1970) in Yugoslavian rivers. They found that the chironomids of the River Trebisnjica were distributed in three basic groups. One group, comprising 41 percent of the species, consisted of those living exclusively on the substrate in the current; a second group, comprising 32 percent, was found exclusively in the hyporheos; and a third group (26%) was composed of larvae living both on the surface and in the hyporheos. Tavcar and Mestrov determined that the larvae belonging to the second and third groups, in contrast to those in the first group, were capable of tolerating greater fluctuations in temperature, dissolved oxygen, and carbon dioxide. There were also differences in substrate preferences among the three groups. The authors concluded that the composition of the hyporheic chironomid communities of their study was the result of natural selection.

Ford (1962) found that chironomids penetrated to only about 5 cm in a mud-bottom stream in England. He concluded that compaction of the substratum may have influenced penetration. However, he did find a few larvae of other insect groups at greater depths, including Ceratopogonidae, Sialidae, and Tipulidae. Wiley (1981) investigated the factors affecting successful penetration of sediments by chironomid larvae. He found that penetration consisted of two components: searching for a suitable site and actual penetration. The time each phase took was dependent on the physical make-up of the substrate. As the ratio of particle diameter to larval head-capsule width approached 1.0, the time spent locating a site increased substantially. Penetration success was negatively correlated with the square root of current velocity and penetration time. The author also suggested that body size influences penetration rates in some sediments, in that many sediment-dwelling species are of small size.

Insect taxa that are seldom found in the hyporheos include the heptageniid mayflies, the Odonata, Hemiptera, Megaloptera, most cased caddisflies and the larger caseless forms such as *Rhyacophila,* and adult Coleoptera. These taxa, however, tend to include large or active forms, or forms that require rapid current for respiration or feeding, characteristics that would preclude interstitial living. Simuliids were never found in the Speed River hyporheos, but they were found together with other more delicate taxa such as *Baetis, Amphinemura,* and *Potamophylax,* in the hyporheic zone of the Afon Hirnant, Wales. The substrate of this river consists of broken slate which creates considerably larger interstices than for many other river bed materials.

The Permanent Hyporheos

Few aquatic insects belong to the permanent hyporheos, although it is possible that for species which no longer need to leave the water to reproduce, their life cycle could be spent interstitially (e.g. some Elmidae and Dytiscidae; Vandel 1965). Species that live more or less permanently in the hyporheic zone belong to the Bacteria, Protozoa, Turbellaria, Rotifera, Nematoda, Oligochaeta, and Gastropoda among the non-Arthropoda, and to the Cladocera, Ostracoda, Copepoda, Malacostraca, Tardigrada, and Acari among the noninsect Arthropoda.

Microorganisms such as protozoans and bacteria cannot be ignored in this context since, in the absence of autotrophic algae, they and the aquatic hyphomycete fungi may be a potential food source for interstitial insects. Lupkes (1976) found that in the Fulda River, Germany, the number of ciliate species and their densities usually decreased with increasing depth. At 50 cm depth, only three species were found. These were related to surface species that were small, long, and thin and, hence, seemed especially adapted to an interstitial life. In marine interstitial zones, the vertical extent of ciliates has been correlated with the depth of the redox-discontinuity layer (Fenchel and Jansson 1966).

By far the most studied arthropod groups are the harpacticoid copepods and the Acari, and much of this work has entailed taxonomy and systematics (e.g. see Noodt 1954; Cook 1969; Gledhill 1973). These two groups appear particularly well adapted to an interstitial existence and are well represented not only in fluvial and lacustrine deposits but also in groundwater. In fact, as we have seen, Husmann (1966) proposed using the distribution of the mites to delimit the boundary between the hyporheic and groundwater zones. Figure 14.7 shows the hyporheic distribution of *Bryocampus zschokkei*, a harpacticoid copepod belonging to the permanent component of the Speed River hyporheos.

Morphological Adaptations of Insects to Living a Hyporheic Existence

The shape of an organism may be important in determining whether it is suited for an interstitial existence. There are two basic shapes that may be adaptive for living in this habitat. The first is for an organism to be long and slender, with a flexible body that allows easy passage between the substrate particles. The second is to be small and blunt with a hard protective shell or exoskeleton that will withstand being crushed as the animal pushes its way through the pore spaces. The first shape seems to be a preadaptation of larvae of some Diptera, Ephemeroptera, Trichoptera, Plecoptera, Elmidae, the Chironomidae, and the Collembola. The second shape is a preadaptation of some of the cased Trichoptera, Elmidae adults and the Acari. Perhaps the Copepoda and Ostracoda should comprise yet a third shape, characterized by extreme smallness of size, which enables these organisms to actually swim in the interstitial water.

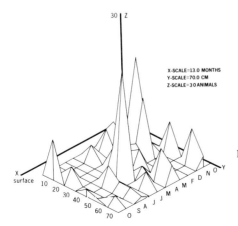

X-SCALE=13.0 MONTHS
Y-SCALE=70.0 CM
Z-SCALE=3.0 ANIMALS

Figure 14.7 Vertical distribution of *Bryocamptus zschokkei* (Copepoda: Harpacticoida), a member of the permanent hyporheos of the Speed River, Ontario.

It is unlikely that any one of the above adaptations to, or characteristics of, the interstitial environment will alone control the number of animals living in it. More likely, it is the combined effect of many of these variables that determines the qualitative and quantitative nature of these sub-benthic populations.

Very little is known about the physiological demands made on an organism by the environment in aquatic interstices although clearly these demands must be of paramount importance.

Advantages and Disadvantages of a Hyporheic Existence

There seem to be several distinct disadvantages to living in the hyporheic zone, notably the limited habitable space (though not in terms of total habitat volume), reduced current, low dissolved oxygen levels, high carbon dioxide levels, the accumulation of nitrogenous wastes, and lack of light. Nevertheless, there also seem to be many advantages to living in such a habitat. Those immediately obvious include safety from large predatory species, safety from fast currents, plentiful food in the form of detritus and microorganisms, and somewhat ameliorated temperature regimes. Species living interstitially also live considerably longer than congeneric epigean species (Henry 1976, cited in Danielopol 1980).

One of the most important advantages of living in the hyporheic zone is that of survival during adverse stream conditions. Several workers have suggested that some of the surface benthos may move down into the substrate during severe droughts in order to follow the retreating water table (e.g. see Clifford 1966; Harrison 1966; Hynes 1968). Imhof and Harrison (1981), for example, found that larvae of the caddisfly *Diplectrona modesta* survived desiccation for at least four weeks in a southern Ontario stream by constructing tubes deep in the substrate. Similarly, Williams and Hynes (1976a) dug up diapausing nymphs of the stonefly *Allocapnia vivipara* from between 5- and 25-cm deep in the dry bed of Kirkland Creek, Ontario. This diapause is an adaptation to surviving high summer temperatures (Harper and Hynes 1970). Nebeker and Lemke (1968) have shown that these nymphs are heat sensitive, particularly in their later instars. There may also be a gradual upward migration of mature larvae and pupae of hyporheic species prior to emergence and this occurs just before the dry phase of cyclical temporary streams (Williams and Hynes 1977).

At the other extreme, several authors have noted animals moving deeper into the substrate during spates, presumably as a protective mechanism against the current increase, which may cause scouring and increased silt loading (e.g. see Williams and Hynes 1974; Poole and Stewart 1976). At other times, the hyporheic fauna may act as a source of recolonizing animals should anything natural or unnatural happen to the surface benthos (Williams and Hynes 1976b; Rosenberg and Wiens 1978). The fast repopulation of the substrate surface, noted by many authors, after application of insecticides or chemical spills may be from hyporheic sources (e.g. see Wallace et al. 1973).

Migrations Within the Hyporheic Zone

It is unlikely that many members of the occasional hyporheos move vertically between the stream surface and the hyporheic zone on a daily basis, as is the case with many lake zooplankters (e.g. *Chaoborus*). Even though the vertical extent through which the latter move is much greater, the complex interconnections of the interstitial spaces would preclude this for the occasional hyporheos. Evidence, at present, (e.g. see Williams and Hynes 1976b) points to vertical (and probably horizontal) migration over a period of several days (as in the case of spates) or, more commonly, on a seasonal basis. Figure 14.8 presents a summary of the ways in which many stream insects make use of the hyporheic zone. Eggs from ovipositing females may be worked down into the interstices by a combination of current action and gravity, as previously discussed. This will be most effective for eggs that are laid singly or egg masses that break up soon after contact with water. A wide variety of egg shapes and sizes are commonly encountered in the hyporheic zone. These hatch in the interstices where the resulting larvae find suitable conditions for growth. Also, larvae or nymphs

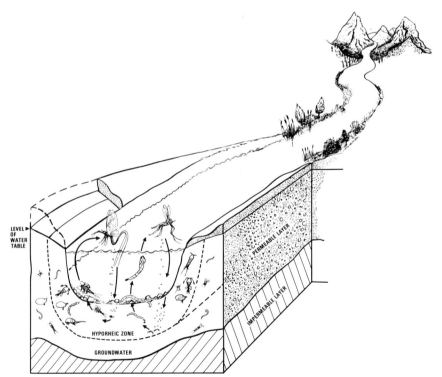

Figure 14.8   Diagramatic summary of the hyporheic zone and its role in the life cycles of stream insects (redrawn from Williams 1981, reprinted with permission of Plenum Publishing Corp.).

hatching from egg masses laid on the streambed surface may actively burrow down into the interstices. In fact, the results of various colonization pot studies (e.g. see Coleman and Hynes 1970; Hynes 1974; O'Connor 1974) seem to substantiate this as a method of entry into the hyporheos. As these larvae grow, the limits of interstitial spaces in some streams may force them towards the surface so that prepupal and pupal stages are more common near the surface. The larvae of some taxa, for example the chironomids, occur deeper in the hyporheic zone than others. This may be purely the result of body size or the consequence of differences in tolerance to environmental variables.

Some noninsect arthropods also seek out the hyporheic zone during their immature stages. Pieper (1978) has shown, for example, that the young of *Gammarus fossarum* migrate into the hyporheic zone, returning to the surface only at the end of their juvenile stage.

EVOLUTIONARY ASPECTS OF THE HYPORHEOS

How the hyporheic community may have developed provides for interesting speculation. Undoubtedly, colonization has occurred from more than one source. Factors that may have promoted colonization include advantages in terms of reduced fluctuations in temperature, current, and other conditions offered by this "middle" zone. Egglishaw (1964) suggested that the large amounts of organic matter which get trapped in the interstices of streambeds may well have been the most important factor leading to the invasion of this habitat by surface detritivores, and small predatory species may have followed in turn. Interestingly, Ladle et al. (1980) have shown from studies on a large artificial stream, that oviposition in the absence of drift or other inputs of animals is capable of establishing a hyporheic fauna.

Obviously, animals preadapted to an interstitial environment stand the best chance of survival, although further adaptations may occur as the species evolves in its new environment. For example, in amphipods, the loss of a pelagic, terminal male stage was a prerequisite to colonization of freshwater epigean, interstitial, and cave habitats (Bousfield 1978). For the mites, Schwoerbel (1961) pointed out that species of the *Atractides* type, together with most other hyporheic mites, must have been preadapted to an interstitial existence. In contrast, mites of the *Hygrobates* type have not become established in the hyporheic zone because they retain a parasitic nymphal phase and are therefore dependent on an epigean host.

Effective post-colonization adaptations may be controlled by the degree of isolation from the surface together with lack of genetic interchange with epigean populations (Schwoerbel 1967). Progressive reduction in body size and loss of eyes is readily observable in interstitial mites. Such changes may even be seen over the relatively short term. Piearce and Cox (1977), for example, have shown marked pigment changes between two parts of a population of *Gammarus pulex* in England, one in a stream above a cave, and the other in the same stream at the point where it emerged from a cave. Similarly for insects, Stanford and Gaufin (1974) found that stonefly nymphs in the hyporheic zone of two Montana rivers were pale, almost transparent, and the eyes of early instar nymphs were poorly developed.

In terms of fauna colonizing from the groundwater habitat, Husmann (1976) showed that such species can be carried out of their primary habitat by rapidly moving groundwater. Some of these animals may become well established in a secondary biotope and, over time, it may become their optimum environment. On a larger scale, Delamare-Deboutteville (1960) and Fenchel (1978) have suggested that some elements of freshwater faunas (e.g. harpacticoids and amphipods) originated from the existence of an uninterrupted interstitial habitat, extending from sandy marine littoral habitats via coastal groundwater to inland groundwater. In fact, representatives of subterranean, including cave, faunas show very close affinities with coastal marine species.

There may also have been contributions to the hyporheic fauna from the soil fauna. Angelier (1953) suggested that mites of the family Stygothrombidiidae are forms which probably originated in soil. Being dependent on moist interstitial conditions, they became easily established in the hyporheic zone. Phylogenetically, their recent ancestors were terrestrial.

The relatively stable ambient conditions offered by the hyporheic habitat may, as in the case of the groundwater habitat, enable survival of sensitive species, some of which may be relics of ancient faunas (Delamare-Deboutteville 1957). The hyporheic zone may therefore have an important function as a refuge not only for species of epigean origin but for those of endogean and phreatic origin as well.

SIGNIFICANCE OF HYPORHEIC POPULATIONS IN STREAM STUDIES

The distribution of benthos in three dimensions has serious implications in our understanding, and measurement, of freshwater invertebrate production. Williams and Hynes (1974) made a detailed study of the hyporheos of an Ontario stream and concluded that large numbers of invertebrates occurred beneath the streambed. They calculated that the total number of animals beneath a 1 m² area of riffle to a depth of 1 m (i.e. a volume of 1 m³) varied seasonally from about 135,000 to 800,000 organisms, with a corresponding dry weight biomass of 31 to 253 g. The vertical distribution of the fauna meant that only a maximum of 20 percent (by numbers) of the total fauna occurred in the top 5 cm of substrate. Other workers report similar findings (e.g. see Poole and Stewart 1976; Godbout and Hynes 1982). This brings to light the inefficiencies of many of the benthic samplers in use today. Although some of these devices are fairly reliable in their quantitative and qualitative estimates of surface benthos populations, with few exceptions they only sample to about 5 cm. Thus, if species in many streams can extend down to depths of 1 m or more in quite large numbers, then most of these samplers fail to give an accurate picture of the total biomass. Perhaps to be accurate, the exact depth to which these devices sample should be stated and, for any kind of accurate production work (see Benke, Chapter 10), they should be used in conjunction with a more suitable quantitative interstitial sampler.

FUTURE NEEDS

Research on hyporheic populations is in its infancy and a great many aspects of this fascinating zone and its community are yet to be unravelled. For example, what is the exact extent of the zone, where does the groundwater zone begin, and what is the detailed nature and importance of the hyporheic-groundwater interface? More data are required on the taxa found in hyporheic communities, with emphasis on examining different types of streams in different countries with a view to identifying common elements in the faunas and their zoogeographical relationships. The morphology and physiology of interstitial species, particularly with regard to how they differ from closely related epigean forms, should be examined as well as the tolerance of hyporheic species to low dissolved oxygen levels. Additional research on comparative growth studies and the migrations of animals within, into, and out of the hyporheic zone would be useful. Finally, there is a need for continued work on the design of more efficient hyporheic samplers to better estimate total benthic invertebrate standing crop and production, by taking into account the hyporheic component.

References

Angelier, E. 1953. Recherches écologiques et biogeographiques sur la faune des sables sub-mergés. Archives de zoologie expérimentale et générale 90:37–162.
Aubert, J. 1959. Plecoptera. Insecta Helvetia Fauna 1. Imprimerie la Concorde, Lausanne, Switzerland. 140 pp.
Berthélemy, C. 1966. Recherches écologiques et biogeographiques sur les Plécoptères et Coléoptères d'eau courante (*Hydraena* et Elminthidae) des Pyrénées. Annales de Limnologie 2:227–458.
Bishop, J. E. 1973. Observations on the vertical distributions of the benthos in a Malaysian stream. Freshwater Biology 3:147–56.
Bou, C. and R. Rouch. 1967. Un nouveau champ de recherches sur la faune aquatique souterraine. Comptes rendus hebdomadaires des séances de l'Académie des sciences 265:369–70.

Bousfield, E. L. 1978. A revised classification and phylogeny of amphipod crustaceans. Transactions of the Royal Society of Canada 16:343–90.

Bruce, J. R. 1928. Physical factors on the sandy beach I. Tidal, climatic and edaphic. Journal of the Marine Biological Association of the United Kingdom 15:535–52.

Burbanck, W. D. and G. P. Burbanck. 1967. Parameters of interstitial water collected by a new sampler from the biotopes of *Cyathura polita* (Isopoda) in six southeastern states. Chesapeake Science 8:14–27.

Chappuis, P. A. 1942. Eine neue Methode zur Untersuchung der Grundwasser-fauna. Acta Scientiarum Mathematicarum 6:3–7.

Clifford, H. F. 1966. The ecology of invertebrates in an intermittent stream. Investigations of Indiana Lakes and Streams 7:57–98.

Coleman, M. J. and H. B. N. Hynes. 1970. The vertical distribution of the invertebrate fauna in the bed of a stream. Limnology and Oceanography 15:31–40.

Cook, D. R. 1969. The zoogeography of interstitial water mites, pp. 81–87. *In*: G. O. Evans (ed.) Proceedings of the 2nd International Congress of Acarology, 1967. Akadémiai Kiadó, Budapest. 652 pp.

Danielopol, D. L. 1976. The distribution of the fauna in the interstitial habitats of riverine sediments of the Danube and the Piesting (Austria). International Journal of Speleogy 8:23–51.

Danielopol, D. L. 1980. The role of the limnologist in groundwater studies. Internationale Revue der gesamten Hydrobiologie 65:777–91.

Danielopol, D. L., R. Ginner, and H. Waidbacher. 1980. Some comments on the freezing core method of Stocker and Williams (1972). Stygo News 3:4–5.

Delamare-Deboutteville, C. 1957. Lignées marines ayant pénétré dans les eaux souterraines continentales. Un probleme de biogéographie actuelle. Compte rendu des séances de la Société de biogéographie 34:53–67.

Delamare-Deboutteville, C. 1960. Biologie des eaux souterraines littorales et continentales. Actualités Scientifiques et Industrielles No. 1280. Hermann, Paris, France. 740 pp.

Egglishaw, H. J. 1964. The distributional relationship between the bottom fauna and plant detritus in streams. Journal of Animal Ecology 33:463–76.

Fenchel, T. M. 1978. The ecology of micro- and meiobenthos. Annual Review of Ecology and Systematics 9:99–121.

Fenchel, T. M. and B. O. Jansson. 1966. On the vertical distribution of the microfauna in the sediments of a brackish-water beach. Ophelia 3:161–77.

Ford, J. B. 1962. The vertical distribution of larval Chironomidae (Dipt.) in the mud of a stream. Hydrobiologia 19:262–72.

Freeze, R. A. and J. A. Cherry. 1979. Groundwater. Prentice-Hall, Englewood Cliffs, NJ. 604 pp.

Gilpin, B. R. and M. A. Brusven. 1976. Subsurface sampler for determining vertical distribution of stream-bed benthos. Progressive Fish-Culturist. 38:192–94.

Gledhill, T. 1971. The genera *Azugofeltria, Vietsaxona, Neoacarus* and *Hungarohydracarus* (Hydrachnellae:Acari) from the interstitial habitat in Britain. Freshwater Biology 1:61–82.

Gledhill, T. 1973. Observations on the numbers of water-mites (Hydrachnellae, Porohalacaridae) in Karaman/Chappuis samples from the interstitial habitat of riverine gravels in Britain, pp. 249–57. *In*: Colloque National de Spéleologie, October 2–11, 1971, Bucureşti-Cluj. Editura Academiei Republicii Socialiste România.

Godbout, L. and H. B. N. Hynes. 1982. The three dimensional distribution of the fauna in a single riffle in a stream in Ontario. Hydrobiologia 97:87–96.

Harper, P. P. and H. B. N. Hynes. 1970. Diapause in the nymphs of Canadian winter stoneflies. Ecology 51:925–27.

Harrison, A. D. 1966. Recolonization of a Rhodesian stream after drought. Archiv für Hydrobiologie 62:405–21.

Henry, J. P. 1976. Recherches sur les Asellidae hypogés de la lignee cavaticus (Crustacea, Isopoda, Asellota). Unpublished D. Sc. thesis. 270 pp. (Available from Centre de Documentation, CNRS A.O. 12, 143 Paris, France).

Husmann, S. 1966. Versuch einer ökologischen Gliederung des interstitiellen Grundwassers in Lebensbereiche eigener Prägung. Archiv für Hydrobiologie 62:231–68.

Husmann, S. 1971. Eine neue Methode zur Entnahme von Interstitialwasser aus subaquatischen Lockergesteinen. Archiv für Hydrobiologie 68:519–27.

Husmann, S. 1976. Studies on subterranean drift of stygobiont crustaceans (Niphargus, Crangonyx, Graeteriella). International Journal of Speleology 8:81–92.

Hynes, H. B. N. 1968. Further studies on the invertebrate fauna of a Welsh mountain stream. Archiv für Hydrobiologie 65:360–79.

Hynes, H. B. N. 1974. Further studies on the distribution of stream animals within the substratum. Limnology and Oceanography 19:92–99.

Hynes, H. B. N. 1975. The stream and its valley. Internationale Vereinigung für Theoretische und Angewandte Limnologie Verhandlungen 19:1–15.

Hynes, H. B. N., D. D. Williams, and N. E. Williams. 1976. Distribution of the benthos within the substratum of a Welsh mountain stream. Oikos 27:307–10.

Imhof, J. G. A. and A. D. Harrison. 1981. Survival of Diplectrona modesta Banks (Trichoptera:Hydropsychidae) during short periods of desiccation. Hydrobiologia 77:61–63.

Krogius, F. V. and E. M. Krokhin. 1948. On the production of young sockeye salmon (Oncorhynchus nerka Walb.). Izvestiia Tikhookeanskovo Nauchno-Issledovatelskovo. Instituta Rybnovo Koziasitva i Okeanografii 28:3–27. (Fisheries Research Board of Canada Translation Series, No. 109).

Kühtreiber, J. 1934. Die Plecopterenfauna Nordtirols. Berichte des naturwissenschaftlichmedizinischen Vereins in Innsbruck 43/44:1–219.

Ladle, M., J. S. Welton, and J. A. B. Bass. 1980. Invertebrate colonization of the gravel substratum of an experimental recirculating channel. Holarctic Ecology 3:116–23.

Lee, D. R. and H. B. N. Hynes. 1977. Identification of groundwater discharge zones in a reach of Hillman Creek in southern Ontario. Water Pollution Research in Canada 13:121–33.

Lupkes, G. 1976. Die vertikale Verteilung von Ciliaten im Stygorhithral der Fulda (Beitrag zur Kenntnis mesopsammaler Ciliaten in Fliessgewässern). International Journal of Speleology 8:127–33.

Mason, W. T., J. B. Anderson, and G. E. Morrison. 1967. A limestone-filled artificial substrate sampler-float unit for collecting macroinvertebrates in large streams. Progressive Fish-Culturist 29:74.

Mestrov, M. and V. Tavcar. 1972. Hyporheic as a selective biotope for some kinds of larvae of Chironomidae. Bulletin Scientifique, Conseil des Academies des Sciences et des Arts de la RSF de Yougoslavie, Section A: Sciences Naturelles Techniques et Medicales 17:7–8.

Mundie, J. H. 1971. Sampling benthos and substrate materials, down to 50 microns in size, in shallow streams. Journal of the Fisheries Research Board of Canada 28: 849–60.

Nebeker, A. V. and A. E. Lemke. 1968. Preliminary studies on the tolerance of aquatic insects to heated waters. Journal of the Kansas Entomological Society 41:413–18.

Neel, J. K. 1948. A limnological investigation of the psammon in Douglas Lake, Michigan, with especial reference to shoal and shoreline dynamics. Transactions of the American Microscopical Society 67:1–53.

Noodt, W. 1954. Die Verbreitung des Genus *Parastenocaris* ein Beispiel einer subterranen Crustaceen-Gruppe. Verhandlungen der Deutschen zoologischen Gesellschaft 19: 429–35.

O'Connor, J. F. 1974. An apparatus for sampling gravel substrate in streams. Limnology and Oceanography 19:1007–11.

Orghidan, T. 1959. Ein neuer Lebensraum des unterirdischen Wassers: Der hyporheische Biotop. Archiv für Hydrobiologie 55:392–414.

Paterson, C. G. and C. H. Fernando. 1969. The effect of winter drainage on reservoir benthic fauna. Canadian Journal of Zoology 47:589–95.

Pennak, R. W. 1940. Ecology of the microscopic metazoa inhabiting the sandy beaches of some Wisconsin lakes. Ecological Monographs 10:537–615.

Piearce, T. G. and M. Cox. 1977. The distribution of unpigmented and pigmented *Gammarus pulex* L. in two streams in northern England. Naturalist (Hull) 102:21–23.

Pieper, H. G. 1978. Ökophysiologische und produktionsbiologische Untersuchungen an Jugendstadien von *Gammarus fossarum* Koch 1835. Archiv für Hydrobiologie Supplement 54:257–327.

Pollard, R. A. 1955. Measuring seepage through salmon spawning gravel. Journal of the Fisheries Research Board of Canada 12:706–41.

Poole, W. C. and K. W. Stewart. 1976. The vertical distribution of macrobenthos within the substratum of the Brazos River, Texas. Hydrobiologia 50:151–60.

Racovitza, E. G. 1907. Essai sur les problèmes biospéléologiques. Biospeologica. I. Archives de zoologie expérimentale et générale 6:371–488.

Rosenberg, D. M. and V. H. Resh. 1982. The use of artificial substrates in the study of freshwater benthic macroinvertebrates, pp. 175–235. *In*: J. Cairns, Jr. (ed.). Artificial substrates. Ann Arbor Science Publishers, Inc. Ann Arbor, MI. 279 pp.

Rosenberg, D. M. and A. P. Wiens. 1978. Effects of sediment addition on macrobenthic invertebrates in a northern Canadian river. Water Research 12:753–63.

Sassuchin, D. N., N. M. Kabanov, and K. Neiswestnova-Shadina. 1927. Über die mikroskopische Pflanzen- und Tierwelt der Sandfläche des Okaufers bei Murom. Russische Hydrobiologische Zeitschrift Saratow 6:59–83.

Schwoerbel, J. 1961. Über die Lebensbedingungen und die Besiedlung des hyporheischen Lebensraumes. Archiv für Hydrobiologie Supplement 25:182–214.

Schwoerbel, J. 1964. Die Bedeutung des Hyporheals für die benthische Lebensgemeinschaft der Fliessgewässer. Internationale Vereinigung für Theoretische and Angewandte Limnologie Verhandlungen 15:215–26.

Schwoerbel, J. 1967. Das hyporheische Interstitial als Grenzbiotop zwischen oberirdischem und subterranem Ökosystem und seine Bedeutung für die Primär-Evolution von Kleinisthöhlenbewohnern. Archiv für Hydrobiologie Supplement 33:1–62.

Schwoerbel, J. 1970. Methods of hydrobiology. Pergamon Press, Oxford, England. 200 pp.

Sheridan, W. L. 1962. Waterflow through a salmon spawning riffle in southeastern Alaska. U.S. Fish and Wildlife Service Special Scientific Report on Fisheries 407:1–20.

Stanford, J. A. and A. R. Gaufin. 1974. Hyporheic communities of two Montana rivers. Science 185:700–2.

Stocker, Z. S. J. and D. D. Williams. 1972. A freezing core method for describing the vertical distribution of sediments in a stream bed. Limnology and Oceanography 17: 136–38.

Tavcar, V. and M. Mestrov. 1970. Larvae of Chironomidae in some flowing waters and hyporheic in Yugoslavia. Ekologiya 5:185–216.

Tsuda, M. 1966. Occurrence of young larvae of *Ephemerella* in sandy bottom of streams. Japanese Journal of Limnology 27:91–93 (In Japanese).

Vandel, A. 1965. Biospeleology. The biology of cavernicolous animals. Pergamon Press, London. 524 pp.

Vaux, W. G. 1966. Intragravel flow and interchange of water in a streambed. U.S. Department of the Interior, Fish and Wildlife Bulletin 66:479–89.

Viré, A. 1904. La biospéléologie. Comptes rendus hebdomadaires des séances de l'Académie des sciences 139:992–95.

Wallace, R. R., A. S. West, A. E. R. Downes, and H. B. N. Hynes. 1973. The effects of experimental blackfly (Diptera: Simuliidae) larviciding with abate, dursban and methoxychlor on stream invertebrates. The Canadian Entomologist 105:817–31.

Waters, T. F. and R. J. Knapp. 1961. An improved stream bottom fauna sampler. Transactions of the American Fisheries Society 90:225–26.

Wetzel, R. G. 1975. Limnology. W.B. Saunders Co., Toronto. 743 pp.

Wiley, M. J. 1981. An analysis of some factors influencing the successful penetration of sediment by chironomid larvae. Oikos 36:296–302.

Williams, D. D. 1981. Migrations and distributions of stream benthos, pp. 155–207. *In*: M. A. Lock and D. D. Williams (eds.). Perspectives in running water ecology. Plenum Press, New York, NY. 430 pp.

Williams, D. D. and H. B. N. Hynes. 1974. The occurrence of benthos deep in the substratum of a stream. Freshwater Biology 4:233–56.

Williams, D. D. and H. B. N. Hynes. 1976a. The ecology of temporary streams. I. The faunas of two Canadian streams. Internationale Revue der gesamten Hydrobiologie 61:761–87.

Williams, D. D. and H. B. N. Hynes. 1976b. The recolonization mechanisms of stream benthos. Oikos 27:265–72.

Williams, D. D. and H. B. N. Hynes. 1977. The ecology of temporary streams. II. General remarks on temporary streams. Internationale Revue der gesamten Hydrobiologie 62:53–61.

Wiszniewski, J. 1934. Remarques sur les conditions de la vie du psammon lacustre. Internationale Vereinigung für Theoretische und Angewandte Limnologie Verhandlungen 6:263–74.

chapter 15

INSECTS OF EXTREMELY SMALL AND EXTREMELY LARGE AQUATIC HABITATS

David R. Barton
Stephen M. Smith

"Extreme habitats are always instructive as to the ecological potentiality of a group" (Brundin 1966, p. 97).

INTRODUCTION

Each of the environmental parameters that combine to make up an aquatic habitat occurs over some range, such as from small to large or from low to high. This applies to virtually every feature, from temperature and environmental chemistry to temporal and physical stability. From this perspective, any single body of water is, strictly speaking, unique. In practical terms, however, most habitats share enough characteristics that we can generalize and speak with a degree of common understanding about what is meant by a river, stream, lake, or pond (i.e. the "typical" aquatic habitats). Such habitats have attracted the attention of most aquatic ecologists and it is the results of studies from these types of habitats that comprise the topics discussed in most of this book. If we consider that such places (lakes, ponds, etc.) are the "typical" habitats for aquatic insects, then this chapter deals with some features of the insect fauna of certain kinds of "atypical" aquatic habitats (i.e. those that lie at the extremes of one or more environmental variables).

There is an enormous variety of these habitats, just as there is an enormous variety of environmental parameters. Naturally, the most atypical habitats, such as hot springs and subterranean waters, are rarely encountered unless specifically searched for; others, such as the sea, could be considered atypical since insects are only a small part of the total fauna.

There is a vast but uneven literature dealing with the insect fauna of atypical aquatic habitats, including some excellent reviews (e.g. caves: Howarth 1983; high-altitude waters: Mani 1968; Brodsky 1980; thermal waters: Wirth and Mathis 1979; marine habitats: Cheng 1976; inland saline waters: Bayly 1972; Krebs 1982; temporary waters: Williams and Hynes 1976; Wiggins et al. 1980;

plant-container habitats: Seifert 1982) and regional treatments such as those of Bayly and Williams (1973) and Beadle (1981). There are, additionally, some habitats at the interface between land and water, such as splash zones, seepage areas on rock faces, and possibly even perpetually wetted vegetation in montane cloud forests, which, except for passing mention (e.g. Brodsky 1980), have been essentially ignored. Rather than attempt brief reviews of this multitude of atypical habitats, we will confine our discussion to examples that illustrate the influence of a single environmental factor, physical size, on species and communities of insects. The specific habitats we will consider are container habitats at the small end of the size spectrum, and very large lakes and rivers at the other.

CONTAINER HABITATS

A large variety of containers of plant, animal, and human origin serve as habitats for the immature and adult stages of a variety of aquatic insects. Plant-container habitats (termed *phytotelmata* by Varga 1928) include tree holes, leaf axils (e.g. as found in the Araceae, Bromeliaceae, and Pandanaceae), the internodes of monocotyledonous plants such as bamboo, cavities in fungi, monkey-discarded fruit husks (especially of the Apocynaceae and Loganiaceae), and the leaves of the insectivorous pitcher plants (especially of the genera *Sarracenia* and *Nepenthes*). Container habitats of animal origin range from empty snail shells (especially of large snails such as *Achatina*), to the living bodies of such animals as starfish that sometimes serve as oviposition sites for caddisflies (Winterbourn and Anderson 1980). Artificial container habitats may be thought of as analogs of the natural habitats and include tin cans, jars, water barrels, discarded tires, steps carved in trees, bird baths, roof gutters, and funeral urns. Artificial container habitats usually contain a subset of the fauna of natural habitats and are often of importance from a medical/veterinary perspective because they serve as breeding sites for mosquitoes that are vectors of disease (e.g. yellow fever, St. Louis encephalitis). Resh and Grodhaus (1983) have recently reviewed the aquatic insect fauna of such artificial containers; these habitats will not be treated further in this chapter.

All naturally occurring container habitats share a number of important features that can have profound effects on the fauna occupying them. At the same time, some of these features have made such habitats attractive to experimentally inclined ecologists. For example, container habitats are small, discrete, patchily distributed microhabitats. The patchy distribution and small size necessitate effective dispersal strategies and oviposition behaviors that will avoid over-crowding the tiny habitats. Their availability may change temporally, either seasonally or capriciously, depending on the weather. Their small size makes them vulnerable to rapid daily and seasonal variations in physical-chemical characteristics, and the insect fauna occupying them must therefore be adapted to an environment that may change considerably over short periods of time. At the same time, however, the small size of containers makes it possible to undertake a complete census of the limnetic fauna, making the habitats attractive for demographic and community studies.

How small can container habitats be? In Tanzania, the small pools of water contained among the bracts of pineapple plants are usually less than 10 ml in volume, and the catchments provided by crudely cut steps in coconut trees are less than 50 ml in volume, yet both habitats are extensively utilized by the mosquitoes *Aedes simpsoni* and *Toxorhynchites brevipalpis* (Smith, unpublished data). Clearly, the small size of such habitats will make them liable to desiccation. They are also susceptible to large-scale fluctuations in the water conditions as might occur, for example, should a single large leaf fall onto the water surface and impede gas exchange, or should a single large insect die and decay.

Rohnert (1951) subdivided the tree-hole fauna into three groups, based mainly on the nature of the temporal and ecological relationships of the fauna and habitat. Her classification is readily extended to all container habitats:

Dendrolimnetoxene.   This category includes a fauna whose occurrence in tree holes is accidental. It includes many small forms such as rotifers and protozoans that have high rates of passive dispersal.

Dendrolimnetophile.   This category includes a variety of aquatic or semi-aquatic species that are found in container habitats as well as in other habitats such as ground pools, marshes, and the forest floor. It includes terrestrial isopods, pulmonate gastropods, and several families of Diptera.

Dendrolimnetobionten.   This category includes species that are obligate inhabitants of container habitats. A large variety of insects, especially Diptera and Coleoptera, are found in this group.

Rohnert's classification is a useful approach to the analysis of the container-habitat fauna, but generally, long-term rather than survey-type studies are required to place a taxon in one of the three categories.

Extensive, but not intensive, survey data are available for the faunas of both leaf-axil and tree-hole environments. Lengthy lists for tree-hole habitats are available in Thienemann (1934), Park et al. (1950), Rohnert (1951), Kitching (1971), Maguire (1971), and Teskey (1976). Lists of the fauna found in plant-axil habitats are available from Picado (1913), Thienemann (1932), Strenzke (1950), Beaver (1979), and Erber (1979). Only a small portion of the aquatic insect fauna is found in container habitats; certain groups are almost always lacking, notably the Ephemeroptera, Plecoptera, Orthoptera, Megaloptera, Neuroptera, and Trichoptera. Other orders common in lentic habitats, such as the Odonata and Hemiptera, are rare in container habitats. The two orders that have been most successful in the container environment are the Diptera and the Coleoptera. Usually, both plant-axil and tree-hole habitats are dominated by larvae and/or adults of either flies or beetles (Seifert 1980). Among the Diptera, families of the Nematocera predominate (as they do in all aquatic environments), whereas among the Coleoptera, the common aquatic groups are rare or absent. Instead, the niche is occupied by a number of interesting and abundant container-habitat groups, notably the families Scirtidae and Pselaphidae.

ECOLOGY OF CONTAINER-INHABITING INSECTS

The following observations suggest that the container-habitat environment is an ancient one for the class Insecta:

1. Much of the fauna comes from the more primitive lines within groups of aquatic insects, or the fauna is of tropical origin. This is particularly true of the Diptera; most of the successful families are drawn from the suborder Nematocera and many temperate-zone forms are derivatives of large, tropical groups. For example, *Wyeomyia smithii*, the North American pitcher-plant mosquito, is a member of the tribe Sabethini, which is a widely distributed and successful group in the tropics.
2. The habit of breeding in container habitats is cosmopolitan, and has evolved independently among many groups and in many different circumstances.
3. Although detailed ecological studies have been carried out for only a few members of the container-habitat fauna (and these mostly on Diptera), a number of similar morphological and behavioral trends are evident, which suggests a long history of association with the container environment.
4. In many cases, specificity for a certain type of container habitat is strikingly rigid. Some species, such as the mosquito *Wyeomyia smithii*, are found in only a single type of container habitat.

Life-history Adaptations

Among habitats that are spatially and temporally heterogeneous, we might expect adaptations such as phenotypic plasticity, iteroparity (repeated reproduction), dispersed oviposition, and genetic polymorphisms (Giesel 1976). Although few detailed studies are available, aspects of each of these predicted adaptational trends have been found among the container-habitat insect fauna.

Phenotypic plasticity is most marked in the ability of many larval stages to withstand lengthy periods of food scarcity (Beaver 1979; Moeur and Istock 1980) or temporary desiccation of the habitat. A study of the reproductive biology of the predatory, tree-hole mosquito *Toxorhynchites rutilus* (Watts and Smith 1978) revealed many features that are clearly adaptive to a predator that occupies spatially variable and temporally unpredictable habitats. Unlike other mosquitoes, *T. rutilus* (and other species of the genus) exhibits a precocious oogenesis so that females emerge from the pupal stage with their oocytes almost mature. In many insects, ovarian precocity is associated with short-lived adults, but in *T. rutilus* the adult life span is considerably longer than in other mosquitoes. Presumably, ovarian precocity insures that at least some eggs will be laid in the tree-hole environment from which the female emerged. In addition to ovarian precocity, the female of *T. rutilus* exhibits an interruptible oviposition behavior resulting in the deposition of only small clutches of eggs (1–7) in each habitat, presumably to prevent the over-utilization of resources in a site. Reproductive effort by *T. rutilus* females is continuous and asynchronous, rather than periodic and synchronous, which is the pattern found in most other Nematocera. This permits the female to have a reserve of eggs so that any sudden availability of habitats can be exploited. Studies of the reproductive biology of other container-inhabiting insects might reveal similar life-history modifications.

There are also interesting differences in the reproductive strategies of closely related species. Lamb and Smith (1980) investigated the reasons for the different egg sizes of two species of the Toxorhynchitini. They found that one species produces many clutches of small eggs, presumably in response to spatial heterogeneity of the habitats, whereas the other species produces a few clutches of large eggs, presumably in response to temporal heterogeneity of the food supply. Lamb and Smith speculated that differences in egg size (and hence, first-instar larval size) were also correlated with mean prey size, but none of their hypotheses has been tested in the field.

Moderate drought resistance is common among many species of the container-habitat environment, particularly in the egg stage (Buxton and Breland 1952; Lounibos 1980). For example, at Point Pelee, Ontario, the larvae of many Diptera crawl into the sediment when the free water in tree holes disappears, and they can survive there for long periods of time (Smith, unpublished data). The larvae of some container-habitat species in the north-temperate zone are resistant to freezing (Paterson 1971), but in some of these the resistance is rather mild, reflecting the tropical origin of many of the container species (Smith and Brust 1971).

Container Habitats as Islands

The patchy distribution and discreteness of the container environment have led several workers to explore the dynamics of community composition using island-biogeography theory (e.g. see Maguire 1971; Seifert 1975). In general, the theory applies quite well; species-area curves resemble those of islands rather than those of mainland areas, but distance relationships are less important. Even within a single type of container such as the inflorescences of *Heliconia*, a variety of habitats is provided by differences in age and size of the container, resulting in significant habitat differences in the community structure (Seifert 1975).

Species Interactions

In such small habitats as provided by containers, competitive interactions among and within the taxa might be anticipated. Indeed, instances of both scramble (Seifert and Seifert 1976) and contest competition (Forsyth and Robertson 1975) have been found. Since small numbers of detritivores or predators can radically alter the community structure of a container habitat (Maguire et al. 1968), resource limitations probably occur in these habitats. Likewise, cannibalism is common among some container-habitat predators (Lamb and Smith 1980), and other species exhibit facultative predation (Lounibos 1980). However, Seifert and Seifert (1976) found only low levels of direct interactions among the fauna of *Heliconia* bracts, and Seifert and Barrera (1981) concluded that competition was not a major factor in structuring leaf-axil communities. The analyses indicate that many container habitats are not equilibrium communities controlled by species interactions, but rather the composition of the communities is the result of processes such as dispersal and local extinction (Seifert and Seifert 1976). Since most container habitats are spatially and temporally varied, dispersal can be an important factor in maintaining inferior competitors in these communities (Seifert and Seifert 1976).

Resource Partitioning

Detailed ecological studies have revealed that the insect fauna of container habitats segregate within these habitats on both a microscale and a macroscale level. For example, there may be spatial and/or temporal separation of faunal elements based on the age of the habitat (Seifert 1980), or even spatial separation within a habitat. The latter occurs between the mosquito *Wyeomyia smithii*, which occupies the free-water areas of *Sarracenia purpurea* leaves, and the chironomid *Metriocnemus knabi*, which occupies the sediment region. On a macroscale level, a rigid separation of habitats usually results from oviposition preferences or phenological differences (Lounibos 1980, 1981). Segregation of species by growth stages may also occur, and it is often achieved by variable patterns of egg hatching that desynchronize larval cohorts in the same container. A common mechanism to achieve this end is the plasticity or polymorphism of hatching responses, as is seen among the eggs of many mosquito species (Buxton and Breland 1952).

Methods for Studying the Fauna of Container Habitats

Most container habitats are so small that no particularly novel or complex collecting techniques are required—whether the sample is to be used for faunal surveys or for chemical analysis. Indeed, access to the habitat is often more difficult than collecting from the container itself as, for example, with container habitats that are formed by epiphytic plants high in the canopy of tropical forests.

If access to the habitat is possible, samples of the free water can be removed with a variety of simple devices such as pipettes or small hand-operated dippers. For tree holes that are in deep cavities in trees, a siphon will successfully sample the free water.

Water removed by pipette or siphon will usually not contain a representative sample of the insect fauna that is normally found in the detritus at the bottom of container habitats. In the case of tree holes, samples of the sediment can be removed by a small scoop or spoon but, in the case of plant-container habitats, reliable sampling of the sediment regions may require the destruction of the habitat. For example, the larvae of the chironomid *Metriocnemus knabi* live in the narrowed petiole of the leaves of the pitcher plant. Extraction of the water from the leaf will often not remove these larvae.

The small size of many container habitats makes it possible to census the entire fauna rather than attempting to take a representative sample of it. This is most easily done by repeatedly flushing the habitat with fresh quantities of water. This is necessary because some of the fauna may cling to the bottom or sides of crevices within the tortuous geometry of many container habitats.

In sampling adult insects from container habitats, an aspirator can be used to collect recently emerged or ovipositing adults, but more representative collections can be obtained by enclosing the entire habitat (or at least its openings) by netting to form an emergence trap (e.g. see Yates 1974). The awkward morphology of many of the habitats makes a standard design impracticable; it is usually sufficient to adapt a make-shift arrangement of netting and wire supports.

Supplementary faunal surveys and data on dispersal patterns can be obtained by the use of artificial container habitats. These need not be complex: in the case of tree holes, sections of bamboo (Corbet 1964) or black-painted jars or tins often suffice; in the case of plant-container habitats, suitably colored analogs of the leaves are easily constructed.

EXTREMELY LARGE LAKES

It is well known that the number of species of benthic invertebrates, especially insects, tends to decrease with increasing distance from the shore of a lake (Brinkhurst 1974). There are two main reasons why this should be so. First, the variety of microhabitats is greatest in shallow water where wave action and plant growth produce a more heterogeneous environment than in the deeper, darker, offshore regions. Second, very few aquatic insects have become completely adapted to life in the water; nearly all retain a terrestrial or aerial stage during which reproduction occurs. The retention of this ancestral trait may be because most freshwater habitats are relatively small, and adults emerging from streams, rivers, and most lakes are rarely far from land. Those emerging from the surface of very large lakes may be tens of kilometers from the shore (e.g. as in the center of Lake Superior in North America, or Lake Victoria in Africa). With only a few exceptions, such lakes are geologically ephemeral; therefore, there have been correspondingly few opportunities for the development of evolutionary adaptations to conditions in remote offshore areas.

A satisfactory definition of a "very large lake" is elusive, if not impossible. Standing waters vary in surface area from a few cm^2 to many thousands of km^2, and there is really no clear demarcation among puddles, ponds, lakes, and very large lakes. The latter are often deep (e.g. Lake Baikal in Russia has an area of about 31,500 km^2 with a maximum depth of 1,741 m), but not always (e.g. the area of Lake Chad in Africa has recently fluctuated between 10,000 and 25,000 km^2 with a maximum depth of about 12 m). The exposed shores of very large lakes typically lack rooted aquatic vegetation because of the scouring action of waves breaking in shallow water, but many small lakes also lack a macrophytic flora for a variety of reasons (e.g. extreme oligotrophy or lack of suitable sediments).

The more-or-less continuous disturbance of the near-shore zone by waves is the most distinctive feature of very large lakes, so perhaps the best definition of this habitat is one based on the biological consequences of wave action; very large lakes are those in which net-spinning caddisflies of the family Hydropsychidae are a regular component of the fauna. These insects require moving water to operate their feeding nets, and sufficiently continuous movement of water occurs in lakes that are large enough to generate "lake breezes" due to temperature differences between the water and surrounding land. The minimum diameter of such lakes seems to be about 20 km (e.g. Lake Simcoe, Ontario: Rawson 1930). Such lakes are found throughout the world, as shown by Herdendorf's (1982) list of 153 freshwater lakes of this size.

Continuous wave action and the resulting high degree of oxygen saturation in the near-shore zone are the only physical or chemical features common to all very large lakes. Surface temperatures in the open waters range from 24 to 26°C at the equator (Lake Victoria in Africa: Beadle 1981) to 0 to 6°C in the far north (Great Bear Lake, Northwest Territories, Canada: Johnson 1975). Bottom temperatures are somewhat less variable, being generally near 4°C in temperate and northern lakes and in the low 20s in the tropics. Thermal stratification develops seasonally in most large lakes, but can vary in both strength and duration. For example, Lake Tanganyika in Africa is meromictic, Lake Chad stratifies only during periods of exceptional calm (Beadle 1981), and, in some years, Great Bear Lake does not stratify at all (Johnson 1975). Chemically, very large lakes are even more diverse (e.g. see Kozhov 1963; Chandler 1964; Beadle 1981; Johnson 1975).

Despite these differences in climatic, physical, and chemical conditions, very large lakes all share two unique habitats: their remote offshore regions, and their wave-swept surf zones. The composition of the fauna in each of these habitats appears to be influenced more by physical factors that are related to the lake's size, wave action, or distance from shore, than by the lake's water chemistry.

Offshore Areas of Large Lakes

Independence from land would be the logical end point in adaptation to life in remote offshore regions. This seems to have been achieved by only a few species of chaoborids and chironomids inhabiting ancient African lakes such as Victoria, Edward, and Malawi (Beadle 1981). MacDonald's (1956) observations of swarms of these "lake flies" on Lake Victoria suggest that adult emergence is linked to the phase of the moon, with abundance peaks occurring a few days after the new moon. Mating and oviposition occur over the open lake, and swarms that are blown ashore are presumably lost. Synchronized emergence increases the probability of successful mating during the short time that the flies can remain in flight over the water. In the absence of the visual cues used by species that swarm over land, orientation of the mating swarm is likely achieved by auditory cues (MacDonald 1956); one could imagine that a cloud of midges that is 50 m or more in height and up to 1 km in length could be quite loud indeed! Swarm formation is also enhanced by the participation of several species at one time, although only one or two species of *Chaoborus, Chironomus, Tanypus,* and *Procladius* usually predominate.

A further step toward independence from land would be the development of a fully aquatic life cycle. This has been achieved by many species of Hemiptera and some Coleoptera, but these have not been recorded offshore in large lakes. However, five species of *Halobates* (Hemiptera) complete their life cycles on the surface of the open ocean (Andersen and Polhemus 1976). There is also a report of apterous adults of a capniid stonefly collected from depths of 60 to 80 m in Lake Tahoe, California and Nevada (Jewett 1963). Both adults and nymphs of *Capnia tahoensis* were found in association with *Chara* and aquatic moss in this exceptionally transparent lake. The only other examples of apterous, aquatic insect adults are the flightless caddisflies recorded from Lakes Baikal (Kozhov 1963), Tanganyika, and Titicaca (Marlier 1962; Beadle 1981). Best known are the Baicalinini (= Thamastini) of the family Limnephilidae, which are characterized by strong sexual dimorphism, primitive genitalic anatomy, reduced wings in some species, reduced larval gills, large size, and long (2–3 year) life cycles (Martynov, in Kozhov 1963). Completely flightless species include *Thamastes dipterus* and *Baicalina reducta*, both of which have reduced hind wings and the mesotarsi and metatarsi modified for swimming. These species do not fly at all; mating and oviposition occur in the water over the larval habitat, which is the rocky substrata at depths of 2 to 4 m. The larvae of *Baicalina bellicosa* also live in the rocky littoral. The adults attempt to fly but usually fail; they then swim to shore where eggs are deposited on exposed rocks that are covered by the annual rise in lake level later in the summer (Kozhov 1963).

The possibility that fully aquatic adults of other insect species exist in deep water of other large lakes cannot be dismissed since these animals would probably live on hard, stony substrata that are virtually impossible to sample with conventional equipment. Larval mayflies, stoneflies, and caddisflies have been taken in dredges and found clinging to gill nets that were set in deep-water areas (32–100 m) in Lake Superior that had a steep and rocky bottom (Selgeby 1974). In contrast to Miron's (1973) observations that sudden increases in atmospheric pressure had adverse effects on the nymphs of the stonefly *Perla,* these records suggest that depth is not necessarily a limiting factor in the distribution of lentic insects.

In contrast to Lake Baikal and the large African lakes mentioned above, most northern and temperate lakes are geologically very young, and specialized forms of aquatic insects are correspondingly rare. In these lakes, the offshore insect fauna consists almost entirely of Chironomidae. Various species of *Heterotrissocladius* and *Monodiamesa* characterize oligotrophic basins, with each species exhibiting somewhat different depth and substrate preferences (Saether 1973, 1975). Species of the tribe Chironomini, especially *Chironomus* spp., generally inhabit warmer, more eutrophic lakes (e.g. see Hilsenhoff 1967). None of the species in either group appears to be restricted to large lakes.

Endemic offshore species of midges do occur in the very ancient Lake Baikal. At depths exceeding 20 to 25 m, the chironomid fauna consists of six endemic species, five of them members of the genus *Sergentia* (= *Phaenopsectra*) (Linevich 1971a). These species apparently partition this habitat by depth and substratum. For example, the deepest-dwelling midge larva, the large (12–20 mm) *S. koschowi,* lives in the fine abyssal ooze at depths of 200 to 1360 m. The behavior of the adults has not attracted special comment (Kozhov 1963; Linevich 1971a), so presumably they swarm over or near the shore in the usual way. Perhaps because this lake is relatively narrow, imagos are seldom likely to emerge far from land; thus there has been little pressure favoring mating over water.

Larval behavior seems to be the key to the success of chironomids in offshore regions of large lakes, especially outside the tropics. First-instar larvae of most, if not all, lentic chironomids are capable of planktonic existence (Lellák 1968; Oliver 1971; Davies 1974), which allows them to be transported throughout the lake by currents that result from wind and thermal processes (see review by Barton 1981). The tendency to reenter the water column persists in later instars (Hilsenhoff 1966; Davies 1974), and distinct migrations to shallow water just prior to emergence have been noted (Kraemer 1980). The small populations, or complete absence, of chironomids in the profundal zone of many deep lakes may be more a reflection of the lake's weak internal circulation than the strictly biological limitations of the insects (Barton 1981).

Methods for Studying the Fauna of Offshore Areas of Large Lakes

Most of the standard samplers for collecting invertebrates from soft lacustrine sediments can be used to collect insects from offshore regions of very large lakes. (One simply needs more rope!) This is not to suggest, of course, that each of the various grabs, corers, and dredges will provide the same, or even readily comparable, results. Since Elliott and Tullett (1978) and Rosenberg (1978) have recently compiled complete bibliographies of samplers and efficiency comparisons, we will emphasize those factors unique to offshore regions of very large lakes that might influence the choice of sampling device used.

Any sampler lowered to the bottom in deep water may be deflected from a vertical descent by currents or drifting of the boat. Such deflections will result in poor penetration of the substratum if the sampler strikes the bottom at an angle. Even if adequate penetration is achieved, drifting of the line or the boat during the descent of a messenger may move the sampler before it is closed. Rawson (1947) recognized these problems and added righting flanges and an automatic closing mechanism to the design of his Ekman grab. Self-triggering mechanisms are essential, and the most commonly used samplers (e.g. Peterson, Ponar, and Shipek grabs, K-B corer) have them. The problem of nonperpendicular contact is minimized by the use of heavy grabs and samplers with streamlined shapes (e.g. many corers) or a righting frame (e.g. large box corers). Each of these is large and heavy, thus necessitating a fairly large boat as a working platform. If only a small craft is available, an Ekman-type grab with a righting frame and the top lid modifications described by Burton and Flannagan (1973) is probably the most efficient sampler.

Even a well-designed sampler must be dropped very rapidly during the last few metres above the bottom to insure vertical contact and proper depth of penetration. Unfortunately, this maximizes the pressure-wave effect (see Brinkhurst 1974, for discussion), which may displace many surficial animals. This results not only in an underestimate of faunal abundance, but also a biased assessment of qualitative composition. The problem is minimized, but not eliminated, by a design that offers minimal resistance to the passage of water through the sampling chamber. Selgeby's (1974) report of Plecoptera and Trichoptera larvae at depths of 32 to 100 m in Lake Superior suggests the possibility that active, crawling species may be much more common at great depths than the results of grab sampling would indicate.

The situation in deep lakes may be analogous to the deep sea where large individual animals account for a large fraction of the total standing stock, despite their low numbers. Deep-sea techniques such as dredging or photography might be very useful in this regard, and direct observations from a deep-diving submersible would be even better. A small submarine might also be ideal for collecting truly undisturbed core samples.

As discussed elsewhere in this chapter, it is unlikely that many insects emerge directly from remote offshore regions. Drifting or planktonic larvae and pupae can be collected in plankton nets (e.g. see Mundie 1959; Wiley and Mozley 1978) or in traps (Davies 1976). Traps have the advantage of sampling continuously over long periods of time, whereas plankton tows provide information about short-term periods of activity. Dredging could be used to document migrations of crawling or sporadically swimming species.

Wave-swept Shores of Large Lakes

The principal requirement for existence in the exposed, shallow, littoral zone of very large lakes is a means of resisting the erosive forces of breaking waves. Many lotic insects, with adaptations such as streamlined or flattened shapes, well-developed tarsal claws, use of silken anchor threads, and cryptic habits, are well prepared for maintaining their position under moving water, and thus form the bulk of the wave-zone community. Characteristic forms include the hydropsychid caddisflies mentioned earlier, heptageneid and baetid mayflies, a variety of stoneflies, and dryopoid beetles (Rawson 1930; Krecker and Lancaster 1933; Fryer 1959; Barton and Hynes 1978a). The composition of the insect fauna of the wave-washed shores of very large lakes is strikingly similar throughout the world, with the notable exception of Lake Baikal. To appreciate the unique character of that ancient lake, let us first consider the shallow-water insect communities of a few other large lakes that have been examined in detail.

Fryer (1959) recognized three distinct types of shore in the African Lake Nyasa (now Lake Malawi): boulder shores, bare sand, and sandy beaches with scattered beds of tape-grass, *Vallisneria*. Boulder shores supported the densest populations and greatest variety of invertebrates, including several species of mayflies, dragonflies, caddisflies, beetles, craneflies, and midges. The presence of Heptageniidae, Hydropsychidae, and Psephenidae emphasized the lotic character of the habitat. The only insects on bare sand were chironomid midges. In beds of *Vallisneria*, several mayflies, odonates, and caddisflies were also present.

A similar list of insect families was reported from Lake Simcoe, Ontario, by Rawson (1930). Exposed rocky shores supported 14 species of Ephemeroptera, one Plecoptera, two Odonata, one Neuroptera, eight Trichoptera, one Coleoptera, and several chironomid species. Sandy beaches were inhabited by three species of mayflies, seven caddisflies, and several chironomid species. The distinctive indicators of continuous wave action, two species of hydropsychids and one species of psephenid, were found at depths of 0 to 2 m.

Many of the earliest studies of the St. Lawrence Great Lakes were undertaken from the Franz Theodore Stone Laboratory on South Bass Island in western Lake Erie. Krecker and Lancaster (1933) studied the animals on six types of substratum in the wave zone (0–1.9 m), including sand, gravel, shelving rock, clay, flat rubble, and angular rubble. Their list of taxa was similar to Rawson's (1930) from Lake Simcoe. The variety and abundance of animals was directly related to the stability and complexity of the substratum: sand and gravel had low-density populations of only a few taxa, whereas angular rubble had the richest fauna. Shelving rock supported a large number of individuals but only a few species. Shelford and Boesel (1942) recognized four communities in western Lake Erie, including a *Goniobasis* snail-*Hydropsyche* caddisfly community on rocky substrata at depths of 0 to 8 m. Other abundant insects in this community were *Tanytarsus, Stenelmis crenata, Stenonema* sp., *Psephenus,* Leptoceridae, and *Helicopsyche.*

In a survey of the exposed Canadian shores of the Great Lakes, Barton and Hynes (1978a) found 208 insect taxa, including 40 species of Ephemeroptera, 10 Plecoptera, 44 Trichoptera, and 82 Chironomidae. As in the studies mentioned above, there was a strong direct relationship between stability and complexity of the substratum, and the variety of invertebrates collected. Mayflies and caddisflies were major components of the fauna in all lakes except Lake Ontario, where a total of only seven specimens of heptageneid mayflies and fewer than 40 hydropsychid caddisflies were taken in all collections combined. Sandy substrata were inhabited by some of the same species from the *Harnischia*-complex of midges that are characteristic of unstable sand in large rivers.

It is apparent from the aforementioned examples that a large variety of insects, particularly mayflies and caddisflies, is a regular feature of the invertebrate communities of wave-swept shores of very large lakes. The species living there are a subset of the rheophilic fauna of the region. This is true for at least some ancient tropical lakes, as well as for relatively young temperate lakes. But what about an ancient temperate lake, such as Lake Baikal?

According to Kozhov (1963), representatives of Ephemeroptera, Odonata, Hemiptera, and Coleoptera do not live on the exposed shores of Lake Baikal, but are found only in sheltered gulfs and bays. Two species of stoneflies occur only in the vicinity of the outfall of the Angara River. Adults of 50 species of Trichoptera have been collected around the shore, but the larvae of only 14 species, all Limnephilidae, inhabit the main basin of the lake. The chironomid fauna seems to be similarly impoverished, with only 22 species having been found in the main basin at all depths (Linevich 1971a). Most of the littoral species in the lake also inhabit streams in the region (Linevich 1971b). Kozhov (1963) suggested that low temperature was the primary factor that favored the development and persistence of an endemic fauna over more widely distributed species of Insecta. He emphasized that water temperatures seldom exceed 12 to 13°C at depths of 3 to 5 m, and are that high for only 10 to 15 days in August. Furthermore, most of the lake has a very narrow littoral region, and frequent upwellings reduce the temperature to 4 to 5°C.

It is unlikely that low temperatures alone would have such a limiting effect on the insect fauna. The annual range of temperature in the littoral zone of Lake Baikal is nearly identical to that of Lake Superior, but Lake Superior has a very diverse insect fauna (Thomas 1966; Barton and Hynes 1978a). Thermal shock resulting from upwellings of cold hypolimnetic water can be fatal to many benthic species (Emery 1970), and Barton and Hynes (1978a) suggested that this might at least partially explain the low diversity of insects along the north shore of Lake Ontario. Total invertebrate standing stocks in shallow-water areas of the Laurentian Great Lakes that lack substantial numbers of mayflies and caddisflies, such as Lake Ontario and offshore shoals in the other lakes (Barton, unpublished data), do not appear to be lower than those at sites where the fauna is more diverse. Amphipods, and to a lesser extent isopods, make up the standing-stock difference in these situations. This suggests that competition may be occurring between Amphipoda and littoral insects.

Of the lakes just discussed, Lake Malawi has no amphipods (a group that is absent from all of tropical Africa: Beadle 1981), and Lake Simcoe has two of the seven amphipod species known from the St. Lawrence Great Lakes; however, Lake Baikal has 240 species with littoral populations reaching 30,000 individuals m^{-2} (Kozhov 1963)! With the possible exception of *Pontoporeia hoyi*, the other amphipods of the Laurentian Great Lakes are eurytopic omnivores. They occur in springs, ponds, and streams, as well as in large lakes. On the other hand, virtually all the baikalian species are endemic to the lake (although a few species have extended their range down the Angara River: Greze 1953), and include specialized herbivores, carnivores, and scavengers.

Although it is thought that Lake Baikal did not freeze completely during the Pleistocene glaciations, the near-shore zone was probably ice covered year round (Kozhov 1963). Freezing of the shallow substratum and reduction of water movements would have eliminated most of the rheophilic, littoral insect fauna. The Gammaridae and Baicalinini, which can complete their life cycles in or over deeper water, have survived. Subsequent postglacial invasion of the exposed littoral zone by rheophilic insects apparently met with limited success in the face of competition from the established amphipod community and the frequent upwellings of cold water that resulted from the generally colder climate.

Methods of Studying the Fauna of Wave-swept Shores

Collecting insects from the wave-swept shallows of very large lakes is complicated by the turbulence of passing or breaking waves, and the coarse texture of the substratum. These conditions usually preclude the use of a boat as a stable working platform, as well as grabs that are normally appropriate for lake work. Depth and the absence of a directed current make several standard lotic-sampling methods ineffective. The most successful collecting techniques, therefore, are combinations of stream and lake methods.

Hand picking of insects from stones is easily done while wading or snorkeling. Kick sampling, similar to the technique used in streams, is useful at wadeable depths but, since the direction and velocity of water movements change rapidly, it is necessary to work the net actively to catch dislodged animals rather than depending on the current to sweep them into the net. In deeper water, a portable airlift is very effective when used to "vacuum" all types of substratum.

Quantitative sampling is always more difficult. On sandy beaches an Ekman- or Peterson-type grab can be forced into the bottom (McKim 1962). Pole-mounted grabs might be successful in deeper water, especially if equipped with a lever or hydraulic system for closing the jaws (Slack 1972; W. Burton, personal communication). Coring tubes of metal or plexiglass are easier to drive into sand and can be used at any depth that can be reached by divers.

Normal grab samplers will not close on stones, even if they can be forced to penetrate rocky substrata. The inverted grab described by Dall (1981) may be used for gravel and shingle in shallow water, and a clumsier, but effective, technique was used by Barton and Hynes (1978b). In deeper water, SCUBA or other types of diving techniques seem to be the only satisfactory methods of taking quantitative samples from hard substrata, although the suction samplers (airlifts) designed by Pearson et al. (1973) and Verollet and Tachet (1978) would probably be effective for animals that are not firmly attached to the stones.

Thomas (1966) used SCUBA to embed metal trays in the bottom of selected bays in Lake Superior. Rotenone was then injected into the enclosed space, and the animals killed by the poison were collected 24 hours later. Other appropriate, diver-operated equipment includes the dome sampler of Gale and Thompson (1975) or a combination of a portable airlift and quadrat sampler used by Barton and Hynes (1978b).

EXTREMELY LARGE RIVERS

Shifting-sand Habitats in Large Rivers

Sand, which is composed of mineral particles that range in diameter from about 0.1 to 1.5 mm, is a component of the substratum of nearly every body of water. Sand grains settle out of the water column over a fairly narrow range of current velocities, but do not pack together as closely as do the finer silts and clays. Since sand grains have greater surface-to-volume ratios than coarser gravels and stones, they can be resuspended and moved by small increases in the speed of the current (Hynes 1970). Thus, sandy substrates are common, but are both temporally and spatially ephemeral in most running-water environments. Sandy substrates persist year-round only in low-gradient areas that typically occur in the lower reaches of large, unregulated rivers where continual fluctuations in discharge and turbulence keep the sand in almost continual motion, producing one of the most severe environments for benthic animals.

Extensive, shifting-sand habitats occur from the arctic to southern Africa, in rivers that vary over almost the complete range of chemical, physical, and climatic conditions. The only features that many of these rivers share are low levels of light penetration (usually due to high turbidity) and fluctuating discharge during much or all of the year. The interaction of these two factors is crucial since it prevents the periodic stabilization of sandy substrata that occurs in small streams as a result of algal or macrophyte growth during periods of low flow. Furthermore, shifting-sand deposits are usually isolated from one another by long expanses of dry ground or salt water. Despite these factors, the insect fauna of physically unstable, shifting sands is remarkably uniform, being characterized by species from only three orders: Ephemeroptera, Odonata, and Diptera.

Species of mayflies whose larvae live only in lotic, shifting sands are listed in Table 15.1. Most species are known from only a small number of localities. These apparently restricted distributions, and perhaps the small number of taxa in these genera, may simply be a reflection of the difficulty involved in collecting animals from shifting substrata located beneath several meters of flowing water, a problem compounded by the strong swimming abilities of many of these species. For example, nymphs of *Analetris* swim about as well as many species of small fish.

The ability to swim is an obvious advantage to an organism that is likely to be displaced from the substratum at unpredictable intervals by the vagaries of turbulence. This capacity is reflected morphologically in the streamlined shape and dorso-ventrally flattened abdomen of such nymphs.

The detritivore *Ametropus* typically lives just under the sand with only the eyes protruding (Clifford and Barton 1979), whereas the predators *Acantha-*

TABLE 15.1:   Genera of Psammophilic Ephemeroptera and Their Global Distributions

FAMILY	GENUS	DISTRIBUTION
Siphlonuridae	*Acanthametropus*	Holarctic
	Analetris	Nearctic
Ametropodidae	*Ametropus*	Holarctic
Baetidae	*Apobaetis*	Nearctic
	Paracloeodes	Neotropical
Oligoneuriidae	*Homoeoneuria*	Neotropical
Heptageniidae	*Pseudiron*	Nearctic
Behningiidae	*Behningia*	Palaearctic
	Dolania	Nearctic
	Protobehningia	Palaearctic

metropus, Analetris, and *Pseudiron* are believed to anchor themselves in the sand using their long, slender thoracic legs (Edmunds et al. 1976). *Apobaetis* probably lives in the shelter of sand ripples (Edmunds et al. 1976), darting among them in typical baetid fashion. Nonswimmers include *Homoeoneuria* and the Behningidae, which are not streamlined but burrow 2 to 30 cm below the surface of the sand where they feed on detritus or chironomids (Peters and Jones 1973; Edmunds et al. 1976). In addition to these morphological and behavioral adaptations, Clifford and Barton (1979) suggested that prolonged emergence periods may also be characteristic of univoltine mayflies in large rivers.

Among the Odonata, several of the Gomphidae (e.g. *Gomphus* sp., *Hagenius brevistylus, Ophiogomphus* sp., *Paragomphus cognatus,* and *Progomphus obscurus*) are consistently found in collections of animals from lotic sand (Walker 1958; Harrison 1959; Russev 1974; Peters and Peters 1977; Benke et al. 1979; Wells and Demas 1979; Barton 1980). All are predaceous, tend to burrow into the substratum, and are capable of high speed movement through the water, at least over short distances. While these features are shared with some of the psammophilic Ephemeroptera, they are also shared with most other Anisoptera, especially congeners with very different habitat preferences. In short, these species have no obvious morphological specializations for life in shifting sand. Perhaps this is the reason why Odonata are rarely found in habitats with the coarsest sand and that are exposed to the full force of the current.

Like the dragonflies just discussed, ceratopogonid larvae that are characteristic of lotic sand habitats do not appear to be any more specialized morphologically than species found in other habitats. With present limitations on larval taxonomy, it is impossible to assess substrate specificity at even the generic level, but larvae of the *Palpomyia-Bezzia* type are obviously well suited to life among sand grains because of their long, thin bodies, lack of protruding appendages, and predatory habits.

The importance of larval chironomids to benthic communities of sandy substrata has been recognized since the first intensive study of a large river system, the Volga (Behning 1928). Chernovskii (1949) recognized a number of species, all grouped in the genus *Cryptochironomus,* as being highly specialized psammorheobionts. These species are members of the *Harnischia*-complex that Saether (1977) and Jackson (1977) revised into 17 different genera. All larvae of this group seem to prefer sandy substrata, and five genera are known only from unstable sand—*Cyphomella, Chernovskiia, Beckidia, Robackia,* and *Saetheria.* At least two species of *Paracladopelma* are also strictly psammophilic (Saether 1979). Besides their extreme habitat specificity, these species all appear to have wide geographic distributions. For example, *Beckidia* is known from Russia, Africa, and North America (Saether 1977); *Saetheria tylus* occurs from the Mackenzie River to Florida in North America (Jackson 1977).

A number of other chironomid genera are frequently encountered in sandy habitats, including *Stictochironomus, Polypedilum, Corynoneura, Cordites,*

Monodiamesa, Cricotopus, and the *Parakiefferiella*-group. Of these, only some of the *Parakiefferiella*-group appear to be restricted to such unstable environments (Benke et al. 1979; Barton 1980; D. Soluk, unpublished data). The other genera are probably opportunistic forms that invade sandy habitats during periods of low discharge.

These psammophilic chironomid larvae share a number of morphological characteristics and habits that enable them to cope with their severe habitat (Chernovskii 1949): (1) carnivory; (2) elongated head and body with the segments sometimes subdivided, which allows greater flexibility; (3) long antennae and maxillary palpi with well-developed sensory organs; (4) posterior prolegs reduced and very slender. Several additional features—such as a thickened body wall that might be more resistant to abrasions, small size (e.g. some species of the *Parakiefferiella*-group, *Cordites*), and the long flexible antennae of at least some species, which may be used for locomotion via a kind of "swimming" through the sand (e.g. *Beckidia, Cordites*)—may also be involved (David R. Lenat, personal communication).

Importance of Psammophilic Insects

In contrast to the fauna of most other "atypical" habitats, insects are the only group of macroinvertebrates that have evolved specialized sand-dwelling forms. The fauna of shifting sand usually includes a small variety of other animals, but these are eurytopic species and their relative importance varies considerably from river to river. Amphipods and mysids are a major component of the benthic communities of some rivers such as the Volga (Zhadin 1956) and the Danube (Russev 1974). Oligochaetes are present in most sandy river beds, but there is considerable variation in the species present and their contribution to the total fauna (Zhadin 1956; Harrison 1959; Barton and Lock 1979; Benke et al. 1979; Barton 1980). Such variation seems to reflect differences in grain size of the sediments, the biogeographical province, and the techniques used in collecting and processing the samples.

Of these factors, the aperture of the sieves used to concentrate the animals is responsible for most of the variability among studies, both quantitatively and qualitatively. From Table 15.2 it is apparent that when fine sieves (100–200 μm apertures) are used, the numbers of organisms obtained are usually two to three orders of magnitude greater than results obtained with coarse sieves, but the biomass is not proportionally greater. This emphasizes the small size of most of the individual animals in shifting sand.

Low standing-stock biomass does not necessarily mean low productivity. Benke et al. (1979) found that the numerically dominant animal on sand in the Satilla River, Georgia, was a small chironomid referred to as a member of the *Parakiefferiella*-group. This species accounted for about 90 percent of the individuals and 50 percent of the biomass, but it dominated production in the

TABLE 15.2: **Estimates of Total Invertebrate Abundance and Biomass from Sand Habitats in Large Rivers**

RIVER	NUMBER m^{-2}	$g\ m^{-2}$	REFERENCE
A. Sieve apertures $\geq 500\ \mu$m:			
Volga[a]	30	ND[b]	Behning 1928
Missouri[c]	>1	>0.1	Berner 1951
Missouri	78	>0.1	Volesky 1969
Fraser[d]	19–979	0.7–8.6	Northcote et al. 1976
Danube[e]	17	0.2	Russev 1974
B. Sieve apertures $\leq 200\ \mu$m:			
Volga	up to 9500	1.4	Zhadin 1956
Angara[a]	14,066	2.2	Greze 1953
Great Berg[f]	7,749	ND	Harrison 1959
Athabasca[d]	1100–40,000	0.2	Barton and Lock 1979

[a]USSR
[b]No data
[c]USA
[d]Canada
[e]Bulgaria
[f]South Africa

sand due to its rapid turnover rate of about once per day. Noting that pupae were present virtually year-round, Benke et al. concluded that rapid turnover of small organisms at relatively high temperatures resulted in a moderately high total production from sand (13.6–28.3 g m^{-2}yr^{-1}). The *Parakiefferiella*-group contributed little to fish production directly, but may have been indirectly significant since drifting individuals were probably an important food of the larger filter-feeding insects that occurred on snags in the river.

Methods for Studying the Fauna of Lotic Sand Habitats

When compared with most other lotic habitats, lotic sands are relatively easy to sample where the water is shallow enough for wading. Quantitative collections can be made using scoop or shovel samplers (e.g. see Dittmar 1955; Prater et al. 1977; Tolkamp 1980), or by forcing an Ekman grab into the sand (Barton and Lock 1979). Coring with tubes that are stopped by hand before being withdrawn should also work well. In any case, animals are most easily separated from the sand by elutriation and concentrated with a sieve of appropriate mesh size. Sieves with very small apertures (100 μm or less) are necessary to retain many of the small species of Chironomidae.

Qualitative collections from shallow water can be obtained by using a kick net held at the level of the sand surface while disturbing the substratum upstream. For larger insects, such as Ephemeroptera and Odonata, which tend to occur in fairly low numbers, a net with meshes of at least 500 μm allows some sand grains to pass through, thus minimizing clogging. Nymphs of *Analetris*, which are capable of swimming very rapidly, seem to be caught only by dragging the net very quickly upstream (Lehmkuhl 1976; Barton 1980) or by herding them into a very large stationary net (Edmunds and Koss 1972).

It is much more difficult to sample in deeper water. The problem of deflection of the sampler by water currents is compounded by the resistance of coarse sand to penetration. Schräder (1932) suggested lowering an Ekman grab to just above the bottom, then allowing the boat to drift while the messenger and grab are dropped simultaneously, but we have had limited success with this technique in muddy sediments of the Athabasca River. Heavier grabs with automatic closing mechanisms reduce these problems. The "Okean" grab described by Lisitsyn and Udintsev (1955, in Elliott and Tullett 1978) may be very effective but a large research vessel is needed to handle its weight of 150 kg. Various dredges might also be effective but, again, weight may be a limiting factor.

From a smaller boat, both qualitative and quantitative collections can be made using an airlift. Mackey (1972) described a device made of plastic tubing that was comparable in efficiency to a Maitland corer on muddy sand, but more versatile in the range of substrata that it could sample. In cohesive sediments, the airlift samples an area the same size as the opening of the tube. For stronger currents and coarser sediments, plastic pipe can be replaced by stronger materials and sample size can be estimated from the volume of sand lifted to the surface (Barton 1980).

Artificial substrates may be of limited value. Although they have been used to collect invertebrates from large rivers (Rosenberg and Resh 1982), we note that when placed over shifting sand (e.g. see McCart et al. 1977), these devices do not collect psammophilic species.

CONCLUSIONS

Since "atypical" habitats, as we have defined them, lie at the extremes of some environmental factor, their faunas tend to reflect the limitations imposed on them by that dominant factor. In this light, such habitats can be viewed as natural experiments that demonstrate the role of individual habitat variables in shaping biological communities. In the preceding discussion, we have attempted to illustrate this point by citing examples of extremes in physical size. In concluding, we should emphasize the general implications of habitat size for the insects living there.

Small habitats are much more numerous than very large ones, but individually they are more unpredictable, being subject to short-term fluctuations in their capacity to support life. Large bodies of water are much rarer but they tend to be stable since their volumes buffer against change. The selective pressures influencing insects that exploit container habitats have been toward strong powers of dispersal, with oviposition at a number of sites to minimize the impact of random catastrophic events and to prevent overexploitation of local resources. For species inhabiting the offshore zone of very large lakes and unstable riverine sands, selection tends to be against dispersal since the probability of reaching another similar habitat is small. In contrast, there seems to be little selection for or against adult dispersal among species living in the wave-washed shallows of very large lakes because these habitats are directly connected to inflowing and outflowing streams. Here, the main pressure seems to have been on the aquatic stages, especially favoring those species that can cope with erratic fluctuations in turbulence.

As a result, it can be predicted that insect communities of individual container habitats will tend to be variable both temporally and between containers; random events will have profound effects on species composition. The remarkable similarity of insect communities in very large aquatic habitats suggests that the physical stresses associated with shifting sand, continuous but variable wave action, and long distances to land are far more important than local differences in climate and water chemistry.

The available literature, as just discussed, suggests that a survey of container habitats within any geographic province should reveal a large total number of species. In a given container, the combination of abiotic conditions and biotic interactions may limit diversity; however, because many containers of markedly different character are available, the total number of container species should be large. High overall insect diversity for very small habitats is a logical consequence of selection pressures favoring dispersed reproductive effort combined with the local and regional proximity of breeding sites.

Since large habitats tend to be isolated from one another, dispersal is maladaptive, and we should expect a high degree of endemicity in the faunas of large habitats. This seems to be the case for some very large lakes, at least very old ones such as Baikal, Victoria, and possibly Titicaca, but not for very large sandy rivers. The apparently global distributions of many psammophilic insects suggests that areas of shifting sand are not so completely isolated. Sandy shores of large lakes offer similar conditions and such lakes would have connected many rivers during periods of glaciation, permitting direct access from one river to another. In the absence of such direct connections, small patches of shifting sand in intervening streams may allow gradual dispersal, and perhaps effectively permanent contact, between large rivers, but the minimum area of suitable sand sufficient to maintain populations of true psammophiles is unknown.

Shifting-sand habitats have been greatly reduced in extent, or eliminated completely, by dams on many large rivers. Edmunds and Koss (1972) noted that all of the then-known habitats of *Analetris eximia* were threatened by dams or pollution. The extreme habitat specificity of strict psammophiles makes them particularly useful indicators of past drainage patterns (e.g. see Lehmkuhl 1976), so it might be profitable to assess their distributions much more completely before even more of their habitats are eliminated. In this regard, we are particularly ignorant of the fauna of South America.

To date, most of the work on insects of extremely small and large aquatic habitats has been in the form of surveys. We have made reference to the few available studies concerned with the dynamic aspects of community structure, and this should stress the need for much more work of that nature. This is not to suggest that additional faunal surveys are unnecessary. With only a few exceptions, reliable faunal lists for extreme habitats in most areas of the world are lacking, and it is certain that many of the taxa are yet to be described.

Integrated, long-term studies of the physical, chemical, and biological characteristics of such "atypical" (and, unfortunately, "typical" as well!) habitats are needed. For example, almost nothing is known of the variability of water chemistry within and among container habitats, especially over time. The physical and chemical aspects of a few very large lakes and rivers have been studied in detail, but biological data are conspicuously lacking. Even in relatively well-known areas, the life histories of many species of aquatic insects are poorly known. Even basic adaptations may be unknown; for example, what cues prompt chironomids to emerge from the profundal depths of large lakes where temperature and light are virtually uniform year-round?

It is clear that for many container-habitat species, dispersal is the *sine qua non* of continued survival. With the exception of a few studies of mosquitoes, nothing is known of the flight behavior, longevity, or range of the container-habitat fauna. This is particularly intriguing since dispersal from such patchily distributed habitats would seem to be maladaptive, yet individual species are often very widely distributed. Such questions offer interesting possibilities to future researchers of such habitats.

References

Andersen, N. M. and J. T. Polhemus. 1976. Water-striders (Hemiptera:Gerridae, Veliidae, etc.), pp. 187–224. *In*: L. Cheng (ed.). Marine insects. North-Holland Publishing Co., Amsterdam, The Netherlands. 581 pp.

Barton, D. R. 1980. Benthic macroinvertebrate communities of the Athabasca River near Ft. Mackay, Alberta. Hydrobiologia 74:151–60.

Barton, D. R. 1981. Effects of hydrodynamics on the distribution of lake benthos, pp. 251–63. *In*: M. A. Lock and D. D. Williams (eds.). Perspectives in running water ecology. Plenum Press, New York, NY. 430 pp.

Barton, D. R. and H. B. N. Hynes. 1978a. Wave-zone macrobenthos of the exposed Canadian shores of the St. Lawrence Great Lakes. Journal of Great Lakes Research 4:27–45.

Barton, D. R. and H. B. N. Hynes. 1978b. Seasonal study of the fauna of bedrock substrates in the wave zones of Lakes Huron and Erie. Canadian Journal of Zoology 56:48–54.

Barton, D. R. and M. A. Lock. 1979. Numerical abundance and biomass of bacteria, algae and macrobenthos of a large northern river, the Athabasca. Internationale Revue der gesamten Hydrobiologie 64:345–59.

Bayly, I. A. E. 1972. Salinity tolerance and osmotic behavior of animals in athalassic saline and marine hypersaline waters. Annual Review of Ecology and Systematics 3:233–68.

Bayly, I. A. E. and W. D. Williams. 1973. Inland waters and their ecology. Longman Australia Pty Ltd., Hawthorn, Australia. 316 pp.

Beaver, R. A. 1979. Biological studies of the fauna of pitcher plants (Nepenthes) in West Malaysia. Annales de la Société entomologique de France (New Series) 15:3–17.

Beadle, L. C. 1981. The inland waters of tropical Africa: an introduction to tropical limnology. 2nd ed. Longman Group Ltd., London. 475 pp.

Behning, A. 1928. Das Leben der Wolga, zugleich eine Einführung in die Flüssbiologie. Die Binnengewässer 5:1–162.

Benke, A. C., D. M. Gillespie, F. K. Parrish, T. C. Van Arsdall, Jr., R. J. Hunter, and R. L. Henry, III. 1979. Biological basis for assessing impacts of channel modifications: invertebrate production, drift, and fish feeding in a southeastern blackwater river. Environmental Resources Center Publication No. ERC 06-79, Georgia Institute of Technology, Atlanta, GA. 187 pp.

Berner, L. M. 1951. Limnology of the lower Missouri River. Ecology 32:1–12.

Brinkhurst, R. O. 1974. The benthos of lakes. Macmillan Press Ltd., London. 190 pp.

Brodsky, K. A. 1980. Mountain torrent of the Tien Shan: a faunistic-ecology essay. Monographiae Biologicae 39. Dr. W. Junk B. V. Publishers, The Hague, The Netherlands. 311 pp.

Brundin, L. 1966. Transantarctic relationships and their significance as evidenced by chironomid midges, with a monograph of the subfamilies Podonominae and Aphroteniinae and the Austral Heptagyiae. Kungliga Svenska Vetenskapsakademiens Handlingar, Fjärde Serien 11:1–472.

Burton, W. and J. F. Flannagan. 1973. An improved Ekman-type grab. Journal of the Fisheries Research Board of Canada 30:287–90.

Buxton, J. A. and O. P. Breland. 1952. Some species of mosquitoes reared from dry materials. Mosquito News 12:209–14.

Chandler, D. C. 1964. The St. Lawrence Great Lakes. Internationale Vereinigung für Theoretische und Angewandte Limnologie Verhandlungen 15:59–75.

Cheng, L. 1976. Marine insects. North-Holland Publishing Co., Amsterdam, The Netherlands. 581 pp.

Chernovskii, A. A. 1949. Identification of larvae of the midge family Tendipedidae. USSR Academy of Sciences, Moscow, Russia. Translation by National Lending Library for Science and Technology, Boston Spa, Yorkshire, England. 302 pp.

Clifford, H. F. and D. R. Barton. 1979. Observations on the biology of Ametropus neavei (Ephemeroptera: Ametropodidae) from a large river in northern Alberta, Canada. The Canadian Entomologist 111:855–58.

Corbet, P. S. 1964. Observations on mosquitoes ovipositing in small containers in Zika Forest, Uganda. Journal of Animal Ecology 33:141–64.

Dall, P. C. 1981. A new grab for the sampling of zoobenthos in the upper stony littoral zone. Archiv für Hydrobiologie 92:396–405.

Davies, B. R. 1974. The planktonic activity of larval Chironomidae in Loch Leven, Kinross. Proceedings of the Royal Society of Edinburgh, Series B 74:275–83.

Davies, B. R. 1976. A trap for capturing planktonic chironomid larvae. Freshwater Biology 6:373–80.

Dittmar, H. 1955. Die quantitative Analyse des Fliesswasser-Benthos. Anregungen zu ihrer methodischen Anwendung und ihre theoretische und praktische Bedeutung. Archiv für Hydrobiologie Supplement 22:295–300.

Edmunds, G. F., Jr. and R. W. Koss. 1972. A review of the Acanthametropodinae with a description of a new genus (Ephemeroptera: Siphlonuridae). Pan-Pacific Entomologist 48:136–44.

Edmunds, G. F., Jr., S. L. Jensen, and L. Berner. 1976. The mayflies of North and Central America. University of Minnesota Press, Minneapolis. 330 pp.

Elliott, J. M. and P. A. Tullett. 1978. A bibliography of samplers for benthic invertebrates. Freshwater Biological Association Occasional Publication No. 4. 61 pp.

Emery, A. R. 1970. Fish and crayfish mortalities due to an internal seiche in Georgian Bay, Lake Huron. Journal of the Fisheries Research Board of Canada 27:1165–68.

Erber, D. 1979. Untersuchungen zur Biozönose und Nekrozönose in Kannenpflanzen auf Sumatra. Archiv für Hyrobiologie 87:37–48.

Forsyth, A. B. and R. J. Robertson. 1975. K reproductive strategy and larval behavior of the pitcher plant sarcophagid fly, *Blaesoxipha fletcheri*. Canadian Journal of Zoology 53:174–79.

Fryer, G. 1959. The trophic interrelationships and ecology of some littoral communities of Lake Nyasa with especial reference to the fishes; and a discussion of the evolution of a group of rock-frequenting Cichlidae. Proceedings of the Zoological Society of London 132:153–281.

Gale, W. F. and J. D. Thompson. 1975. A suction sampler for quantitatively sampling benthos on rocky substrates in rivers. Transactions of the American Fisheries Society 104:398–405.

Giesel, J. T. 1976. Reproductive strategies as adaptations to life in temporally heterogeneous environments. Annual Review of Ecology and Systematics 7:57–79.

Greze, I. I. 1953. Hydrobiology of the lower part of the River Angara. Trudȳ Vsesoyuznogo gidrobiologicheskogo obshchestva (Moscow) 5:203–11. (In Russian. Translation by Mrs. A. M. L. Hynes).

Harrison, A. D. 1959. Hydrobiological studies on the Great Berg River, Western Cape Province. Part 2: Quantitative studies on sandy bottoms, notes on tributaries and further information on the fauna, arranged systematically. Transactions of the Royal Society of South Africa 35:227–76.

Herdendorf, C. E. 1982. Large lakes of the world. Journal of Great Lakes Research 8:379–412.

Hilsenhoff, W. L. 1966. The biology of *Chironomus plumosus* (Diptera: Chironomidae) in Lake Winnebago, Wisconsin. Annals of the Entomological Society of America 59:465–73.

Hilsenhoff, W. L. 1967. Ecology and population dynamics of *Chironomus plumosus*

(Diptera: Chironomidae) in Lake Winnebago, Wisconsin. Annals of the Entomological Society of America 60:1183–94.

Howarth, F. G. 1983. Ecology of cave arthropods. Annual Review of Entomology 28:365–89.

Hynes, H. B. N. 1970. The ecology of running waters. University of Toronto Press, Toronto. 555 pp.

Jackson, G. A. 1977. Nearctic and Palearctic *Paracladopelma* Harnisch and *Saetheria* n.gen. (Diptera: Chironomidae). Journal of the Fisheries Research Board of Canada 34:1321–59.

Jewett, S. G., Jr. 1963. A stonefly aquatic in the adult stage. Science 139:484–85.

Johnson, L. 1975. Physical and chemical characteristics of Great Bear Lake, Northwest Territories. Journal of the Fisheries Research Board of Canada 32:1971–87.

Kitching, R. L. 1971. An ecological study of water-filled tree-holes and their position in the woodland ecosystem. Journal of Animal Ecology 40:281–302.

Kozhov, M. 1963. Lake Baikal and its life. Monographiae Biologicae Vol. XI. Dr. W. Junk Publishers, The Hague, The Netherlands. 344 pp.

Kraemer, G. 1980. The macrozoobenthos of a deep, ultraoligotrophic, humic acid lake in the eastern Canadian shield with special reference to the Chironomidae. Unpublished M.S. thesis, University of Waterloo, Waterloo, Canada. 273 pp.

Krebs, B. P. M. 1982. Chironomid communities of brackish inland waters. Chironomus 2:19–23.

Krecker, F. H. and L. Y. Lancaster. 1933. Bottom shore fauna of western Lake Erie: a population study to a depth of six feet. Ecology 14:79–93.

Lamb, R. J. and S. M. Smith. 1980. Comparison of egg size and related life-history characteristics for two predaceous tree-hole mosquitoes (*Toxorhynchites*). Canadian Journal of Zoology 58:2065–70.

Lehmkuhl, D. M. 1976. Additions to the taxonomy, zoogeography, and biology of *Analetris eximia* (Acanthametropodinae: Siphlonuridae: Ephemeroptera). The Canadian Entomologist 108:199–207.

Lellák, J. 1968. Positive Phototaxis der Chironomiden-Larvulae als regulierender Faktor ihrer Verteilung in stehenden Gewässern. Annales Zoologici Fennici 5:84–87.

Linevich, A. A. 1971a. The Chironomidae of Lake Baikal. Limnologica (Berlin) 8: 51–52.

Linevich, A. A. 1971b. Rheophil chironomids of the trans-Baikal area and their association with the littoral chironomid fauna of Lake Baikal. Limnologica (Berlin) 8: 99–101.

Lounibos, L. P. 1980. The bionomics of three sympatric *Eretmapodites* (Diptera: Culicidae) at the Kenya coast. Bulletin of Entomological Research 70:309–20.

Lounibos, L. P. 1981. Habitat segregation among African treehole mosquitoes. Ecological Entomology 6:129–54.

MacDonald, W. W. 1956. Observations on the biology of chaoborids and chironomids in Lake Victoria and on the feeding habits of the 'elephant-snout fish' (*Mormyrus kannume* Forsk.). Journal of Animal Ecology 25:36–53.

Mackey, A. P. 1972. An air-lift for sampling freshwater benthos. Oikos 23:413–15.

Maguire, B., Jr. 1971. Phytotelmata: biota and community structure determination in plant-held waters. Annual Review of Ecology and Systematics 2:439–64.

Maguire, B., Jr., D. Belk, and G. Wells. 1968. Control of community structure by mosquito larvae. Ecology 49:207–10.

Mani, M. S. 1968. The ecology and biogeography of high altitude insects. Series Entomologica 4. Dr. W. Junk B. V. Publishers, The Hague, The Netherlands. 527 pp.

Marlier, G. 1962. Genera des Trichoptères de l'Afrique. Annales du Musée royale de l'Afrique centrale, Sciences Zoologiques 109:1–261.

McCart, P., P. Tsui, W. Grant, and R. Green. 1977. Baseline studies of aquatic environments in the Athabasca River near Lease 17. Syncrude Canada Limited, Environmental Research Monograph 1977-2. 205 pp.

McKim, J. M. 1962. The inshore benthos of Michigan waters of southwestern Lake Huron. Unpublished M.S. thesis, University of Michigan, Ann Arbor. 69 pp.

Miron, I. 1973. Réponse des larves de *Perla burmeisteriana* Claassen (Plecoptera) aux variations de la pression hydrostatique. Hydrobiologia 42:345–54.

Moeur, J. E. and C. A. Istock. 1980. Ecology and evolution of the pitcher-plant mosquito. IV. Larval influence over adult reproductive performance and longevity. Journal of Animal Ecology 49:775–92.

Mundie, J. H. 1959. The diurnal activity of the larger invertebrates at the surface of Lac La Ronge, Saskatchewan. Canadian Journal of Zoology 37:945–56.

Northcote, T. G., N. J. Johnston, and K. Tsumura. 1976. Benthic, epibenthic and drift fauna of the Lower Fraser River. Technical Report 11. Westwater Research Centre, Vancouver, Canada. 227 pp.

Oliver, D. R. 1971. Life history of the Chironomidae. Annual Review of Entomology 16:211–30.

Park, O., S. Auerbach, and G. Corley. 1950. The tree-hole habitat with emphasis on the pselaphid beetle fauna. Bulletin of the Chicago Academy of Sciences 9:19–45.

Paterson, C. G. 1971. Overwintering ecology of the aquatic fauna associated with the pitcher plant *Sarracenia purpurea* L. Canadian Journal of Zoology 49:1455–59.

Pearson, R. G., M. R. Litterick, and N. V. Jones. 1973. An air-lift for quantitative sampling of the benthos. Freshwater Biology 3:309–15.

Peters, W. L. and J. Jones. 1973. Historical and biological aspects of the Blackwater River in northwestern Florida, pp. 242–53. *In*: W. L. Peters and J. G. Peters (eds.). Proceedings of the First International Conference on Ephemeroptera. E. J. Brill, Leiden, The Netherlands. 312 pp.

Peters, W. L. and J. G. Peters. 1977. Adult life and emergence of *Dolania americana* in northwestern Florida (Ephemeroptera: Behningiidae). Internationale Revue der gesamten Hydrobiologie 62:409–38.

Picado, C. 1913. Les bromeliacées epiphytes, considerées comme milieu biologique (1). Bulletin scientifique de la France et de la Belgique 47:315–60.

Prater, B. L., D. R. Barton, and J. H. Olive. 1977. New sampler for shallow-water benthic invertebrates. Progressive Fish-Culturist 39:57–58.

Rawson, D. S. 1930. The bottom fauna of Lake Simcoe and its role in the ecology of the lake. University of Toronto Studies, Biological Series 40:1–183.

Rawson, D. S. 1947. An automatic-closing Ekman dredge and other equipment for use in extremely deep water. Special Publications of the Limnological Society of America 18:1–8.

Resh, V. H. and G. Grodhaus. 1983. Aquatic insects in urban environments, pp. 247–76.

In: G. W. Frankie and K. W. Koehler (eds.). Urban entomology: interdisciplinary perspectives. Praeger Publishing Co., New York, NY. 493 pp.

Rohnert, U. 1951. Wassererfullte Baumhöhlen und ihre Besiedlung. Ein Beitrag zur Fauna dendrolimnetica. Archiv für Hydrobiologie 44:472-516.

Rosenberg, D. M. 1978. Practical sampling of freshwater macrozoobenthos: a bibliography of useful texts, reviews, and recent papers. Canadian Fisheries and Marine Service Technical Report 790. 15 pp.

Rosenberg, D. M. and V. H. Resh. 1982. The use of artificial substrates in the study of freshwater benthic macroinvertebrates, pp. 175-235. *In*: J. Cairns, Jr. (ed.). Artificial substrates. Ann Arbor Science Publishers Inc., Ann Arbor. 279 pp.

Russev, B. 1974. Das Zoobenthos der Donau zwischen dem 845ten und dem 375ten Stromkilometer. III. Dichte und Biomasse. Izvestiya na Zoologeseskiya Institut s Muzei (Sofiya) 40:175-94.

Saether, O. A. 1973. Taxonomy and ecology of three new species of *Monodiamesa* Kieffer, with keys to Nearctic and Palaearctic species of the genus (Diptera: Chironomidae). Journal of the Fisheries Research Board of Canada 30:665-79.

Saether, O. A. 1975. Nearctic and Palaearctic *Heterotrissocladius* (Diptera: Chironomidae). Bulletin of the Fisheries Research Board of Canada 193:1-68.

Saether, O. A. 1977. Taxonomic studies on Chironomidae: *Nanocladius, Pseudochironomus*, and the *Harnischia* complex. Bulletin of the Fisheries Research Board of Canada 196:1-143.

Saether, O. A. 1979. *Paracladopelma doris* (Townes) [syn. *"Cryptochironomus"* near *rollei* (Saether 1977) n.syn.] and *P. rollei* (Kirpichenko) n. comb. (Diptera: Chironomidae). Entomologica Scandinavica Supplementum 10:117-18.

Schräder, T. 1932. Über die Möglichkeit einer quantitativen Untersuchungen der Boden- und Ufertierwelt fliessender Gewässer zugleich: Fischereibiologische Untersuchungen im Wesergebiet I. Zeitschrift für Fischerei 30:105-27.

Seifert, R. P. 1975. Clumps of *Heliconia* inflorescences as ecological islands. Ecology 56:1416-22.

Seifert, R. P. 1980. Mosquito fauna of *Heliconia aurea*. Journal of Animal Ecology 49:687-97.

Seifert, R. P. 1982. Neotropical *Heliconia* insect communities. Quarterly Review of Biology 57:1-28.

Seifert, R. P. and R. Barrera. 1981. Cohort studies on mosquito (Diptera: Culicidae) larvae living in the water-filled floral bracts of *Heliconia aurea* (Zingiberales: Musaceae). Ecological Entomology 6:191-97.

Seifert, R. P. and F. H. Seifert. 1976. A community matrix analysis of *Heliconia* insect communities. The American Naturalist 110:461-83.

Selgeby, J. H. 1974. Immature insects (Plecoptera, Trichoptera, and Ephemeroptera) collected from deep water in western Lake Superior. Journal of the Fisheries Research Board of Canada 31:109-11.

Shelford, V. E. and M. W. Boesel. 1942. Bottom animal communities of the island area of western Lake Erie in the summer of 1937. Ohio Journal of Science 42:179-90.

Slack, H. D. 1972. A lever-operated Ekman grab. Freshwater Biology 2:401-5.

Smith, S. M. and R. A. Brust. 1971. Photoperiodic control of the maintenance and termination of larval diapause in *Wyeomyia smithii* (Coq.) (Diptera: Culicidae) with notes on oogenesis in the adult female. Canadian Journal of Zoology 49:1065-73.

Strenzke, K. 1950. Die Pflanzengewässer von *Scirpus silvaticus* und ihre Tierwelt. Archiv für Hydrobiologie 44:123–70.

Teskey, H. J. 1976. Diptera larvae associated with trees in North America. Memoirs of the Entomological Society of Canada 100:1–53.

Thienemann, A. 1932. Die Tierwelt der *Nepenthes*-Kannen. Archiv für Hydrobiologie Supplement 11:1–54.

Thienemann, A. 1934. Die Tierwelt der tropischen Pflanzengewässer. Archiv für Hydrobiologie Supplement 13:1–91.

Thomas, M. L. H. 1966. Benthos of four Lake Superior bays. The Canadian Field-Naturalist 80:200–12.

Tolkamp, H. H. 1980. Organism-substrate relationships in lowland streams. Agricultural Research Report 907. Agricultural University, Wageningen, The Netherlands. 211 pp.

Varga, L. 1928. Ein interessanter Biotrop der Biocönose von Wasserorganismen. Biologisches Zentralblatt 48:143–62.

Verollet, G. and H. Tachet. 1978. Un échantillonneur à succion pour le prélèvement du zoobenthos fluvial. Archiv für Hydrobiologie 84:55–64.

Volesky, D. F. 1969. A comparison of the macrobenthos from selected habitats in cattail marshes of the Missouri River. Unpublished M.A. thesis, University of South Dakota, Vermillion, SD. 43 pp.

Walker, E. M. 1958. The Odonata of Canada and Alaska. Volume 2. Part III: The Anisoptera—four families. University of Toronto Press, Toronto. 318 pp.

Watts, R. B. and S. M. Smith. 1978. Oogenesis in *Toxorhynchites rutilus* (Diptera: Culicidae). Canadian Journal of Zoology 56:136–39.

Wells, F. C. and C. R. Demas. 1979. Benthic invertebrates of the lower Mississippi River. Water Resources Bulletin 15:1565–77.

Wiggins, G. B., R. J. Mackay, and I. M. Smith. 1980. Evolutionary and ecological strategies of animals in annual temporary pools. Archiv für Hydrobiologie Supplement 58:97–206.

Wiley, M. J. and S. C. Mozley. 1978. Pelagic occurrence of benthic animals near shore in Lake Michigan. Journal of Great Lakes Research 4:201–5.

Williams, D. D. and H. B. N. Hynes. 1976. The ecology of temporary streams I. The fauna of two Canadian streams. Internationale Revue der gesamten Hydrobiologie 61:761–87.

Winterbourn, M. J. and N. H. Anderson. 1980. The life history of *Philanisus plebeius* (Trichoptera: Chathamiidae), a caddisfly whose eggs were found in a starfish. Ecological Entomology 5:293–303.

Wirth, W. W. and W. N. Mathis. 1979. A review of the Ephydridae living in thermal springs, pp. 21–45. *In*: D. L. Deonier (ed.). First symposium on the systematics and ecology of Ephydridae. North American Benthological Society, Springfield, IL. 147 pp.

Yates, M. G. 1974. An emergence trap for sampling adult tree-hole mosquitoes. Entomologist's Monthly Magazine 109:99–101.

Zhadin, V. I. 1956. Life in rivers. Juzni preznih vod SSSR (Moscow, Russia) 3:113–256. (In Russian. Translation by Mrs. A. M. L. Hynes).

chapter 16

HYPOTHESIS TESTING IN ECOLOGICAL STUDIES OF AQUATIC INSECTS

J. David Allan

INTRODUCTION

Ecological research in general is currently in transition from a classical descriptive approach to a more experimental one that emphasizes the testing of hypotheses (Simberloff 1980; Willson 1981). Aquatic insect studies have tended to lag behind those in some other fields (e.g. research in the marine intertidal zone, plankton ecology, and certain areas of terrestrial ecology) that have used field manipulations and an explicitly experimental context to initiate an era of hypothesis testing. Another—some might say intermediate—approach is one that falls between description and experimentation. It concerns the use of intriguing comparisons when it appears that only one factor varies between two areas; for example, in one such comparison, differences may be attributed to the presence or absence of a competitor. Termed a "natural experiment," this approach has been widely used where manipulations are impractical, and it has been advocated by some (e.g. see Cody 1974) as preferable to experimental manipulation.

No attempt is made here to review the history of manipulative approaches in ecology, but it would be parochial to view them as a modern innovation. Darwin (1859) commented on plant species composition in meadows with and without "cattle exclosure cages" (i.e. fences), and researchers at Rothamsted, England, have studied the effects of fertilizer treatment on plant communities since the 1860s (Brenchley and Warington 1958). However, the landmark experiments of Connell (1961) concerning competition between species of barnacles introduced an increased emphasis on manipulations to test specific hypotheses. This has resulted in many excellent studies and improved insights into ecological relationships. It has also provided healthy debate concerning the design and interpretation of such studies, and some realization of their difficulties and limits.

The main goal of this chapter is to distinguish clearly between descriptive and experimental approaches, and to center this discussion around studies of aquatic insects. Of course, quantification is fundamental to either approach, and sampling variability due to the extreme patchiness of benthic invertebrates has been a major hindrance to quantitative analysis. Therefore, this chapter includes some discussion of the constraints that this variability places on interpretation.

Finally, a set of recommendations is presented for conducting field studies of aquatic insects, based on the ten principles proposed by Green (1979).

DESCRIPTIVE VERSUS EXPERIMENTAL APPROACHES

Examples of the descriptive approach might include studies of seasonal changes in numbers of aquatic insects, their distribution in various habitats, or their abundance and composition in polluted compared with unpolluted areas. In contrast, an experimental study might compare aquatic insect abundances in a section of stream that has been experimentally acidified to abundances in a control section, or compare species preferences for different, experimentally constructed substrates. Both descriptive and experimental studies may use statistical analysis, perhaps even the same statistics. They differ in that the experimental study generally poses a more specific question, and always uses a planned, manipulative approach. The comparison of, for example, a polluted and an unpolluted arm of some lake could be made experimental by sampling prior to the pollution event as well as subsequent to it. Simply comparing two areas after the fact, however, is most like a natural experiment in that we hope the differences are due only to pollution.

Unfortunately, there is a tendency to equate descriptive with "bad," and experimental with "good." These are misplaced values. Descriptive studies are necessary to collect basic information on phenomena such as life cycles or distributions; in general, they are necessary to describe patterns. The results may suggest hypotheses to test (e.g. if one species is rare where another is common, and vice-versa, this could indicate competition). In fact, when experimental studies are conducted without sufficient knowledge of the natural history of the system, they may be a waste of time.

A hypothesis results from rewording some question in such a way that an experiment will refute or support it. For example, we could ask the question: Does the presence of fish affect the numbers of benthic invertebrates? This may be restated as a null hypothesis (H_0): removal (or addition) of fish will not affect the numbers of benthic invertebrates. The experiment then is performed and the hypothesis is refuted or supported. There is always an alternative hypothesis (H_A), which may be exploratory (i.e. removal of fish will affect the numbers . . .) or specific (i.e. removal of fish will lead to an increase in numbers . . .). Generally, a probability level is associated with rejecting the null hypothesis (e.g. an $\alpha = 0.05$ level of rejection describes a result sufficiently unusual that it would be observed by chance only 5 times in 100 trials). This amounts to setting an acceptable level of error since, on average, 5 out of 100 results will lead us to reject the null hypothesis when it is true. Such an event is called a type I error. A significant result means that the experimenter has rejected the null hypothesis with some stated probability of being wrong. Rejecting the null hypothesis does not prove the alternative hypothesis, since some as yet unconsidered explanation may be the true cause of the significant result. Similarly, acceptance of the null hypothesis does not prove it correct. There is the possibility that the null hypothesis actually is false, but by chance this is not detected. Acceptance of the null hypothesis when it is false is termed a type II error, which occurs with probability β. Figure 16.1 illustrates these two kinds of errors. Field experiments, and especially natural experiments, risk type I errors because the comparison areas may differ in additional, unknown ways. The variability of field data and logistical constraints on adequate replication make type II errors a real possibility, as results may be too variable to distinguish an effect.

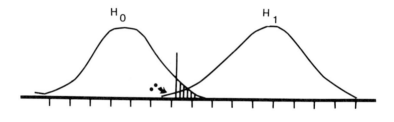

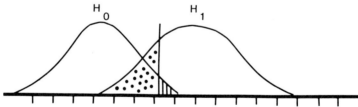

Figure 16.1 Illustration of type I (α) and type II (β) errors, simplified to consider only the one-tailed alternative hypothesis ($H_1 > H_0$). Vertical hatching shows a type I error (i.e. 5% of the observations are large enough to cause the null hypothesis to be rejected when it is true). In a two-tailed test, 2.5 percent of the observations will be sufficiently large, and 2.5 percent sufficiently small, to falsely reject the null hypothesis. The *dotted region* illustrates a type II error. Some observations from the "small end" of the alternative (H_1) distribution are sufficiently small to be accepted as coming from distribution H_0, when that is not true. Comparison of the bottom to the top pair of curves shows that, as the alternative hypothesis becomes more similar to the null hypothesis, β increases while α remains unchanged. Similarly, if both curves are broadened (have higher variances), β increases. Thus the probability of a type II error is increased when small differences must be detected and when variances are large.

QUANTIFICATION

The degree of effort put into measurement and quantification depends on the question being asked. If one wishes to be able to detect a small change in numbers between habitats, or between polluted and unpolluted areas, much more effort is required than if one wishes to detect only gross changes. Unless one specifies the methods of data collection and analysis, and provides some measure of variability of the data, other investigators will be unable to assess how much confidence to place in the results. It is useful to distinguish between precision and accuracy. Most analyses are concerned with precision, which is a measure of the similarity of repeated (i.e. replicate) measurements. Standard deviations, standard errors, and 95 percent confidence limits (95% CL) are all measures of precision, and say nothing about the accuracy of a measurement. Accuracy refers to the closeness of a measurement to its true value. In sampling stream benthos, for example, Hynes et al. (1976) have demonstrated that surface-substrate samplers may collect only 10 percent of total numbers present as revealed by sampling the whole substratum profile (see also Williams, Chapter 14). This is a problem of accuracy, not precision.

Sampling devices and variables affecting sampling methods are discussed by Resh (1979). Methods of statistical analysis and design of sampling programs are discussed in detail by Elliott (1977) and Green (1979), and these two books should be read thoroughly by every benthic ecologist. In this section on quantification, I will discuss two topics that persistently provide difficulties for quantitative analysis: (1) the need for data transformation; and (2) the relationship between sample size and precision, which determines our ability to detect differences.

Transformation

Species counts from field collections generally show a dependency of the variance on the mean. Widely used tests to compare means (e.g. analysis of variance [ANOVA], t-tests) make several assumptions, the most critical of which is approximate equality of variances among the samples to be compared (Sokal and Rohlf 1981). If the ratio of the largest to smallest variance is large (e.g. see Sokal and Rohlf 1981, p. 402), this assumption probably is not met and the test will be in error. Unequal variances often may be caused by a dependence of sample variance on sample mean, and require appropriate variance-stabilizing transformation prior to statistical analysis.

Analysis by nonparametric methods (e.g. see Siegel 1956; Hollander and Wolfe 1973; Conover 1980) is an alternative. Such tests often are simpler to perform, make fewer assumptions, and may be nearly as efficient as their parametric equivalents. However, in many instances one can proceed with standard (parametric) analyses after suitable transformation (Green 1979).

Log(x + 1) is often recommended as a good general transformation because it acts like the square root for small values and like the log transformation for larger values (Steel and Torrie 1960; Green 1979). Alternatively, one may choose the square root transformation for data that best correspond to a random (Poisson) distribution and the log transformation for data that are more highly clumped. Further alternatives include use of the negative binomial or Taylor's (1961) power law to find an exact transformation for a particular data set (see Elliott 1977; Southwood 1978; Green 1979 for further discussion) or the Box-Cox transformation (see Sokal and Rohlf 1981, p. 423). Recently, Downing (1979) suggested that the most appropriate general transformation for species counts obtained in benthic sampling is the fourth root.

Many investigators simply assume a transformation is needed, use log x (or log [x + 1] if zero counts occur), and then assume success. A better approach is to examine the data to see what transformation is indicated, and reexamine the data after transformation to determine if it was effective. Thus, a series of collections is made, each consisting of a number of replicates, and the mean and variance are computed for each collection. The variances are regressed against their respective means for all collections, and if the regression is significant, a transformation is indicated. If no relationship for such a regression (or a greatly reduced one) is found after transformation, the transformation is viewed as successful.

A log-log regression using the original data generally is linear, and Taylor (1961) suggested the following equation:

$$s^2 = am^b \qquad (1)$$

where a and b are the intercept and slope parameters fit by regression, s^2 is the sample variance, and m is the sample mean for the individual collections. The slope b quantifies the dependency of sample variance on sample mean, and usually takes values between 1 and 2. Taylor (1965) suggested that an exact transformation could be found using the value b from equation (1), as follows:

$$x^p = x^{1-b/2} \qquad (2)$$

where x is an observation from a single sample which is transformed by being raised to the power $p = 1 - b/2$. If $b = 1$, this is identical to the square root transformation; if b is close to 2, this is similar to the log transformation. Downing (1979) observed that collections from lakes and large rivers tended to have b values near 1.5, and therefore recommended the fourth root transformation.

Allan (1982) collected a series of Surber samples (0.093 m^2 surface area) from the stony bottom of a Rocky Mountain stream. Twelve replicate samples were collected at each of several sites over two years, for a total of 31 collections. The mean and variance of numbers for species a,b,c . . . were computed for each collection, resulting in up to 31 pairs of mean and variance estimates for each species.

These data, fit to equation (1), also showed a tendency for b to fall near 1.5 (Table 16.1). While this would appear to support Downing's advocacy of the $\sqrt[4]{x}$ transformation, three of the 21 estimates of b were significantly larger and

TABLE 16.1: Value of b from Equation (1), Taylor's Power Law (See Text), Based on Variance-Mean Regression for 21 Taxa. Each Regression is Based on up to 31 Collections of 12 Replicates from a Rocky Mountain Stream

SPECIES	b	95% CL Lower	95% CL Upper	n	r^2
Ephemeroptera					
Baetis bicaudatus	1.60	1.35	1.84	31	.86
Cinygmula sp.	1.49	1.35	1.63	31	.86
Epeorus longimanus	1.34	1.20	1.49	31	.92
Rhithrogena spp.	1.40	1.16	1.65	31	.83
Ephemerella infrequens	1.51	1.37	1.66	30	.97
E. coloradensis	1.30[a]	1.13	1.47	31	.90
E. doddsi	1.27[a]	1.07	1.47	28	.93
Ameletus velox	1.38	1.20	1.57	24	.96
Plecoptera					
Alloperla spp.	1.47	1.24	1.70	31	.86
Zapada haysi	1.84[b]	1.61	2.08	31	.90
Trichoptera					
Brachycentrus americanus	1.31[a]	1.18	1.43	23	.98
Rhyacophila acropedes	1.20[a]	1.04	1.36	31	.89
R. valuma	1.39	1.08	1.69	31	.75
Diptera					
Simuliidae	1.57	1.29	1.85	26	.80
Chironomidae	1.81[b]	1.52	2.10	31	.85
Coleoptera					
Heterlimnius sp.	1.37	.90	1.83	31	.55
Taxonomic groupings					
Ephemeroptera	1.84[b]	1.51	2.17	31	.82
Plecoptera	1.79	1.49	2.09	31	.84
Trichoptera	1.51	1.19	1.83	31	.76
Diptera	1.69	1.35	2.03	31	.78
Total collection	1.90	1.43	2.37	31	.70

[a]Slope significantly less than 1.50.
[b]Slope significantly greater than 1.50.

four were significantly smaller than 1.5 (see 95% CL on slope, Table 16.1). This parallels the observation by W. D. Taylor (1980) and L. R. Taylor (1980) that, in some instances, the fourth root will not work as well as either a square root or a log transformation.

From examination of the data sets of Table 16.1 after various transformations had been made, it was apparent that when many zero observations occurred among the 12 replicate samples (defined as $\bar{x} < 0.5$ individuals/sample), a significant variance-mean relationship generally could not be eliminated with any transformation. Downing (1979, 1981) compared the adequacy of various transformations only for means >3/sample, citing Anderson (1965) as stating that variance-stabilizing transformations could not be found for collections with $\bar{x} < 3$. For the data set of Table 16.1, excluding low means always allowed a satisfactory transformation to be found and usually favored $\sqrt[4]{x}$ and x^p transformations over square root or log. However, when low means were included, log $(x + 1)$ generally achieved the greatest reduction in the dependency of the variance on the mean, but did not completely eliminate it.

Downing's (1979) original article has generated much useful debate (W. D. Taylor 1980; L. R. Taylor 1980; Downing 1980, 1981; Chang and Winnell 1981), but it should be emphasized that Downing's examples all were higher taxonomic groupings (e.g. family or order) from which data sets with low means had been eliminated. Yet in examining many questions, the investigator will need to work with particular species, which may occasionally or perenially be rare. In these instances, $\sqrt[4]{x}$ may work less well than $\log(x + 1)$. Reliance on a single, general transformation may be unwise unless the data set is too limited to permit examination by equation (1).

Sample Size and Precision

Benthic invertebrates are clumped in their distribution, as evidenced by the considerable range of species counts obtained in replicate samples, and the wide applicability of equation (1). This clumping, or aggregation, is the basis for the significant dependence of sample variance on sample mean. Higher values of b indicate greater aggregation. This high variability in replicate samples results in low precision of estimates of means that are based on modest numbers of samples.

In their pioneering effort to address the problem of precision in sampling a stream riffle, Needham and Usinger (1956) concluded that 73 replicate samples were required for a precise estimate of mean total numbers, and 194 replicates for mean wet weight. Chutter and Noble (1966) corrected Needham and Usinger's estimate, and showed that the problem was even worse than originally thought (if one accepts that 95% CL ± 5% of the mean is the necessary level of precision). Resh (1979) reanalyzed these earlier data sets, as well as some of his own, and basically confirmed the pessimistic view that led Needham and Usinger (1956, p. 338) to say, "We then can conclude from these data that purely quantitative routine sampling in streams to determine weights and numerical data is impractical."

Undoubtedly, this question of sampling precision is one of the most fundamental problems limiting quantitative studies of the benthos. However, it is clear that the problems associated with sampling precision have to be faced, and the gloomy prognosis of Needham and Usinger should not be an excuse for bypassing quantitative analysis.

The Effect of Number of Samples on 95 Percent CL. The 95 percent CL are necessary to specify the precision of the estimate of the mean. In addition, one can make some inferences concerning whether two means differ significantly by inspection of their 95 percent CL. If the limits on one mean include the mean of comparison sample, they are not significantly different. If the 95 percent CL of two samples do not overlap, they differ significantly. If the 95 percent CL overlap one another but do not include either mean, a further test is required.

The dependence of precision on sample size is illustrated with the stream benthic samples described previously. The effect is most easily seen if confidence limits are expressed as some multiple of the mean. The log(x) transformation has a unique advantage in that $\pm$ limits of transformed data become a times-divide ($\underset{\div}{\times}$) factor on back-transformation, because adding in the log scale is the same as multiplication in the arithmetic scale. For example, if the count of numbers of insects in each of n samples is transformed using $\log_{10}(x)$, and the mean and standard error of the transformed data are found to be 2.0 and 0.15 respectively, then 95 percent CL = t $\times$ standard error $\approx 2 \times 0.15 = 0.30$ (see Elliott 1977, p. 91). The mean and 95 percent CL can be expressed in one of several ways:

in the logarithmic scale with upper and lower CL shown	2.0 (1.7 − 2.3)
as above, but in the arithmetic scale by back-transformation	100 (50 − 200)
in the logarithmic scale with a $\pm$ factor	2.0 ($\pm$0.30)
in the arithmetic scale with a $\underset{\div}{\times}$ factor by back-transformation	100 ($\underset{\div}{\times}$ 2)

A $\underset{\div}{\times}$ factor of 2 corresponds to a range of 100 percent (confidence limits are double and half the mean), whereas 1.5 corresponds to 50 percent, and so on. Since the transformation presumably has stabilized the variance, rendering it independent of the mean, this $\underset{\div}{\times}$ factor can be applied to the arithmetic mean as well as the mean obtained by back-transformation, which will be a different number (Elliott 1977). Confidence limits obtained from transformed data are very asymmetrical around, and in extreme circumstances may not even include, the arithmetic mean, so use of a $\underset{\div}{\times}$ factor with an arithmetic mean allows one to avoid this troubling feature.

Since many collections will include at least some zeros, $\log(x + 1)$ can always be used, whereas the log transformation cannot. The 95 percent CL obtained with $\log(x + 1)$ are slightly asymmetrical, but $\times$ factors can be calculated as $UTD = UCL/\bar{x}$, and $LTD = \bar{x}/LCL$, where UTD and LTD are upper and lower times-divide terms; and UCL, LCL, and x are upper and lower confidence limits, and the mean after back-transformation.

Some 31 data sets, each with 12 replicates, were analyzed after $\log (x + 1)$ transformation, and the average variance from these was used to calculate 95 percent CL as a function of sample size and the (back-transformed) mean. The results for the mayfly *Baetis bicaudatus* (Fig. 16.2, *top*) show that for means much less than 5 per sample, precision is so poor that quantitative interpretation is virtually useless. A $\times$ factor of 1.25, or 95 percent CL of ± 25 percent of the mean, requires 50 samples and means > 10 per sample. A similar analysis for the total fauna (Fig. 16.2, *bottom*) shows virtually no sensitivity to mean abundance, because the lowest mean was 44.7 per sample. Overall precision was only slightly better than for *B. bicaudatus* alone, due to a somewhat lower average variance. Twenty replicates would give 95 percent CL of ± 30 percent, and 10 replicates would give 95 percent CL of ± 53 percent of the mean.

A general recommendation from this analysis is to collect 10 to 20 replicates for modest precision in quantitative analysis, and at least 50 replicates if high precision is needed. Although it generally is preferable to take many small samples rather than few larger samples in terms of both precision (Elliott 1977; Green 1979) and quantity of benthic substrate that must be processed (Downing 1979), samples must not be so small that counts give low means and frequent zeros. Sampling devices that collect fewer than five individuals per sample of the study organism are, on average, too small for most quantitative work. Either the device should sample a larger area, or very many samples must be collected. Any particular collection may have a variance of transformed counts considerably larger or smaller than the average variance used here, with corresponding effects on CL. The variance of $\log(x + 1)$ transformed counts from other stream benthos collections were of the same general range as the data just analyzed (Needham and Usinger 1956; Chutter and Noble 1966; Resh 1979). Benthic samples from lakes and large rivers are somewhat less variable than those from stony streams (Downing 1979). It appears likely, however, that a relatively high level of variability will prove to be characteristic of field sampling regardless of the system. For example, Eberhardt (1978) compared coefficients of variation (= standard deviation $\times 100/\bar{x}$) for a series of aquatic and terrestrial field counts, and found that a similar range of values generally occurred.

Power of the Test.   A planned comparison between two or more means constitutes a test of a hypothesis. We may find that there is no difference between means. This may be real, or the variability of the system may limit the ability of

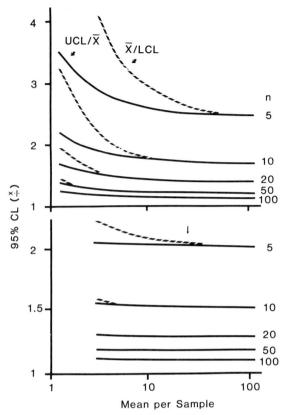

Figure 16.2 Relationship between 95 percent confidence limits (95% CL) around the mean and the magnitude of the mean, for various sample sizes (n). A times-divide ($\overset{\times}{\div}$) factor of 1.5 corresponds to ± 50 percent of the mean. *Top panel:* data for a single species of mayfly, *Baetis bicaudatus. Bottom panel:* data for total aquatic invertebrates. The arrow in the bottom panel denotes the lowest mean observed (~44/sample). Curves are truncated on the left because no densities this low were observed, and to stay within the vertical scale.

the test to detect a difference (type II error). When no significant differences are observed, one wishes to know the power of the test to detect a difference. This, of course, depends on replication.

Green (1979, pp. 42–43) provides an example using a 2×2 factorial ANOVA. Two sites are chosen (upstream and downstream from a potentially toxic input), and samples are collected before and after the event. For three faunal groups selected for their different variances, Table 16.2 gives the number of replicates required per cell to detect a change in numbers of 25 percent, 50 percent, or 100 percent at the 5 percent level. As with 95 percent CL, low means require enormous sample sizes; but for means greater than 5 to 10 per sample, the

TABLE 16.2: The Number of Replicates Required in each of Four Area ×Time Cells (See Diagram of ANOVA Design, Below) in Order to Distinguish Means that Differ by 25%, 50%, or 100%, at the 0.05 Significance Level, for Several Values of the Mean*

AREA

		Control	Impact
	Before	—— —— ——	—— —— ——
SAMPLING TIME			
	After	—— —— ——	—— —— ——

		TOTAL FAUNA	BAETIS BICAUDATUS	EPHEMERELLA COLORADENSIS
Variance (log [x + 1])		0.059	0.105	0.079
Highest mean		330.1	119.2	68.2
Lowest mean		43.7	1.5	0.12
	Difference to be detected (%)	*Number of replicates required*		
Lowest observed mean	100	11	39	624
	50	32	124	2461
	25	106	451	9591
Mean = 10	100	12	20	16
	50	36	60	48
	25	119	202	158
Mean = 50	100	11	18	14
	50	31	54	42
	25	104	178	140
Highest observed mean	100	11	18	14
	50	31	52	42
	25	102	173	140

*Computations were performed for the total fauna and for two mayfly species using the data of Allan (1982) described in the text. The highest and lowest observed mean density (no./m^2), and the average variance of log-transformed counts, were as shown.

required sample size for a given level of precision is fairly independent of the mean. It appears that it would always be unreasonable to use fewer than 15 to 20 replicates per cell in this design, since this would just detect a doubling or halving of densities. Probably few investigators would have the resources to collect sufficient replicates to detect less than a 25 to 50 percent change.

Allan (1982) conducted an experimental study of the effects of trout predation on the aquatic insect community of a mountain stream. An experimental section 1.2 km in length was fenced at both ends, and trout were removed by electrofishing. Two control sections (normal numbers of trout), one upstream and one downstream, were sampled for comparison. Twelve replicate benthic samples were collected from each of these three treatment areas and analyzed by ANOVA. Collections made approximately four times per year for four years allowed many such ANOVAs to be done comparing abundances of a number of faunal groups. With few exceptions (which were expected as type I errors), significant differences were not found between experimental and control treatments. This prompted the following questions: (1) Did trout removal really have no effect on insect abundances? (2) By how much must the experimental area differ from one of the control areas to be significant at the 0.05 level?

If a significant difference had been found, the *a posteriori* Student-Newman-Keuls test (Sokal and Rohlf 1981, p. 262) could be used to determine which two of the three means differed significantly. A difference value, which must be exceeded for significance, can be used to determine by what amount the experimental treatment must differ from the controls. This value (least significant range [LSR]) is based on the average within-treatment variance of the samples (the error mean square [EMS] value of the ANOVA). Since data were log transformed, the back-transformed LSR can be expressed as a times-divide factor. Its dependence on n (Fig. 16.3) is of the same general shape as for 95 percent CL. The choice of 12 replicates appears reasonable, as a reduction in replication would lead to substantial loss of precision, whereas a corresponding increase in replication would bring about rather small gains. However, only means which differed by a factor of 2 would be significantly different. In other words, removal of trout would have to cause at least a 100 percent increase or decrease in numbers of aquatic insects; a 50 percent change would go undetected. Unfortunately, the only way to improve the power of the test is to collect many more replicates per treatment area.

THE DESIGN OF FIELD EXPERIMENTS

Observation of the nonoverlapping distribution of two potentially competing species might be taken as evidence that competition determines where each species occurs. More properly, it is the basis for a hypothesis to be tested, and the least equivocal test is a field manipulation. Since this is where the field

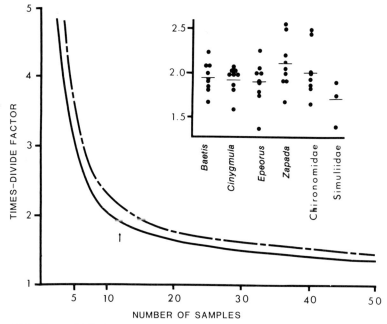

Figure 16.3 The multiple that one mean must be of a second mean (times-divide [×] factor) in order to differ significantly ($\alpha = 0.05$), as a function of sample size. Result applies when three treatments are compared and the goal is to distinguish one mean from the other two. For 12 replicates per treatment (*arrow*), roughly a doubling or halving is required. *Solid line = Baetis bicaudatus, Cinygmula* sp., and *Epeorus longimanus; dashed line = Zapada haysi.* The curve for Chironomidae was between the two curves depicted. Inset shows actual × factors from multiple samplings. Figure from Allan (1982), copyright © Ecological Society of America, reprinted with permission.

experimental approach has the greatest potential, some strengths and weaknesses are discussed below.

Field experiments differ from both laboratory and natural experiments (Connell 1974). In the laboratory, all variables are believed to be controlled except one, the experimental variable, which is allowed to vary. In the field, one variable is manipulated, and all the rest are allowed to vary naturally and, hopefully, similarly. One can never be certain that some naturally varying factor has not caused the observed result, or cancelled an expected result. Adequate replication of experimental and control treatments minimizes this risk, however. In a natural experiment, two areas are compared that the investigator assumes to differ only in a single variable of interest, such as the presence or absence of a competitor. The likelihood that additional factors differ and may contribute to observed differences between areas is considerably greater than in a field manipulation (Hairston 1980).

Research in the marine intertidal zone has progressed perhaps the furthest in exploiting experimental manipulations. In addition to Connell (1980), Paine (1977), Virnstein (1978), and Peterson (1980) provide valuable summaries of methodology. Paine (1977) lists four reasons why the experimental approach has worked so well in the rocky intertidal zone. These include the following: (1) many species are large, relatively accessible and relatively immobile; (2) the habitat allows construction of cages, barriers, and so forth; (3) the length of time for response is moderate (i.e. a few years); and (4) for many species the limiting resource is space, which is stationary and easily measured.

Before freshwater benthic ecologists become too enthusiastic about the transfer of experimental marine benthic technology to lakes and streams, it would be wise to consider these four points. Only (3) seems clearly true in freshwater, while (1) typically is not valid. With regard to (2), running water poses special problems of fouling and damage of cages. How often (4) is true in freshwater has yet to be decided. Peterson (1980) has also discussed the fact that soft bottom marine communities have proven less responsive to a manipulative approach than their hard bottom counterparts.

Previous studies also reveal some of the difficulties with field experiments. First, cage effects can cause changes not anticipated in designing the experiment. For example, a cage of 1-mm mesh placed in a clear mountain stream resulted in a layer of silt over the usually clear substrate within hours, making the caged habitat unsuitable for the fauna that typically lived there (J. D. Allan, personal observation). Open plots have been suggested as an alternative (e.g. see Hairston 1980), but they are not feasible with highly mobile animals. Cage controls are needed to explore the unintended effects of cages. For example, shading can be tested by constructing a cage with a roof but no walls. A two- or three-sided cage that allows free movement of organisms can be used to test for cage effects that result from microhabitat changes (e.g. changing current patterns, cage walls serving as substrates, etc.). Second, the high variability of natural systems increases the likelihood of falsely concluding that there are no differences between the experimental and control treatments (a type II error). Negative results are often and unjustifiably considered uninteresting. This view can only be intensified by lack of confidence in the validity of the decision not to reject the hypothesis of no difference among treatments (Wise, in press). Third, spatial heterogeneity in nature makes it difficult to choose experimental and control plots that are satisfactorily similar (Wise, in press). If, for example, one wishes to place replicate cages in the littoral zone of a pond, we might expect cages to differ in exposure to direct sunlight, substrate type, macrophyte density, and so on (e.g. see Benke 1978). A block design (see Sokal and Rohlf 1981, p. 348) may be used if several treatments can be placed in each habitat type, but one still must assume that all treatments within each block differ only in the experimental manipulation. Fourth, decisions concerning size of plot often are an uncomfortable compromise. The labor involved in a manipulation (e.g. removing a certain species from an area) may preclude the investigator from using plots that are spatially large, or from adequately replicating the design. Small plots suffer from edge effects and only the middle of a caged plot may be free of artifacts due to the cage itself, such as fouling. Immigration and emigration of mobile organisms can swamp an experimental effect if the plot is too small; this is all the more likely in open plots that avoid cage-induced edge effects.

A fifth problem with field experiments deserves more detailed discussion, because it has frequently been overlooked. The basic principle in experimental design is replication of experimental and control treatments, so that variation between plots treated alike can be compared with variation between plots treated differently (Eberhardt 1978; Hurlbert, 1984). Often, however, a field experiment consists of one experimental plot and one control plot, both of which are subsampled but not replicated. Statistical analysis applied to such experimental data must be viewed cautiously, at best. Hurlbert argues that such an approach simply is wrong, and terms it *pseudo-replication*.

There are two faults resulting from pseudo-replication. First, one does not have an estimate of variation among several experimental plots and several control plots, which is the appropriate basis for judging if differences between experimental and control plots exceed average differences between plots that are treated alike. If one finds a significant difference between one experimental and one control plot, based on pseudo-replication, one cannot infer that it is due to the experiment. The two areas do differ, but it may be for some other reason. Second, one has an inflated estimate of degrees of freedom, hence a reduced estimate of the error mean square and a greater likelihood of finding a difference between the experimental and control plot, which results in a type I error. Hurlbert points out that no two plots are ever truly identical. Thus if one goes on subsampling indefinitely, eventually two plots will be found to differ significantly even without an experimental manipulation of one plot. This is a sure sign of pseudo-replication. However, unlimited true replication of treatment and control plots can lead falsely to the conclusion of a significant treatment effect only 5 times in 100, the typical type I error, because spatial heterogeneity is accounted for by both replicate control and replicate experimental plots. Some examples of acceptable and pseudo-replicated designs are presented in Figure 16.4.

SOME EXAMPLES OF FIELD EXPERIMENTS IN AQUATIC INSECT ECOLOGY

A selective review of the freshwater benthic literature provided few satisfactory examples of experimental manipulations. The design used by Allan (1982, see preceding discussion) illustrates a number of problems typical of field experiments. First, the experiment was pseudo-replicated. Pseudo-replication would have been more of a problem if a significant difference had been found, because it could have resulted from environmental heterogeneity (a difference

Arrangement of Replicates	Schema

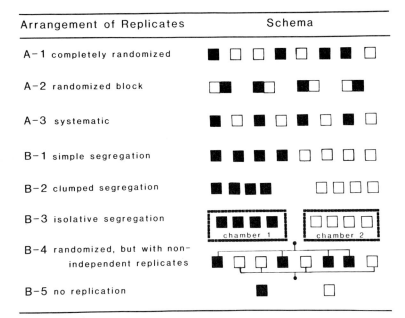

A-1 completely randomized

A-2 randomized block

A-3 systematic

B-1 simple segregation

B-2 clumped segregation

B-3 isolative segregation

chamber 1   chamber 2

B-4 randomized, but with non-
    independent replicates

B-5 no replication

Figure 16.4 Schematic representation of various acceptable modes (*A*) of interspersing the replicates (boxes) of two treatments (shaded, unshaded) and various ways (*B*) in which the principle of interspersion can be violated, leading to pseudo–replication (from "Pseudo–replication and the design of ecological field studies," by S. H. Hurlbert, *Ecological Monographs* 1984, 54. Copyright 1984 by the Ecological Society of America. Reprinted by permission).

between areas) and the fraudulently increased power due to inflated degrees of freedom. Second, the conclusion of no significant effect of trout on insect abundance may be valid, or it may be a type II error. The estimate of the least detectable difference (Fig. 16.3) also poses a problem of inference. It probably is true that the mean of 12 replicates from each of two areas would have to differ two-fold to be significantly different. On this basis, neither the predator removal zone differed (for most taxa, on most sampling dates, from either control section), nor did the controls differ from one another. It is not correct, however, to say that there were 12 replicates of the predator removal section, and 12 replicates of each of the control sections. The error mean square (EMS) value used in Fig. 16.3 had inflated degrees of freedom, and may have underestimated the required difference between means for statistical significance.

Third, this study illustrates the realistic limits of conducting field experiments. The exclusion of trout from 12 replicate 1.2 km sections of stream would require prohibitive effort. Perhaps smaller sections of stream could be used, but this would intensify edge effects. One cannot consider the 1.2 km section to consist of a series of separate experimental treatments, because they are spatially contiguous (see B-1 and B-2, Fig. 16.4). One must use some interspersion scheme of experimental and control sections (A-1 through A-3, Fig. 16.4). Strictly speaking, this would allow inference about the effects of fish removal for one particular stream, but one must be cautious about extending the inference to other streams, unless streams themselves are the unit of replication.

Reice (1983) conducted a predation experiment in which baskets (25 × 25 × 10 cm) of substrate were placed in a stream riffle to permit invertebrate colonization. Some baskets had mesh tops to exclude fish, whereas other baskets were open. True replication of open and enclosed baskets with complete census of invertebrate densities is an excellent design, completely avoiding pseudo-replication. Reice also observed no differences due to fish predation but here the criticism of edge effects must be considered, because movement of aquatic insects in and out of baskets may have been great enough to swamp any effect of predation in the topless cages.

Hall et al. (1980) experimentally acidified a section of a New Hampshire stream to mimic the effects of acid rain on stream communities. Drift increased significantly in the treatment compared to the control area, and benthic densities declined greatly. Some differences in benthic densities were significant, but many were not, despite dramatic declines in numbers. This appears to be an example where the lack of statistical power prevented an unequivocal demonstration of an effect of acidification that almost certainly did occur. It also illustrates pseudo-replication but, again, logistic constraints make true replication impractical.

Macan (1966, 1977) conducted a long term (>20-year) study of the fauna of a small pond in the English Lake District. Predation by fish (*Salmo trutta*) was first reduced to zero by removal of fish, then fish were added, and later they were removed again. Concomitantly, much of the aquatic vegetation declined, and then recovered. Some changes in abundance of species of aquatic insects were attributed primarily to the introduction of fish, whereas others were attributed to the concomitant decline and recovery of rooted plants. Macan (1977) clearly recognized the possibility for changes unrelated to either obvious causal factor. He argued that year-to-year continuation of a trend when one variable remained constant, and reversal of that trend when the effect of that variable changed, increased the likelihood of a causal relationship. This is reasonable and the study is admirable but, again, one wishes for a stronger basis from which to infer causality.

These examples show some of the limitations of the field experimental approach. The lack of statistical power limits our ability to demonstrate differences when they appear real (the acid addition), and to argue convincingly that the experimental manipulation had no effect when that appears to be the case (the trout removal). Benthic ecologists who wish to test hypotheses about factors affecting aquatic insect abundance must redouble their efforts in terms of sample numbers, and also set realistic goals about the magnitude of effects that may be detected.

Environmental impact studies may be viewed as analogous to field manipulations, and Green's (1979) text is an excellent basis for their analysis. It is worth noting that such studies suffer at least as many problems as the field manipulations just discussed. If no preimpact data are available to document similarity of control and impacted areas, the study is no different from a natural experiment in which two areas are compared that are believed to differ only in the factor of interest. Even if preimpact and postimpact data are available, there is typically only one impact and one control area. The ideal impact design presented by Green (1979, pp. 42–43), and used in Table 16.2 as an example of statistical power, actually is pseudo-replicated. If a difference is found between the areas, the inference that it is due to the impact is statistically unwarranted. Yet here, the logistical problem of proper replication of treatments is truly insurmountable.

Community studies (especially those based on counts of individuals in benthic samples) are most difficult to replicate and sample adequately. Substrate preferences of individual species are reasonably amenable to experimental study, because one may utilize baskets of substrate that are sorted into particular substrate sizes or combinations, and placed on the stream bottom. Cummins and Lauff (1969) conducted one of the first such studies in an experimental stream. Subsequent studies in the field (Allan 1975; Minshall and Minshall 1977; Rabeni and Minshall 1977; Reice 1980) have used appropriate experimental designs in examining substrate preferences and in separating the effects of substrate, current, and silt on these preferences. Peckarsky (1979) introduced a cage (10 X 30 X 40 cm) that has proven effective in studying the effects of invertebrate density on colonization rate, and predator-prey interactions (Peckarsky 1980; Peckarsky and Dodson 1980a, 1980b). Proper experimental replication of treatments is feasible because each unit is small. Benke (1978) investigated competition among dragonfly larvae in a small pond, using 16-m^2 cages; Thorp and Bergey (1981) excluded predators from regions of the littoral zone of a reservoir, using 4-m^2 cages.

Recognition of the importance of heterotrophic inputs has led stream biologists to focus on leaf processing (Kaushik and Hynes 1968; Petersen and Cummins 1974). The usual procedure is to pick leaves from trees prior to abscission, construct leaf packs of a single species and given size, and place these on the stream bottom tied to bricks. Weight losses, chemical changes in the leaves, and differences among leaf type, leaf age, substrate type, and colonization by invertebrates are easily measured for each replicate. Also, the small size of each experimental unit permits reasonable replication. A number of studies have used three or four replicates per treatment or observation date, including those by Petersen and Cummins (1974), Davis and Winterbourn (1977), Reice (1974, 1977), Short et al. (1980), and Sedell et al. (1975). However, few of these used an explicit experimental design or performed statistical analysis of their replicated data. Reice's (1980) examination of leaf decomposition in various substrates is an excellent example of the power obtainable by appropriate experimental design.

One promising future avenue for comparisons of aquatic insect communities is the use of artificial substrates that are readily colonized and, on average, yield greater precision that do direct substrate samples (Rosenberg and Resh 1982). Thus, in some instances it may be possible to detect smaller differences than would otherwise be the case. However, as Rosenberg and Resh point out, if artificial substrate samples do not adequately reflect population trends in the natural substrate, gains in precision may not repay the sacrifice of lost accuracy.

This brief and selective review of experimental studies in the freshwater benthos indicates that some strong beginnings have been made toward improved quantification, and incorporation of a field experimental approach. However, environmental heterogeneity and the patchy distribution of organisms continue to pose serious challenges. Only an increased effort, coupled with continued advances in techniques of field manipulations, will lead to further maturation of the hypothesis-testing approach.

SUMMARY AND RECOMMENDATIONS

Green (1979) presents ten principles (Table 16.3) to guide sampling design and statistical analysis in environmental studies, and again I recommend strongly that his book be read carefully. Below, I add some additional recommendations.

1. Researchers should distinguish between collecting data that may be useful in formulating a hypothesis, and collecting data to test a hypothesis. Only an experimental manipulation can test a hypothesis under field conditions. While this approach is not always feasible, and certainly is fraught with difficulties, it is probably the single most powerful technique for understanding ecological interactions. Attention to principle 1 (Table 16.3) will resolve whether the approach is descriptive or experimental, exploratory or an explicit test of a hypothesis.
2. Whenever possible, treatments should be replicated in an experimental design. However, logistical constraints may limit one's options here, and impact studies almost always provide only one treatment. Be aware that using subsamples from a single treatment (principle 4) is not true replication, weakening the inference that differences between areas are due to the treatment or impact.
3. It is always preferable to examine the data before choosing a transformation and to test the data after transformation to determine its appropriateness (principle 9). Although the fourth root is a good general transformation, so is log $(x + 1)$, and it may be better than $\sqrt[4]{x}$ when low means are included. Log(x) has the attractive property of providing 95 percent CL as a $\times$ factor on back-transformation, but it cannot be used with zero observations.
4. It is always preferable to decide what level of precision is necessary to answer a particular question before choosing a sample size. Although this requires a preliminary estimate of sample variance (principle 5), analysis of various benthic data suggests that certain generalities can be made. Samples with means <3 to $5/$sample and fewer than 10 replicates generally are inadequate for quantitative analysis. The advice to take many small samples rather than few large samples reaches a limit when sample dimension is so small that zero counts are common.

TABLE 16.3:   Ten Principles to Guide Field Environmental Studies*

1. Be able to state concisely to someone else what question you are asking. Your results will be as coherent and as comprehensible as your initial conception of the problem.

2. Take replicate samples within each combination of time, location, and any other controlled variable. Differences among can only be demonstrated by comparison to differences within.

3. Take an equal number of randomly allocated replicate samples for each combination of controlled variables. Putting samples in "representative" or "typical" places is *not* random sampling.

4. To test whether a condition has an effect, collect samples both where the condition is present and where the condition is absent but all else is the same. An effect can only be demonstrated by comparison with a control.

5. Carry out some preliminary sampling to provide a basis for evaluation of sampling design and statistical analysis options. Those who skip this step because they do not have enough time usually end up losing time.

6. Verify that your sampling device or method is sampling the population you think you are sampling, and with equal and adequate efficiency over the entire range of sampling conditions to be encountered. Variation in efficiency of sampling from area to area biases among-area comparisons.

7. If the area to be sampled has a large-scale environmental pattern, break the area up into relatively homogeneous subareas and allocate samples to each in proportion to the size of the subarea. If it is an estimate of total abundance over the entire area that is desired, make the allocation proportional to the number of organisms in the subarea.

8. Verify that your sample unit size is appropriate to the sizes, densities, and spatial distributions of the organisms you are sampling. Then estimate the number of replicate samples required to obtain the precision you want.

9. Test your data to determine whether the error variation is homogeneous, normally distributed, and independent of the mean. If it is not, as will be the case for most field data, then (a) appropriately transform the data, (b) use a distribution-free (nonparametric) procedure, (c) use an appropriate sequential sampling design, or (d) test against simulated H_0 data.

10. Having chosen the best statistical methods to test your hypothesis, stick with the result. An unexpected or undesired result is not a valid reason for rejecting the method and hunting for a "better" one.

*Reprinted from Green (1979) with permission of John Wiley & Sons, Inc.

Modest precision (e.g. 95% CL $\pm$ 50% of the mean) likely requires at least 10 replicates, and high precision (e.g. 95% CL $\pm$ 15–25% of the mean) likely requires >50 replicates.

5. Similarly, in benthic sampling, means must differ by about 100 percent (two-fold) to be distinguished in an ANOVA design that uses 10 to 20 replicates per

treatment. A difference of 50 percent may be the smallest that is practical to distinguish, as this requires 30 to 50 replicates per treatment. It is likely that few benthic investigators would have the resources to collect sufficient replicates to distinguish between means that differ by less than 50 to 100 percent, and that may be the realistic limit to statistical power for all but the most ambitious study.

Acknowledgments

I benefitted from discussions with J. M. Elliott, T. T. Macan, and C. M. Drake, and from reviews of earlier drafts by D. D. Hart, B. L. Peckarsky, M. L. Reaka, S. A. Woodin, D. Wise, and anonymous reviewers. Clearly this chapter presents a personal perspective, however, and the final responsibility for it is mine. I am grateful to the Freshwater Biological Laboratory, Windermere, for space and library facilities. Research was supported by National Science Foundation grant DEB-77-11131, and the University of Maryland Computer Center.

References

Allan, J. D. 1975. The distributional ecology and diversity of benthic insects in Cement Creek, Colorado. Ecology 56:1040–53.

Allan, J. D. 1982. The effects of reduction in trout density on the invertebrate community of a mountain stream. Ecology 63:1444–55.

Anderson, F. S. 1965. The negative binomial distribution and the sampling of insect populations, p. 395. In: Proceedings of the XIIth International Congress of Entomology, London. 842 pp.

Benke, A. C. 1978. Interactions among coexisting predators—a field experiment with dragonfly larvae. Journal of Animal Ecology 47:335–50.

Brenchley, W. and K. Warington. 1958. The Park Grass Plots at Rothamsted. Rothamsted Experimental Station, Harpenden, England. 271 pp.

Chang, W. Y. B. and M. H. Winnell. 1981. Comment on the fourth-root transformation. Canadian Journal of Fisheries and Aquatic Sciences 38:126–27.

Chutter, F. M. and R. G. Noble. 1966. The reliability of a method of sampling stream invertebrates. Archiv für Hydrobiologie 62:95–103.

Cody, M. L. 1974. Competition and the structure of bird communities. Princeton University Press, Princeton. 318 pp.

Connell, J. H. 1961. The influence of interspecific competition and other factors on the distribution of the barnacle Chthamalus stellatus. Ecology 42:710–23.

Connell, J. H. 1974. Field experiments in marine ecology, pp. 21–54. In: R. Mariscal (ed.). Experimental marine biology. Academic Press, New York, NY. 373 pp.

Connell, J. H. 1980. Diversity and the coevolution of competitors, or the ghost of competition past. Oikos 35:131–38.

Conover, W. J. 1980. Practical nonparametric statistics. 2nd ed. John Wiley and Sons, New York, NY. 493 pp.

Cummins, K. W. and G. H. Lauff. 1969. The influence of substrate particle size on the microdistribution of stream macrobenthos. Hydrobiologia 34:145–81.

Darwin, C. 1859. The origin of species. John Murray, London. 502 pp.

Davis, S. F. and M. J. Winterbourn. 1977. Breakdown and colonization of *Nothofagus* leaves in a New Zealand stream. Oikos 28:250–55.

Downing, J. A. 1979. Aggregation, transformation, and the design of benthos sampling programs. Journal of the Fisheries Research Board of Canada 36:1454–63.

Downing, J. A. 1980. Precision vs. generality: a reply. Canadian Journal of Fisheries and Aquatic Sciences 37:1329–30.

Downing, J. A. 1981. How well does the fourth-root transformation work? Canadian Journal of Fisheries and Aquatic Sciences 38:127–29.

Eberhardt, L. L. 1978. Appraising variability in population studies. Journal of Wildlife Management 42:207–38.

Elliott, J. M. 1977. Some methods for the statistical analysis of samples of benthic invertebrates. 2nd ed. Freshwater Biological Association Scientific Publication No. 25. 160 pp.

Green, R. H. 1979. Sampling design and statistical methods for environmental biologists. John Wiley and Sons, New York, NY. 257 pp.

Hairston, N. G. 1980. The experimental test of an analysis of field distributions: competition in terrestrial salamanders. Ecology 61:817–26.

Hall, R. J., G. E. Likens, S. B. Fiance, and G. R. Hendrey. 1980. Experimental acidification of a stream in the Hubbard Brook Experimental Forest, New Hampshire. Ecology 61:976–89.

Hollander, M. and D. A. Wolfe. 1973. Nonparametric statistical methods. John Wiley and Sons, New York, NY. 503 pp.

Hurlbert, S. H. 1984. Pseudo-replication and the design of ecological field studies. Ecological Monographs. (In press.)

Hynes, H. B. N., D. D. Williams, and N. E. Williams. 1976. Distribution of the benthos within the substratum of a Welsh mountain stream. Oikos 27:307–10.

Kaushik, N. K. and H. B. N. Hynes. 1968. Experimental study of the role of autumn-shed leaves in aquatic environments. Journal of Ecology 56:229–43.

Macan, T. T. 1966. The influence of predation on the fauna of a moorland fishpond. Archiv für Hydrobiologie 61:432–52.

Macan, T. T. 1977. A twenty-year study of the fauna in the vegetation of a moorland fishpond. Archiv für Hydrobiologie 81:1–24.

Minshall, G. W. and J. N. Minshall. 1977. Microdistribution of benthic invertebrates in a Rocky Mountain (U.S.A.) stream. Hydrobiologia 55:231–49.

Needham, P. R. and R. L. Usinger. 1956. Variability in the macrofauna of a single riffle in Prosser Creek, California, as indicated by the Surber sampler. Hilgardia 24:383–409.

Paine, R. T. 1977. Controlled manipulations in the marine intertidal zone, and their contributions to ecological theory, pp. 245–70. *In:* C. E. Goulden (ed.). Changing scenes in natural sciences 1776–1976. Philadelphia Academy of Natural Sciences, Special Publication 12. 312 pp.

Peckarsky, B. L. 1979. Biological interactions as determinants of distributions of benthic invertebrates within the substrate of stony streams. Limnology and Oceanography 24:59–68.

Peckarsky, B. L. 1980. Predator-prey interactions between stoneflies and mayflies: behavioral observations. Ecology 61:932–43.

Peckarsky, B. L. and S. I. Dodson. 1980a. Do stonefly predators influence benthic distributions in streams? Ecology 61:1275–82.

Peckarsky, B. L. and S. I. Dodson. 1980b. An experimental analysis of biological factors contributing to stream community structure. Ecology 61:1283–90.

Petersen, R. C. and K. W. Cummins. 1974. Leaf processing in a woodland stream. Freshwater Biology 4:343–68.

Peterson, C. H. 1980. Approaches to the study of competition in benthic communities in soft sediments, pp. 291–302. In: V. S. Kennedy (ed.). Estuarine perspectives. Academic Press, New York, NY. 533 pp.

Rabeni, C. F. and G. W. Minshall. 1977. Factors affecting microdistribution of stream benthic insects. Oikos 29:33–43.

Reice, S. R. 1974. Environmental patchiness and the breakdown of leaf litter in a woodland stream. Ecology 55:1271–82.

Reice, S. R. 1977. The role of animal associations and current velocity in sediment-specific leaf litter decomposition. Oikos 29:357–65.

Reice, S. R. 1980. The role of substratum in benthic macroinvertebrate microdistribution and litter decomposition in a woodland stream. Ecology 61:580–90.

Reice, S. R. 1983. Predation and substratum: factors in lotic community structure, pp. 325–45. In: T. D. Fontaine, III, and S. M. Bartell (eds.). Dynamics of lotic ecosystems. Ann Arbor Science Publishers, Ann Arbor, MI. 494 pp.

Resh, V. H. 1979. Sampling variability and life history features: basic considerations in the design of aquatic insect studies. Journal of the Fisheries Research Board of Canada 36:290–311.

Rosenberg, D. M. and V. H. Resh. 1982. The use of artificial substrates in the study of freshwater benthic macroinvertebrates, pp. 175–235. In: J. Cairns, Jr. (ed.). Artificial substrates. Ann Arbor Science Publishers, Inc., Ann Arbor. 279 pp.

Sedell, J. R., F. J. Triska, and N. S. Triska. 1975. The processing of conifer and hardwood leaves in two coniferous forest streams: I. Weight loss and associated invertebrates. Internationale Vereinigung für Theoretische und Angewandte Limnologie Verhandlungen 19:1617–27.

Short, R. A., S. P. Canton, and J. V. Ward. 1980. Detrital processing and associated macroinvertebrates in a Colorado mountain stream. Ecology 61:727–32.

Siegel, S. 1956. Nonparametric statistics for the behavioral sciences. McGraw-Hill Book Company Ltd., New York, NY. 312 pp.

Simberloff, D. S. 1980. A succession of paradigms in ecology: essentialism to materialism and probabilism. Synthese 43:3–39.

Sokal, R. R. and F. J. Rohlf. 1981. Biometry. 2nd ed. W. H. Freeman and Company, San Francisco. 859 pp.

Southwood, T. R. E. 1978. Ecological methods. 2nd ed. Chapman and Hall, London. 524 pp.

Steel, R. G. D. and J. H. Torrie. 1960. Principles and procedures of statistics with special reference to the biological sciences. McGraw-Hill Book Company Ltd., New York, NY. 481 pp.

Taylor, L. R. 1961. Aggregation, variance and the mean. Nature (London) 189:732–35.

Taylor, L. R. 1965. A natural law for the spatial disposition of insects, pp. 396–97. *In*: Proceedings of the XIIth International Congress of Entomology, London. 842 pp.

Taylor, L. R. 1980. New light on the variance/mean view of aggregation and transformation: comment. Canadian Journal of Fisheries and Aquatic Sciences 37:1330–32.

Taylor, W. D. 1980. Comment on "Aggregation, transformation, and the design of benthos sampling programs." Canadian Journal of Fisheries and Aquatic Sciences 37:1328–29.

Thorp, J. H. and E. A. Bergey. 1981. Field experiments on responses of a freshwater, benthic macroinvertebrate community to vertebrate predators. Ecology 62:365–75.

Virnstein, R. W. 1978. Predator caging experiments in soft sediments: caution advised, pp. 261–73. *In*: M. L. Wiley (ed.). Estuarine interactions. Academic Press, New York, NY. 603 pp.

Willson, M. F. 1981. Ecology and science. Bulletin of the Ecological Society of America 62:4–12.

Wise, D. H. In press. The role of competition in spider communities: insights from field experiments with a model organism. *In*: L. Abele, D. Simberloff, D. Strong, and A. Thistle (eds.). Ecological communities: conceptual issues and the evidence. Princeton University Press, Princeton, NJ.

chapter 17

RESPONSES OF AQUATIC INSECTS TO ENVIRONMENTAL POLLUTION

Torgny Wiederholm

INTRODUCTION

Pollution may be defined as the introduction into the environment, by humans, of substances or energy that are likely to cause harm to living resources and ecological systems or to interfere with legitimate uses of the environment (Holdgate 1979; see also Warren 1971). More simply stated, pollution is something in the wrong place, at the wrong time, and in the wrong quantity.

From the earliest traces of civilization losses of plant nutrients and erosion of soil undoubtedly resulted when the protective plant cover was altered by clear-cutting of trees, burning, and plowing for the cultivation of land. Likewise, massive discharges of wastes into surface waters during industrial development and urbanization had a deleterious impact that spread from local areas to whole watersheds and, finally, to receiving coastal areas. Although many of the problems associated with these activities remain today, another dimension has been added; that is, the increasing transport of airborne pollutants such as acids, heavy metals, and organic compounds over large distances. Pollution is no longer confined to individual watersheds, nations, or even continents; increasingly it has become a matter of global concern.

Streams and lakes that are entirely free of pollution are becoming rare, and in many areas they no longer exist. Otherwise pristine lakes in central and northern Norway, for example, have background concentrations of metals that are above natural levels, due to the atmospheric supply of pollutants (Henriksen and Wright 1978). Similar conditions probably exist in other supposedly unpolluted areas within the Northern Hemisphere.

Although the early, subtle influence of a pollutant may well be significant to an undisturbed aquatic ecosystem, the present review will focus on more obvious situations of pollution. Even then, and despite the vast volume of literature available, one is faced with an embarrassing scarcity of hard facts on cause-effect relationships. Particularly disturbing is the lack of relevant data on the effects of pollutants on life-history features of aquatic insects (Lehmkuhl 1979; Waters 1979). Most studies have dealt with effects on community parameters, and conclusions on causal relationships have generally been based on circumstantial evidence rather than a procedure involving the testing of explicitly formulated hypotheses (see Allan, Chapter 16).

A major reason for the lack of conclusiveness in many pollution studies is the fact that pollution often involves several potentially interacting agents. For example, industrial wastewater may contain some substances that promote plant growth and others that are toxic to plant or animal life. Razing of forests and agricultural activities may increase nutrient availability to fresh waters but may decrease habitat heterogeneity. Thus, it is often difficult to resolve and quantify the contribution from each factor or, conversely, to predict the combined effect of several factors acting together.

Numerous articles and books have dealt with the subject of aquatic pollution and its biological effects. However, as the field of pollution ecology has broadened, it has become increasingly more difficult to find the necessary width and depth of coverage in treatments of this subject in a single volume. Hynes (1960) is probably still the best single work on the biology of polluted waters, particularly if read concurrently with his later book on running waters (Hynes 1970). Other useful texts are those of Warren (1971), Holdgate (1979), and Welch (1980). More technical aspects of pollution studies, including sampling and statistical analysis appropriate for benthic pollution studies, are included in a series of American Society for Testing and Materials publications (e.g. see Cairns and Dickson 1973; Mayer and Hamelink 1977; Dickson et al. 1978; Buikema and Cairns 1980) and books by Alabaster (1977), Hellawell (1978), and Green (1979).

The effects of pollution on aquatic insects will be treated here under headings that cover a broad overview of the field. The discussion will parallel the historical development of both pollution and pollution analysis as the focus of coverage shifts from inorganic and organic enrichment to siltation and thermal alteration, and, finally, to the addition of toxic materials and acidification.

EUTROPHICATION AND ORGANIC ENRICHMENT

Eutrophication is the enrichment of a water system by inorganic nutrients, primarily nitrogen and phosphorus (Hutchinson 1969). Eutrophication influences aquatic insects by increasing the production of algae and other vegetation that provide these organisms with substrate, shelter, and food (the latter may be either in a fresh form or as detritus). Organic matter that enters the system can be used similarly, although the bacterial decomposition of organic material may reduce the amount of oxygen available to animals for respiration, with serious consequences to many species of aquatic insects.

Although the effects of eutrophication on lakes have been the subject of much study, far less is known about the effects of this process on the fauna of running-water habitats (Hynes 1969). It is possible that the effects of eutrophication are not perceived to be as disturbing in running waters as they are in lakes, probably because extensive plant production, a characteristic of eutrophic conditions, is often prevented by light inhibition. This usually results from the high turbidity caused by the heavy sediment loads that often occur under eutrophic conditions.

Functional Zonation and the Influence of Nutrients in Lotic Environments

Many lotic systems have a longitudinal gradient that includes the whole range from nutrient-poor, or oligotrophic, to nutrient-rich, or eutrophic, conditions (e.g. see Hawkes 1975). In addition, many lotic systems are largely heterotrophic in their upper parts, depending on organic material that is produced in the terrestrial portion of their watersheds (e.g. Hynes 1975; Merritt et al., Chapter 6). Under natural conditions, the shift from predominance of heterotrophy to autotrophy is gradual and is related to both stream size and geographical region. It usually occurs in the range of third to fourth-order streams but, when riparian vegetation is restricted (as occurs at high elevations, high latitudes, and in open areas), autotrophy may be predominant even in first-order streams. Heterotrophic processes tend to be predominant again in downstream reaches (Minshall 1978; Anderson and Sedell 1979; Cummins 1980; Vannote et al. 1980).

The relative contribution of heterotrophy compared to autotrophy is important because different functional groups of invertebrates are involved in the processing of organic matter, depending on the character of this material (Cummins and Klug 1979). For example, leaf-shredding insect larvae are characteristic of heterotrophic headwater communities (Cummins 1973, 1978; Cummins and Klug 1979; Williams 1981). Attached algae, with associated microfauna and detritus (i.e. the Aufwuchs or periphyton in a broad sense), are an important food resource for the grazing insects in autotrophic mid-reaches of a lotic system. Macroinvertebrates that function as collectors are more abundant in downstream reaches (see Merritt et al., Chapter 6, for a definition of functional groups).

The general effect of eutrophication in nutrient-poor heterotrophic streams is to shift the longitudinal succession of both heterotrophic and autotrophic processes upstream (Odum 1956; Hynes 1969). Thus, insects that are typical of the predominantly autotrophic, unpolluted midstream zone will occur further upstream, provided that factors other than nutrient concentrations are favorable for plant growth. The heterotrophic character of the lower, "potamon" region of a stream likewise will extend further upstream as the formation of detritus, the degree of turbidity, and other factors are accelerated by eutrophication.

Mild eutrophication will favor organisms such as grazing mayflies, filter-feeding black flies and hydropsychid caddisflies, and deposit-feeding Chironomidae. In contrast, when eutrophication results in substantial amounts of detritus being formed and accumulated, pronounced oxygen deficiency may result, which may adversely affect many insects. At this stage, however, the environment will differ from one having nutrient-poor conditions in many other ways as well. Silt may smother hard surfaces and fill interstices in the stream bed. As a result, attachment and normal feeding activities will be difficult for many insects, and their preferred microhabitat will be entirely eliminated. Thus, the effects of excessive eutrophication will be similar to those that result from organic pollution (see following discussion).

Lake Outlets

A rich community of filter-feeders that use the organic material produced in the lentic environment is often found in lake outlets and below reservoirs (e.g. see Ward 1976; Armitage 1978; Mackay and Wiggins 1979). Such communities may have a more eutrophic character than those downstream, but the enrichment effects usually do not persist very far below these outlets (Armitage 1978; Wotton 1979). Thus, eutrophication of lakes, besides affecting the lentic fauna, can influence stream communities as well (see Ward, Chapter 18).

Organic Pollution in Streams and Rivers

Organic pollution generally reduces the number of insect species; however, since some species benefit from the increased supply of food, the total numbers of animals may increase (Hynes 1960). Few common river insects survive at very high loads of organic material where complete deoxygenation occurs, but some air-breathing species (e.g. *Eristalis, Psychoda,* and *Culex*) may occur (Hynes 1960).

The pattern of succession downstream from a zone of gross organic pollution is familiar (e.g. see Hynes 1960; Hawkes 1979; Wielgosz 1979) and oligochaetes (*Tubifex tubifex* and *Limnodrilus hoffmeisteri*) are the only macroinvertebrates that occur in high numbers in the most polluted zone where minimal amounts of oxygen occur. A region of rich algal growth (e.g. *Cladophora*) occurs further downstream, and autotrophic conditions predominate there; the crustacean *Asellus* is characteristic of this region. Increased numbers of insects such as the European species *Sialis lutaria* and *Baetis rhodani*, along with *Tanytarsus* sp. and other Chironomidae, also occur in this region. Dense mats of *Cladophora* often harbor several species of the chironomid *Cricotopus* (e.g. see LeSage and Harrison 1980).

Further downstream, filter-feeders such as *Hydropsyche* and some species of Simuliidae may be found, often in great numbers. The assemblage of insects at this stage is numerous and fairly rich in species. Species of Plecoptera are usually among the last insects to reappear following organic pollution.

The last effect to be observed in the recovery of a stream from organic pollution is increased numbers of animals, which is a condition that is also typical of streams receiving very mild organic pollution. Most species that are present in unpolluted conditions are represented in these situations, but the balance between species may be altered, possibly because of different responses to changes in the types of algae or detritus available (Hynes 1960) or due to sedimentation of the substrate. As in an unpolluted environment, the ability to use the various components of available food may play a key role in determining fitness and species distribution, but the relative importance of algae compared to detritus is often difficult to assess (McCullough et al. 1979), and the importance of algal community structure to consumer abundance and composition is virtually unknown (Hawkins and Sedell 1981; see also Lamberti and Moore, Chapter 7).

The Littoral Zone of Lentic Habitats

Most lakes are small and shallow, and a major part of the total primary production occurs in the littoral vegetation of these lakes where the Aufwuchs, or periphyton, grows on macrophytes (Wetzel 1975; Pieczyńska 1976; Saunders 1980; Westlake 1980; see also Lamberti and Moore, Chapter 7). The littoral zone of lakes has been largely ignored faunistically, even though a rich animal life occurs both on solid substrates and in stands of macrophytes (e.g. see Pieczyńska 1972; Glowacka et al. 1976; Prejs 1976; Barton and Hynes 1978).

Presumably, eutrophication affects insect life in wave-exposed littoral areas in a way similar to that in riffle areas of lotic habitats (i.e. by increasing food availability, but also by modifying the character of the substrate). However, few observations on the responses of insects exist. Streit and Schröder (1978) reported increased abundances of the trichopterans *Agraylea multipunctata* and *Hydroptila* sp. in the stony littoral zone of Lake Constance, central Europe, since the 1940s, which has been a period of increased eutrophication. Observations in Lake Geneva, Switzerland, by Lang (1974) indicated that three species of Trichoptera were less abundant than reported earlier, or were completely absent. These changes were attributed to increased algal growth resulting from eutrophication.

The littoral benthos that occurred among macrophytes in a more protected part of Lake Constance also changed over 20 years of increased eutrophication (Reavell and Frenzel 1981). The burrowing mayfly *Ephemera* sp. was abundant in 1948–1949 but had disappeared by 1968. The periodic occurrence of low oxygen levels and the increased access of vertebrate predators due to alterations in macrophyte density were probable causes. In contrast to *Ephemera*, the biomass of invertebrates in the lake increased approximately ten-fold, although Oligochaeta replaced Chironomidae as the dominant organisms (Frenzel 1983).

Gross organic pollution impoverishes the littoral fauna of lakes, just as it does in lotic environments. Species of Chironomidae, Psychodidae, Culicidae, and Oligochaeta often predominate under such conditions, whereas species of Trichoptera, Ephemeroptera, and other groups typical of unpolluted conditions are usually absent. The Aufwuchs assemblage may recover and actually become more diverse within a rather short distance from the outfall due to rapid dilution of sewage water; in contrast, effects on benthic animals may persist further out into the lake (Pieczyńska et al. 1975; Ozimek and Sikorska 1976).

Insects living in the sublittoral zone of lakes experience the same deleterious effects from eutrophication as those in littoral sediments. The increased oxygen demand of water and sediments may result in heavy oxygen depletion during periods of calm weather. Probably the best example of this effect is the catastrophic decline of the mayfly *Hexagenia* in western Lake Erie, Ohio, which occurred during a period of calm weather in the summer of 1952. As much as 30,000 metric tons of mayflies may have been lost; populations of *Hexagenia* never recovered, but they were replaced by worms, midges, and other tolerant organisms (Wood 1973; Cook and Johnson 1974).

The Profundal Zone of Lentic Habitats

Although the profundal environment is more homogeneous than that of the littoral, different zones of erosion or deposition and the occurrence of temperature and oxygen gradients often result in large-scale habitat heterogeneity (Brinkhurst 1974; Sly 1978; Håkanson 1981). The number of animals per surface area may be quite high within the profundal zone but, except for the Chironomidae, few groups of insects become abundant.

The predominance of Chironomidae in the lake profundal, and their response to trophic conditions, have made them useful for classifying lakes (Thienemann 1922; Brundin 1949, 1958; Sæther 1975). Several lake types are named after some of the characteristic members of their chironomid assemblages (Table 17.1). Sæther (1979) recognized 15 different profundal chironomid associations, which were later given trophic index numbers by Warwick (1980a). Such trophic indices have been correlated with mean concentrations of total phosphorus/mean depth or with total chlorophyll/mean depth in many North American and European lakes (Wiederholm 1979, 1980b; Sæther 1980).

Eutrophication results in rather predictable changes in chironomid species composition, and different chironomid assemblages generally succeed each other (Fig. 17.1). Eutrophication changes the species composition most rapidly under initially nutrient-poor conditions, but shallow lakes have a more eutrophic assemblage of benthic insects than deep lakes with similar concentrations of phosphorus. In other words, the nutrient addition and primary production that is needed to change the species composition of the benthic insect fauna is larger in a deeper lake. Likewise, there is a larger discrepancy between the trophic status that is apparent from algal biomass and that apparent from the composition of the bottom fauna in a deeper lake (Sæther 1980; Wiederholm 1980a).

The effects of eutrophication are generally similar in different geographical areas (e.g. see Johnson and Brinkhurst 1971; Wiederholm 1974; Jumppanen

TABLE 17.1: Lake Classification Based on the Profundal Chironomid Fauna*

LAKE TYPE	EUROPE	NORTH AMERICA	TROPHIC STATUS
I	*Heterotrissocladius subpilosus:* lakes	*Heterotrissocladius oliveri:* lakes	Ultraoligotrophic
II	*Tanytarsus lugens:* lakes (with *Heterotrisso- cladius grimshawi* or *H. scutellatus*)	*Tanytarsus* sp.: lakes (with *Monodiamesa tuberculata* and *Heterotrissocladius changi*)	Oligotrophic
II/III	*Stictochironomus rosenschoeldi* and *Sergentia coracina:* lakes	*Chironomus atritibia* and *Sergentia cora- cina:* lakes	Mesotrophic
III	a) *Chironomus anthra- cinus:* lakes	a) *Chironomus de- corus:* lakes	Moderately eutrophic
	b) *Chironomus plumosus:* lakes	b) *Chironomus plumosus:* lakes	Strongly eutrophic
IV	*Chironomus tenuistylus:* lakes (with *Zalutschia zalutschicola*)	*Chironomus* sp.: lakes (with *Zalutschia zalutschicola*)	Dystrophic

*From Brundin 1958; Saether 1975

1976; Saether 1980; Kansanen and Aho 1981). In addition to changes in species composition, slight eutrophication or mild organic pollution usually increases the abundance and biomass of benthic organisms and also the total number of species, as it does in lotic habitats. Some free-living, predatory species of the chironomid *Procladius* and the phantom midge *Chaoborus* are generally the last insects that remain in the lake profundal zone if hypereutrophic conditions develop.

The structure of profundal benthic communities was initially thought to be determined by differences in tolerance to low oxygen concentrations (Thiene- mann 1922; Brundin 1949). Following this, the observed pattern of increasing mean size of benthic species in increasingly eutrophic lakes was suggested as important, since larger size would enable larvae to break through the layer of oxygen microstratification that occurs near the bottom of lakes (Brundin 1951). Later, the significance of food supply was emphasized more (Warwick 1975; Wiederholm 1976; Saether 1979) but, unfortunately, the quantity and quality of

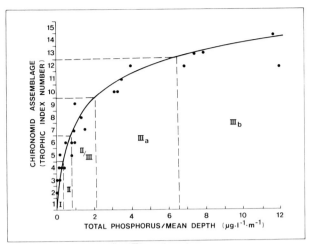

Figure 17.1 Profundal chironomid assemblages relative to lake fertility. Modified after Saether (1979), see also Saether (1980). *Roman figures* refer to the lake type system of Brundin (1958) and Saether (1975) (see Table 17.1); trophic index numbers are from Warwick (1980a), and lake names are given in Saether (1979).

organic matter that reaches the sediments in different types of lakes is poorly known, and this prevents a further analysis of food as a regulating factor. However, in view of the environmental differences at the sediment/water interface among lakes of different trophic state, it is apparent that different physiological, morphological, and behavioral traits may be important in the different stages of a trophic gradient. For example, *Chironomus anthracinus,* which occurs in eutrophic lakes, is adapted to seasonal variations in its habitat by its ability to survive the long periods without oxygen (Nagell and Landahl 1978) that may occur in deep eutrophic lakes in summer. The larvae remain inactive in the absence of oxygen, but grow and mature in fall and in spring and early summer when phytoplankton production is high and oxygen is abundantly available (Jónasson 1972). In contrast, tolerance to poor oxygen conditions has no survival value in oligotrophic lakes where the bottom water usually is near saturation all year around. Here the lack of food is the main problem, and those insect species that live in strongly oligotrophic lakes must be particularly efficient in obtaining and utilizing what food there is. It is perhaps significant that six of the seven major chironomid species in the oligotrophic Char lake, which is in the Canadian high arctic, were reported to be free living (Davies 1975), because such vagility should increase their range of foraging. In eutrophic lakes, where food is more abundant, the majority of Chironomidae are sessile, feeding by filtration or browsing around the tube in which they live.

ADDITION OF SEDIMENT AND OTHER SOLIDS

Erosion of undisturbed watersheds usually releases rather small amounts of particulate material (Bormann et al. 1969), whereas certain farming, forestry, or mining practices—along with dredging, industrial, or construction activities—often result in the introduction of substantial amounts of such solids. The effects of these introductions on aquatic insects, which have largely evolved in fairly clear water (Hynes 1973), may be quite serious. The organisms can be affected directly, such as when food collection or respiration is obstructed, or indirectly, such as when depletion of the resources on which they depend occurs. In the latter case, suspended or sedimented material can be deleterious because these substances reduce light penetration and consequently plant growth, smother hard surfaces, and fill interstices within the substrate. Specific pollutants associated with sediment can create problems in addition to those caused by the inert material.

Lotic Environments

Experimental studies have demonstrated that the primary effect of sediment addition to a stream is to initiate drift of animals from the affected site (White and Gammon 1976; Rosenberg and Wiens 1978, 1980). Studies on the mayfly *Ephemera subvaria* in artificial streams indicated that minor spates may initiate significant increases in drift when siltation is involved (Ciborowski et al. 1977). In general, the effect of a prolonged heavy sediment load on streams is to reduce the number of species and the density of animals, although some groups may take advantage of the altered habitat conditions. Size, morphology, and behavior are important in determining tolerance to sediment load (McClelland and Brusven 1980). For example, burrowing and deposit-feeding forms such as Oligochaeta and some Chironomidae tend to be favored by such alteration of habitat (Nuttall and Bielby 1973).

Several studies have reported severe effects of logging on stream invertebrates (Iwamoto et al. 1978). Streams subjected to the effects of logging had lower invertebrate diversity and reduced numbers of some organisms (e.g. Plecoptera) but higher overall densities of others (e.g. Oligochaeta, Chironomidae, Ephemeroptera) within one to five years after logging (Graynoth 1979; Newbold et al. 1980). Buffer strips 30 to 60 m wide have been proposed as a method of protecting streams from adverse effects of logging (see Ward, Chapter 18).

Drainage from mining activities may cause water quality problems in several ways (Greenfield and Ireland 1978; Letterman and Mitsch 1978; Moon and Lucostic 1979; Aanes 1980; Scullion and Edwards 1980; see also later sections). In terms of sediments, precipitation of ferric hydroxide under neutral and oxidized conditions may result in complete blanketing of the stream bottom, which detrimentally affects both invertebrates and fish. The deposit may support a restricted fauna that is dominated by Oligochaeta and chironomid larvae, but sometimes contains a few species of other insect orders. Generally, the Ephemeroptera, Plecoptera, Trichoptera, and Coleoptera are eliminated or greatly reduced in both abundance and number of species. Among those taxa that remained in some numbers in a Welsh stream that was subjected to mining impacts were the mayfly *Baetis rhodani*, the ceratopogonid *Bezzia* sp., the empidid *Chelifera flavella*, and the coleopteran *Limnius volkmari* (Scullion and Edwards 1980). The net-spinning caddisfly *Hydropsyche* sp. occurred in an impacted Pennsylvania stream (Letterman and Mitsch 1978).

Suspended coal particles may also deleteriously affect Ephemeroptera, Plecoptera, and Trichoptera, although species such as the mayfly *Baetis rhodani*, the stonefly *Amphinemura sulcicollis*, and the caddisflies *Hydropsyche pellucidula* and *Rhyacophila dorsalis* appear to be more resistant than others (Scullion and Edwards 1980). Siltation by heavy coal ash may eliminate a variety of insects (e.g. see Learner et al. 1971), but Chironomidae and Odonata are less susceptible than others (Cherry et al. 1979).

Nutrients, pesticides, increased turbidity, and other factors may influence the quality of waters that receive agricultural runoff; thus, the impact of sediment alone is often difficult to isolate. When streams that drain areas with different land use have been compared, grazers and detritivores dominate under silty conditions when there is a low allochthonous input of organic material (e.g. see Dance and Hynes 1980). However, the effects of land-use disturbances in upstream areas are not necessarily detectable in downstream reaches when the degree of impact changes along the course of a stream (Marsh and Waters 1980).

Sediments do not remain where they are originally deposited, except under very stable hydrodynamic conditions (see Newbury, Chapter 11), so the stream fauna may recover when the deposited material is flushed out of the system. Recovery may occur after spates, or as a result of normal seasonal variation in current velocity but, for that to occur, recolonization by drift from upstream areas is important (Tebo 1955; Hynes 1973; Rosenberg and Snow 1977; see also Sheldon, Chapter 13).

Lentic Environments

The effects of increased sediment load in lakes and reservoirs are poorly known. However, head capsules of chironomid larvae from sediment cores have been used to evaluate a lake's response to environmental conditions, including climatic change and anthropogenic influence, sometimes over periods of 10,000 years or more (e.g. see Frey 1964, 1977; Stahl 1969; Hofmann 1971a, 1971b; Warwick 1975, 1980a, 1980b). Sediment cores from the Bay of Quinte, Lake Ontario, Canada, indicated that in the stratum corresponding to the period of most intensive logging in the watershed (about 1850–1860), the relative abundance of Chironomidae representative of oligotrophic conditions increased relative to the abundance of Chironomidae that are more typical of eutrophic conditions (Fig.

17.2). The rapid accumulation of mineral sediments during this period of large scale deforestation and erosion of the watershed (which resulted in a high input of clay to the lake) reduced the availability of food materials, as represented by the increase in the organic matter:organic carbon profile (Fig. 17.2) Some organisms declined as a result of this (e.g. *Ablabesmyia peleensis, Polypedilum simulans* gr., and *Chironomus*), and a more oligotrophic fauna was able to persist or even increase in number (e.g. *Microspectra* and Tanytarsini type 7). After mineral sedimentation declined, food materials became more available again, and probably even increased due to eutrophication, and Tanytarsini and *Microspectra* were replaced by *Chironomus*. This development continued into the depleted *Chironomus-Procladius* fauna tolerant of present-day eutrophic conditions.

Observations from other studies support the idea of inorganic siltation resulting in effects similar to an oligotrophication process (e.g. see Nursall 1952; Grimås 1961). Whether the addition of sediment reduces the lake productivity in general (e.g. by reducing light transmission and primary production), or directly affects the bottom fauna (e.g. by diluting the nutrient content of seston and sediments), is not known.

Interestingly, the two species of Chironomidae that were apparently favored by increased sedimentation in the Bay of Quinte, *Stempellina* sp. and *Abisko-*

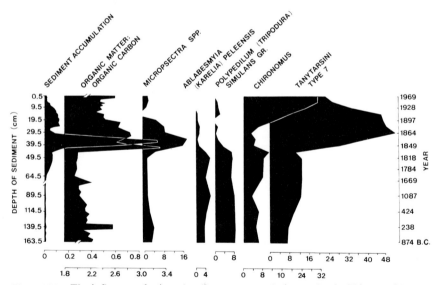

Figure 17.2   The influence of mineral sediment accumulation on lentic Chironomidae, as reflected in subfossil material from the Bay of Quinte, Lake Ontario. Sediment accumulation is expressed as g dry sediment per cm^2 per yr, and relative abundance of Chironomidae is presented as percent of total individuals (horizontal scale). Modified from Warwick (1980a).

myia sp., both inhabit cases that they can transport across the bottom (Warwick 1980a). Two case-dwelling species of Chironomidae that were favored by increased inorganic sedimentation in a regulated lake in northern Sweden, *Constempellina brevicosta* and *Abiskomyia virgo* (Grimås 1961), are also able to move their cases freely about the lake bottom. Possibly, a transportable case enables these larvae to exploit food resources in the sediment/water interface, whereas the tubes of more sedentary chironomid species would be clogged by sedimentation deposits (Grimås and Wiederholm 1979). In general, a high efficiency in food uptake and metabolism, and an ability to avoid siltation, should be favorable traits for surviving conditions of high inorganic sediment deposition.

TEMPERATURE ALTERATIONS

Temperature is of overriding importance to nearly all functions within an ecosystem. The discussion here will be limited to the direct influence of temperature alterations on insects, but other effects may result if, for instance, flow conditions or the rates of production and mineralization processes change due to thermal impact. Ward and Stanford (1982) and Sweeney (Chapter 4) consider various aspects of thermal influence on the ecology of aquatic insects in more detail.

Alterations of the thermal regime in a water body occur in many ways. The effects of industrial discharges, particularly in the form of cooling water from power plants, are immediately obvious; however, agricultural and forestry activities and urbanization probably are more important because they may influence large areas or even whole watersheds (Wurtz 1969). Removal of riparian vegetation and the channelization and regulation of streams may change not only mean temperature, but also the diurnal and seasonal pattern of temperature fluctuation, and this may have profound biological consequences.

Tolerance Limits

Definitive temperature-tolerance limits of aquatic insects are difficult to establish. The lethal levels for some species of Plecoptera and Ephemeroptera are around 20°C (Whitney 1939; Nebeker and Lemke 1968). Several species of Chironomidae have been shown to be somewhat more hardy (Walshe 1948). Critical or lethal temperatures around 40°C have been demonstrated for some Odonata (Garten and Gentry 1976; Cherry et al. 1979). Variation in temperature tolerance occurs within aquatic insect groups. For example, temperatures just below 30°C were fatal to most mayflies in a Czechoslovakian river, but one mayfly species, *Potamanthus luteus,* was slightly more resistant (Obrdlík et al. 1979).

In terms of community responses, field studies have indicated that temperatures above 30°C cause a reduction in species number, abundance, biomass, and production, whereas moderate heating (below 25°C) may result in an increase in biomass and production without limiting diversity (Dusoge and Wieśniewski 1976; see also Russian studies referred to therein). In two examples, temperatures of 35 to 41°C eliminated virtually all animals from cooling-water channels (Durrett and Pearson 1975; Parkin and Stahl 1981).

The tolerance of aquatic insects to high temperatures seems to be related to the kind of environment in which the species normally live. Greater tolerance to high temperatures has been demonstrated for species of Ephemeroptera, Odonata, and Chironomidae that live in stagnant waters than for those species that live in streams (e.g. see Whitney 1939; Walshe 1948; Garten and Gentry 1976). Odonata collected in streams and lakes receiving cooling water were more heat resistant than those from unaffected habitats (Garten and Gentry 1976). Physiological acclimation and selection for high thermal tolerance are probably important adaptive strategies (Gentry et al. 1975). Whether selection for tolerant strains occurs in the rather short time perspective of pollution activities needs further study, but evidence from toxicological studies indicates that this is a strong possibility (e.g. see Bryan and Hummerstone 1973; Fraser et al. 1978; Metcalf 1980).

Effects on Growth and Development

The sublethal effects of temperature on metabolism and growth may be more important than the absolute tolerance to high temperature that an aquatic insect may exhibit. The rates of most biological functions increase with higher temperatures, although many aquatic insects are able to adjust their rates of metabolism as the environmental temperature changes (Wigglesworth 1965). Such an increase may be manifested in a higher growth rate (Markarian 1980) and a shorter developmental time (Harper 1973; Parkin and Stahl 1981). Thus, temperature may directly or indirectly influence the competitive ability and distribution of species both within drainage systems and over large geographic areas (Hynes 1970; Sweeney and Vannote 1978). For example, it has been suggested that a change in body size and fecundity, induced by thermal pollution, may cause elimination or replacement of species, which result in a community with altered or lower diversity at the polluted site (Sweeney and Vannote 1978; see also Sweeney, Chapter 4). Adult size and fecundity, then, may be important variables to use in monitoring the sublethal effects of temperature alterations.

Effects on Emergence

Increased temperatures may also disrupt the seasonal emergence pattern of aquatic insects. Unseasonally high water temperatures in winter caused emergence to occur up to five months early, and a greater time lag occurred between emergence of males and females in laboratory experiments on ten species of aquatic insects (Nebeker 1971). Similar observations were made by Rupprecht (1975), who also reported gill damage in *Perla* stonefly larvae kept at high temperatures. Temperature increases could have serious effects in natural communities because insects emerging too early in the season might be killed or inactivated by low ambient temperatures (Nebeker 1971), and an extended period of time between emergence of males and females might prevent mating, thereby eliminating the species from the impacted habitat.

These predicted effects of increased temperatures on emergence patterns have been verified in some field studies but not in others. A two-year study in the River Severn, England, revealed no detectable effects on emergence, abundance, or diversity of insects when mean weekly temperature increased up to 8°C, or degree hours increased by 25 percent in the early growing season (Langford 1975). Langford concluded that factors such as photoperiod, flow, and meteorological conditions may override the effects of temperature and/or that a temperature threshold must be exceeded before control of emergence by other factors is released (see also Hynes 1970). Other field studies have shown that earlier emergence does occur when water temperature is elevated (Fey 1977; Mattice and Dye 1978). For example, *Hydropsyche pellucidula* emerged up to four months earlier than normal immediately below a power station in the River Lenne, Germany (Fey 1977).

Low Temperature Effects

The lowering of water temperatures, such as may occur at the release of cold, hypolimnetic water from reservoirs into streams during summer, may be detrimental to aquatic insects if their metabolism and growth decrease to the point that the insects become competitively inferior to others that are not similarly affected (Rupprecht 1975; Ward and Stanford 1979; Ward, Chapter 18). Lowered fecundity, apparently because of smaller size of adults, has been related to reduced temperature (Sweeney 1978; Sweeney and Vannote 1978; Ward and Stanford 1979), and may place a species at a competitive disadvantage.

Disturbance of Diurnal or Seasonal Patterns

Natural fluctuations are important in regulating the temporal pattern of insect life histories. Diurnal or seasonal disruptions may prevent the animals from optimally adapting their development to environmental conditions, although it is difficult to separate the influence of temperature *per se* from that of temperature fluctuations (see Sweeney, Chapter 4). Diurnal temperature fluctuations and growth seem to be linked, possibly because different physiological processes have different thermal optima. Diurnal and seasonal temperature fluctuations also may be effective in preventing competitive exclusion, and hence in promoting coexistence of more species (Ward and Stanford 1979). In this sense, constant temperatures could have a negative effect on species richness and diversity.

The depleted fauna downstream of many reservoir outflows in North America has been partly attributed to elevated winter temperatures caused by outflowing hypolimnetic water (Lehmkuhl 1972; Ward 1974; Ward and Stanford 1979). Many species in regions where the temperature normally drops to near 0°C in winter require these decreased temperatures to complete their development, and are adversely affected by constantly higher winter water temperatures from the reservoir outflows.

HEAVY METALS

Interest in the effects of heavy metals on aquatic insects is a result of the wide use and release of these metals into the aquatic environment, and their documented toxicity to both aquatic organisms and humans (e.g. see Nriagu 1978a, 1978b, 1979a, 1979b, 1979c, 1980a, 1980b; Förstner and Wittmann 1979; Webb 1979).

Toxic Effects

Field studies have shown that heavy metal pollution often reduces the abundance and species richness of aquatic insects and changes the proportional abundance of different groups (e.g. see research summarized in Hynes 1960; Eyres and Pugh-Thomas 1978). Evidence suggests that there is increasing tolerance to heavy metal pollution in the sequence from mayflies, to caddisflies, to midges (e.g. see Savage and Rabe 1973; Winner et al. 1975, 1980; Solbé 1977; Armitage 1980). Of course, it should not be assumed that taxonomic groups above the species level behave in a uniform way (e.g. see Resh and Unzicker 1975) but, rather, that some species of these groups do. The chance that some species of a particular group will be tolerant or intolerant is also related to the number of species in that group. For example, in the European fauna (Illies 1978), the number of species in the family Chironomidae outnumber those of the orders Ephemeroptera and Trichoptera by about ten to one, and three to one, respectively. Thus, because of such numerical dominance, the Chironomidae will appear as more highly tolerant to extreme conditions than these other groups.

Short-term toxicity tests have indicated that insects may be more tolerant than fish and other invertebrates to heavy metals (Warnick and Bell 1969). However, although such tests may be useful in demonstrating relative differences in the sensitivity of various insects to heavy metals, the ecological significance of these results is unclear because of the highly unnatural experimental conditions used. Lethal effect levels in such tests are often three or four orders of magnitude above the concentrations that occur naturally in fresh waters (Warnick and Bell

1969; Rehwoldt et al. 1973; Clubb et al. 1975a), and heavy metals are rarely found in these amounts even under severely polluted conditions.

In contrast to the results of short-term toxicity tests, when longer exposure times were used the results indicated that some insects may be equally or even more sensitive than fish (Spehar et al. 1978). Such tests also underline the importance of sublethal effects and the need to focus tests on susceptible stages in the insect's development to achieve ecologically meaningful results (Elder and Gaufin 1974; Spehar et al. 1978). For example, long-term exposure of several insects to cadmium (Cd) concentrations well below those that are acutely toxic reduced the incidence of molting and inhibited emergence (Clubb et al. 1975a). Also, small, immature insects were killed at lower concentrations of Cd than were larger, mature insects. Water with cobalt (Co) at a concentration of 5.2 ppb contained nymphs of *Ephemerella ignita* that weighed less, took longer to emerge, and sometimes produced nymphs that lacked the ability to fly after emergence (Fig. 17.3) (Södergren 1976). This concentration is about 10 to 20 times the natural background level, but it is nearly three orders of magnitude lower than the LC_{50} reported for related species of mayflies (e.g. *Ephemerella mucronata*, Södergren 1976; and *E. subvaria*, Warnick and Bell 1969). Decreased

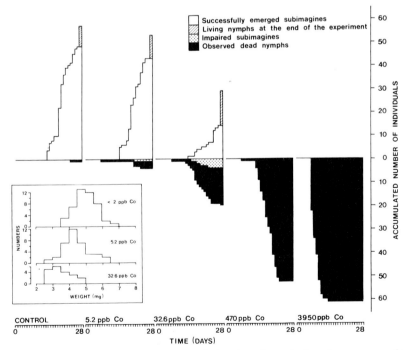

Figure 17.3   Effects of cobalt (Co) on growth (see *insert*), survival, and emergence of the mayfly *Ephemerella ignita* in running water tests with Co added as Co(NO₃)₂. Modified after Södergren (1976).

weight (e.g. as much as 90%), and both delays and reductions in successful emergence were also found for larvae of the midge *Chironomus tentans* that were laboratory reared in sediments that were excessively polluted by chromium (Cr), zinc (Zn), and particularly Cd (Wentsel et al. 1977b, 1978). Given a choice, larvae avoided these polluted sediments (Wentsel et al. 1977c). In a field study, *C. tentans* was absent from the most polluted lake sediment, with the oligochaete *Limnodrilus hoffmeisteri* the only survivor there (Wentsel et al. 1977a).

Behavioral disturbances may influence the overall fitness of an organism in a polluted environment and may also serve as a useful index of environmental quality. For example, the ability of *Hydropsyche* caddisfly larvae to construct nets can be affected by high concentrations of copper (Cu) effluents from a chloralkali plant, or a mixture of heavy metals at stream sites polluted by surface-plating and other industrial activities (Besch et al. 1979; Petersen and Petersen 1983). However, the pesticide Fenethcarb had the same effect (Besch et al. 1977), therefore, the behavioral disturbance is apparently not specifically related to the influence of heavy metals.

Secondary effects of heavy metals, which influence other components of the ecosystem and consequently affect aquatic insects, may also prove to be significant. For example, higher numbers of Oligochaeta and Chironomidae occurred within lake enclosures to which mercury (Hg), Cu, Cd, Zn, and lead (Pb) were added (Lang and Lang-Dobler 1979). This may have resulted from an increase in food supply to the benthos because the zooplankton were adversely affected by heavy metals and, consequently, their grazing was reduced (Gächter and Máreš 1979; Urech 1979).

Life-history features of aquatic insects often determine much of their susceptibility to heavy metals and other toxicants. For example, since concentrations of Co were highest at low flows in late fall and winter, winter species were more severely affected by Co pollution in the River Rickleån, northern Sweden, than were summer species (Södergren 1976). Winter generations of mayflies and black flies were almost eliminated, and this had detrimental effects on juvenile salmon, which fed on these organisms during the winter. Likewise, organisms living on *Fontinalis* and feeding on its periphyton assemblage, which was where Co accumulated, were most affected.

Insects accumulate heavy metals in varying degrees, which is a characteristic that is useful for monitoring purposes. This aspect of heavy metal pollution will not be considered here, but further discussions of this appear in Prosi (1979) and Phillips (1980).

Habitat Factors Influencing Toxicity of Heavy Metals

The varying responses of benthic invertebrates to heavy metals, including the apparent lack of response on some occasions (e.g. see Brooker and Morris 1980), is the result of metals being less toxic under certain circumstances and/or the organisms apparently adapting to live at high heavy metal concentrations. It is well known that the toxic effects of heavy metals vary with their chemical properties and with habitat variables such as temperature, pH, oxygen content, alkalinity or hardness, and the amount of complexing substances (e.g. organic material) in the water (Prosi 1979; Kaiser 1980). For example, since clear, oligotrophic waters have lower amounts of complexing material than eutrophic waters, the biota of oligotrophic waters would be more exposed to the toxic action of metals. Heavy metals and other pollutants also may act synergistically or, in fact, they may actually reduce the individual effect of each other. Therefore, considerable variation in the effects of interacting pollutants can be expected under field conditions.

In general, dissolved metals are more toxic than metals in other forms (e.g. as precipitates). Since pH affects metal solubility, the pH of a polluted body of water is important. This may be one reason for the negative influence of acid-mine water and acid precipitation on freshwater invertebrates (e.g. see Roback and Richardson 1969; Tomkiewicz and Dunson 1977), in addition to the direct effect of low pH on the organisms. In water courses with high levels of precipitated and sediment-bound metals, toxic pulses may result from periodic acid discharges and the consequent increases in dissolved metals—even though the concentration of dissolved metals in water is low at other times (Eyres and Pugh-Thomas 1978).

Higher temperatures often increase the effects of chemicals on organisms, and heavy metals are no exception in this respect. Little pertinent research has been conducted on invertebrates (Cairns et al. 1975, 1978), although Fey (1977) demonstrated increased toxicity of lead to larvae of *Hydropsyche* caddisflies at higher temperatures.

Fish are more sensitive to heavy metal toxicity at low oxygen concentrations (e.g. see Lloyd 1965), but the response of insects is unclear. For example, the effect of Cd on insects decreased at lower oxygen concentrations, presumably because of lower metabolism and consequently lower Cd uptake (Clubb et al. 1975b).

Adaptation

Some populations of terrestrial plants, aquatic plants, and aquatic invertebrates that live in contaminated environments are tolerant to heavy metals (e.g. see Antonovics et al. 1971; Bryan and Hummerstone 1971; Stokes et al. 1973; Brown 1977a; Wentsel et al. 1978), but this adaptation seems to be of limited scope and its genetic basis is unclear. However, studies on fish indicate that some of the variation in toxicant sensitivity is genetically based, and that fish can develop increased resistance to metal toxicants (Rahel 1981). Presumably, since insects have shorter life cycles and higher numbers than fish, they should be able to develop metal-resistant strains faster.

OIL AND RELATED PRODUCTS

Oil can enter the aquatic environment by accidental spills, chronic additions from industrial waste water, and surface runoff from paved land. Approximately 6.1 million metric tons per year or as much as 0.25 percent of annual crude oil production has been estimated to enter the aquatic environment from natural and anthropogenic sources (National Academy of Sciences 1975; Neff 1979). Although marine waters receive the major share of this burden, the impact on freshwater habitats can be substantial.

Oil is a complex mixture of many organic and inorganic substances. Its effect on aquatic biota is likely to depend on both the quantity and the quality of the particular mixture involved. The deleterious effects of oil on animals are caused by chemical toxicity, due mainly to the lighter fractions of the oil, and by mechanical clogging of body surfaces and covering of bottom substrates.

The consequences of oil pollution have been less studied in freshwater than marine environments, and very little is actually known about its effects on freshwater organisms (Parker et al. 1976; Rosenberg and Wiens 1976). The effects of gross pollution, such as from a massive spill, are obviously negative to most insects. Large amounts of oil, whether present as a surface slick shortly after a spill or as a sludge on the stream or lake bottom, will effectively eliminate most insects and other invertebrates (McCauley 1966; Jahn 1972; Bugbee and Walter 1973; Meynell 1973).

Similarly, low or moderate levels of contamination may result in decreased species richness and abundance of some insects, but these levels may also have the opposite effect (Rosenberg and Snow 1975). Increased abundance probably results from the proliferation of attached algae, which serve as food and substrate for some insects (Rosenberg and Wiens 1976; Lock et al. 1981a). For example, the herbivorous larvae of two species of *Cricotopus* midges apparently produced part of another generation on oiled compared to on nonoiled artificial substrates (Rosenberg et al. 1977). Decreased abundance probably results from the persistence of oil residues on the substrate and the large amount of organic matter that results from algal growth (e.g. see Rosenberg et al. 1980 for effects on mayfly nymphs). Some of the factors that determine the type of response to oil contamination that will occur include the amount of oil added, the discharge or water-exchange time at the polluted site, the degree of weathering of the oil, and the specific tolerances of individual species (Rosenberg and Wiens 1976; Lock et al. 1981b).

Chronic exposure to petroleum hydrocarbons can reduce both diversity and abundance of insects (Barton and Wallace 1979), and can impair the life functions of those that exist under polluted conditions. The gills of *Cheumatopsyche* caddisfly larvae from a chronically oil-contaminated Pennsylvania stream were smeared with a tarlike substance, and the larvae were considerably smaller than those living under normal conditions (Simpson 1980). Accumulation of oil in lake sediments, especially in combination with oxygen deficiency and/or other potential toxicants such as heavy metals, may have particularly serious consequences for the benthos (Bengtsson and Berggren 1972; Wiederholm, unpublished data).

Finally, it should be emphasized that although there is increased interest in manufacture and use of coal-derived synthetic fuels and oils, these mixtures may involve a greater risk to aquatic insects than do the petroleum-derived products. Recent studies have already indicated a higher toxicity of these new products to some aquatic organisms (Giddings et al. 1980; Cushman and McKamey 1981).

PESTICIDES

Pesticides are used to control unwanted aquatic insects such as black flies, midges, and mosquitoes, and unwanted aquatic plants such as pondweed (*Elodea canadensis*) and water milfoil (*Myriophyllum*). These chemicals may also inadvertently reach the aquatic environment when applied to forests or crops. In fact, with the common use of chemicals in agriculture and forestry, this source is probably more significant than direct application to watercourses for control purposes.

Direct Effects of Pesticides

Intuitively, one would expect that the effects of pesticides on aquatic ecosystems should be better known and easier to quantify than the effects of other toxicants, because there has been considerable attention paid to dose-effect relationships that result in both safe and economical levels of application. However, even if selective, controlled use of a pesticide is intended, it is extremely difficult to foresee unwanted consequences to nontarget organisms.

Rosenberg (1975, p. 106) stated that the uptake and accumulation of chlorinated hydrocarbon pesticides such as DDT and dieldrin in (and one might add "their effect on") aquatic invertebrates is " . . . more likely a function of habitat, mode of life, and exchange equilibria than food but is also affected by size of the organism, pharmacokinetics, physical and chemical properties of the pesticides, and various extrinsic factors." Further, "until adequate research is done, the relative contributions of the factors listed above to pesticide levels in invertebrates will remain unknown." In view of such a statement, the outlook for making any simple, valid generalization on the effects of pesticides on aquatic insects seems rather unlikely. Yet, among the large volume of facts on specific toxicants, some may be of more general interest and importance than others.

Aquatic insect species differ greatly in their susceptibility to pesticides, just as they do to other toxicants. However, no pattern has been established that relates susceptibility to specific physiological processes, body configuration, or mode of life. Lower resistance of younger or smaller specimens than of older or larger ones has been observed (Jensen and Gaufin 1964a; Fredeen 1972; Eidt 1975), and a correlation with higher metabolic rate has been suggested (Jensen and Gaufin 1964b). This might be true for interspecific differences as well, although clear evidence is lacking. Besides the lower resistance of young specimens, it seems likely that insects would be particularly susceptible during physiologically active periods, such as during molting and emergence. In fact, disturbance at these stages has been reported (e.g. Jensen and Gaufin 1964b).

Food selection and microhabitat may influence the susceptibility of aquatic insects to toxicants. For example, it has been suggested that pesticides may be used selectively for control of black flies and filter-feeding caddisflies because, either in particulate form or as a result of their adherence to suspended particles, these pesticides would be ingested by filter-feeding organisms and "filtered out" of the environment (Fredeen et al. 1953; Kershaw et al. 1965; Fredeen 1974). Unfortunately the validity of this approach is contradicted by observations that a number of nontarget insects that have diverse food sources are also adversely affected (Wallace and Hynes 1975; Flannagan et al. 1979). Even if selective results are achieved in the short term, the effects of such selectivity over a longer term can hardly be predicted with our present knowledge of toxic effects and ecosystem dynamics (Helson and West 1978; Flannagan et al. 1979).

Microhabitat can be an important determinant of pesticide effects. For example, animals that live on the upper surfaces of stones or rocks in running-water habitats are more likely to be exposed to a toxicant during short- to moderate-term exposures than those that dwell deeper down in the substrate. Surface dwellers are also more easily swept away by the current when affected. The successive physiological disturbances induced by pesticides, which include hyperactivity, loss of equilibrium, tremors, and convulsions (Jensen and Gaufin 1964a; Sanders and Cope 1968; Muirhead-Thomson 1978; Anderson and DeFoe 1980), will tend to increase the amount of drifting, but drift may also be a passive defense mechanism whereby an insect avoids prolonged exposure to the irritant (see also Wiley and Kohler, Chapter 5). Pesticide-induced drift may reach catastrophic proportions; in fact, portions of the near-surface benthic community may be eliminated (e.g. see Coutant 1964; Welch and Spindler 1964; Wallace and Hynes 1975; Courtemanch and Gibbs 1980). Flannagan et al. (1980) found that 50,000 specimens entered a 15-cm diameter drift sampler during a four-hour period after treatment of the Athabasca River in Alberta with the pesticide methoxychlor. An estimated 2.5 billion animals drifted past a particular site during that period of time.

Duration of exposure may be of foremost importance among the various extrinsic factors that influence the toxicity of a pesticide. Toxicant effects are often more pronounced at lower concentrations with long-term exposure. This may be due to the organism's inability to evade, excrete, or detoxify a harmful agent over a prolonged period of time, or to the higher probability that more sensitive stages of development will be exposed. However, a short pulse of high pesticide concentration sometimes has a greater effect than a long exposure to low pesticide concentration, even though the product of the concentration and the period of exposure is the same (Muirhead-Thomson 1973).

Indirect Effects of Pesticides

Toxic effects of pesticides on aquatic organisms will also influence more tolerant species in that their numbers and types of competitors, predators, and prey organisms may expand or retract as niche space is vacated or eliminated. For example, numerous pesticide-induced population increases appear to have resulted from predator removal or from increased food supply through elimination of competitors, although reasons for the change are usually difficult to identify with certainty (Hurlbert 1975).

There are several examples of the effect of top predator elimination. Following treatment of a California mountain stream by rotenone, and the resulting fish kill, a several-fold increase in density of some major insect orders occurred (Cook and Moore 1969). Elimination of fish from a Wisconsin lake by toxaphene was followed by a 200-fold increase in the *Chironomus* midge population, which declined again when fish were restocked into the lake (Hilsenhoff 1965). However, depending on concentration, toxaphene may also eliminate Chironomidae (e.g. see Cushing and Olive 1956). Reduction or elimination of invertebrate predators was also the likely cause of several observed increases in prey densities that followed pesticide treatment. For example, numbers of black flies increased greatly following the treatment of streams with DDT and the subsequent reduction in numbers of predatory Plecoptera and Trichoptera (see review by Hurlbert 1975). Likewise, several cases of massive increases in attached algae in streams have been attributed to the elimination of grazer invertebrates by pesticides (e.g. see Hurlbert 1975; Yasuno et al. 1982). On other occasions, the removal of competitors may have caused the expansion of species with similar niche requirements. For example, in a malaria eradication program, pesticide-resistant populations of the *Anopheles* mosquito may have benefited from the pesticide-caused mortality of species that compete for their preferred prey (Hurlbert 1975).

Insecticides Versus Herbicides

Unless used in unusually high concentrations, herbicides exert a less dramatic primary effect on aquatic ecosystems than do insecticides (Hurlbert 1975), but their secondary effects may be substantial. For example, increased density of benthic invertebrates after herbicide treatment, which presumably resulted from the increased availability of detritus associated with the decaying plants, has been reported by several workers; however, in other studies, little change has been noted (see review in Brooker and Edwards 1975). Species closely associated with macrophytes (e.g. some Trichoptera, Ephemeroptera, and Chironomidae) can be affected strongly when substrate availability and complexity is reduced (Brooker and Edwards 1975). The decay of large amounts of plant material can reduce the amount of dissolved oxygen and alter pH conditions, thereby affecting many organisms. As a result, fewer species, lower numbers of individuals, and a predominance of more sluggish, deposit-feeding forms often occur (see review in Hurlbert 1975). The release of nutrients from decaying macrophytes may also give rise to higher phytoplankton production before the macrophytes begin to reappear and, thus, a period of restructuring of the whole ecosystem can follow herbicide application (Hurlbert 1975).

ACIDIFICATION

Acid precipitation, which is caused by combustion of fossil fuels and the long-distance transport of pollutants that result from this process, occurs in large parts of Europe and North America. Areas with slowly weathering bedrock and poorly buffered soils are particularly sensitive, and acid precipitation has become a serious environmental problem in Scandinavia, southeastern Canada, and the northeastern United States (Haines 1981; Cowling 1982). For example, by 1980 approximately 13,000 lakes and 15,000 to 20,000 km of running waters in southern and central Sweden were reported to be acidified to a level at which the risk of damage to plant and animal life was acute, or to a level where damage had already occurred (Jóhansson and Nyberg 1981). This amounts to almost half of the total number of lakes and length of streams in that part of the country.

Chemical Changes from Acid Precipitation

Three major phases may be distinguished in the acidification of surface waters (Almer et al. 1978; Last et al. 1980; Johansson and Nyberg 1981). During the first phase, excessive acidity is buffered by the various natural systems in soil and water than entrain hydrogen ions while releasing calcium (Ca^{++}), magnesium (Mg^{++}), and other cations (primarily the carbonic acid/carbonate system). Although the alkalinity of water decreases, the pH remains fairly stable, and biological changes do not yet appear. In the second phase, the buffering capacity of the carbonic acid/carbonate system is eventually exhausted, and aluminum (Al) is released from silicate minerals in the soil, first as polymeric hydroxocations but, at very low pH, as Al^{+++}. Alkalinity decreases below 0.1 meqv/l, and pH becomes unstable and decreases more rapidly. Consequently, water with higher acid content reaches streams and lakes. This occurs particularly during periods of high flows and may result in pronounced acid shocks, and damage to plant and animal life may appear in receiving waters. In the third phase, pH values stabilize around 4.5 due to buffering by humic materials. During this stage, numbers of species of plants and animals are strongly reduced, and production and decomposition processes are considerably slowed down.

Characteristics of Acidified Communities

Mayflies are among the most sensitive groups of aquatic insects to acidification. Not only does the number of species decrease under increasingly acid conditions (i.e. lower pH), but mayflies can become rare or absent under such circumstances even though other groups may still be flourishing (e.g. see Morgan and Egglishaw 1965; Minshall and Kuehne 1969; Bell 1971; Sutcliffe and Carrick 1973; Leivestad et al. 1976; Nilssen 1980). In a survey of 600 otherwise unpolluted Swedish streams, the maximum number of mayfly species occurred at a pH range of 6.0 to 7.5; far fewer species occurred at pH 5.0 than at pH 6.0; only three species, *Leptophlebia vespertina*, *L. marginata*, and *Heptagenia fuscogrisea*, were found in the range 4.0 to 5.0; and only the first of these latter three, *L. vespertina*, remained at pH < 4.0 (P. Lingdell, unpublished data; cited in Johansson and Nyberg 1981).

In contrast to the Ephemeroptera, insects in the orders Odonata, Megaloptera, Coleoptera, and Hemiptera, and species of the phantom midge *Chaoborus*, are sometimes abundant in acidified lakes (Grahn et al. 1974; Mossberg and Nyberg 1979; Nilssen 1980; Henrikson and Oscarson 1981). The predominance in the sediments of one or more species of the midge *Chironomus* is also characteristic of acidic conditions (Wiederholm and Eriksson 1977; Mossberg 1979; Raddum and Sæther 1981); this occurs in oligotrophic lakes, even though the genus is typical of harmonic eutrophic lakes (Brundin 1949; Sæther 1975; Table 17.1). In fact, *Chironomus riparius* is one of the few insect species that has been reported from natural, highly acidic conditions (Harp and Campbell 1967; Jernelöv et al. 1981).

Acid waters typically have fewer species, and also lower abundance and biomass, of benthic invertebrates than have nonacidic waters (Hendrey et al. 1976; Leivestad et al. 1976; Arnold et al. 1981). However, biomass and abundance may sometimes be unchanged (Mossberg 1979; Collins et al. 1981), which indicates that factors other than pH may also be important.

Reduced overall abundance of benthic invertebrates, and a greater sensitivity of individuals in certain functional groups, may influence the metabolism of acidified ecosystems. For example, Hall et al. (1980) noted that experimental acidification, which resulted in increased drift of organisms (Fig. 17.4A), affected the drift rate of the collector functional group more than that of the scraper group, and that the scrapers were affected more than the predators (see also

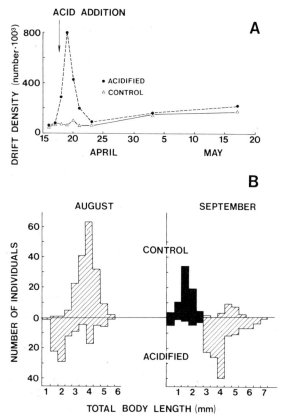

Figure 17.4   Effects of experimental acidification of a New Hampshire stream on (*A*) drift of invertebrates, and (*B*) growth and recruitment of the mayfly *Ephemerella funeralis*. Black histograms indicate the newly recruited cohort, striped histograms indicate the cohort recruited during the previous year. (*A*) redrawn from Hall et al. (1980), copright © 1980, the Ecological Society of America. (*B*) redrawn from Fiance (1978) with permission of OIKOS.

Friberg et al. 1980). These changes might have been responsible for the observed increase in attached periphyton biomass. In addition, these changes decreased the storage of nutrients and, thus, increased the downstream transport distance of particulate organic matter (Hall et al. 1980).

Sensitivity of Different Life Stages

There is little information on the sensitivity of different life stages to acidification, although Bell (1971) found that the period around emergence was a critical time. For example, dead pupae of the chironomids *Sergentia albescens* and *Chironomus* sp. have been reported from acidic lakes (Mossberg 1979); however, the mayfly *Ephemerella funeralis* did not show any alteration in emergence phenology after exposure to experimental acidification for two months (Fiance 1978). Since all other mayflies were eliminated or severely reduced, *E. funeralis* may be a resistant species; however, after a three-month exposure to a pH of 4, exposed larvae of this species were about 30 percent smaller than those at a reference site (Fig. 17.4*B*). Fiance (1978) suggested that this might reflect greater allocation of energy resources to the maintenance of internal ionic balance during acid stress. The reduction in the growth and subsequent size of the adult is ecologically significant, because of the positive correlation that exists between individual size of an insect and its fecundity (see Sweeney, Chapter 4).

A further observation of the aforementioned study (Fiance 1978) was that the recruitment of *E. funeralis* was reduced in the acid zone (Fig. 17.4*B*). Sutcliffe and Carrick (1973) contended that adult females of the mayfly *Baetis* are selective in choosing oviposition sites, and they apparently avoid laying eggs in water with pH <6.0; this would explain the observed distribution of *Baetis* eggs and larvae relative to pH conditions in the River Duddon, England. Hudson (1956) and Strenzke (1960) reported a similar capacity for selection of oviposition sites in mosquitoes and midges. These individual observations suggest that some insects may be able to assess water quality, possibly via the chemoreceptors on their legs or mouth parts (Wigglesworth 1965), and such a trait would be of obvious advantage to them in distinguishing between a polluted and a nonpolluted environment.

If the life stages involved in reproduction are the most sensitive to acid stress, as some of the observations just cited indicate, species that reproduce in the spring and live in areas that have a snow cover during winter will be particularly affected since this is when the effects of acid precipitation are most pronounced. For example, snow melt and runoff frequently result in a pulse of low pH water, which produces an acid shock; in stratified lakes this pulse is distributed almost entirely in the near surface layer (Hultberg 1976; Hendrey et al. 1980). Hence, at this time the emerging or egg-laying insect is directly exposed to acidic conditions.

Mechanisms of Damage

How does acidification affect aquatic insects and other aquatic animals? Hall et al. (1980) proposed the following three explanations: (1) low pH affects the physiology of organisms; (2) metals are released at low pH in a state and concentration that is toxic to many organisms; and (3) indirect effects occur through reduced primary production, and/or reduced bacterial decomposition, or altered patterns of competition and predation.

Physiological Disturbances. Four major physiological functions of aquatic animals are affected at low pH; they are Ca regulation, sodium (Na) regulation, respiration, and acid-base balance (see review by Havas 1981). At sublethal pH levels (i.e. <5.5), aquatic animals have problems with Ca regulation, which may result in disturbance of reproduction, growth, and molting processes. At lower pHs (i.e. <5.0), in addition to Ca stress, aquatic animals may also experience problems with Na regulation. More energy is needed to maintain constant Na concentration and less energy is available for growth and reproduction; disruption of the Na balance usually results in death. If the pH decreases further (i.e. to < 4.0), the ability of organisms either to transport oxygen internally or remove oxygen from the water is affected, and mortality occurs rapidly at this stage. Penetration of hydrogen (H^+) ions across body membranes occurs over a wide range of pH levels and results in problems with acid-base balance, with consequences to all the aforementioned physiological functions.

Insects that persist under increasingly acid conditions must be particularly well adapted in the above respects. For example, the midge *Chironomus riparius*, which has been collected from a pond with a pH of 2.8, has more buffering capacity in its hemolymph, because it has more hemoglobin than have several other species of *Chironomus* (Havas 1981; Jernelöv et al. 1981). This also benefits respiration, and *C. riparius* is one of the few truly aquatic insects that can occur in nearly oxygen-free, organically polluted bodies of water (see above).

Mobilization of Toxic Metals. Interest has focused on Al, which occurs at increased concentrations in acidified waters. Elevated Al concentrations are toxic to fish and may also be toxic to invertebrates. The matter is complicated by the interacting effects of H^+ ions and Al. For example, experiments with brook trout (*Salvelinus fontinalis*) and white sucker (*Catostomus commersoni*) showed that, with increasing age from eggs to fry, both species of fish were less sensitive to low pH but more sensitive to Al; thus, the factor controlling survival tended to be reversed with age (Baker and Schofield 1980). Furthermore, the presence of Al mitigated the toxic effects of low pH levels to fish eggs; in fact, egg survival at low pH was better with Al present!

Heavy metals may also be a threat to aquatic insects under acidic conditions. Low pH increases the mobility of heavy metals from soils and sediment. Observations from Swedish waters show higher concentrations and greater seasonal variations of Cd, Zn, and Pb in acidic lakes. Fish from these waters had elevated levels of muscle mercury that were near or above the levels permitted for commercial marketing, and this occurred in the absence of any point sources of mercury pollution (Johansson and Nyberg 1981).

Indirect Effects on Insects. Changes in primary production and patterns of competition and predation undoubtedly contribute to the overall effects of acidification on invertebrate community structure. However, to what extent acidification reduces primary production in streams and lakes has been a matter of dispute. Grahn et al. (1974) used the decrease in rooted macrophytes, the expansion of *Sphagnum* mats, and the accumulation of coarse debris in acidified lakes to imply that nutrient cycling, and hence production and decomposition processes, were retarded. Grahn et al. labeled this process "oligotrophication" (see also Hendrey et al. 1976; Gahnström et al. 1980; Stokes 1981). However, other observations showed that production and decomposition were not affected, at least at moderately low pH (Almer et al. 1978; Schindler 1980). In fact, increases in algal biomass have been observed in acidic waters, and are presumably related to reduced grazing by invertebrates or to changes in algal species composition (Hendrey 1976; Hall et al. 1980; Müller 1980). Hence, food shortage should not become critical to invertebrates in acidified waters, unless changes in algal species composition, and perhaps reduction of microbial decomposers, eliminate the most valuable food items. The effect of acidification on competition among invertebrates is presently unclear, but effects of top predator removal are more apparent. Increased abundance of the phantom midge *Chaoborus* and surface-dwelling aquatic Hemiptera and Coleoptera occurs in acidified lakes from which fish have been eliminated, and a similar fauna resulted from experimental removal of fish from a nonacidified lake (Eriksson et al. 1980). Thus, the damage done to fish by acid precipitation would in itself be enough to cause substantial changes in the benthic fauna, but the depauperate invertebrate communities that eventually result from acidification are more likely the result of direct physiological disturbance caused by low pH and/or the toxic action of metals.

Acid-Mine Wastes

The preceding discussion has been centered exclusively around the effects of acid precipitation, but other types of acidification may occur as well. For example, acid-mine waste water has been a widespread cause of disturbance. The fauna in streams affected by mine wastes is often similar to that found in streams subjected to airborne acidification (i.e. it is depauperate and low in diversity). Megaloptera (e.g. *Sialis lutaria*), some Trichoptera (e.g. *Hydropsyche pellucidula, Rhyacophila*), Plecoptera (*Leuctra hippopus, Amphinemura sulcicollis*), and Chironomidae are among the few insects that remain under such conditions (Roback and Richardson 1969; Scullion and Edwards 1980). However, the effects of low pH in acid-mine waste water are often difficult to separate from the effects of suspended solids and heavy metals (see earlier discussion).

CONCLUDING REMARKS

Ecological studies of freshwater communities have frequently been more descriptive than explanatory (Wiggins and Mackay 1978), and this is particularly true for pollution-oriented investigations. However, successful management of freshwater ecosystems requires that effects of disturbance be predictable. Such predictions must be based on a genuine understanding of the habitat, the natural dynamics of its biota, and the response of its biota to disturbance (Södergren et al. 1972; Flannagan et al. 1979; Muirhead-Thomson 1979; Cairns 1981). Unfortunately, our current state of knowledge is insufficient to enable us to make predictions about how a lake or stream will respond to disturbance, except in general, qualitative terms.

To improve this situation, clearer insights are needed into how pollutants are distributed and transformed in the environment, because these factors determine the exposure conditions or the effective dose of a particular pollutant that an organism will encounter. For example, temperature, pH, hardness, and the occurrence of chelating substances can influence the fate of a pollutant and be instrumental in determining whether biological effects will or will not occur.

There is also a need for research that is directed toward the response of populations (particularly at the sublethal effects of disturbance), rather than at community parameters such as species composition, diversity, and so forth, which has been the approach used in most investigations. Changes in community structure and function will not be understood and explainable without insight into the response mechanisms of the individual populations.

The significance of both natural environmental factors and pollutants lies in their influence on the basic biological functions that an organism must fulfill, namely, (1) occupying and maintaining a place to live, (2) acquiring nutrients and oxygen, (3) maintaining chemical homeostasis, (4) eluding predation, and (5) avoiding or overcoming parasites and diseases. Failure or success in fulfilling these functions is reflected in the fitness of individual organisms and the long-term growth rate of populations. The development of populations will affect community structure and, depending on the species involved, the functioning of the whole ecosystem.

Recruitment and removal are the two balancing elements of population growth rate. Lethal effects of pollutants, the determination of which has long preoccupied aquatic toxicologists, clearly have a bearing on the elimination of organisms. In contrast, sublethal effects impinge on recruitment, and this is the topic that should interest the investigator who strives for ecologically meaningful results. Effects on parameters such as rate of maturation, brood size, longevity, and frequency of reproduction then become a matter of major concern. Daniels and Allan (1981) and Gentile et al. (1982) have demonstrated how life-table data can be used to measure the effect of sublethal stress on the reproductive success of organisms and the growth rate of populations. In the latter study, the effect of a pollutant, nickel (Ni), was combined with the additional stress of predation to indicate the critical concentrations of Ni that would cause a population of the crustacean *Mysidopsis bahia* to decline and be eliminated (Fig. 17.5). This approach seems to have considerable potential for development and use with aquatic insects.

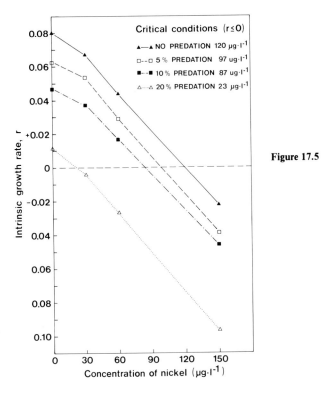

Figure 17.5 The combined effects of nickel (Ni) concentration and predation on population development (as measured by r, the intrinsic growth rate) of the crustacean *Mysidopsis bahia*. Critical concentrations of metal that would cause the population to decline and be eliminated are indicated in the upper right corner of the figure. Redrawn from Gentile et al. (1982), with permission of Hydrobiologia.

The mechanisms that make some organisms able to proliferate in a disturbed environment, while others are eliminated, also need further exploration. Several adaptive characteristics have proved to be of survival value for aquatic insects that are exposed to stress (Table 17.2). Further studies will eventually demonstrate to what extent resistance to such disturbances is related to the possession of a few particularly robust characteristics in the tolerant species, such as a capacity for osmoregulation, detoxification of toxicants, or the use of alternative metabolic pathways and rates. It may be, however, that the life-history strategy of an organism is equally or even more important. The best example to illustrate this point is the polychaete *Capitella capitata*. Grassle and Grassle (1974) demonstrated that the predominance and high abundance of *C. capitata* in highly polluted marine environments is due to particular life-history traits that result in an ability to quickly colonize and reproduce in disturbed areas, rather than to a particular tolerance to pollution. Such a life-history trait is typical of an ecologically opportunistic species. Frenzel (1983) pointed to the shorter turnover time among some chironomid species that have become more abundant following eutrophication of Lake Constance in Germany. Facultative partheno-
(text continues on page 541)

TABLE 17.2: Examples of Adaptations Enabling Aquatic Insects to Cope with Pollution

STRESS	HYPOTHESIZED MECHANISMS OF ADAPTATION	TYPES OF AQUATIC INSECTS SHOWING ADAPTATION	REMARKS	REFERENCES
O_2 deficiency	Possession of various respiratory structures (e.g. tracheal gills, rectal folds)	Odonata		Hynes (1970), Gaufin (1973)
	Possession of gills; adjustment of gill beat frequency; undulating body movements	Various Plecoptera, Ephemeroptera, Trichoptera		Gaufin and Gaufin (1961), Knight and Gaufin (1966), Gaufin et al. (1974)
	Adjustment of body position relative to flow and O_2 gradient; inactivity	Some Plecoptera (e.g. *Acroneuria pacifica*, *Pteronarcys californica*)		Gaufin and Gaufin (1961)
	Orientation towards high flow sites	Simuliidae		Gaufin et al. (1974)
	Surface breathing	Various Hemiptera, Coleoptera, Diptera		Gaufin (1973)
	Comparatively low metabolic rate	Various insect groups from standing or slowly running waters		Hynes (1970)
	Short-term anaerobic metabolism	Plecoptera, Ephemeroptera, Trichoptera		Gaufin et al. (1974)

538

	Response	Taxa	Comments	References
	Increased capacity of oxygen storage and transport by possession of hemoglobin; capacity of long-term anaerobic metabolism; cessation of feeding; inactivity	Chironomidae (e.g. *Chironomus anthracinus, C. plumosus, C. riparius*)		Walshe (1950), Neumann (1962), Jónasson (1972), Frank (1980), Nagell and Landahl (1978)
Sediment	Protection by movable cases	Chironomidae (e.g. *Constempellina brevicosta, Abiskomyia virgo*)		Grimås and Wiederholm (1979)
	Burrowing to avoid effects of sediment addition	Chironomidae (e.g. *Constempellina, Stempellinella*)	Rigid cases of sand may also give protection	Rosenberg and Wiens (1980)
	Burrowing behavior; hairiness; dorsal position of gills	Several Ephemeroptera and Plecoptera (e.g. *Ephemera, Leuctra nigra, Caenis*)		Hynes (1970)
	Capacity to inhabit alternative microhabitat as streambed interstices become filled with sand	Ephemeroptera (e.g. *Ephemerella grandis, E. doddsi*)	Ability to tolerate higher current velocity necessary	McClelland and Brusven (1980)
Heavy metals				
Cd, Cr, Zn	Avoidance of polluted sediments	Chironomidae (e.g. *Chironomus tentans*)		Wentsel et al. (1977c)
Cu, Zn	Protection by cases	Case-bearing Trichoptera	Decreased permeability of body cuticle to heavy metals; food choice, habitat selection, behavioral responses, and physiological regulatory mechanisms possibly also important	Brown (1977b)

539

TABLE 17.2: Concluded

STRESS	HYPOTHESIZED MECHANISMS OF ADAPTATION	TYPES OF AQUATIC INSECTS SHOWING ADAPTATION	REMARKS	REFERENCES
Cd	Excretion of Cd	Plecoptera (e.g. *Pteronarcys californica, Pteronarcella badia*)	Excretion of Cd may enable insects to survive and recover after pulse of high metal concentration in ambient water	Clubb et al. (1975a)
Co	Life cycle permitting development during seasonal periods with low pollution pressure	Ephemeroptera (e.g. *Baetis, Ephemerella ignita*)	High water flow and low concentration of metal favorable to species with summer development	Södergren (1976)
Low pH	Possession of large quantities of hemoglobin gives high buffering capacity and helps maintain internal acid-base balance; large anal papillae may improve osmo-regulation	Chironomidae (e.g. *Chironomus riparius*)	Organisms tolerant to low pH should have relative impermeability to water, well-buffered body fluids, and capacity to utilize large organic compounds for osmoregulation; hemoglobin improves tolerance to oxygen deficiency as well	Havas (1981), Jernelöv et al. (1981)
Various pollutants (e.g. sediment, low O_2)	Short generation time	Various Chironomidae, Ephemeroptera (e.g. *Baetis*), Plecoptera (e.g. *Nemoura*)	Results in rapid population increase and/or rapid colonization rate (i.e. opportunistic species)	Newbold et al. (1980)

540

genesis may be another strategy for rapid population expansion and exploitation of suddenly, or intermittently, available resources (Cuellar 1977; Wiederholm et al. 1977; Fiance 1978).

Finally, attention should be directed to the genetic control of life histories and consequent responses to habitat conditions (see also Butler, Chapter 3; Sweeney, Chapter 4). It is essential that we obtain a clearer picture of how and to what extent observed cases of resistance to pollutants depend on acclimation or selection for tolerant strains. Tolerance to heavy metals in populations of the polychaete *Nereis diversicolor* and the freshwater isopods *Asellus aquaticus* and *A. meridianus* from areas with a history of pollution seems to be genetically based (Bryan and Hummerstone 1971, 1973; Brown 1976; Fraser et al. 1978), but although resistance to metals and other toxicants occurs among several other organisms, it is not a universally observed phenomenon (Rahel 1981). In contrast, the number of documented cases of resistance to insecticides has, since about 1950, increased exponentially among insect pests, and cross- and multiple-resistance (i.e. resistance to related, and to a broad spectrum of pesticides, respectively) have become increasingly widespread (e.g. see Metcalf 1980). Luoma (1977) noted that resistance to toxins reduces the overall fitness of the population when compared to a nonresistant one, and that when the pressure of a toxicant is removed the population reverts to dominance by intolerant genotypes in just a few generations. However, once resistant genes are selected for, they may persist for a long time in insect populations, even though the resistant allele may decrease in frequency (Metcalf 1980). Crow (1957), however, suggested that reversion to susceptibility would be slow and that resistance would increase rapidly if the population was once again exposed to the selective agent.

Since ecologically opportunistic species have a high degree of genetic variability (Ayala 1968; Grassle and Grassle 1974; Luoma 1977), their potential for short-term selection in response to habitat factors and pollutants is increased. Opportunistic species, when compared to more specialized ones, are also favored by unpredictable environmental conditions (e.g. see Pianka 1978). Hence, habitat variability, life-history patterns, and adaptive success are closely interrelated in the final response of insect communities to environmental disturbance.

References

Aanes, K. J. 1980. A preliminary report from a study on the environmental impact of pyrite mining and dressing in a mountain stream in Norway, pp. 419–42. *In*: J. F. Flannagan and K. E. Marshall (eds.). Advances in Ephemeroptera biology. Proceedings of the Third International Conference on Ephemeroptera. Plenum Press, New York, NY. 552 pp.

Alabaster, J. S. (ed.). 1977. Biological monitoring of inland fisheries. Applied Science Publishers Ltd., London. 226 pp.

Almer, B., W. Dickson, C. Ekström, and E. Hörnström. 1978. Sulphur pollution and the aquatic ecosystem, pp. 271–311. *In:* J. O. Nriagu (ed.). Sulphur in the environment. Part II. Ecological impacts. John Wiley and Sons Inc., New York, NY. 482 pp.

Anderson, N. H. and J. R. Sedell. 1979. Detritus processing by macroinvertebrates in stream ecosystems. Annual Review of Entomology 24:351–77.

Anderson, R. L. and D. L. DeFoe. 1980. Toxicity and bioaccumulation of endrin and methoxychlor in aquatic invertebrates and fish. Environmental Pollution (Series A) 22:111–21.

Antonovics, J., A. D. Bradshaw, and R. G. Turner. 1971. Heavy metal tolerance in plants. Advances in Ecological Research 7:1–85.

Armitage, P. D. 1978. Downstream changes in the composition, numbers and biomass of bottom fauna in the Tees below Cow Green reservoir and in an unregulated tributary Maize Beck, in the first five years after impoundment. Hydrobiologia 58:145–56.

Armitage, P. D. 1980. The effects of mine drainage and organic enrichment on benthos in the River Nent system, Northern Pennines. Hydrobiologia 74:119–28.

Arnold, D. E., P. M. Bender, A. B. Hale, and R. W. Light. 1981. Studies on infertile, acidic Pennsylvania streams and their benthic communities, pp. 15–33. *In:* R. Singer (ed.). Effects of acidic precipitation on benthos. North American Benthological Society, Springfield, IL. 154 pp.

Ayala, F. J. 1968. Genotype, environment, and population numbers. Science 162: 1453–59.

Baker, J. P. and C. L. Schofield. 1980. Aluminum toxicity to fish as related to acid precipitation and Adirondack surface water quality, pp. 292–93. *In:* D. Drabløs and A. Tollan (eds.). Ecological impact of acid precipitation. Proceedings of an international conference, Sandefjord, Norway, March 11–14, 1980. SNSF Project, Olso, Norway. 383 pp.

Barton, D. R. and H. B. N. Hynes. 1978. Seasonal study of the fauna of bedrock substrates in the wave zones of lakes Huron and Erie. Canadian Journal of Zoology 56: 48–54.

Barton, D. R. and R. R. Wallace. 1979. Effects of eroding oil sand and periodic flooding on benthic macroinvertebrate communities in a brown-water stream in Northeastern Alberta, Canada. Canadian Journal of Zoology 57:533–41.

Bell, H. L. 1971. Effect of low pH on the survival and emergence of aquatic insects. Water Research 5:313–19.

Bengtsson, L. and H. Berggren. 1972. The bottom fauna in an oil-contaminated lake. Ambio 1:141–44.

Besch, W. K., I. Schreiber, and D. Herbst. 1977. Der *Hydropsyche*-Toxizitätstest, erprobt an Fenethcarb. Schweizerische Zeitschrift für Hydrologie 39:69–85.

Besch, W. K., I. Schreiber, and E. Magnin. 1979. Influence du sulfate de cuivre sur la structure du filet des larves d'*Hydropsyche* (Insecta, Trichoptera). Annales de Limnologie 15:123–38.

Bormann, F. H., G. E. Likens, and J. S. Eaton. 1969. Biotic regulation of particulate and solution losses from a forest ecosystem. BioScience 19:600–10.

Brinkhurst, R. O. 1974. The benthos of lakes. The Macmillan Press Ltd., London. 190 pp.

Brooker, M. P. and R. W. Edwards. 1975. Review paper: Aquatic herbicides and the control of water weeds. Water Research 9:1–15.

Brooker, M. P. and D. L. Morris. 1980. A survey of the macro-invertebrate riffle fauna of the rivers Ystwyth and Rheidol, Wales. Freshwater Biology 10:459–74.

Brown, B. E. 1976. Observations on the tolerance of the isopod *Asellus meridianus* Rac. to copper and lead. Water Research 10:555–59.

Brown, B. E. 1977a. Uptake of copper and lead by a metal-tolerant isopod *Asellus meridianus* Rac. Freshwater Biology 7:235–44.

Brown, B. E. 1977b. Effects of mine drainage on the river Hayle, Cornwall. A) Factors affecting concentrations of copper, zinc and iron in water, sediments and dominant invertebrate fauna. Hydrobiologia 52:221–33.

Brundin, L. 1949. Chironomiden und andere Bodentiere der südschwedischen Urgebirgsseen. Institute of Freshwater Research Drottningholm Report 30:1–914.

Brundin, L. 1951. The relation of O_2-microstratification at the mud surface to the ecology of the profundal bottom fauna. Institute of Freshwater Research Drottningholm Report 32:32–42.

Brundin, L. 1958. The bottom faunistical lake type system and its application to the southern hemisphere. Moreover a theory of glacial erosion as a factor of productivity in lakes and oceans. Internationale Vereinigung für Theoretische und Angewandte Limnologie Verhandlungen 13:288–97.

Bryan, G. W. and L. G. Hummerstone. 1971. Adaptation of the polychaete *Nereis diversicolor* to estuarine sediments containing high concentrations of heavy metals. 1. General observations and adaptation to copper. Journal of the Marine Biological Association of the United Kingdom 51:845–63.

Bryan, G. W. and L. G. Hummerstone. 1973. Adaptation of the polychaete *Nereis diversicolor* to estuarine sediments containing high concentrations of zinc and cadmium. Journal of the Marine Biological Association of the United Kingdom 53:839–57.

Bugbee, S. L. and C. M. Walter. 1973. The response of macroinvertebrates to gasoline pollution in a mountain stream, pp. 725–31. *In*: Proceedings of Joint Conference on Prevention and Control of Oil Spills, March 13–15, 1973, Washington DC. American Petroleum Institute, Washington DC. 834 pp.

Buikema, A. L., Jr. and J. Cairns, Jr. (eds.). 1980. Aquatic invertebrate bioassays. American Society for Testing and Materials Special Technical Publication 715. American Society for Testing and Materials, Philadelphia, PA. 209 pp.

Cairns, J., Jr. 1981. Review paper: Biological monitoring. Part VI—Future needs. Water Research 15:941–52.

Cairns, J., Jr., A. L. Buikema, Jr., A. G. Heath, and B. C. Parker. 1978. Effects of temperature on aquatic organism sensitivity to selected chemicals. Virginia Water Resources Research Center Bulletin 106. 88 pp.

Cairns, J., Jr. and K. L. Dickson (eds.). 1973. Biological methods for the assessment of water quality. American Society for Testing and Materials Special Technical Publication 528. American Society for Testing and Materials, Philadelphia, PA. 256 pp.

Cairns, J., Jr., A. G. Heath, and B. C. Parker. 1975. The effects of temperature upon the toxicity of chemicals to aquatic organisms. Hydrobiologia 47:135–71.

Cherry, D. S., S. R. Larrick, R. K. Guthrie, E. M. Davis, and F. F. Sherberger. 1979. Recovery of invertebrate and vertebrate populations in a coal ash stressed drainage system. Journal of the Fisheries Research Board of Canada 36:1089–96.

Ciborowski, J. J. H., P. J. Pointing, and L. D. Corkum. 1977. The effect of current velocity and sediment on the drift of the mayfly *Ephemerella subvaria* McDunnough. Freshwater Biology 7:567–72.

Clubb, R. W., A. R. Gaufin, and J. L. Lords. 1975a. Acute cadmium toxicity studies upon nine species of aquatic insects. Environmental Research 9:332-41.

Clubb, R. W., A. R. Gaufin, and J. L. Lords. 1975b. Synergism between dissolved oxygen and cadmium toxicity in five species of aquatic insects. Environmental Research 9:285-89.

Collins, N. C., A. P. Zimmerman, and R. Knoechel. 1981. Comparisons of benthic infauna and epifauna biomasses in acidified and nonacidified Ontario lakes, pp. 35-48. *In*: R. Singer (ed.). Effects of acidic precipitation on benthos. North American Benthological Society, Springfield IL. 154 pp.

Cook, D. G. and M. G. Johnson. 1974. Benthic macroinvertebrates of the St. Lawrence Great Lakes. Journal of the Fisheries Research Board of Canada 31:763-82.

Cook, S. F., Jr. and R. L. Moore. 1969. The effects of a rotenone treatment on the insect fauna of a California stream. Transactions of the American Fisheries Society 98: 539-44.

Courtemanch, D. L. and K. E. Gibbs. 1980. Short- and long-term effects of forest spraying of Carbaryl (Sevin-4-Oil^R) on stream invertebrates. The Canadian Entomologist 112:271-76.

Coutant, C. C. 1964. Insecticide Sevin: effect of aerial spraying on drift of stream insects. Science 146:420-21.

Cowling, E. B. 1982. Acid precipitation in historical perspective. Environmental Science and Technology 16:110A-23A.

Crow, J. F. 1957. Genetics of insect resistance to chemicals. Annual Review of Entomology 2:227-46.

Cuellar, O. 1977. Animal parthenogenesis. Science 197:837-43.

Cummins, K. W. 1973. Trophic relations of aquatic insects. Annual Review of Entomology 18:183-206.

Cummins, K. W. 1978. Ecology and distribution of aquatic insects, pp. 29-31. *In*: R. W. Merritt and K. W. Cummins (eds.). An introduction to the aquatic insects of North America. Kendall/Hunt Publishing Company, Dubuque, IA. 441 pp.

Cummins, K. W. 1980. The multiple linkages of forests to streams, pp. 191-98. *In*: R. H. Waring (ed.). Forests: fresh perspectives from ecosystem analysis. Oregon State University Biological Colloquium 40. 198 pp.

Cummins, K. W. and M. J. Klug. 1979. Feeding ecology of stream invertebrates. Annual Review of Ecology and Systematics 10:147-72.

Cushing, C. E., Jr. and J. R. Olive. 1956. Effects of toxaphene and rotenone upon the macroscopic bottom fauna of two northern Colorado reservoirs. Transactions of the American Fisheries Society 86:294-301.

Cushman, R. M. and M. I. McKamey. 1981. A *Chironomus tentans* bioassay for testing synthetic fuel products and effluents, with data on acridine and quinoline. Bulletin of Environmental Contamination and Toxicology 26:601-5.

Dance, K. W. and H. B. N. Hynes. 1980. Some effects of agricultural land use on stream insect communities. Environmental Pollution (Series A) 22:19-28.

Daniels, R. E. and J. D. Allan. 1981. Life table evaluation of chronic exposure to a pesticide. Canadian Journal of Fisheries and Aquatic Sciences 38:485-94.

Davies, I. J. 1975. Selective feeding in some arctic Chironomidae. Internationale Vereinigung für Theoretische und Angewandte Limnologie Verhandlungen 19:3149-54.

Dickson, K. L., J. Cairns, Jr., and R. J. Livingstone (eds.). 1978. Biological data in water pollution assessment: quantitative and statistical analyses. American Society for Testing and Materials Special Technical Publication 652. American Society for Testing and Materials, Philadelphia, PA. 184 pp.

Durrett, C. W. and W. D. Pearson. 1975. Drift of macroinvertebrates in a channel carrying heated water from a power plant. Hydrobiologia 46:33–43.

Dusoge, K. and R. J. Wiśniewski. 1976. Effect of heated waters on biocenosis of the moderately polluted Narew River. Macrobenthos. Polskie Archiwum Hydrobiologii 23:539–54.

Eidt, D. C. 1975. The effect of fenitrothion from large-scale forest spraying on benthos in New Brunswick headwaters streams. The Canadian Entomologist 107:743–60.

Elder, J. A. and A. R. Gaufin. 1974. The toxicity of three mercurials to *Pteronarcys californica* Newport, and some possible physiological effects which influence the toxicities. Environmental Research 7:169–75.

Eriksson, M. O. G., L. Henrikson, B.-I. Nilsson, G. Nyman, H. G. Oscarson, and A. E. Stenson. 1980. Predator-prey relations important for the biotic changes in acidified lakes. Ambio 9:248–49.

Eyres, J. P. and M. Pugh-Thomas. 1978. Heavy metal pollution of the river Irwell (Lancashire, UK) demonstrated by analysis of substrate materials and macroinvertebrate tissue. Environmental Pollution 16:129–36.

Fey, J. M. 1977. Die Aufheizung eines Mittelgebirgsflusses und ihre Auswirkungen auf die Zoozönose—dargestellt an der Lenne (Sauerland). Archiv für Hydrobiologie Supplement 53:307–63.

Fiance, S. B. 1978. Effects of pH on the biology and distribution of *Ephemerella funeralis* (Ephemeroptera). Oikos 31:332–39.

Flannagan, J. F., B. E. Townsend, and B. G. E. de March. 1980. Acute and long term effects of methoxychlor larviciding on the aquatic invertebrates of the Athabasca River, Alberta, pp. 151–58. *In*: W. O. Haufe and G. C. R. Croome (eds.). Control of black flies in the Athabasca River. Technical report. An interdisciplinary study for the chemical control of *Simulium arcticum* Malloch in relation to the bionomics of biting flies in the protection of human, animal, and industrial resources and its impact on the aquatic environment. Alberta Environment, Edmonton, Canada. 241 pp.

Flannagan, J. F., B. E. Townsend, B. G. E. de March, M. K. Friesen, and S. L. Leonhard. 1979. The effects of an experimental injection of methoxychlor on aquatic invertebrates: accumulation, standing crop, and drift. The Canadian Entomologist 111:73–89.

Förstner, U. and G. T. W. Wittmann. 1979. Metal pollution in the aquatic environment. Springer-Verlag, Berlin, West Germany. 486 pp.

Frank, C. 1980. Lactate determinations in *Chironomus plumosus* L. larva after anaerobiosis. Acta Universitatis Carolinae-Biologica 1978:59–62.

Fraser, J., D. T. Parkin, and E. Verspoor. 1978. Tolerance to lead in the freshwater isopod *Asellus aquaticus*. Water Research 12:637–41.

Fredeen, F. J. H. 1972. Reactions of the larvae of three rheophilic species of Trichoptera to selected insecticides. The Canadian Entomologist 104:945–53.

Fredeen, F. J. H. 1974. Tests with single injections of methoxychlor black fly (Diptera: Simuliidae) larvicides in large rivers. The Canadian Entomologist 106:285–305.

Fredeen, F. J. H., A. P. Arnason, and B. Berck. 1953. Adsorption of DDT on suspended solids in river water and its role in black-fly control. Nature (London) 171:700–1.

Frenzel, P. 1983. Eutrophierung und Zoobenthos im Bodensee. Mit besonderer Berück sichtigung der litoralen Lebensgemeinschaften. Verhandlungen der Gesellschaft für Ökologie 10 (Mainz 1981):375–391.

Frey, D. G. 1964. Remains of animals in Quaternary lake and bog sediments and their interpretation. Archiv für Hydrobiologie Beiheft Ergebnisse der Limnologie 2:1–114.

Frey, D. G. 1977. Biological integrity of water—an historical approach, pp. 127–40. In: The integrity of water. Proceedings of a symposium, March 10–12, 1975, Washington, DC. U.S. Environmental Protection Agency, Washington, DC. 230 pp.

Friberg, F., C. Otto, and B. S. Svensson. 1980. Effects of acidification on the dynamics of allochthonous leaf material and benthic invertebrate communities in running waters, pp. 304–5. In: D. Drabløs and A. Tollan (eds.). Ecological impact of acid precipitation. Proceedings of an international conference, Sandefjord, Norway, March 11–14, 1980. SNSF Project, Oslo, Norway. 383 pp.

Gächter, R. and A. Máreš. 1979. MELIMEX, an experimental heavy metal pollution study: Effects of increased heavy metal loads on phytoplankton communities. Schweizerische Zeitschrift für Hydrologie 41:228–46.

Gahnström, G., G. Andersson, and S. Fleischer. 1980. Decomposition and exchange processes in acidified lake sediment, pp. 306–7. In: D. Drabløs and A. Tollan (eds.). Ecological impact of acid precipitation. Proceedings of an international conference, Sandefjord, Norway, March 11–14, 1980. SNSF Project, Oslo, Norway. 383 pp.

Garten, C. T., Jr. and J. B. Gentry. 1976. Thermal tolerance of dragonfly nymphs. II. Comparison of nymphs from control and thermally altered environments. Physiological Zoology 49:206–13.

Gaufin, A. R. 1973. Use of aquatic invertebrates in the assessment of water quality, pp. 96–116. In: J. Cairns, Jr. and K. L. Dickson (eds.). Biological methods for the assessment of water quality. American Society for Testing and Materials Special Technical Publication 528. American Society for Testing and Materials, Philadelphia, PA. 256 pp.

Gaufin, A. R., R. Clubb, and R. Newell. 1974. Studies on the tolerance of aquatic insects to low oxygen concentrations. Great Basin Naturalist 34:45–59.

Gaufin, R. F. and A. R. Gaufin. 1961. The effects of low oxygen concentrations on stoneflies. Proceedings of the Utah Academy of Sciences 38:57–64.

Gentile, J. H., S. M. Gentile, N. G. Hairston, Jr., and B. K. Sullivan. 1982. The use of life-tables for evaluating the chronic toxicity of pollutants to *Mysidopsis bahia*. Hydrobiologia 93:179–87.

Gentry, J. B., C. T. Garten, Jr., F. G. Howell, and M. H. Smith. 1975. Thermal ecology of dragonflies in habitats receiving reactor effluent, pp. 563–74. In: Environmental effects of cooling systems at nuclear power plants. Proceedings of a symposium on the physical and biological effects on the environment of cooling systems and thermal discharges at nuclear power stations, Oslo, Norway, August 26–30, 1974. International Atomic Energy Agency, Vienna, Austria. 832 pp.

Giddings, J. M., B. R. Parkhurst, C. W. Gehrs, and R. E. Millemann. 1980. Toxicity of a coal liquefaction product to aquatic organisms. Bulletin of Environmental Contamination and Toxicology 25:1–6.

Glowacka, I., G. J. Soszka, and H. Soszka. 1976. Invertebrates associated with macrophytes, pp. 97–122. In: E. Pieczyńska (ed.). Selected problems of lake littoral ecology. Warsaw University Press, Warsaw, Poland. 238 pp.

Grahn, O., H. Hultberg, and L. Landner. 1974. Oligotrophication—a self-accelerating process in lakes subjected to excessive supply of acid substances. Ambio 3:93–94.

Grassle, J. F. and J. P. Grassle. 1974. Opportunistic life histories and genetic systems in marine benthic polychaetes. Journal of Marine Research 32:253–84.

Graynoth, E. 1979. Effects of logging on stream environments and faunas in Nelson. New Zealand Journal of Marine and Freshwater Research 13:79–109.

Green, R. H. 1979. Sampling design and statistical methods for environmental biologists. John Wiley and Sons, Inc. New York, NY. 257 pp.

Greenfield, J. P. and M. P. Ireland. 1978. A survey of the macrofauna of a coal-waste polluted Lancashire fluvial system. Environmental Pollution 16:105–22.

Grimås, U. 1961. The bottom fauna of natural and impounded lakes in northern Sweden (Ankarvattnet and Blåsjön). Institute of Freshwater Research Drottningholm Report 42:183–237.

Grimås, U. and T. Wiederholm. 1979. Biometry and biology of Constempellina brevicosta (Chironomidae) in a subarctic lake. Holarctic Ecology 2:119–24.

Haines, T. A. 1981. Acidic precipitation and its consequences for aquatic ecosystems: a review. Transactions of the American Fisheries Society 110:669–707.

Håkanson, L. 1981. On lake bottom dynamics—the energy-topography factor. Canadian Journal of Earth Sciences 18:899–909.

Hall, R. J., G. E. Likens, S. B. Fiance, and G. R. Hendrey. 1980. Experimental acidification of a stream in the Hubbard Brook Experimental Forest, New Hampshire. Ecology 61:976–89.

Harp, G. and R. S. Campbell. 1967. The distribution of Tendipes plumosus (Linné) in mineral acid water. Limnology and Oceanography 12:260–63.

Harper, P. P. 1973. Life histories of Nemouridae and Leuctridae in Southern Ontario (Plecoptera). Hydrobiologia 41:309–56.

Havas, M. 1981. Physiological response of aquatic animals to low pH, pp. 49–65. In: R. Singer (ed.). Effects of acidic precipitation on benthos. North American Benthological Society, Springfield, IL. 154 pp.

Hawkes, H. A. 1975. River zonation and classification, pp. 312–74. In: B. A. Whitton (ed.). River ecology. Studies in ecology. Vol. 2. Blackwell Scientific Publications, Oxford. 725 pp.

Hawkes, H. A. 1979. Invertebrates as indicators of river water quality, pp. 2–1 to 2–45. In: A. James and L. Evison (eds.). Biological indicators of water quality. John Wiley and Sons, Inc. New York, NY. 597 pp.

Hawkins, C. P. and J. R. Sedell. 1981. Longitudinal and seasonal changes in functional organization of macroinvertebrate communities in four Oregon streams. Ecology 62:387–97.

Hellawell, J. M. 1978. Biological surveillance of rivers. Water Research Centre, Stevenage, England. 332 pp.

Helson, B. V. and A. S. West. 1978. Particulate formulations of Abate[R] and methoxychlor as black fly larvicides: their selective effects on stream fauna. The Canadian Entomologist 110:591–602.

Hendrey, G. R. 1976. Effects of pH on the growth of periphytic algae in artificial stream channels. Internal Report 25/76. SNSF Project, Oslo, Norway. 50 pp.

Hendrey, G. R., K. Baalsrud, T. S. Traaen, M. Laake, and G. Raddum. 1976. Acid precipitation: some hydrobiological changes. Ambio 5:224–27.

Hendrey, G. R., J. N. Galloway, and C. L. Schofield. 1980. Temporal and spatial trends in the chemistry of acidified lakes under ice cover, pp. 266–67. *In*: D. Drabløs and A. Tollan (eds.). Ecological impact of acid precipitation. Proceedings of an international conference, Sandefjord, Norway, March 11–14, 1980. SNSF Project, Oslo, Norway. 383 pp.

Henriksen, A. and R. F. Wright. 1978. Concentrations of heavy metals in small Norwegian lakes. Water Research 12:101–12.

Henrikson, L. and H. G. Oscarson. 1981. Corixids (Hemiptera-Heteroptera), the new top predators in acidified lakes. Internationale Vereinigung für Theoretische und Angewandte Limnologie Verhandlungen 21:1616–20.

Hilsenhoff, W. L. 1965. The effect of toxaphene on the benthos in a thermally-stratified lake. Transactions of the American Fisheries Society 94:210–13.

Hofmann, W. 1971a. Die postglaziale Entwicklung der Chironomiden- und *Chaoborus*-Fauna (Dipt.) des Schöhsees. Archiv für Hydrobiologie Supplement 40:1–74.

Hofmann, W. 1971b. Zur Taxonomie und Palökologie subfossiler Chironomiden (Dipt.) in Seesedimenten. Archiv für Hydrobiologie Beiheft Ergebnisse der Limnologie 6:1–50.

Holdgate, M. W. 1979. A perspective of environmental pollution. Cambridge University Press, Cambridge. 278 pp.

Hudson, B. N. A. 1956. The behaviour of the female mosquito in selecting water for oviposition. Journal of Experimental Biology 33:478–92.

Hultberg, H. 1976. Thermally-stratified acid water in late winter—a key factor inducing self-accelerating processes which increase the acidification process, pp. 503–17. *In*: Proceedings of the First International Symposium on Acid Precipitation and the Forest Ecosystem, Columbus, Ohio, May 12–15, 1975. U.S.D.A. Forest Service General Technical Report NE, 23. U.S. Department of Agriculture, Northeastern Forest Experiment Station, Upper Darby, PA. 1074 pp.

Hurlbert, S. H. 1975. Secondary effects of pesticides on aquatic ecosystems. Residue Reviews 57:81–148.

Hutchinson, G. E. 1969. Eutrophication, past and present, pp. 17–26. *In*: Eutrophication: causes, consequences, correctives. Proceedings of a symposium. National Academy of Sciences, Washington DC. 661 pp.

Hynes, H. B. N. 1960. The biology of polluted waters. Liverpool University Press, Liverpool, England. 202 pp.

Hynes, H. B. N. 1969. The enrichment of streams, pp. 188–96. *In*: Eutrophication: causes, consequences, correctives. Proceedings of a symposium. National Academy of Sciences, Washington DC. 661 pp.

Hynes, H. B. N. 1970. The ecology of running waters. University of Toronto Press, Toronto. 555 pp.

Hynes, H. B. N. 1973. The effects of sediment on the biota in running water, pp. 653–63. *In*: Fluvial processes and sedimentation. Proceedings of Hydrology Symposium held at University of Alberta, Edmonton, May 8 and 9, 1973. Canada Department of the Environment, Ottawa, Canada. 759 pp.

Hynes, H. B. N. 1975. The stream and its valley. Internationale Vereinigung für Theoretische und Angewandte Limnologie Verhandlungen 19:1–15.

Illies, J. (ed.). 1978. Limnofauna Europaea. A checklist of the animals inhabiting European inland waters, with accounts of their distribution and ecology (except Protozoa). 2nd ed. Gustav Fischer Verlag, Stuttgart, West Germany. 532 pp.

Iwamoto, R. N., E. O. Salo, M. A. Madej, R. L. McComas, and R. L. Rulifson. 1978. Sediment and water quality: a review of the literature including a suggested approach for water quality criteria with summary of workshop and conclusions and recommendations. U.S. Environmental Protection Agency, Region 10, 910/9-78-048. U.S. Environmental Protection Agency, Seattle, WA. 253 pp.

Jahn, W. 1972. Ökologische Untersuchungen an Tümpeln unter besonderer Berüchsichtigung der Folgen von Wasserverschmutzung durch Öl. Archiv für Hydrobiologie 70: 442–83.

Jensen, L. D. and A. R. Gaufin. 1964a. Effects of ten organic insecticides on two species of stonefly naiads. Transactions of the American Fisheries Society 93:27–34.

Jensen, L. D. and A. R. Gaufin. 1964b. Long-term effects of organic insecticides on two species of stonefly naiads. Transactions of the American Fisheries Society 93:357–63.

Jernelöv, A., B. Nagell, and A. Svensson. 1981. Adaptation to an acid environment in *Chironomus riparius* (Diptera, Chironomidae) from Smoking Hills, NWT, Canada. Holarctic Ecology 4:116–19.

Johansson, K. and P. Nyberg. 1981. Acidification of surface waters in Sweden—effects and extent 1980. Institute of Freshwater Research Drottningholm Information 6:1–118. (In Swedish with English summary).

Johnson, M. G. and R. O. Brinkhurst. 1971. Associations and species diversity in benthic macroinvertebrates of Bay of Quinte and Lake Ontario. Journal of the Fisheries Research Board of Canada 28:1683–97.

Jónasson, P. M. 1972. Ecology and production of the profundal benthos in relation to phytoplankton in Lake Esrom. Oikos Supplementum 14:1–148.

Jumppanen, K. 1976. Effects of waste waters on a lake ecosystem. Annales Zoologici Fennici 13:85–138.

Kaiser, K. L. E. 1980. Correlation and prediction of metal toxicity to aquatic biota. Canadian Journal of Fisheries and Aquatic Sciences 37:211–18.

Kansanen, P. H. and J. Aho. 1981. Changes in the macrozoobenthos associations of polluted Lake Vanajavesi, Southern Finland, over a period of 50 years. Annales Zoologici Fennici 18:73–101.

Kershaw, W. E., T. R. Williams, S. Frost, and H. B. N. Hynes. 1965. Selective effect of particulate insecticides on *Simulium* among stream fauna. Nature (London) 208:199.

Knight, A. W. and A. R. Gaufin. 1966. Oxygen consumption of several species of stoneflies (Plecoptera). Journal of Insect Physiology 12:347–55.

Koehn, T. and C. Frank. 1980. Effect of thermal pollution on the chironomid fauna in an urban channel, pp. 187–96. *In*: D. A. Murray (ed.). Chironomidae. Ecology, systematics, cytology, and physiology. Pergamon Press, Oxford, England.

Lang, C. 1974. Macrofaune des fonds des cailloux du Léman. Schweizerische Zeitschrift für Hydrologie 36:301–50.

Lang, C. and B. Lang-Dobler. 1979. MELIMEX, an experimental heavy metal-pollution study: Oligochaetes and chironomid larvae in heavy metal loaded and control limnocorrals. Schweizerische Zeitschrift für Hydrologie 41:271–76.

Langford, T. E. 1975. The emergence of insects from a British river warmed by power station cooling-water. Part II. The emergence patterns of some species of Ephemeroptera, Trichoptera and Megaloptera in relation to water temperature and river flow, upstream and downstream of the cooling-water outfalls. Hydrobiologia 47:91–133.

Last, F. T., G. E. Likens, B. Ulrich, and L. Walløe. 1980. Acid precipitation—progress and problems. Conference summary, pp. 10–12. *In*: D. Drabløs and A. Tollan (eds.).

Ecological impact of acid precipitation. Proceedings of an international conference, Sandefjord, Norway, March 11–14, 1980. SNSF Project, Oslo, Norway. 383 pp.

Learner, M. A., R. Williams, M. Harcup, and B. D. Hughes. 1971. A survey of the macrofauna of the River Cynon, a polluted tributary of the River Taff (South Wales). Freshwater Biology 1:339–67.

Lehmkuhl, D. M. 1972. Change in thermal regime as a cause of reduction of benthic fauna downstream of a reservoir. Journal of the Fisheries Research Board of Canada 29:1329–32.

Lehmkuhl, D. M. 1979. Environmental disturbance and life histories: principles and examples. Journal of the Fisheries Research Board of Canada 36:329–34.

Leivestad, H., G. Hendrey, I. P. Muniz, and E. Snekvik. 1976. Effects of acid precipitation on freshwater organisms, pp. 87–111. In: F. H. Braekke (ed.). Impact of acid precipitation on forest and freshwater ecosystems in Norway. Research Report 6/76. SNSF Project, Oslo, Norway. 111 pp.

LeSage, L. and A. D. Harrison. 1980. The biology of *Cricotopus* (Chironomidae: Orthocladiinae) in an algal-enriched stream: Part I. Normal biology. Archiv für Hydrobiologie Supplement 57:375–418.

Letterman, R. D. and W. J. Mitsch. 1978. Impact of mine drainage on a mountain stream in Pennsylvania. Environmental Pollution 17:53–73.

Lloyd, R. 1965. Factors that affect the tolerance of fish to heavy metal poisoning, pp. 181–87. In: C. M. Tarzwell (ed.). Biological problems in water pollution, 3rd seminar, 13–17 August, 1962. U.S. Public Health Service Publication No. 999-WP-25. U.S. Department of Health, Education, and Welfare, Cincinnati, OH. 424 pp.

Lock, M. A., R. R. Wallace, D. R. Barton, and S. Charlton. 1981a. The effects of synthetic crude oil on microbial and macroinvertebrate benthic river communities—Part I: Colonization of synthetic crude oil contaminated substrata. Environmental Pollution (Series A) 24:207–17.

Lock, M. A., R. R. Wallace, D. R. Barton, and S. Charlton. 1981b. The effects of synthetic crude oil on microbial and macroinvertebrate benthic river communities—Part II: The response of an established community to contamination by synthetic crude oil. Environmental Pollution (Series A) 24:263–75.

Luoma, S. N. 1977. Detection of trace contaminant effects in aquatic ecosystems. Journal of the Fisheries Research Board of Canada 34:436–39.

Mackay, R. J. and G. B. Wiggins. 1979. Ecological diversity in Trichoptera. Annual Review of Entomology 24:185–208.

Markarian, R. K. 1980. A study of the relationship between aquatic insect growth and water temperature in a small stream. Hydrobiologia 75:81–95.

Marsh, P. C. and T. F. Waters. 1980. Effects of agricultural drainage development on benthic invertebrates in undisturbed downstream reaches. Transactions of the American Fisheries Society 109:213–23.

Mattice, J. S. and L. L. Dye. 1978. Effect of a steam electric generating station on the emergence timing of the mayfly, *Hexagenia bilineata* (Say). Internationale Vereinigung für Theoretische und Angewandte Limnologie Verhandlungen 20:1752–58.

Mayer, F. L. and J. L. Hamelink (eds.). 1977. Aquatic toxicology and hazard evaluation. American Society for Testing and Materials Special Technical Publication 634. American Society for Testing and Materials, Philadelphia, PA. 307 pp.

McCauley, R. N. 1966. The biological effects of oil pollution in a river. Limnology and Oceanography 11:475–86.

McClelland, W. T. and M. A. Brusven. 1980. Effects of sedimentation on the behavior and distribution of riffle insects in a laboratory stream. Aquatic Insects 2:161–69.

McCullough, D. A., G. W. Minshall, and C. E. Cushing. 1979. Bioenergetics of lotic filter-feeding insects *Simulium* spp. (Diptera) and *Hydropsyche occidentalis* (Trichoptera) and their function in controlling organic transport in streams. Ecology 60:585–96.

Metcalf, R. L. 1980. Changing role of insecticides in crop protection. Annual Review of Entomology 25:219–56.

Meynell, P. J. 1973. A hydrobiological survey of a small Spanish river grossly polluted by oil refinery and petrochemical wastes. Freshwater Biology 3:503–20.

Minshall, G. W. 1978. Autotrophy in stream ecosystems. BioScience 28:767–71.

Minshall, G. W. and R. A. Kuehne. 1969. An ecological study of invertebrates of the Duddon, an English mountain stream. Archiv für Hydrobiologie 66:169–91.

Moon, T. C. and C. M. Lucostic. 1979. Effects of acid mine drainage on a southwestern Pennsylvania stream. Water, Air, and Soil Pollution 11:377–90.

Morgan, N. C. and H. J. Egglishaw. 1965. A survey of the bottom fauna of streams in the Scottish Highlands. Part 1. Composition of the fauna. Hydrobiologia 25:181–211.

Mossberg, P. 1979. Benthos of oligotrophic and acid lakes. Institute of Freshwater Research Information 11:1–40. (In Swedish with English summary).

Mossberg, P. and P. Nyberg. 1979. Bottom fauna of small and acid forest lakes. Institute of Freshwater Research Drottningholm Report 58:77–87.

Muirhead-Thomson, R. C. 1973. Laboratory evaluation of pesticide impact on stream invertebrates. Freshwater Biology 3:479–98.

Muirhead-Thomson, R. C. 1978. Lethal and behavioral impact of permethrin (NRDC 143) on selected stream macroinvertebrates. Mosquito News 38:185–90.

Muirhead-Thomson, R. C. 1979. Experimental studies on macroinvertebrate predator-prey impact of pesticides. The reactions of *Rhyacophila* and *Hydropsyche* (Trichoptera) larvae to *Simulium* larvicides. Canadian Journal of Zoology 57:2264–70.

Müller, P. 1980. Effects of artificial acidification on the growth of periphyton. Canadian Journal of Fisheries and Aquatic Sciences 37:355–63.

Nagell, B. and C.-C. Landahl. 1978. Resistance to anoxia of *Chironomus plumosus* and *Chironomus anthracinus* (Diptera) larvae. Holaractic Ecology 1:333–36.

National Academy of Sciences. 1975. Petroleum in the marine environment. National Academy of Sciences, Washington DC. 107 pp.

Nebeker, A. V. 1971. Effect of high winter water temperatures on adult emergence of aquatic insects. Water Research 5:777–83.

Nebeker, A. V. and A. E. Lemke. 1968. Preliminary studies on the tolerance of aquatic insects to heated waters. Journal of the Kansas Entomological Society 41:413–18.

Neff, J. M. 1979. Polycyclic aromatic hydrocarbons in the aquatic environment. Sources, fates and biological effects. Applied Science Publishers Ltd., London. 262 pp.

Neumann, D. 1962. Die anaerobiose-toleranz der Larven zweier Subspecies von *Chironomus thummi*. Zeitschrift für vergleichende Physiologie 46:150–62.

Newbold, J. D., D. C. Erman, and K. B. Roby. 1980. Effects of logging on macroinvertebrates in streams with and without buffer strips. Canadian Journal of Fisheries and Aquatic Sciences 37:1076–85.

Nilssen, J. P. 1980. Acidification of a small watershed in southern Norway and some characteristics of acidic aquatic environments. Internationale Revue der gesamten Hydrobiologie 65:177-207.

Nriagu, J. O. (ed.). 1978a. The biogeochemistry of lead in the environment. Part A. Ecological cycles. Elsevier/North-Holland Biomedical Press, New York, NY. 422 pp.

Nriagu, J. O. 1978b. The biogeochemistry of lead in the environment. Part B. Biological effects. Elsevier/North-Holland Biomedical Press, New York, NY. 397 pp.

Nriagu, J. O. 1979a. Copper in the environment. Part I: Ecological cycling. John Wiley and Sons, Inc. New York, NY. 522 pp.

Nriagu, J. O. 1979b. Copper in the environment. Part II: Health effects. John Wiley and Sons, Inc. New York, NY. 489 pp.

Nriagu, J. O. 1979c. The biogeochemistry of mercury in the environment. Elsevier/North-Holland Biomedical Press, New York, NY. 696 pp.

Nriagu, J. O. 1980a. Zinc in the environment. Part I: Ecological cycling. John Wiley and Sons, Inc. New York, NY. 453 pp.

Nriagu, J. O. 1980b. Zinc in the environment. Part II: Health effects. John Wiley and Sons, Inc. New York, NY. 480 pp.

Nursall, J. R. 1952. The early development of a bottom fauna in a new power reservoir in the Rocky Mountains of Alberta. Canadian Journal of Zoology 30:387-409.

Nuttall, P. M. and G. H. Bielby. 1973. The effect of china-clay wastes on stream invertebrates. Environmental Pollution 5:77-86.

Obrdlík, P., Z. Adámek, and J. Zahrádka. 1979. Mayfly fauna (Ephemeroptera) and the biology of the species *Potamanthus luteus* (L.) in a warmed stretch of the Oslava River. Hydrobiologia 67:129-40.

Odum, H. T. 1956. Primary production in flowing waters. Limnology and Oceanography 1:102-17.

Ozimek, T. and U. Sikorska. 1976. Influence of municipal sewage on the littoral biocenosis, pp. 181-93. *In*: E. Pieczyńska (ed.). Selected problems of lake littoral ecology. Warsaw University Press, Warsaw, Poland. 238 pp.

Parker, B. L., J. D. Brammer, M. E. Whalon, and W. O. Berry. 1976. Chronic oil contamination and aquatic organisms with emphasis on Diptera: status and bibliography. Water Resources Bulletin 12:291-305.

Parkin, R. B. and J. B. Stahl. 1981. Chironomidae (Diptera) of Baldwin Lake, Illinois, a cooling reservoir. Hydrobiologia 76:119-28.

Petersen, L. B.-M. and R. C. Petersen, Jr. 1983. Anomalies in hydropsychid capture nets from polluted streams. Freshwater Biology 13:185-91.

Phillips, D. J. H. 1980. Quantitative aquatic biological indicators. Their use to monitor trace metal and organochlorine pollution. Applied Science Publishers Ltd., London. 488 pp.

Pianka, E. R. 1978. Evolutionary ecology. 2nd ed. Harper and Row Publishers, New York, NY. 397 pp.

Pieczyńska, E. 1972. Ecology of the eulittoral zone of lakes. Ekologia Polska 20:637-732.

Pieczyńska, E. 1976. Some regularities in the functioning of lake littoral, pp. 211-28. *In*: E. Pieczyńska (ed.). Selected problems of lake littoral ecology. Warsaw University Press, Warsaw, Poland. 238 pp.

Pieczyńska, E., U. Sikorska, and T. Ozimek. 1975. The influence of domestic sewage on the littoral zone of lakes. Polskie Archiwum Hydrobiologii 22:141-56.

Prejs, K. 1976. Bottom fauna, pp. 123-44. *In*: E. Pieczyńska (ed.). Selected problems of lake littoral ecology. Warsaw University Press, Warsaw, Poland. 238 pp.

Prosi, F. 1979. Heavy metals in aquatic organisms, pp. 271–323. *In*: Förstner and G. T.W. Wittmann. Metal pollution in the aquatic environment. Springer-Verlag, Berlin, West Germany. 486 pp.

Raddum, G. G. and O. A. Sæther. 1981. Chironomid communities in Norwegian lakes with different degrees of acidification. Internationale Vereinigung für Theoretische und Angewandte Limnologie Verhandlungen 21:399–405.

Rahel, F. J. 1981. Selection for zinc tolerance in fish: results from laboratory and wild populations. Transactions of the American Fisheries Society 110:19–28.

Reavell, P. E. and P. Frenzel. 1981. The structure and some recent changes of the zoobenthic community in the Ermatinger Becken, a shallow littoral part of Lake Constance. Archiv für Hydrobiologie 92:44–52.

Rehwoldt, R., L. Lasko, C. Shaw, and E. Wirhowski. 1973. The acute toxicity of some heavy metal ions toward benthic organisms. Bulletin of Environmental Contamination and Toxicology 10:291–94.

Resh, V. H. and J. D. Unzicker. 1975. Water quality monitoring and aquatic organisms: the importance of species identification. Journal of the Water Pollution Control Federation 47:9–19.

Roback, S. S. and J. W. Richardson. 1969. The effects of acid mine drainage on aquatic insects. Proceedings of the Academy of Natural Sciences of Philadelphia 121:81–107.

Rosenberg, D. M. 1975. Food chain concentration of chlorinated hydrocarbon pesticides in invertebrate communities: a re-evaluation. Quaestiones Entomologicae 11:97–110.

Rosenberg, D. M. and N. B. Snow. 1975. Effect of crude oil on zoobenthos colonization of artificial substrates in subarctic ecosystems. Internationale Vereinigung für Theoretische und Angewandte Limnologie Verhandlungen 19:2172–77.

Rosenberg, D. M. and N. B. Snow. 1977. A design for environmental impact studies with special reference to sedimentation in aquatic systems of the Mackenzie and Porcupine river drainages, pp. III 65–III 78. *In*: Proceedings of the Circumpolar Conference on Northern Ecology, September 15–18, 1975, Ottawa. National Research Council of Canada, Ottawa, Canada. 9 sections.

Rosenberg, D. M. and A. P. Wiens. 1976. Community and species responses of Chironomidae (Diptera) to contamination of fresh waters by crude oil and petroleum products, with special reference to the Trail River, Northwest Territories. Journal of the Fisheries Research Board of Canada 33:1955–63.

Rosenberg, D. M. and A. P. Wiens. 1978. Effects of sediment addition on macrobenthic invertebrates in a northern Canadian river. Water Research 12:753–63.

Rosenberg, D. M. and A. P. Wiens. 1980. Responses of Chironomidae (Diptera) to short-term experimental sediment additions in the Harris River, Northwest Territories, Canada. Acta Universitatis Carolinae-Biologica 1978:181–92.

Rosenberg, D. M., A. P. Wiens, and J. F. Flannagan. 1980. Effects of crude oil contamination on Ephemeroptera in the Trail River, Northwest Territories, Canada, pp. 443–55. *In*: J. F. Flannagan and K. E. Marshall (eds.). Advances in Ephemeroptera biology. Proceedings of the Third International Conference on Ephemeroptera. Plenum Press, New York, NY. 552 pp.

Rosenberg, D. M., A. P. Wiens, and O. A. Sæther. 1977. Responses to crude oil contamination by *Cricotopus (Cricotopus) bicinctus* and *C. (C.) mackenziensis* (Diptera: Chironomidae) in the Fort Simpson area, Northwest Territories. Journal of the Fisheries Research Board of Canada 34:254–61.

Rupprecht, R. 1975. The dependence of emergence-period in insect larvae on water temperature. Internationale Vereinigung für Theoretische und Angewandte Limnologie Verhandlungen 19:3057–63.

Saether, O. A. 1975. Nearctic chironomids as indicators of lake typology. Internationale Vereinigung für Theoretische und Angewandte Limnologie Verhandlungen 19: 3127–33.

Saether, O. A. 1979. Chironomid communities as water quality indicators. Holarctic Ecology 2:65–74.

Saether, O. A. 1980. The influence of eutrophication on deep lake benthic invertebrate communities. Progress in Water Technology 12:161–80.

Sanders, H. O. and O. B. Cope. 1968. The relative toxicities of several pesticides to naiads of three species of stoneflies. Limnology and Oceanography 13:112–17.

Saunders, G. W. (coordinator). 1980. Organic matter and decomposers, pp. 341–92. In: E. D. LeCren and R. H. Lowe-McConnell (eds.). The functioning of freshwater ecosystems. International Biological Programme 22. Cambridge University Press, Cambridge. 588 pp.

Savage, N. L. and F. W. Rabe. 1973. The effects of mine and domestic wastes on macroinvertebrate community structure in the Coeur d'Alene River. Northwest Science 47:159–68.

Schindler, D. W. 1980. Experimental acidification of a whole lake: A test of the oligotrophication hypothesis, pp. 370–74. In: D. Drabløs and A. Tollan (eds.). Ecological impact of acid precipitation. Proceedings of an international conference, Sandefjord, Norway, March 11–14, 1980. SNSF Project, Oslo, Norway. 383 pp.

Scullion, J. and R. W. Edwards. 1980. The effects of coal industry pollutants on the macroinvertebrate fauna of a small river in the South Wales coalfield. Freshwater Biology 10:141–62.

Simpson, K. W. 1980. Abnormalities in the tracheal gills of aquatic insects collected from streams receiving chlorinated or crude oil wastes. Freshwater Biology 10:581–83.

Sly, P. G. 1978. Sedimentary processes in lakes, pp. 65–89. In: A. Lerman (ed.). Lakes. Chemistry, geology, physics. Springer-Verlag, New York, NY. 375 pp.

Södergren, A., Bj. Svensson, and S. Ulfstrand. 1972. DDT and PCB in South Swedish streams. Environmental Pollution 3:25–36.

Södergren, S. 1976. Ecological effects of heavy metal discharge in a salmon river. Institute of Freshwater Research Drottningholm Report 55:91–131.

Solbé, J. F. de L. G. 1977. Water quality, fish and invertebrates in a zinc-polluted stream, pp. 97–105. In: J. S. Alabaster (ed.). Biological monitoring of inland fisheries. Applied Science Publishers Ltd., London. 226 pp.

Spehar, R. L., R. L. Anderson, and J. T. Fiandt. 1978. Toxicity and bioaccumulation of cadmium and lead in aquatic invertebrates. Environmental Pollution 15: 195–208.

Stahl, J. B. 1969. The uses of chironomids and other midges in interpreting lake histories. Internationale Vereinigung für Theoretische und Angewandte Limnologie Mitteilungen 17:111–25.

Stokes, P. M. 1981. Benthic algal communities in acidic lakes, pp. 119–38. In: R. Singer (ed.). Effects of acidic precipitation on benthos. North American Benthological Society, Springfield, IL. 154 pp.

Stokes, P. M., T. C. Hutchinson, and K. Krauter. 1973. Heavy-metal tolerance in algae isolated from contaminated lakes near Sudbury, Ontario. Canadian Journal of Botany 51:2155–68.

Streit, B. and P. Schröder. 1978. Dominierende Benthosinvertebraten in der Geröll-brandungszone des Bodensees: Phänologie, Nahrungsökologie und Biomasse. Archiv für Hydrobiologie Supplement 55:211–34.

Strenzke, K. 1960. Die systematische und ökologische Differenzierung der Gattung *Chironomus*. Annales Entomologici Fennici 26:111–38.

Sutcliffe, D. W. and T. R. Carrick. 1973. Studies on mountain streams in the English Lake District. I. pH, calcium and the distribution of invertebrates in the River Duddon. Freshwater Biology 3:437–62.

Sweeney, B. W. 1978. Bioenergetic and developmental response of a mayfly to thermal variation. Limnology and Oceanography 23:461–77.

Sweeney, B. W. and R. L. Vannote. 1978. Size variation and distribution of hemimeta-bolous aquatic insects: two thermal equilibrium hypotheses. Science 200:444–46.

Tebo, L. B., Jr. 1955. Effects of siltation, resulting from improper logging, on the bottom fauna of a small trout stream in the Southern Appalachians. The Progressive Fish-Culturist 17:64–70.

Thienemann, A. 1922. Die beiden Chironomusarten der Tiefenfauna der norddeutschen Seen. Ein hydrobiologisches Problem. Archiv für Hydrobiologie 13:609–46.

Tomkiewicz, S. M., Jr. and W. A. Dunson. 1977. Aquatic insect diversity and biomass in a stream marginally polluted by acid strip mine drainage. Water Research 11:397–402.

Urech, J. 1979. MELIMEX, an experimental heavy metal pollution study: Effects of increased heavy metal load on crustacea plankton. Schweizerische Zeitschrift für Hydrologie 41:247–60.

Vannote, R. L., G. W. Minshall, K. W. Cummins, J. R. Sedell, and C. E. Cushing. 1980. The river continuum concept. Canadian Journal of Fisheries and Aquatic Sciences 37:130–37.

Wallace, R. R. and H. B. N. Hynes. 1975. The catastrophic drift of stream insects after treatments with methoxychlor (1, 1, 1-trichloro-2, 2-bis [*p*-methoxyphenyl] ethane). Environmental Pollution 8:255–68.

Walshe, B. M. 1948. The oxygen requirements and thermal resistance of chironomid larvae from flowing and from still waters. Journal of Experimental Biology 25:35–44.

Walshe, B. M. 1950. The function of hemoglobin in *Chironomus plumosus* under natural conditions. Journal of Experimental Biology 27:73–95.

Ward, J. V. 1974. A temperature-stressed stream ecosystem below a hypolimnial release mountain reservoir. Archiv für Hydrobiologie 74:247–75.

Ward, J. V. 1976. Effects of flow patterns below large dams on stream benthos: a review, pp. 235–53. *In:* J. F. Orsborn and G. H. Allman (eds.). Instream flow needs, solutions to technical, legal, and social problems caused by increasing competition for limited stream flow. Symposium and speciality conference, Boise, ID, May 1976. American Fisheries Society, Bethesda, MD. 1208 pp.

Ward, J. V. and J. A. Stanford. 1979. Ecological factors controlling stream zoobenthos with emphasis on thermal modification of regulated streams, pp. 35–55. *In:* J. V. Ward and J. A. Stanford (eds.). The ecology of regulated streams. Plenum Press, New York, NY. 398 pp.

Ward, J. V. and J. A. Stanford. 1982. Thermal responses in the evolutionary ecology of aquatic insects. Annual Review of Entomology 27:97–117.

Warnick, S. L. and H. L. Bell. 1969. The acute toxicity of some heavy metals to different species of aquatic insects. Journal of the Water Pollution Control Federation 41:280–84.

Warren, C. E. 1971. Biology and water pollution control. W. B. Saunders Co., Philadelphia, PA. 434 pp.

Warwick, W. F. 1975. The impact of man on the Bay of Quinte, Lake Ontario, as shown by the subfossil chironomid succession (Chironomidae, Diptera). Internationale Vereinigung für Theoretische und Angewandte Limnologie Verhandlungen 19: 3134–41.

Warwick, W. F. 1980a. Palaeolimnology of the Bay of Quinte, Lake Ontario: 2800 years of cultural influence. Canadian Bulletin of Fisheries and Aquatic Sciences 206:1–118.

Warwick, W. F. 1980b. Chironomidae (Diptera) responses to 2800 years of cultural influence; a palaeolimnological study with special reference to sedimentation, eutrophication, and contamination processes. The Canadian Entomologist 112:1193–1238.

Waters, T. F. 1979. Benthic life histories: summary and future needs. Journal of the Fisheries Research Board of Canada 36:342–45.

Webb, M. (ed.). 1979. The chemistry, biochemistry and biology of cadmium. Elsevier/ North-Holland Biomedical Press, Amsterdam, The Netherlands, 465 pp.

Welch, E. B. 1980. Ecological effects of waste water. Cambridge University Press, Cambridge. 337 pp.

Welch, E. B. and J. C. Spindler. 1964. DDT persistence and its effect on aquatic insects and fish after an aerial application. Journal of the Water Pollution Control Federation 36:1285–92.

Wentsel, R., A. McIntosh, and V. Anderson. 1977a. Sediment contamination and benthic macroinvertebrate distribution in a metal-impacted lake. Environmental Pollution 14:187–93.

Wentsel, R., A. McIntosh, and G. Atchison. 1977b. Sublethal effects of heavy metal contaminated sediment on midge larvae (*Chironomus tentans*). Hydrobiologia 56: 153–56.

Wentsel, R., A. McIntosh, W. P. McCafferty, G. Atchison, and V. Anderson. 1977c. Avoidance response of midge larvae (*Chironomus tentans*) to sediments containing heavy metals. Hydrobiologia 55:171–75.

Wentsel, R., A. McIntosh, and W. P. McCafferty. 1978. Emergence of the midge *Chironomus tentans* when exposed to heavy metal contaminated sediment. Hydrobiologia 57:195–96.

Westlake, D. F. (coordinator). 1980. Primary production, pp. 141–246. *In*: E. D. LeCren and R. H. Lowe-McConnell (eds.). The functioning of freshwater ecosystems. International Biological Programme 22. Cambridge University Press, Cambridge. 588 pp.

Wetzel, R. G. 1975. Limnology. W. B. Saunders Co., Philadelphia, PA. 743 pp.

White, D. S. and J. R. Gammon. 1976. The effect of suspended solids on macroinvertebrate drift in an Indiana creek. Proceedings of the Indiana Academy of Science 86: 182–88.

Whitney, R. J. 1939. The thermal resistance of mayfly nymphs from ponds and streams. Journal of Experimental Biology 16:374–85.

Wiederholm, T. 1974. Bottom fauna and eutrophication in the large lakes of Sweden. Acta Universitatis Upsaliensis. Abstracts of Uppsala Dissertations from the Faculty of Sciences 270:1–15.

Wiederholm, T. 1976. A survey of the bottom fauna of Lake Sammamish. Northwest Science 50:23–31.

Wiederholm, T. 1979. Use of benthic communities in lake monitoring, pp. 196–211. *In*:

The use of ecological variables in environmental monitoring. The National Swedish Environment Protection Board, Report PM 1151. 348 pp.

Wiederholm, T. 1980a. Chironomids as indicators of water quality in Swedish lakes. Acta Universitatis Carolinae-Biologica 1978:275–83.

Wiederholm, T. 1980b. Use of benthos in lake monitoring. Journal of the Water Pollution Control Federation 52:537–47.

Wiederholm, T., K. Danell, and K. Sjöberg. 1977. Emergence of chironomids from a small man-made lake in northern Sweden. Norwegian Journal of Entomology 24:99–105.

Wiederholm, T. and L. Eriksson. 1977. Benthos of an acid lake. Oikos 29:261–67.

Wielgosz, S. 1979. The effect of wastes from the town of Olsztyn on invertebrate communities in the bottom of the River Łyna. Acta Hydrobiologica 21:149–65.

Wiggins, G. B. and R. J. Mackay. 1978. Some relationships between systematics and trophic ecology in Nearctic aquatic insects, with special reference to Trichoptera. Ecology 59:1211–20.

Wigglesworth, V. B. 1965. The principles of insect physiology. 6th ed. Methuen and Co. Ltd., London. 741 pp.

Williams, D. D. 1981. Migrations and distributions of stream benthos, pp. 155–207. In: M. A. Lock and D. D. Williams (eds.). Perspectives in running water ecology. Plenum Press, New York, NY. 430 pp.

Winner, R. W., M. W. Boesel, and M. P. Farrell. 1980. Insect community structure as an index of heavy-metal pollution in lotic ecosystems. Canadian Journal of Fisheries and Aquatic Sciences 37:647–55.

Winner, R. W., J. Scott Van Dyke, N. Caris, and M. P. Farrel. 1975. Response of the macroinvertebrate fauna to a copper gradient in an experimentally-polluted stream. Internationale Vereinigung für Theoretische und Angewandte Limnologie Verhandlungen 19:2121–27.

Wood, K. G. 1973. Decline of *Hexagenia* (Ephemeroptera) nymphs in western Lake Erie, pp. 26–32. In: W. L. Peters and J. G. Peters (eds.). Proceedings of the First International Conference on Ephemeroptera. E. J. Brill, Leiden, The Netherlands. 312 pp.

Wotton, R. S. 1979. The influence of a lake on the distribution of blackfly species (Diptera: Simuliidae) along a river. Oikos 32:368–72.

Wurtz, C. B. 1969. The effects of heated discharges on freshwater benthos, pp. 199–213. In: P. A. Krenkel and F. L. Parker (eds.). Biological aspects of thermal pollution. Proceedings of the national symposium on thermal pollution, Portland, OR, June 3–5, 1968. Vanderbilt University Press, Nashville, TN. 407 pp.

Yasuno, M., J. Ohkita, and S. Hatakeyama. 1982. Effects of temephos on macrobenthos in a stream of Mt. Tsukuba. Japanese Journal of Ecology 32:29–38.

chapter 18

ECOLOGICAL PERSPECTIVES IN THE MANAGEMENT OF AQUATIC INSECT HABITAT

James V. Ward

INTRODUCTION

Despite the major ecological roles played by insects in aquatic habitats, management strategies have generally ignored or given only cursory consideration to their requirements. In this context, aquatic insects have been viewed as fish food organisms or water quality indicators; only rarely are they considered to be an integral part of habitat management (Hynes 1970). Although knowledge of the environmental requirements of aquatic insects is far from complete, what is known should be more fully integrated with information on other biotic and abiotic components of aquatic ecosystems in the development of management plans. It is only by maintaining a broad ecological perspective, including consideration of land-water interactions, that management strategies can optimize aquatic resources, particularly when these resources are under constraints posed by multiple-use practices within the watershed.

This chapter addresses some of the ways by which aquatic insect habitats may be influenced by management strategies. Consideration is given to ameliorative practices, as well as to management strategies specifically designed to enhance aquatic resources.

RIPARIAN HABITAT MANAGEMENT

Human modification of watersheds has often resulted in the removal of vegetation along streams (Hynes 1969). As a result, ecological conditions that originally characterized lower reaches of rivers (high summer temperatures, illumination, and nutrient levels) have been shifted upstream. In addition, because downstream conditions are dependent upon upstream functions (Vannote et al. 1980), alteration of the riparian habitats of headwater streams will be reflected in lower reaches.

Thermal conditions in natural waters have played a major role in the ecology and evolution of aquatic insects (Ward and Stanford 1982). Temperature serves as the cue for a variety of life-cycle responses, and it often determines the distribution patterns of species, the outcomes of competitive interactions, and the temporal segregation of species. Consequently, altered thermal conditions may have serious consequences for aquatic insects. Compared with streams with well-developed riparian vegetation, open streams generally exhibit greater diel and annual temperature ranges, higher daily and seasonal maxima, and lower temperature minima. Heavily canopied streams also warm more slowly in the spring and cool less rapidly in autumn than do open streams. Removal of riparian vegetation may more than double the annual maximum stream temperature, with maximum diel fluctuations exhibiting an order of magnitude increase (Brown and Krygier 1970).

The recovery of aquatic insect communities from alterations engendered by watershed disturbances may be directly dependent upon the rate of recovery of the terrestrial vegetation (Haefner and Wallace 1981). The use of buffer strips (i.e. narrow rows of natural vegetation along stream banks) is an effective management strategy for protecting the stream environment from severe alterations in temperature and other variables soon to be discussed. In a clearcut watershed, no stream temperature changes attributable to logging were detected where 15 to 30 m-wide buffer strips were retained (Brown and Krygier 1970). Ten years after an exclosure was constructed along an Oregon stream that was heavily grazed by cattle, the maximum summer temperature of the stream within the exclosure was lower (18.9°C versus 25.6°C) and maximum diel fluctuations were less (7.2°C versus 14.4°C) in the reach from which cattle were excluded than in the grazed section (Claire and Storch, in press). The density of the canopy in the path of the sun, called angular canopy density (Brown and Brazier 1972), is the most important buffer strip variable for water temperature control.

Although nutrients have considerable mobility within terrestrial ecosystems, undisturbed forests lose only small amounts of vital nutrients to adjacent aquatic systems (Likens and Bormann 1974). In contrast, watersheds affected by disturbances such as logging may exhibit dramatic increases in nutrient output. Since nutrients are adsorbed by sediment particles, management strategies that minimize erosion will also reduce nutrient losses from the watershed. Protection of riparian habitats from grazing, logging, and other potential disruptions will, therefore, do much to protect adjacent aquatic habitats.

Allochthonous detritus may be arranged along a continuum from highly labile inputs that are rapidly utilized by aquatic insects and microorganisms, to very resistant materials requiring many years for complete decomposition. The diversity of organic inputs (i.e. differences in resistance to decomposition, particle size, food quality, and timing of input) is used by a diverse community of aquatic insects that is adapted in various ways to exploit this energy resource. More resistant detritus (logs and branches) may provide important physical structure in streams as well as habitat niches and food for certain aquatic insects (Dudley and Anderson 1982).

Watershed practices that reduce the allochthonous organic input, modify the detrital structure within the stream, or alter the composition and diversity of riparian flora, will be reflected by changes in the aquatic insect community. For example, a stream in a drainage basin planted with a white pine monoculture supported only 20 to 50 percent as much benthic invertebrate biomass as streams in nearby watersheds that were more diverse (Woodall and Wallace 1972). Regrowth of riparian vegetation in a previously clearcut watershed increased amounts and food quality of allochthonous detritus within the stream, and reduced in situ primary production by shading (Haefner and Wallace 1981). This was reflected by associated reductions in grazer densities (e.g. elmid beetles) and increases in numbers of shredders (e.g. *Peltoperla maria*).

A survey of small watersheds in northern California was conducted to assess the protection that buffer strips of vegetation provided to stream invertebrates (Newbold et al. 1980). Macroinvertebrate communities of streams in undisturbed control watersheds differed significantly in total density, Shannon diversity, and Euclidean distance (a dissimilarity index) from communities residing in streams affected by logging without buffer strips. No detectable differences were found between control streams and streams with wide buffer strips (≥30 m). Shannon diversity index values exhibited a positive correlation with buffer strip width (range: 3–60 m) up to 30 m, beyond which the invertebrate communities were similar to control streams. Density values were highest at stations logged without buffer strips primarily due to increases in the mayfly *Baetis,* the stonefly *Nemoura,* and chironomid midges. The overall effects of logging without protection by buffer strips were likened to mild organic enrichment.

STREAM REGULATION

The impoundment of a previously free-flowing river by damming (i.e. stream regulation) creates an artificial lake and an altered lotic environment downstream from the dam. Management strategies for protecting or restoring the aquatic insect habitat in the receiving stream will be considered here and implications for the reservoir will be dealt with in the section on Lake Management.

The altered temperature regime below dams may be responsible for the elimination of many species of aquatic insects (Ward and Stanford 1979). This can occur in several ways (e.g. species that require a winter chill to break diapause may be eliminated by the winter-warm conditions below deep-release dams; the rapid vernal rise in temperature required by some species for hatching of eggs may not occur in regulated streams; and the attainment of an absolute level of temperature, or the accumulation of a certain number of degree days, which is necessary for maturation and emergence, may not occur in the summer-cool waters below deep-release dams). Species that require diel thermal fluctuations for maximum metabolic efficiency will be at a competitive disadvantage because of constant temperatures in many regulated streams. In addition, thermal alterations may result in emergence during winter when air temperatures are lethal to adults.

High dams with multiple-level outlets that impound large stratified reservoirs allow considerable managerial control over downstream water temperature. The thermal regimes of such regulated streams could be managed to mimic the temperature pattern required by a given species or assemblage of aquatic insects. For example, a certain number of degree days (see Sweeney, Chapter 4) could be programmed in an annual cycle. Such systems also offer virtually unlimited potential for experimental manipulations designed to investigate the ecological role of temperature in stream ecosystems. Although models have been developed to predict downstream temperatures or to control thermal conditions below dams, only specific portions of the total thermal regime have been considered. Water bodies with very different thermal patterns may not exhibit discernible differences in total degree days or annual mean temperatures (Ward and Stanford 1982). To be biologically meaningful, temperature management should consider the influence of the total thermal regime upon all biotic components of the habitat and also (within the constraints of present knowledge) their interactions.

The flow regime below dams may be considerably different from that of unregulated streams. The discharge pattern engendered by stream regulation is primarily determined by the function of the reservoir (see Ward 1976a, for a conceptual model relating the various flow regime alterations to aquatic insect habitat parameters).

Short-term flow fluctuations may modify aquatic insect communities in several ways. Aquatic insects may be stranded in pockets of standing water or along exposed shorelines due to fluctuating water levels. Differential tolerance to stranding eliminates certain groups while allowing others to predominate. Mayflies are particularly susceptible to stranding and are relatively intolerant of exposure, whereas many chironomids tolerate considerable exposure, especially in cool weather. The majority of stream insects can apparently tolerate brief periods of exposure, and some species migrate as the water level recedes. This, of course, also varies with the life-cycle stage (e.g. larvae are generally more mobile than eggs or pupae). Management strategies to reduce stranding losses should consider not only the duration of exposure and the rate of change in water levels, but also time of year (atmospheric conditions), channel configuration, and substrate heterogeneity.

Both increasing and decreasing discharge induce drift of aquatic insects. Populations of certain running water forms may thus be depleted in streams below dams because drift from upstream lotic reaches is unavailable to replenish the individuals lost from the regulated stream segment. Research is needed to assess the relationship between drift rates and altered discharge patterns for different species of aquatic insects. Severe flow fluctuations may greatly alter debris dams and remove accumulations of detritus, thereby modifying spatial and trophic characteristics. Reduction of physical retention structures lengthens nutrient spirals, thus altering the ability of the stream ecosystem to withstand and recover from perturbations (Webster et al. 1975). Stream reaches below deep-release hydropower dams may exhibit extreme short-term thermal variations in summer associated with daily fluctuations in discharge. In addition, the diel thermal maxima in regulated segments may coincide with daily minima in natural streams (Ward and Stanford 1982). The ecological implications for aquatic insects of such thermal phase shifts certainly warrant further investigation.

Flow constancy, especially of reduced flow, results in a different set of habitat modifications. The quality and quantity of the allochthonous detritus available for aquatic insects may be altered if nonriparian plant species invade the shore zone in the absence of periodic flooding. A constant flow regime may not provide the proper signals for some species of aquatic insects. For example, the annual migration of the mayfly *Leptophlebia cupida* from the main stream into small tributaries appears to be initiated by rising water during spring runoff (Hayden and Clifford 1974).

The effects of siltation, including reduction or elimination of the hyporheic community, create severe problems in some regulated streams. A transmountain diversion scheme reduced the total annual discharge from Lake Granby on the upper Colorado River, Colorado, to only 11 percent of its historical flow, and resulted in a constant flow regime for several km below the dam. The regulated flow regime was not sufficient to remove sediment that accumulated during dam construction; some areas of streambed had silt deposits 30 cm deep with the benthos consisting primarily of Oligochaeta and Chironomidae. To ameliorate the sediment buildup, a large volume of water ($1,224,677 \text{ m}^3$) was released from the dam during a four-day period in April (Eustis and Hillen 1954). The flushing action of the simulated spring runoff removed nearly all accumulated fine sediment from riffle areas, enabling recolonization by insects requiring a clean rocky substrate. Although the aquatic insect community remains altered by other effects of stream regulation (J. V. Ward, unpublished data), sedimentation is no longer a major problem.

Regulation of the Trinity River in northern California has reduced the flow below Lewiston Dam to approximately 20 percent of historical levels (Boles 1981). To mitigate loss of fish spawning habitat, a series of riffles was restored by removing the heterogeneous, sediment-laden gravel and adding homogenous, clean gravel. However, the aquatic insect communities that colonized these replacement gravels were less diverse, had lower biomass, and exhibited more severe population fluctuations than did insects of clean natural riffles. Boles (1981) concluded that the substrate used to restore the riffles did not have sufficient particle size heterogeneity to support a diverse and stable biotic community.

Dense masses of algae such as *Cladophora glomerata* and *Hydrurus foetidus* often occur in regulated streams. Beds of submerged angiosperms (e.g. water buttercup, *Ranunculus aquatilis*) may develop in regulated streams in regions where such plants are normally absent from lotic biotopes (Ward 1976b). In addition to changing the food base, the enhanced aquatic flora may virtually eliminate clean rock surfaces, but it may also provide sites that serve as refugia from the current. As a result, some species are eliminated, whereas others, which were previously absent, can now invade the regulated segment of stream (Ward 1976b). In eutrophic waters, aquatic plants may reach nuisance proportions below dams; however, the additional habitat niches and the increased amount of food provided for aquatic insects may be viewed as a beneficial change in an unproductive stream system. Management schemes may control aquatic flora by regulating the intensity and frequency of flood events, and by releasing water from reservoir strata with certain nutrient levels.

Downstream losses of plankton from an upstream reservoir may account for the dense populations of filter-feeding insects such as Simuliidae and Hydropsychidae that may develop in streams below dams (e.g. see Ward and Short 1978). Control of plankton concentrations released from the dam, based on monitoring the depth distribution of reservoir plankton, also offers potential as an effective management technique, as described in the section on Fisheries Enhancement.

Since aquatic insects of cool rocky streams have an evolutionary history associated with oxygen-saturated waters (e.g. see Ross 1967), few species have developed mechanisms to cope with oxygen deficits. Therefore, discharge of water with low oxygen concentrations may greatly alter and reduce aquatic fauna in regulated streams. Reduced compounds such as hydrogen sulfide, which may be directly toxic to aquatic insects, are associated with anaerobic conditions in a reservoir. However, severe problems caused by such toxic compounds are generally limited to streams below deep-release eutrophic reservoirs during the period of stratification. Even if anaerobic hypolimnia develop, water may be drawn from upper layers of a reservoir, or air drafts may be installed in the water release ports of a dam.

Excessive amounts of dissolved gases (gas supersaturation) cause gas bubble disease in aquatic organisms (see Weitkamp and Katz 1980 for a review). Although gas supersaturation may occur under natural conditions, it poses the most serious problems where water falling from high dams is mixed with air that subsequently dissolves under the hydrostatic pressures in deep plunge pools. While aquatic insects are relatively more tolerant of gas supersaturation than fishes (Montgomery and Fickeisen 1979), they apparently lack the latter's ability to adjust their depth to compensate for gas supersaturation levels. However, Montgomery and Fickeisen (1979) reported increased buoyancy of benthic insects in gas-supersaturated waters. If this increased buoyancy increases drift rates, populations may be depleted from lotic habitats that have low recolonization potential due to the lentic water body upstream. Spillway deflectors, devices that direct the spilled flow along the surface of the tailrace (rather than allowing the water to be carried deep into the plunge basin), and other mechanisms are available to alleviate problems of gas-supersaturation below dams.

Operators of high dams are provided with powerful tools to manage habitat conditions in the receiving stream because of the myriad ecological interactions engendered by control of release depth and discharge. Although the potential exists for ecologically sound management of aquatic insect habitats in regulated streams, mitigative strategies have largely dealt with protecting sport fishes. The requirements of fishes may (as in the case of silt-free substrate interstices) or may not (as in the case of thermal requirements) correspond to those of aquatic insects. Management objectives for protection of aquatic biota should include a consideration of the habitat requirements of aquatic insects as well.

LAKE MANAGEMENT

A variety of lake and reservoir management techniques, especially regarding reversal of eutrophication, has been developed and assessed in recent years. However, relatively few studies have considered, even cursorily, the implications of various lake management schemes on aquatic insect communities. This is somewhat surprising since the development of lake typologies has been based primarily on the presence or absence of *Chaoborus* and the species composition of profundal Chironomidae (see Wiederholm, Chapter 17, and Brinkhurst 1974, for an historical account). In addition, because littoral insects form a major portion of the diets of some lake fishes and are an essential part of the diet of many species of waterfowl, it is surprising that more consideration has not been directed toward management techniques to enhance aquatic insect productivity.

The removal of trees from the area to be inundated by an impoundment is a common management practice. For example, prior to filling, trees were removed from prospective fishing areas in Volta Lake in Ghana (Petr 1973). As inundation progressed there was a massive development of the wood-burrowing mayfly *Povilla adusta* on the remaining drowned trees. Since *Povilla* is an important food for some lake fishes, the presence of trees greatly increased fish production in Volta Lake (Petr 1973). However, although a high benthic biomass develops rapidly when softwood trees are submerged, *P. adusta* is unable to bore in hardwood trees. When hardwood trees are exposed during drawdown, terrestrial wood-boring beetles create burrows that can later be utilized by *P. adusta*, the caddisfly *Amphipsyche senegalensis*, and other insects after water levels increase. Although wood-burrowing mayflies are restricted to the tropics, dense populations of benthic invertebrates may also be associated with submerged trees in temperate reservoirs. For example, densities of Chironomidae were 11 times greater on submerged trees than on the bottom substrate in a Missouri River reservoir (Cowell and Hudson 1967).

Compared with natural lakes, many reservoirs exhibit considerable fluctuations in water level. In addition to direct effects resulting from desiccation and exposure to temperature extremes in the drawdown zone, water-level fluctuations indirectly influence aquatic insects by altering sediment composition and by the virtual elimination of higher aquatic plants from certain areas of the littoral zone. Although precise water level management schedules have been devised for mosquito control (e.g. see Hess and Kiker 1944), few investigations have specifically addressed effects of managing water-level fluctuations on other aquatic insects.

A case-history study of Llyn Tegid, Wales, offers an exceptional opportunity to examine short-term effects of water-level fluctuations and recovery of the littoral insect fauna following restabilization (Hunt and Jones 1972). In 1955 a controlled outflow scheme decreased the minimum lake level and increased water level fluctuations. The maximum permissible rate of change (15 cm/day) was often exceeded, resulting in rapid and irregular fluctuations for several years after dam construction. However, because of the construction of an additional reservoir, fluctuations lessened, and by 1967 the water level amplitude was similar to pre-1955 levels. What changes occurred over this period? Sampling in 1957 (two years after conversion of Llyn Tegid to a reservoir) revealed that many major groups of invertebrates (e.g. Ephemeroptera, Plecoptera, Trichoptera) had been either eliminated or greatly reduced. Although loss of some species was due to stranding, the disappearance of more mobile forms was attributed to destruction of littoral vegetation or increased deposition of fine sediments. Chironomidae and Oligochaeta adapted to the muddy littoral zone and became the predominant macroinvertebrates. By 1959, despite continued high fluctuations in water level, some of the Plecoptera, Ephemeroptera, and Trichoptera began to recolonize the littoral zone, which had by then restabilized at the lower mean water level. In 1969, following a reduction of water-level fluctuations to preimpoundment values, all major groups and many species present before 1955 were again found, although the community composition remained altered (Hunt and Jones 1972). At that time, faunal recovery was attributed to silt removal by wave action and the reestablishment of littoral vegetation.

Aquatic insects may also play a role in waterfowl management in habitats with controlled water levels. Since egg-laying ducks, ducklings, and molting adults cannot obtain all their protein requirements from aquatic plants (Krull 1970), the abundance and availability of benthic invertebrates should be a major consideration in managing wetlands for waterfowl production. For example, some plants (e.g. *Elodea*) that are poor waterfowl food in themselves harbor large numbers of insects. Conversely, some of the pondweeds (*Potamogeton* spp.) are excellent waterfowl food, although they are poorly colonized by insects (Krull 1970). Drawdown enhances some species of macrophytes while adversely affecting others, and such effects vary with the timing and duration of drawdown. The seasonal availability of insects also varies as a function of the plant species present, hence habitat manipulations designed to provide large numbers of

invertebrates during the waterfowl breeding season must be based on consideration of a variety of interrelated factors.

Drawdown was used as a management tool to enhance the sport fishery in a
Florida lake after regulation of water levels had decreased the natural historical
fluctuations by 71 percent (Wegener et al. 1975). After reflooding, insect
abundance greatly increased, especially the phytomacrofauna, which exhibited
over a three-fold increase in numbers. As a result, the sport-fish population nearly
doubled. Wegener et al. (1975) concluded that the most effective and economical
method to enhance sport fishing was accomplished by increasing water level
fluctuation as a mechanism to stimulate macroinvertebrate production. More
typically, drawdown of a Michigan waterfowl impoundment decimated aquatic
invertebrates, which recolonized reflooded areas very slowly (Kadlec 1962).

The depth(s) at which water is released from a dam allows managerial
control over temperature, nutrients, oxygen, and turbidity within a stratified
reservoir. In addition, the timing and depth of release may directly influence lentic
insect populations. For example, migratory movements involving incursions into
the limnetic zone may be a significant factor in the dynamics of benthic insects in
reservoirs. Cowell and Hudson (1967) recorded 0.5 to 8 benthic invertebrates per
m^3 in the water column of Lewis and Clark Lake, South Dakota, and estimated
an annual loss of 44 metric tons of insects (24 tons of *Hexagenia* nymphs, and
20 tons of Chironomidae and Ceratopogonidae) from the reservoir to the Missouri River. Such losses do not necessarily reduce total reservoir productivity
(Hudson and Lorenzen 1981), but they may provide the managerial potential
for fishery enhancement of downstream lotic habitats as discussed later in
this chapter.

Although various techniques have been used to aerate lakes, they may be
grouped in two general categories (Fast 1979): 1) techniques that disrupt thermal
stratification (i.e. destratification), and 2) techniques that increase the oxygen
content of hypolimnetic waters while maintaining thermal stratification (i.e.
hypolimnetic aeration). The latter technique is used to provide conditions
suitable for a "two-story" fishery with warmwater fishes in shallow waters and
coldwater species in deep waters. Few data exist regarding the effects that the
myriad alterations in abiotic and biotic variables induced by lake aeration have
on aquatic insects. The abundance of benthic invertebrates greatly increased
following aeration of reservoirs in Wisconsin (Wirth et al. 1970) and California
(Fast 1979), but declined in a Colorado montane lake (Lackey 1973). Following
artificial destratification of highly eutrophic reservoirs, benthic insects may
invade deep waters previously devoid of macrobenthos, and a shift in the
dominant species of Chaoboridae may result as fishes begin feeding in deeper
strata (Wirth et al. 1970).

Shapiro (1979, p. 161), in reference to lake restoration schemes such as
artificial aeration, cautions us to remember that lakes are ecosystems, not
"phosphorus-driven generators of algae." A preoccupation with nutrients by

those engaged in research on lake restoration may indeed be partly responsible for our general ignorance of the ecological implications of the various lake restoration schemes. There is a great need for research to elucidate the direct and indirect effects of various lake renewal techniques on aquatic insect communities and the resulting implications for the total lentic ecosystem.

RIVER RESTORATION

The unidirectional flow of running waters greatly enhances the recovery potential of damaged lotic systems (Cairns and Dickson 1977). The presence of unimpacted upper reaches and entering tributaries may provide direct routes for recolonization by aquatic insects and other organisms. Aquatic insects are generally highly vagile within a free-flowing river system because of the aerial adult stage and the propensity of immatures to drift.

The effects of impact on abiotic habitat conditions may be specific to certain types of activities and these effects in turn directly influence recovery. For example, a chemical spill that alters pH and eliminates the aquatic biota from a river reach may not engender much permanent change in the habitat. In contrast, channelization alters the physical characteristics of the affected stream reach for many years. The presence of residual toxicants, such as certain pesticides and heavy metals, slows the recovery process and reduces the recovery potential of an ecosystem. A chemical spill in the Roanoke River, Virginia, resulted in prolonged toxicity to aquatic insects because of the retention of ethyl benzene-creosote residues by the sediment (Cairns and Dickson 1977). The rate at which physical-chemical environmental conditions return to an approximation of prestressed values depends on the duration, severity, and type of perturbation. Many of the management strategies discussed in this chapter (e.g. reintroductions of indigenous species, substrate restoration) may speed the recovery of damaged ecosystems. A thorough understanding of the structure and function of the ecosystem prior to perturbation is necessary to maximize such ameliorative practices.

The Thames River in England demonstrates the remarkable ability of running waters to at least partially recover from severe environmental deterioration. In the seventeenth century "the waters at Oxford were revolting" due to large amounts of untreated sewage entering the river (Mann 1972, p. 229). However, fishing, boating, and swimming are again possible because of strict control of pollution in the Thames basin. There is no longer evidence of serious oxygen deficits. Fish biomass is extremely high (2000 kg/ha) as is production of aquatic invertebrates (331 kcal/m^2/yr), over 60 percent of which is contributed by Chironomidae. A productive riverine environment now exists because of the management practices of the Thames Conservancy (Mann 1972).

In Norway, river waters are diverted through tunnels and flow to power stations at lower elevations, leaving the intervening reach of river with reduced discharge. A project involving the use of weir basins, the Weir Project (Terskelprosjektet), is an attempt to partially mitigate the adverse effects of low flow by constructing small on-stream impoundments to maintain perennial aquatic habitats. A variety of studies have been conducted on organic matter budgets, physical-chemical characteristics, benthos, and fishes associated with the weir basins. Black flies, previously making up only a small portion of the lotic insect fauna, greatly increased downstream from the weir basins because of increased seston (Raastad 1979). Mean densities of total benthic invertebrates up to 160,000 organisms/m^2 have been recorded at outlet stations, 80 to 90 percent of which were black flies. There appears to be a succession of black fly species over time, and, because older weir systems are dominated by Chironomidae and Ephemeroptera, there is some indication that the predominance of Simuliidae may represent an early successional stage.

Many colonization pathways are available to stream insects (Williams and Hynes 1976). Although drift constitutes a major recolonization mechanism, other sources of colonizers (aerial, hyporheic, and upstream aquatic migration) may also contribute significantly to stream recovery (e.g. see Resh et al. 1981). Blocking one or more of these four major sources of colonization (e.g. by siltation of the hyporheic zone) will likely result in the reestablishment of a different benthic community than if all pathways remain open.

MINING AND ROAD CONSTRUCTION

Surface Coal Mining

Low pH, high concentrations of heavy metals, and ferric hydroxide precipitation associated with acid-mine drainage may eliminate aquatic insects from affected areas for many years following cessation of mining (see Roback and Richardson 1969 for a review). Recovery of the insect community may take decades, especially in lentic habitats. To be most effective, environmental control of surface mining must be a part of the mine-planning stage. Sedimentation ponds should be established below mining areas prior to commencement of mining operations, and methods to reduce oxidation of pyritic material, a major source of acid, and to prevent formation of acid seeps should be used (Hill and Grim 1977).

Mining in the Western Energy Region of the United States (e.g. Colorado, Wyoming) engenders a somewhat different set of problems for aquatic habitats and their inhabitants (Ward et al. 1978). Although coal mining adjacent to a mountain stream in northwestern Colorado began over 30 years prior to the initiation of a year-round study, no detrimental effects on the extant aquatic insect community were discernible. Community composition was generally similar at stream locations above, adjacent to, and below the mine spoils; species diversity and richness values were comparable at all locations; in fact, abundance and biomass increased, rather than decreased, in the downstream direction. The lack of discernible detrimental effects was attributed to the following: (1) the absence of acid-mine drainage; (2) the low solubility of most heavy metals; (3) the buffer zone between the mine spoils and the stream; (4) the short period of soluble salt input; and (5) the large, relatively undisturbed watershed above the mine spoils. While some of these conditions may be peculiar to the study stream, many generally characterize the states of this Energy Region.

The insect communities of spoils ponds that differed in regard to age and differences in input of coal-mine drainage were also examined, but the extreme detrimental effects associated with acid-mine drainage were not observed (Canton and Ward 1981). Diversity values were lower in the ponds affected by mine drainage, but Ephemeroptera, Hemiptera, Odonata, Coleoptera, and Diptera occurred in all ponds examined. Trichoptera were absent only from the youngest pond. Similarity coefficients suggested that colonization phenomena such as age and distance from a source of colonizers may be responsible, in part, for faunal differences among ponds. Colonization of spoils ponds may, therefore, be accelerated by direct transplants of aquatic fauna and flora from established ponds in xeric regions of the western U.S. where natural lentic habitats are generally small and widely spaced. It appears that, with proper environmental considerations, coal mining in the western U.S. may not always have the localized severe impacts on aquatic insects that characterize many other regions. However, the cumulative effects of increased salinity, sedimentation, and water depletion may adversely affect aquatic habitats downstream. As with forest buffer strips, maintenance of even a short distance between mine spoils and running waters may do much to protect the habitat of lotic insects.

Road Construction

Logging roads, and not timber harvest practices *per se*, may be responsible for the majority of sediment inputs to streams associated with logging. Compared to an undisturbed control watershed, the mean annual sediment yield was increased only 3.3 times in a clearcut watershed without roads, but was increased 109 times in a patch-cut watershed with roads (Fredriksen 1970). Cline et al. (1982) examined the immediate and residual effects of localized highway construction activities, such as bridge construction, on aquatic insects of a high mountain stream. Contrary to expectations, effects were minimal and, where discernible changes occurred, recovery was rapid, despite 10- to 100-fold elevations in suspended solids. The relatively high inertia (ability to resist disturbance) and resilience (ability to recover from disturbance) of the lotic insect community was attributed to the following: (1) the rapid and persistent return of suspended solids to background levels following cessation of construction activities; (2) the absence of sedimentation during spring runoff when highest concentrations of suspended solids from construction occurred; (3) the steep gradient and virtual absence of pools in the study segment, which allowed the system to be flushed; (4) the presence of unimpacted upstream reaches; and (5) the relatively short duration and localized nature of each construction disturbance. Mountain stream insects have evolved to withstand periods of high runoff with associated high levels of suspended solids. If, as was true in the study by Cline et al. (1982), the major increases in suspended sediment due to construction activities occur immediately preceding or during spring runoff, it is perhaps not surprising to find only minimal effects on aquatic insects. In contrast, insect communities of high mountain streams may indeed be extremely sensitive to anthropogenic inputs of toxic substances or organic wastes.

FISHERIES ENHANCEMENT

Artificial Enrichment

When mineral fertilizer was added to unproductive lakes in Russia, the summer biomass of benthos, previously 9 to 83 kg/ha, increased to 56 to 268 kg/ha within two to three years (Baranov et al. 1973). The increase in invertebrates was accompanied by higher production of lake fishes. However, because fertilization of lakes may speed the eutrophication process and result in oxygen depletion, such practices may be more compatible with running waters, at least where salmonid fisheries are concerned. Sections of an Oregon stream enriched with sucrose developed large populations of sewage fungus, *Sphaerotilus natans,* which is actually a bacterium associated with organic pollution (Warren et al. 1964). The annual mean biomass of benthic insects was nine times greater in an enriched section than in the control reach, primarily due to increases in Chironomidae, which fed on *Sphaerotilus.* Trout production exhibited a seven-fold increase. A mixture of cereal grain and cattle feed was added to a 55-m stretch of a north Swedish stream five times during the summer as an experiment to enhance salmonid production by increasing bottom fauna (Henricson and Sjöberg 1981). After fertilization, the mean biomass of total benthos was 2.6 times greater in the enriched section than at an upstream control station. Significant increases in biomass of Chironomidae, Ephemeroptera, and Plecoptera resulted in the enriched section. Increases in biomass were exhibited by gatherers, shredders, and filter-feeders. No species or other taxonomic group occurred with lower abundance or biomass at the fertilized location. Number of taxa, Shannon diversity, equitability, and dominance values were not significantly different in enriched and control reaches.

Semi-natural Channels

Side channels connected to the main stream offer several advantages over standard hatchery channels for incubation and rearing of salmonids (Mundie 1979). For example, the ratio of riffles, where greatest production of benthos occurs, to pools, which are used as fish habitat, can be optimized when such semi-natural channels are constructed. The stream gradient and flow regime in the side channel can be regulated to increase production of aquatic insects and fishes. Floods and drought, which limit fish and insect production, can be eliminated, yet fluctuations in discharge can be retained and substrate heterogeneity can be controlled. The smolts produced are indistinguishable from wild fish with respect to fat content, and they exhibit natural immune responses and feeding modes. A semi-natural rearing channel, 396-m long, has been operating for several years along the Big Qualicum River in British Columbia (Mundie 1979). This approach is economically feasible because natural food supplements reduce commercial food requirements to <50 percent of those in standard hatchery rearing procedures. Savings in commercial food are possible because drifting aquatic insects from the main river continuously recolonize riffles in the side channel, and insects and other invertebrates assimilate fish wastes and uneaten commercial food, thus enhancing benthic production in the semi-natural channel.

In the Big Qualicum River experimental side channel, invertebrate biomass was 20-fold greater in downstream than upstream pools, and four times greater in downstream than upstream riffles, which suggests that even greater benthic production could be attained by further enriching the upper end of the channel. Small quantities of fish wastes and food pellets added to delineated areas of substrate increased benthic standing crop as much as three times over control values (Williams et al. 1977). However, because young salmonids feed primarily on drifting invertebrates, and since drift occurs mainly at night, much of the increased production is unavailable to the fish. This could be remedied if riffles are disturbed to induce catastrophic drift during the day, but to what extent can recolonization phenomena such as behavioral drift compensate for population depletions? Williams et al. (1977) experimentally determined that small areas of gravel in the side channel could be depleted of benthic fauna every 15 days, based on the recovery rate of aquatic invertebrate densities. About four weeks was required to allow complete colonization of all taxa. The authors suggest that water jets or compressed air may be suitable for routinely inducing drift in rearing channels. It is not known, however, what proportion of the total riffle area can be simultaneously disturbed and still maintain benthic communities.

Reservoirs, especially those impounded by dams with multidepth water release capabilities, offer an untapped potential for downstream fish rearing facilities. Both the main stream and the reservoir could be used to optimize habitat conditions and fish production in semi-natural side channels (Fig. 18.1).

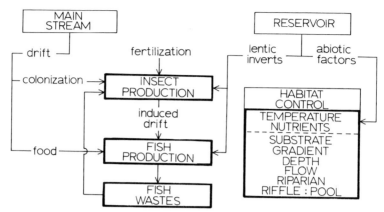

Figure 18.1 Management strategies for potential enhancement of salmonid production in semi-natural side channels (*heavy lines*) based on optimization of habitat conditions, and increased production and availability of lentic and lotic invertebrates. Modified and expanded from Mundie (1979); see text for details.

Temperature and nutrient control are made possible by withdrawing water from different reservoir strata. For example, winter-warm conditions may be maintained in the rearing channels by introducing water from near the bottom of the reservoir. Lentic plankton and benthos released from dams could be utilized to enhance production of filter-feeding insects in the rearing channels, and to provide a direct dietary supplement for fishes. For example, each year 13,000 to 30,000 metric tons (wet weight) of zooplanktonic crustaceans and about 44 metric tons of insects are lost from Lewis and Clark Reservoir (Benson and Cowell 1967; Cowell and Hudson 1967). Multidepth outlets would allow selective withdrawal of reservoir strata rich in food organisms as a function of diel and seasonal patterns of depth distribution. Lotic invertebrates from the main stream could be used directly as food for fishes and as a source of colonizers to assure that benthic populations are maintained in the side channels.

Introduction of Exotic Invertebrates

Large-scale introductions of invertebrates have been undertaken, especially in Russia, as a means to increase fish production. Organisms of low vagility such as Mollusca and malacostracan Crustacea, rather than insects, are most commonly used in transplants. From 1947 to 1967, 49 species of invertebrates were introduced into 51 reservoirs in the U.S.S.R. (Ioffe 1972). Twenty-six of the species became acclimatized to the reservoir environment, and 19 of those became common items in the diets of fishes. The exotic species did not appear to reduce the established communities of Chironomidae and Oligochaeta.

Aquatic insects have also been introduced, with limited success, for biological control of mosquitoes (Bay 1974) and macrophytes (Balciunas and Center 1981). However, it should be emphasized that the introduction of any exotic species, whether as prey organisms for fishes, or as control agents, is fraught with uncertainties and should be approached with extreme caution.

CONCLUSIONS

The broad ecological perspective essential for sound management of aquatic habitats and their inhabitants has not always been apparent. In this chapter I have attempted to relate a few of the ecological implications of management practices to the habitat requirements of aquatic insects. Many aspects of resource management remain *terra incognita* with respect to their effects on aquatic insects. For example, most of the myriad lake renewal techniques have not been even superficially examined in the context of aquatic insect ecology; those for which data are available (e.g. lake aeration) have generally received only limited study in this regard. Clearly, the implications of resource management on the ecology of aquatic insects provide fruitful areas for further research. The responses of aquatic insect communities to habitat alterations, including mitigative measures, may allow considerable insight into structural and functional attributes of aquatic ecosystems, in addition to providing a fuller understanding of aquatic insect ecology. Since insects play a central role in most aquatic habitats and act as integrators of environmental conditions, they should be considered an essential component of any ecologically sound management strategy.

Acknowledgments

The author is grateful to Drs. R. W. Pennak, University of Colorado; W. D. Fronk, Colorado State University; J. H. Mundie, Pacific Biological Station; T. Wiederholm, The National Swedish Environment Protection Board; and an anonymous reviewer for improving the manuscript. Miss Nancy Flaming and Mrs. Nancy Heisler typed the manuscript. This chapter was written while the author was supported by a research grant from the Colorado Experiment Station.

References

Balciunas, J. K. and T. D. Center. 1981. Preliminary host specificity tests of a Panamanian *Parapoynx* as a potential biological control agent for *Hydrilla*. Environmental Entomology 10:462–67.

Baranov, I. V., O. N. Bauer, and V. V. Pokrovskii. 1973. Biological aspects of increased fish productivity of the USSR lakes. Internationale Vereinigung für Theoretische und Angewandte Limnologie Verhandlungen 18:1851–57.

Bay, E. C. 1974. Predator-prey relationships among aquatic insects. Annual Review of Entomology 19:441–53.

Benson, N. G. and B. C. Cowell. 1967. The environment and plankton density in Missouri River reservoirs, pp. 358–73. In: Reservoir fishery resources symposium. American Fisheries Society, Washington DC. 569 pp.

Boles, G. L. 1981. Macroinvertebrate colonization of replacement substrate below a hypo-limnial release reservoir. Hydrobiologia 78:133–46.

Brinkhurst, R. O. 1974. The benthos of lakes. St. Martin's Press, New York, NY. 190 pp.

Brown, G. W. and J. R. Brazier. 1972. Controlling thermal pollution in small streams. EPA-R2-72-083. U.S. Environmental Protection Agency, Washington DC. 64 pp.

Brown, G. W. and J. T. Krygier. 1970. Effects of clear-cutting on stream temperature. Water Resources Research 6:1133–39.

Cairns, J., Jr and K. L. Dickson. 1977. Recovery of streams from spills of hazardous materials, pp. 24–42. In: J. Cairns, Jr., K. L. Dickson, and E. E. Herricks (eds.). Recovery and restoration of damaged ecosystems. University of Virginia Press, Charlottesville, VA. 531 pp.

Canton, S. P. and J. V. Ward. 1981. Benthos and zooplankton of coal strip mine ponds in the mountains of northwestern Colorado, U.S.A. Hydrobiologia 85:23–31.

Claire, E. and R. Storch. in press. Streamside management and livestock grazing: an objective look at the situation. In: J. Menke (ed.). Symposium on livestock inter-actions with wildlife, fish and their environments. U.S. Forest Service, Pacific South-west Forest and Range Experiment Station, Berkeley, CA.

Cline, L. D., R. A. Short, and J. V. Ward. 1981. The influence of highway construction on the macroinvertebrates and epilithic algae of a high mountain stream. Hydrobiologia 96:149–59.

Cowell, B. C. and P. L. Hudson. 1967. Some environmental factors influencing benthic invertebrates in two Missouri River reservoirs, pp. 541–55. In: Reservoir fishery resources symposium. American Fisheries Society, Washington DC. 569 pp.

Dudley, T. and N. H. Anderson. 1982. A survey of invertebrates associated with wood debris in aquatic habitats. Melanderia 39:1–21.

Eustis, A. B. and R. H. Hillen. 1954. Stream sediment removal by controlled reservoir releases. Progressive Fish-Culturist 16:30–35.

Fast, A. W. 1979. Artificial aeration as a lake restoration technique, pp. 121–31. In: Lake restoration. EPA 440/5-79-001. U.S. Environmental Protection Agency, Washing-ton DC. 254 pp.

Fredriksen, R. L. 1970. Erosion and sedimentation following road construction and timber harvest on unstable soils in three small western Oregon watersheds. U.S. Forest Service Research Paper PNW-104, Pacific Northwest Forest and Range Experiment Station, Portland, OR. 15 pp.

Haefner, J. D. and J. B. Wallace. 1981. Shifts in aquatic insect populations in a first-order southern Appalachian stream following a decade of old field succession. Canadian Journal of Fisheries and Aquatic Sciences 38:353–59.

Hayden, W. and H. F. Clifford. 1974. Seasonal movements of the mayfly *Leptophlebia cupida* (Say) in a brown-water stream of Alberta, Canada. American Midland Naturalist 91:90–102.

Henricson, J. and G. Sjöberg. 1981. Fishery management in tributaries—a fertilization experiment. The bottom fauna investigation in 1980. FÅK Informerar No. 10:17–36. National Board of Fisheries, Härnösand, Sweden (In Swedish). 39 pp.

Hess, A. D. and C. C. Kiker. 1944. Water level management for malaria control on impounded waters. The Journal of the National Malaria Society 3:181–96.

Hill, R. D. and E. C. Grim. 1977. Environmental factors in surface mine recovery, pp. 290–302. In: J. Cairns, Jr., K. L. Dickson, and E. E. Herricks (eds.). Recovery and restoration of damaged ecosystems. University of Virginia Press, Charlottesville, VA. 531 pp.

Hudson, P. L. and W. E. Lorenzen. 1981. Manipulation of reservoir discharge to enhance tailwater fisheries, pp. 568–79. In: R. M. North, L. B. Dworsky, and D. J. Allee (eds.). Unified river basin management. American Water Resources Association, Minneapolis. 654 pp.

Hunt, P. C. and J. W. Jones. 1972. The effect of water level fluctuations on a littoral fauna. Journal of Fish Biology 4:385–94.

Hynes, H. B. N. 1969. The enrichment of streams, pp. 188–96. In: Eutrophication: causes, consequences, correctives. National Academy of Sciences, Washington DC. 661 pp.

Hynes, H. B. N. 1970. The ecology of flowing waters in relation to management. Journal of the Water Pollution Control Federation 42:418–24.

Ioffe, T. I. 1972. The improvement of reservoir productivity through acclimatization of invertebrates. Internationale Vereinigung für Theoretische und Angewandte Limnologie Verhandlungen 18:818–21.

Kadlec, J. A. 1962. Effects of a drawdown on a waterfowl impoundment. Ecology 43: 267–81.

Krull, J. N. 1970. Aquatic plant-macroinvertebrate associations and waterfowl. Journal of Wildlife Management 34:707–18.

Lackey, R. T. 1973. Bottom fauna changes during artificial reservoir destratification. Water Research 7:1349–56.

Likens, G. E. and F. H. Bormann. 1974. Linkages between terrestrial and aquatic ecosystems. BioScience 24:447–56.

Mann, K. H. 1972. Case history: the River Thames, pp. 215–32. In: R. T. Oglesby, C. A. Carlson, and J. A. McCann (eds.). River ecology and man. Academic Press, New York, NY. 465 pp.

Montgomery, J. C. and D. H. Fickeisen. 1979. Tolerance and buoyancy of aquatic insect larvae exposed to gas supersaturated water. Environmental Entomology 8:655–57.

Mundie, J. H. 1979. The regulated stream and salmon management, pp. 307–19. In: J. V. Ward and J. A. Stanford (eds.). The ecology of regulated streams. Plenum Press, New York, NY. 398 pp.

Newbold, J. D., D. C. Erman, and K. B. Roby. 1980. Effects of logging on macroinvertebrates in streams with and without buffer strips. Canadian Journal of Fisheries and Aquatic Sciences 37:1076–85.

Petr, T. 1973. Some factors limiting the distribution of Povilla adusta Navas (Ephemeroptera, Polymitarcidae) in African lakes, pp. 223–30. In: W. L. Peters and J. G. Peters (eds.). Proceedings of the First International Conference on Ephemeroptera. E. J. Brill, Leiden, The Netherlands. 312 pp.

Raastad, J. E. 1979. Investigations of the bottom fauna of regulated rivers, with special emphasis on black-flies, (Diptera, Simuliidae). Terskelprosjektet Informasjon Nr. 8:1–60.

Resh, V. H., T. S. Flynn, G. A. Lamberti, E. P. McElravy, K. L. Sorg and J. R. Wood. 1981. Responses of the sericostomatid caddisfly *Gumaga nigricula* (McL.) to environmental disruption, pp. 311–18. *In*: G. P. Moretti (ed.). Proceedings of the Third International Symposium on Trichoptera. Series Entomologica, Vol. 20. Dr. W. Junk Publishers, The Hague, The Netherlands. 472 pp.

Roback, S. S. and J. W. Richardson. 1969. The effects of acid mine drainage on aquatic insects. Proceedings of the Academy of Natural Sciences of Philadelphia 121:81–107.

Ross, H. H. 1967. The evolution and past dispersal of the Trichoptera. Annual Review of Entomology 12:169–206.

Shapiro, J. 1979. The need for more biology in lake restoration, pp. 161–67. *In*: Lake restoration. EPA 440/5-79-001. U.S. Environmental Protection Agency, Washington, DC. 254 pp.

Vannote, R. L., G. W. Minshall, K. W. Cummins, J. R. Sedell, and C. E. Cushing. 1980. The river continuum concept. Canadian Journal of Fisheries and Aquatic Sciences 37:130–37.

Ward, J. V. 1967a. Effects of flow patterns below large dams on stream benthos: a review, pp. 235–53. *In*: J. F. Orsborn and C. H. Allman (eds.). Instream flow needs, Vol. II. American Fisheries Society, Bethesda, MD. 657 pp.

Ward, J. V. 1976b. Comparative limnology of differentially regulated sections of a Colorado mountain river. Archiv für Hydrobiologie 78:319–42.

Ward, J. V., S. P. Canton, and L. J. Gray. 1978. The stream environment and macroinvertebrate communities: contrasting effects of mining in Colorado and the eastern United States, pp. 176–87. *In*: J. H. Thorp and J. W. Gibbons (eds.). Energy and environmental stress in aquatic systems. Department of Energy Symposium Series, Springfield, VA. 854 pp.

Ward, J. V. and R. A. Short. 1978. Macroinvertebrate community structure of four special lotic habitats in Colorado, U.S.A. Internationale Vereinigung für Theoretische und Angewandte Limnologie Verhandlungen 20:1382–87.

Ward, J. V. and J. A. Stanford. 1979. Ecological factors controlling stream zoobenthos with emphasis on thermal modification of regulated streams, pp. 35–55. *In*: J. V. Ward and J. A. Stanford (eds.). The ecology of regulated streams. Plenum Press, New York, NY. 398 pp.

Ward, J. V. and J. A. Stanford. 1982. Thermal responses in the evolutionary ecology of aquatic insects. Annual Review of Entomology 27:97–117.

Warren, C. E., J. H. Wales, G. E. Davis, and P. Doudoroff. 1964. Trout production in an experimental stream enriched with sucrose. Journal of Wildlife Management 28:617–60.

Webster, J. R., J. B. Waide, and B. C. Patten. 1975. Nutrient recycling and the stability of ecosystems, pp. 1–27. *In*: F. G. Howell, J. B. Gentry, and M. H. Smith (eds.). Mineral cycling in southeastern ecosystems. Energy Research and Development Administration Symposium Series, Technical Information Center, Oak Ridge, TN. 898 pp.

Wegener, W., V. Williams, and T. D. McCall. 1975. Aquatic macroinvertebrate responses to an extreme drawdown. Proceedings of the Southeastern Association of Game and Fish Commissioners 28:126–44.

Weitkamp, D. E. and M. Katz. 1980. A review of dissolved gas supersaturation literature. Transactions of the American Fisheries Society 109:659–702.

Williams, D. D. and H. B. N. Hynes. 1976. The recolonization mechanisms of stream benthos. Oikos 27:265–72.

Williams, D. D., J. H. Mundie, and D. E. Mounce. 1977. Some aspects of benthic production in a salmonid rearing channel. Journal of the Fisheries Research Board of Canada 34:2133–41.

Wirth, T. L., R. C. Dunst, P. D. Uttormark, and W. Hilsenhoff. 1970. Manipulation of reservoir waters for improved quality and fish population response. Wisconsin Department of Natural Resources Research Report No. 62. Madison, WI. 23 pp.

Woodall, W. R. and J. B. Wallace. 1972. The benthic fauna in four small southern Appalachian streams. American Midland Naturalist 88:393–407.

chapter 19

AQUATIC INSECTS AND MANKIND

H. B. N. Hynes

It is impossible to consider the use of fresh water for any purpose involving quality, in the sense of the civil engineer, or yield, in the sense of the angler, without very soon coming upon topics that involve aquatic insects. This is not only because aquatic insects are ubiquitous, but they also make up a major part of the biomass in water. As major converters of energy, they play an important role in the spiraling of materials in streams, and their diversity and often close adaptation to fairly limited ecological niches makes them excellent indicators of local conditions. These topics have all been discussed in various contexts in previous chapters. Aquatic insects, however, impinge upon humans in many other ways, several of which will be considered here.

Mass emergences of adult insects frequently lead to severe nuisance conditions. The lake flies of Lake Victoria in East Africa were mentioned in Chapter 2, and more detailed account of them is given in Beadle (1974) and Rźoska (1976). Further downstream on the River Nile, at Khartoum and Wadi Halfa, mass emergences of Chironomidae cause problems involving allergic reactions such as asthma (Rźoska 1976). Indeed, even in the absence of any allergic effects, mass emergences cause adverse public reaction, and there are accounts of such problems from many parts of the world. Particularly common are great clouds of Chironomidae, which emerge from lakes and rivers that have been enriched by organic pollution, and which then move into residential areas. Somewhat similar swarms of moth flies (Psychodidae) and window flies (Rhyphidae) can be produced by sewage treatment works that use the trickling-filter process. Other examples are the mass emergences of large mayflies (*Hexagenia*) and caddisflies (Hydropsychidae) from the Mississippi River reported by Fremling (1960a, 1960b). The mayfly swarms are sometimes dense enough to impede traffic along the river and on nearby roads. The second produces annoying swarms at street lights with piles of dead insects beneath them, allergic reactions, and hysterical disgust.

It is noteworthy that at least some of these plagues of insects result from human interference with the habitat. Many are due to organic enrichment of one sort or another, and are thus really part of the effluent disposal problem. Others are caused by construction work. For example, dense concentrations of hydropsychid caddisflies, frequently called shadflies by the suffering public, are often encouraged by impoundments where production of food, which is used by populations of these filter-feeders living below the dam, is enhanced. In the

Mississippi, Hydropsychidae benefit from the construction of solid groins and other anti-erosional structures that serve as ideal substrata for the larvae. To some extent then the avoidance of such maddening swarms is a matter of management.

A second important impingement upon mankind is biting flies. Three of the important families of blood-sucking Diptera, the Culicidae, the Simuliidae, and the Tabanidae (mosquitoes, black flies, and horse flies) have aquatic larvae. The other blood-sucking flies, mostly species of Ceratopogonidae (biting midges) and Psychodidae (Phlebotominae, sand flies), insofar as species that bite man or domestic animals are concerned, have larvae that are associated only with damp soil, and so do not concern us further here.

The first three of the biting-fly families just mentioned contain species that are well known as vectors of disease. It is common knowledge that mosquitoes of the genus *Anopheles* transmit malaria, that *Aedes* species are vectors of yellow fever and dengue, and that other mosquitoes transmit the various nematodes known collectively as filarias. Similarly, black flies and some horse flies transmit filarial diseases to humans. Well-known examples include onchocerciasis (caused by the nematode *Onchocerca*), which leads to river blindness and is carried by *Simulium* species in Africa and South America, and the eye-worm, or calabar swellings, transmitted by the deer-fly *Chrysops* in West Africa. Such diseases are transmitted similarly to domestic animals (e.g. onchocerciasis and blue tongue in cattle, the latter also occurring in sheep; and malaria in domestic waterfowl, which is caused by the protozoan *Leucocytozoon*). The various viral diseases known as equine encephalitis are carried among horses by mosquitoes and are occasionally transmitted to humans. Thus, aquatic biting flies are of great medical and veterinary importance, and it is because of this that we know so much more about the taxonomy of mosquitoes and black flies (and of their larvae) than we do of other groups. We also know more about their way of life—an important factor in the efficient control of these insects. It is economically, and also environmentally, more desirable to know just where a species breeds so that control measures can be applied precisely, with minimal effects on other organisms.

In this regard it is interesting to note that while major control efforts have been directed against the larvae of mosquitoes and black flies, the larvae of horse flies and deer flies live in soft sediments and among plants where they are hard to find and control. As a result, our knowledge of the taxonomy and biology of larval Tabanidae lags far behind that of the other two families. The same applies to the soil-dwelling larvae of the Ceratopogonidae and Phlebotominae mentioned previously. As in all scientific endeavor the amount of study undertaken is very much controlled by economic considerations. It soon became clear that the cryptic habits of the larvae of horse flies, midges, and sand flies would make their control difficult; therefore, research efforts were directed elsewhere.

Apart from the transmission of disease, the sheer harassment by biting flies can have important economic consequences. Anyone who has visited the Canadian northlands in springtime will be aware of the hordes of mosquitoes and black flies that can make life almost intolerable; and massive attack by black flies may lead to collapse, the so-called black fly syndrome. Biting flies are indeed credited with having seriously inhibited development in many parts of North America, as has recently been documented by Laird et al. (1982). It is also well known that caribou and reindeer desert the forests and lowlands in summer to avoid biting flies. Similarly, visitors to tropical Africa become impressed by the fact that *Simulium damnosum* is well named, quite apart from its role as a vector of onchocerciasis. *Simulium metallicum* of tropical South America, also a vector of *Onchocerca,* can similarly drive humans out of its territory by the mere persistence of its biting; and elsewhere, also in a hot climate, horse flies are believed to be a controlling factor in the migrations of nomadic herdsmen in the Sudan (Rźoska 1976). These are just a few of the more spectacular examples, but nearly everywhere on the planet swarms of biting flies can make life miserable for both man and beast.

Generally speaking though, humans can cope with this menace by using screens, nets, suitable clothing, mosquito boots, smoky fires, and repellants. There are, however, reports of deaths caused by excessive exposure (Laird et al. 1982), and many people experience severe allergic reactions to bites. Domestic stock also suffer from these attacks but lack the protective amenities available to humans. Cattle can be seen to mill around or to shelter in barns, and horses become wild and unmanageable during biting-fly attacks. Little is really known of the economic consequences of this harassment, but recent Canadian work indicates that it is very great (Laird et al. 1982). Undoubtedly, over the planet as a whole, the loss of meat, milk, and general productivity must be stupendous; even indirectly, therefore, the negative impact of aquatic insects on mankind is very great indeed.

On the other hand, the aquatic insects have many positive contributions to make. As we have seen (Lamberti and Moore, Chapter 7; Healey, Chapter 9), aquatic insects are the principle converters of the photosynthetic production of plants into food for fishes; and fishes are undoubtedly among the major economic benefits to mankind that inland waters provide. In many parts of the Third World, fishes are a primary source of protein, and hence are essential to human survival.

Insects are also important to many types of angling, because species of freshwater fishes feed, at least in part, on insects, and many of the most palatable fish species, especially in temperate latitudes, feed in mid-water or at the surface where they catch drifting or emerging insects, or females that have returned to oviposit. It is natural therefore that anglers should look to such insects as bait, and around this idea there has developed, starting in classical times, the manufacture of lures that resemble insects and can be used to capture fish (Williams 1980). This practice can be traced back in Europe to at least the twelfth century. In the seventeenth century, in *The Compleat Angler*, Izaak Walton discussed and illustrated many lures made of pieces of various feathers, wool, hare and rabbit fur, wax and colored silk tied to a hook, and designed to resemble various types, often at least generically identifiable, of aquatic and terrestrial insects. This activity developed into a fishermen's lore that became much refined during the following centuries. It is expounded at some length by Schwiebert (1973) and also by Harris (1952), who supply details of the various European insect species involved, together with color photographs of the lures and of the insects they are designed to mimic. Clearly, in conjunction with the technique of making lures to resemble insects, anglers also learned a great deal about the insects themselves. They not only learned to distinguish some species, but in the Ephemeroptera, which are the most widely mimicked order for lures, the various developmental stages of the insects are clearly distinguished. Thus nymphs are fished as "wet" flies below the surface, whereas subimagos and sometimes females are fished on the surface. Other lures, representing flying insects, are touched lightly on the surface in a series of casts.

Notably, most of these lures represent adult insects, or specimens that are very close to emergence. This is because most fly fishing is done during the spring and summer when adult insects are on the wing, and many insectivorous fishes do indeed tend to feed at or near the surface at these seasons. There is also a tendency for individual fishes to concentrate for certain periods on a particular type of prey, and part of the mystique of fly fishing is to choose a lure that resembles the particular insect species that is emerging, or falling onto the water, at that time. This strategy is often successful, but fishes will often take a lure that does not apparently resemble any live insect that is present in the water at the time. Doubtless there is far more to the art of fishing than simple physical resemblance between lure and fly. The skilled angler endows the lure with movement and behavior, and these are probably just as important as color and form in deceiving the fish. The angler needs, in fact, to be a close observer of insects. Indeed, this matter of behavior must be of primary importance in the capture of adult salmon migrating upstream. Such fish are regularly caught on large, rather improbable-looking, artificial flies even though they are not feeding. One wonders whether these fishes perceive the lures as insects or as the marine crustaceans that constituted their diet before they began their migratory fast. Perhaps a more shrimplike lure would be even more successful.

We should also note that many vertebrates other than fishes feed at least partly upon aquatic insects. This applies to newts and salamanders during their aquatic phase, to small crocodiles before they turn to a diet of fish, to many freshwater turtles, and to some aquatic snakes and mammals. For instance, the Australian eastern water rat, *Hydromys chrysogaster*, eats many large stream insects (Woollard et al. 1978), and about half the diet of the platypus, *Ornithorhynchus*, is composed of aquatic insects (Faragher et al. 1979). However, not all the insectivorous mammals associated with water necessarily feed there. For example, Churchfield (1979) found that the European water shrew, *Neomys fodiens,* feeds mostly on terrestrial insects and eats comparatively few aquatic invertebrates.

Many birds also eat aquatic insects. The dippers of the thrush genus *Cinclus* are specialists in the capture of stream insects, for which they search by propelling themselves under water with their wings. They seem to be selective feeders, taking insects rather than other invertebrates, and when they are breeding in the spring they feed great numbers of newly emerged adults to their nestlings (Shaw 1979). Terrestrial perching birds have also been observed feeding on black fly larvae at a waterfall in Ontario and a dam in Alabama (Snoddy 1967; James 1968). Examination of specimens and fecal material showed that Trichoptera larvae had also been eaten.

Some specialized ducks hunt, like the dippers, for insects in turbulent streams. Such are the (now rare) blue duck of New Zealand, and the torrent ducks of New Guinea and the Andes. It is said of the South American species that it feeds very largely on stoneflies. However, many other ducks eat large numbers of insects gathered from still or slowly flowing water. Swanson et al. (1974) record the consumption of large numbers of insects by blue-winged teal feeding on wetlands in North Dakota, rising to a peak in the use of this food at the breeding season. Swanson (1977) also showed that this species and several other ducks were attracted to waste stabilization ponds as soon as the Chironomidae began to emerge. Clearly, then, aquatic insects are important to these birds and such insects probably supply a large proportion of the protein needed for avian breeding.

Whether or not the nurture of baby crocodiles can be considered of benefit to mankind (except perhaps to those who hunt them for their leather), we all derive aesthetic enjoyment from water birds or from the turtles sunning on the logs near the bank. Some of us also enjoy sporting and culinary pleasures; wild duck makes a splendid dish, and freshwater turtles were widely eaten in pre-Columbian America. The latter are said to be excellent eating.

We also derive aesthetic pleasure from the swallows, many species of which hunt over lakes, rivers, and marshes where they catch emerging insects in great numbers. Their prey almost certainly consists largely of adult Chironomidae, although the notion persists that great numbers of mosquitoes are killed by these birds. Indeed, it is a common practice in North America to put up nesting houses for purple martins so that they will hunt near the homestead and reduce the mosquito population. In all probability, though, the crepuscular mosquitoes, which fly rarely in daylight, are not really menaced by these strictly diurnal birds. One might do better with a belfry full of bats; but there is nevertheless much contentment to be derived from the chatter and activity of a houseful of purple martins.

Humans, to a limited extent, take part in this eating of aquatic insects. Terrestrial insects form, or have formed, part of the human diet in many lands, although our western culture now often seems to regard the practice with distaste. This is a remarkable prejudice among people who delightedly, and often at considerable cost, feast on arthropods such as shrimps, crayfishes, lobsters, and crabs. In Africa, caterpillars are crushed to a tasty paste; termites are often eaten, and their queens are regarded as a delicacy by many tribes. In the Middle East, locusts are widely eaten, and large bags of dried locusts are carried to the markets of Arabia. We all know about the locusts and wild honey that sustained John the Baptist (Mark 1:6). The nomadic aborigines of Australia eat many kinds of insects, and they particularly prize wichita grubs, the larvae of Scarabaeidae, which are similar to those of the May beetle of Europe and the erroneously named June bug of North America. In the past the aborigines organized special wintertime trips into the high mountains to feast on the great assemblages of bogong moths that flew up there to hibernate. To this day, one of the higher mountains in the Australian Alps bears the name of the moth as a memorial to this insectivory.

Consumption of aquatic insects by humans is also recorded in several countries. Aldrich (1972) reports that ovipositing adults of the genus *Atherix* (Athericidae) were beaten by the Indians from bushes along the Pitt River in northeastern California, gathered up, and baked or steamed into cakes of material resembling headcheese, which were then used as winter food. He also refers to the collecting of *Ephydra* along lake shores by other Californian Indians for food as being a well-known phenomenon.

These are practices that have long since disappeared, but in central Africa and Mexico aquatic insects are still eaten by humans. The great clouds of lake flies that emerge from Lake Malawi (Nyasa) in Africa, which are appropriately named *Chaoborus edulis,* are attracted to lights under which they accumulate in piles. These are gathered up, boiled, and made into cakes 10 to 20 cm in diameter and about 3-cm high. The nineteenth century explorer, Dr. Livingstone, reported that they tasted like caviar or salted locusts, but Beadle (1974), from whom this account is quoted, does not record that he tried them.

In Mexico, the gastronomic victims are species of Corixidae, two of which are also appropriately named *Corisella edulis* and *C. mercenaria,* the latter in reference to the fact that they are sold in large numbers as dried produce in the markets (Hungerford 1948). Not only are the adults collected and dried, and either later cooked and eaten or exported to serve as food for pet fishes and birds in the affluent north, but the eggs are also collected for food. For this purpose, bundles of reeds are submerged in the water, and the insects oviposit on them. The eggs are then shaken out of the bundles. One can only assume that these insects are exceptionally abundant in the localities where they are harvested for this activity to be worthwhile. Indeed, in all these examples, unusual abundance or gregarious behavior of aquatic insects is associated with their exploitation as human food.

The same applies to terrestrial insects, except that a few of the species that are eaten, such as termite queens and wichita grubs, are exceptionally large and rich in calories. In this connection, it is worthy of note that the uninitiated can eat only very few wichita grubs without stomach upset, and even the Australian aborigines had to feast on bogong moths with great circumspection, presumably because the fat bodies of the insects are very enriched in preparation for hibernation.

It is also of interest that one of the genera that is extensively eaten by humans (*Corisella*) is a member of the Heteroptera. Nearly all representatives of this order (or suborder) secrete repellant substances, and many are referred to as stink bugs. It is odd that this malodorous group should include a species named *edulis*, and be a species that has remained on the menu into modern times.

The consumption of aquatic insects whether by natural predators or by human ones has probably not had any devastating effects upon their populations, but undoubtedly many other human practices have been catastrophic for them in recent decades. The general effects of pollution are dealt with by Wiederholm (Chapter 17), but in several areas of the world they, and possibly other activities, have led to the loss of or great reduction of species. Thus Nielsen (1976) points out that in Denmark the farming of rainbow trout has eliminated, by downstream pollution, many species of caddisflies, as well as other rheophilic insects, and as a result these species no longer occur in the country. Botosaneanu (1981) has reviewed at length the steady decline of the Trichoptera in many areas of the planet. He stresses that species that are confined to springs, especially in arid areas, or to temporary waters, are at particular risk because of engineering and agricultural activities, and that over-enthusiastic use of pesticides threatens the caddisfly fauna of some tropical islands. However, it is among the inhabitants of large rivers that total extinctions of species have been fairly certainly documented. Botosaneanu gives several examples among the Trichoptera, caused perhaps by pollution or navigational engineering. To these we can add the stonefly *Oemopteryx loewii*, which used to inhabit the Danube, and probably the Rhine, and is now almost certainly extinct (Zwick 1980). According to Zwick (personal communication), several other species of stoneflies have apparently disappeared from West Germany, including *Isoperla obscura* (which in Britain also was last seen in 1911), and *Isogenus nubecula* now confined in Britain to a small area near the northern end of the Welsh border (Hynes 1977). Similarly, Pyle et al. (1981) report that two North American dragonflies of the family Gomphidae are now probably extinct as the result of human alteration of their habitats. They also state that the widespread Hawaiian damselfly *Megalagrion pacificum* was eliminated by the introduction of predatory freshwater fish to the archipelago.

The documented examples of severe reductions or extinctions of species are undoubtedly only a small fraction of those that have actually occurred; it is certain that modern humans are having a marked effect on aquatic insects. Some insects, probably many, are becoming rare or extinct, while others, as mentioned earlier, are so favored by human activity that they become pests.

The value of these losses is, of course, a matter of opinion, and many people would be delighted if disease-carrying biting flies were all to become extinct. However, reference has several times been made to aesthetic considerations, and there is little doubt that many aquatic biologists derive pleasure from their contact with aquatic insects, not only because of their elegance of adaption, but also because of the inherent beauty of many of them. This is readily apparent to readers of older works, such as Miall (1895) or Wesenberg-Lund (1943). Modern scientific writers, particularly in the professional journals, are discouraged from expressing such feelings on the grounds that they introduce unnecessary words. However, the appreciation of beauty is often apparent in illustrations, as for instance in the previously mentioned photographs by Harris (1952), and the even more attractively illustrated book of Leonard and Leonard (1962), which has photographs of many species of mayflies.

Adult mayflies, with their elegant shape, long graceful tails, pleated iridescent wings, and delicate pastel colors are among the most beautiful of all the insects. They are followed in this by the dragonflies, which also have pleated, and hence glistening, wings. It is perhaps significant that the mayflies and dragonflies, which are primitive and do not fold and hide their wings, share with the butterflies, which have lost the ability to fold their wings, an honorable place in art. Dragonflies and butterflies, and more rarely mayflies, are frequently depicted in Far Eastern paintings, and it is possibly because they display their wings so well that they have attracted the attention of artists.

These insects have received less attention in western art, perhaps because it tends to be more heavy-handed than the delicate painting of China and Japan. Thus, the gaudy butterfly and the brightly colored beetle fit better into the western scheme of things than the elegant network of the archaeopteran wing. However, Lord Tennyson did describe a dragonfly as a living flash of light adorned with sapphire mail, and some modern sculptors are using aquatic insects as themes. For example, David Allred of California, whose work is described by Fong (1980), has made some striking sculptures of metal and wood, capturing the elegance of the adult mayfly and even making the rather ugly damselfly nymph resemble an ecclesiastical figure from the Dark Ages. At last the intrinsic beauty of aquatic insects is being appreciated by western artists. Perhaps this reflects the fact that they no longer spend so much of their time in urban environments, nor are they, like the outdoor painters of the turn of the century such as the Canadian Group of Seven, so impressed by the magnificence of the landscape that they do not, as it were, see the trees for the woods. Let us hope that they will now begin to share the delights that people such as readers of this book get from observing the smaller details of nature.

Lastly, the more we come to know about insects the more useful they become to us. This is because each species has a fairly well-defined ecological niche; as a result, when one finds a population one can infer something about the habitat where it occurs. Thus, the more one knows about the species, and the more species one knows about, the more one can say about the habitat by making a simple collection and determining what species are present. Wiederholm has described in Chapter 17 how this general principle can be applied in the assessment of pollution, but its usefulness goes much further than that.

Such an approach has the potential for development into a major tool for environmental monitoring, because aquatic insects, being confined to the water, are nearly always residents of the localities in which they are found. The only exceptions are a few stray adults of beetles and bugs that may be engaged in dispersal. The aquatic insect is, therefore, in place as a sentinel animal (to use the veterinary term for a beast that is set out to test the environment), and the fact that it is there indicates that the habitat is suitable for its development. One can be far less certain about the signficance of the presence of a terrestrial insect, or of the flying adult of an aquatic one, because its range is not bounded by the water surface and it may be far from its true habitat. Moreover, the aquatic habitat integrates most of the parameters of the environment because it accumulates water from both air and land. It is no accident that the effects of acid rain first manifested themselves to the public through lakes, even though biologists had long known of the adverse effects of sulphur dioxide on lichens and its beneficial effects on potato plants afflicted with blight.

Thus the more precisely one can identify the insects, especially in their young stages (as has been stressed by Resh and Unzicker 1975), the more one can learn about a habitat and its neighborhood simply by collecting them. At present, this activity is perhaps more art than science. A skilled malaria control officer can tell at a glance if a pool or ditch will contain the local vector species of *Anopheles,* and an experienced student of stoneflies or mayflies can sit on the bank of a stream in a familiar area and list the species that would be found there. Often these specialists cannot describe precisely how they do this except in general terms, but, as knowledge increases, the terms will become more precise, more scientific, and hence more communicable.

Already, the types of Chironomidae inhabiting deep-water lacustrine deposits have been used to construct a classification of the tophic status of lakes (Thienemann 1954). Similarly, insects have been used in classifying the various zones of streams and rivers (Illies and Botosaneanu 1963). Both these approaches have weaknesses. The first approach can be misleading because the Chironomidae reflect only the extremely local conditions in the sediment, and so, because of other factors in the lake such as form and depth, they may give the wrong impression about the lake as a whole (Wetzel 1975). The second can lead to confusion because different species react to different factors and so produce a less clear-cut zonation than was originally supposed to exist (Hynes 1970). Nevertheless, the concepts behind these ideas are sound, and the classifications fail only because they were too simply based. In the more refined river continuum hypothesis (Vannote et al. 1980), insects still occupy an important place, and the anomalous findings from some lakes are now fairly well understood.

In this computer age when we can handle and evaluate enormous amounts of data, it is not too fanciful to predict a time when it will be possible to collect aquatic invertebrates, most of which will be insects, and from the collection derive a great deal of information about the habitat itself and the basin that drains into it. To do this we shall need to improve our knowledge of the aquatic stages of insects, and of their environmental and toxicological limits. This seems like an enormous amount of work but, when one considers how far we have come during the present writer's half century of intimate contact with aquatic insects, it does not seem a very distant goal. Anyone who doubts this should read Miall (1895) or Wesenberg-Lund (1943) in the light of the contents of this book and of the fact that an enormously greater number of people are now working in this field than were during their time. Both these men would probably have been astonished to be told that one day their beloved insects would be considered to have great environmental importance.

References

Aldrich, J. M. 1972. Flies of the Leptid genus *Atherix* used as food by California Indians (Dipt.). Entomological News 23:159–63.

Beadle, L. C. 1974. The inland waters of tropical Africa. An introduction to tropical limnology. Longman Group Ltd., London. 365 pp.

Botosaneanu, L. 1981. Ordo Trichoptera et Homo insapiens. pp. 11–19. *In*: G. P. Moretti (ed.). Proceedings of the Third International Symposium on Trichoptera. Series Entomologica, Volume 20. Dr. W. Junk Publishers, The Hague, The Netherlands. 472 pp.

Churchfield, J. S. 1979. A note on the diet of the European water shrew, *Neomys fodiens fodiens*. Journal of Zoology 188:294–96.

Faragher, R. A., T. R. Grant, and F. N. Carrick. 1979. Food of the platypus (*Ornithorhynchus anatinus*) with notes on the food of the brown trout (*Salmo trutta*) in the Shoalhaven River N.S.W. Australian Journal of Ecology 4:171–79.

Fong, M. 1980. Artist/angler—Dave Allred. The Flyfisher, Summer 1980:16–19.

Fremling, C. R. 1960a. Biology of a large mayfly, *Hexagenia bilineata* (Say), of the upper Mississippi River. Iowa State University Agricultural and Home Economics Experiment Station Research Bulletin 482:841–52.

Fremling, C. R. 1960b. Biology and possible control of nuisance caddisflies of the upper Mississippi River. Iowa State University Agricultural and Home Economics Experiment Station Research Bulletin 483:856–79.

Harris, J. R. 1952. An angler's entomology. Collins, London. 268 pp.

Hungerford, H. B. 1948. The Corixidae of the western hemisphere (Hemiptera). University of Kansas Science Bulletin 32:1–827.

Hynes, H. B. N. 1970. The ecology of running waters. University of Liverpool Press, Liverpool, England. 555 pp.

Hynes, H. B. N. 1977. A key to the adults and nymphs of the British stoneflies (Plecoptera). Freshwater Biological Association Scientific Publication 17. 92 pp.

Illies, J. and L. Botosaneanu. 1963. Problèmes et méthodes de la classification et de la zonation écologique des eaux courantes, considérées surtout du point de vue faunistique. Internationale Vereinigung für Theoretische und Angewandte Limnologie Mitteilungen 12:1–57.

James, H. G. 1968. Bird predation on black fly larvae and pupae in Ontario. Canadian Journal of Zoology 46:106–7.

Laird, M., A. Aubin, P. Belton, M. M. Chance, F. J. H. Fredeen, W. O. Haufe, H. B. N. Hynes, D. J. Lewis, I. S. Lindsay, D. M. McLean, G. A. Surgeoner, D. M. Wood, and M. D. Sutton. 1982. Biting flies in Canada: health effects and economic consequences. National Research Council of Canada, Ottawa, Publication No. 19248. 157 pp.

Leonard, J. W. and F. A. Leonard. 1962. Mayflies of Michigan trout streams. Cranbrook Institute Science Bulletin 43:1–139.

Miall, L. C. 1895. The natural history of aquatic insects. Macmillan and Co., London. 395 pp.

Nielsen, A. 1976. Pollution and caddis-fly fauna, pp. 159–61. In: H. Malicky (ed.). Proceedings of the First International Symposium on Trichoptera. Dr. W. Junk Publishers, The Hague, The Netherlands. 213 pp.

Pyle, R., M. Bentzien, and P. Opler. 1981. Insect conservation. Annual Review of Entomology 26:233–58.

Resh, V. H. and J. D. Unzicker. 1975. Water quality monitoring and aquatic organisms: the importance of species identification. Journal of the Water Pollution Control Federation 47:9–19.

Rźoska, J. (ed.). 1976. The Nile, biology of an ancient river. Monographiae Biologicae, Vol. 29. Dr. W. Junk Publishers, The Hague, The Netherlands, 417 pp.

Schwiebert, E. 1973. Nymphs: a complete guide to naturals and their imitations. Winchester Press, New York, NY. 339 pp.

Shaw, G. 1979. Prey selection by breeding dippers, Cinclus cinclus. Bird Study 26:66–67.

Snoddy, E. L. 1967. The common grackle, Quiscalus quiscula L. (Aves: Icteridae), a predator of Simulium pictipes Hagen larvae. Journal of the Georgia Entomological Society 2:45–46.

Swanson, G. A. 1977. Diel food selection by Anatinae on a waste stabilization system. Journal of Wildlife Management 41:226–31.

Swanson, G. A., M. I. Meyer, and J. R. Serie. 1974. Feeding ecology of blue-winged teals. Journal of Wildlife Management 38:396–407.

Thienemann, A. 1954. Chironomus. Leben, Verbreitung und wirtschaftliche Bedeutung der Chironomiden. Die Binnengewässer 20:1–834.

Vannote, R. L., G. W. Minshall, K. W. Cummins, J. R. Sedell, and C. E. Cushing. 1980. The river continuum concept. Canadian Journal of Fisheries and Aquatic Sciences 37:130–37.

Wesenberg-Lund, C. 1943. Biologie der Süsswasserinsekten. Verlag von Julius Springer, Berlin, Germany. 682 pp.

Wetzel, R. G. 1975. Limnology. W. B. Saunders Company, Philadelphia. 743 pp.

Williams, D. D. 1980. Applied aspects of mayfly biology, pp. 1–17. In: J. F. Flannagan and K. E. Marshall (eds.). Advances in Ephemeroptera biology. Plenum Press, New York, NY. 552 pp.

Woollard, P., W. J. M. Vestjens, and L. MacLean. 1978. The ecology of the eastern water rat Hydromys chrysogaster at Griffith, N.S.W.; food and feeding habits. Australian Wildlife Research 5:59–73.

Zwick, P. 1980. Plecoptera (Steinfliegen). Handbuch der Zoologie. Walter de Gruyter, Berlin, 4(2) 2/7: Lief. 26. 111 pp.

AUTHOR INDEX

TAXONOMIC INDEX

SUBJECT INDEX

abdominal undulation, 105
abundance: on aquatic plant vs.
   mineral substrates; 363; effect of silt on, 377;
   relation to substrate particle size, 363–365;
   on sand habitats in large rivers, 473–474
accommodation, physiological, 46
accuracy, sampling, 487
acidification, 530–535; management strategies
   for, 568–569; mechanisms of damage, 533–535
acid-mine wastes, 535, 568–569
activity rates: increases in, 123; temperature
   and, 57–59
actual cohort measurement methods, 291–298
adaptation: to cope with pollution, 537–540;
   crypsis, 227–228; environmental factors and,
   85; functional, for fish predation, 256–263; to
   heavy metal pollution, 525; of higher beetles,
   17–18; of hyporheos, 438–449; larval dispersal
   and colonization and, 410–411; life histories
   and, 25–26, 459–460; morphological, for
   hyporheic existence, 446–447; post-
   colonization, 449–450. *See also* behavioral
   adaptations
adaptive coevolution, 46
adult emergence. *See* emergence
adults: colonization by, 405–407; feeding
   retention/loss, 26–28; terrestrial life of, 14;
   variance in size, 64
aeration of lakes, 566–567
aerial swarming, 115–116; regulation on, 563
angling, importance of insects for, 580–581
annual P/B ratio, 298–299, 303, 305–308
apneustic insects, 12–13
apomixis, 37
aposematism, 228
area-restricted searching, 107–109, 206, 207–208
arthropod groups, in hyporheic zone, 445–446
artificial container habitats, 457
artificial enrichment of fisheries, 570
age, relative, 34–35
agricultural runoff, 517
airlift, 475
algae: effect of stream regulation on, 562–563;
   nutritional value of, 172, 174–175; primary
   productivity, 179–182; reproductive ability of,
   183; responses to herbivory, 177–186; selective
   feeding, 182–183; standing crop, 178–179;
   succession and diversity, 183–184
alkalinity: acid precipitation and, 530–531; in
   hyporheic zone, 436

Allen curve method, 291–298
Allen paradox, 307–308
allochthonous detritus, 559–560
allochthonous organic matter
   production, 142, 143–144, 177
anachoresis, 226–227
Analysis of Variance (ANOVA), 487,
   493, 495
angiosperms, effect of stream artificial
   substrates: samplers,
   432, 433; use of, 475, 502
assembly, ecological, 46–47
assimilation: effeciency, 63,
   147–148; in primary consumers,
   173–177
association, defensive, 231
attachment sites, 361
attack, deflection of, 230–231
autochthonous organic matter production,
   142–143, 177
automixis, 37
autotomy, 230
autotrophy, 510

bacteria, nutritional value of, 172, 174–175
bankfull stage: channel dimensions, 348;
   distribution of power at, 352–353; flow
   conditions, 349–350
baseflow, 329, 435
beetles, higher, habitat and adaptation variety,
   17–18
behavioral adaptations, 101–133; behavioral
   drift, 119–124, 125; foraging behavior,
   106–110; regulatory behavior, 101,
   102–106, 124; reproductive behavior,
   115–119, 125; territoriality and
   competition, 110–114
behavioral adaptations, regulatory behavior,
   101, 124
behavioral disturbances, 524
benthic insects: annual production of, 307;
   effects of fish predation on, 276–278;
   limitation in secondary production of,
   310–312, 316; sediment-water exchange of
   nutrients and, 150–151
bioenergetic model, 271–273
biological functions, 536
biological interactions, colonization and,
   417–419